Dynamics and Control of Autonomous Space Vehicles and Robotics

This book presents the established principles underpinning space robotics (conservation of momentum and energy, stability) in a thorough and modern fashion. Chapters build from general physical foundations through an extensive treatment of kinematics of multi-body systems, and then to conservation principles in dynamics. The latter part of the book focuses on real-life applications related to space systems.

Drawing upon years of practical experience and using numerous solved examples, illustrative applications, and MATLAB, Ranjan Vepa discusses:

- Basic space mechanics and the dynamics of space vehicles.
- Conservation and variational principles in dynamics and in control theory that can be applied to a range of space vehicles and robotic systems.
- A systematic presentation of the application of dynamics and control theory to real spacecraft systems.

Dr. Ranjan Vepa is currently a senior lecturer at Queen Mary University of London. He is the author of five books: *Biomimetic Robotics*; *Dynamics of Smart Structures*; *Dynamic Modeling, Simulation and Control of Energy Generation*; *Flight Dynamics Simulation and Control of Aircraft*; and *Nonlinear Control of Robots and UAVs*. His research interests include applications in space robotics, electric aircraft, and autonomous vehicles. He teaches advanced courses on robotics, aeroelasticity, advanced flight control and simulation, and spacecraft design, maneuvering, and orbital mechanics.

Dynamics and Control of Autonomous Space Vehicles and Robotics

RANJAN VEPA

Queen Mary University of London

CAMBRIDGE UNIVERSITY PRESS

CAMBRIDGE
UNIVERSITY PRESS

University Printing House, Cambridge CB2 8BS, United Kingdom

One Liberty Plaza, 20th Floor, New York, NY 10006, USA

477 Williamstown Road, Port Melbourne, VIC 3207, Australia

314–321, 3rd Floor, Plot 3, Splendor Forum, Jasola District Centre, New Delhi – 110025, India

79 Anson Road, #06-04/06, Singapore 079906

Cambridge University Press is part of the University of Cambridge.

It furthers the University's mission by disseminating knowledge in the pursuit of education, learning, and research at the highest international levels of excellence.

www.cambridge.org
Information on this title: www.cambridge.org/9781108422840
DOI: 10.1017/9781108525404

© Ranjan Vepa 2019

This publication is in copyright. Subject to statutory exception and to the provisions of relevant collective licensing agreements, no reproduction of any part may take place without the written permission of Cambridge University Press.

First published 2019

Printed in the United Kingdom by TJ International Ltd. Padstow Cornwall

A catalogue record for this publication is available from the British Library.

Library of Congress Cataloging-in-Publication Data
Names: Vepa, Ranjan, author.
Title: Dynamics and control of autonomous space vehicles and robotics / Ranjan Vepa.
Description: Cambridge, United Kingdom ; New York, NY, USA : University Printing House, 2019. | Includes bibliographical references and index.
Identifiers: LCCN 2018046546 | ISBN 9781108422840 (hardback)
Subjects: LCSH: Roving vehicles (Astronautics) | Autonomous robots. | Space vehicles–Dynamics. | Space vehicles–Control systems.
Classification: LCC TL475 .V47 2019 | DDC 629.43–dc23
LC record available at https://lccn.loc.gov/2018046546

ISBN 978-1-108-42284-0 Hardback

Cambridge University Press has no responsibility for the persistence or accuracy of URLs for external or third-party internet websites referred to in this publication and does not guarantee that any content on such websites is, or will remain, accurate or appropriate.

To my parents, Narasimha Row and Annapurna, and to my family

Contents

Preface		page xv
List of Acronyms		xvii

1 Introduction to Autonomous Space Vehicles and Robotics — 1
- 1.1 Space Exploration: The Unmanned Spacecraft That Ventured into Space — 1
- 1.2 Exploring Mars — 7
- 1.3 Robotic Spacecraft for Planetary Landing and Exploration — 9
- 1.4 Exploring a Comet — 10
- 1.5 Grabbing an Asteroid — 11
- 1.6 Routing Space Debris — 13
- 1.7 Venturing into Deep Space: Spacecraft with Endurance — 15
- 1.8 Planetary Rovers and Robot Walkers, Hoppers, and Crawlers for Exploration — 15
- 1.9 Underwater Rovers and Aquanauts — 16
- 1.10 Humanoid Space Robots and Robonauts — 16
- 1.11 Robot Arms for Tele-Robotic Servicing — 18
- 1.12 Tumbling Cubes — 23
- 1.13 Collaborative Robotic Systems — 23
- 1.14 The Meaning of Autonomy — 24
- 1.15 Dynamics and Control of Space Vehicles — 26
- 1.16 The Future — 26
- References — 27

2 Space Vehicle Orbit Dynamics — 28
- 2.1 Orbit Dynamics: An Introduction — 28
- 2.2 Planetary Motion: The Two-Body Problem — 28
  - 2.2.1 Kepler's Laws — 28
  - 2.2.2 Keplerian Motion of Two Bodies — 29
  - 2.2.3 Orbital Elements — 34
  - 2.2.4 Two-Body Problem in a Plane: Position and Velocity in an Elliptic Orbit — 35
  - 2.2.5 Orbital Energy: The Visa-Viva Equation — 39
  - 2.2.6 Position and Time in Elliptic Orbit — 41
  - 2.2.7 Lambert's Theorem — 42

	2.2.8	Orbit Inclination, Argument of the Ascending Node, Argument of the Perigee, and True Anomaly	43
	2.2.9	The f and g Functions	46
2.3	Types of Orbits		47
	2.3.1	Geosynchronous Earth Orbits	47
	2.3.2	Geostationary Orbits	47
	2.3.3	Geosynchronous Transfer Orbit	47
	2.3.4	Polar Orbits	48
	2.3.5	Walking Orbits	48
	2.3.6	Sun Synchronous Orbits	48
2.4	Impulsive Orbit Transfer		48
	2.4.1	Co-Planar Hohmann Transfer	49
	2.4.2	Non-Planar Hohmann Transfer	51
2.5	Preliminary Orbit Determination		53
	2.5.1	Two Position Vectors of the Satellite	53
	2.5.2	Three Position Vectors of the Satellite	54
	2.5.3	Two Sets of Observations of the Range at Three Locations	55
	2.5.4	Range and Range Rates Measured at Three Locations	56
2.6	Lambert's Problem		56
2.7	Third Body and Other Orbit Perturbations		58
	2.7.1	Circular Restricted Three-Body Problem	59
2.8	Lagrange Planetary Equations		62
	2.8.1	Geostationary Satellites	65
2.9	Gauss' Planetary Equations: Force Perturbations		65
	2.9.1	Effect of Atmospheric Drag	67
	2.9.2	Space Shuttle in a Low Earth Orbit	68
	2.9.3	Lunar Orbits	69
	2.9.4	Third-Body Perturbation and Orbital Elements in Earth Orbit	71
2.10	Spacecraft Relative Motion		71
	2.10.1	Hill-Clohessy-Wiltshire Equations	71
	2.10.2	Linear Orbit Theory with Perturbations	74
	2.10.3	Nonlinear Equations of Relative Motion with Perturbations	75
	2.10.4	Nonlinear Equations of Relative Motion with Reference to an Elliptic Orbit	77
	2.10.5	The Extended Nonlinear Tschauner-Hempel Equations	81
2.11	Orbit Control		85
	2.11.1	Delaunay Elements	86
	2.11.2	Non-Singular Element Sets	86
	2.11.3	Equinoctial Elements	87
	2.11.4	Orbital Elements with the Orbit Plane Quaternion Replacing the Euler Angles in the 3–1–3 Sequence	88
	2.11.5	Gauss Planetary Equations in Terms of Orbit Quaternion Parameters	91
	2.11.6	Other Nonclassical Elements	92

	2.12	Orbit Maneuvers	93
		2.12.1	Feedback Control Laws for Low-Thrust Transfers Based on the GPE	94
		2.12.2	Feedback Control Laws with Constraints on the Control Accelerations	98
	2.13	Interception and Rendezvous	100
	2.14	Advanced Orbit Perturbations	102
		2.14.1	Gravitational Potential of a Perfect Oblate Spheroid Model of the Central Body	102
		2.14.2	Gravitational Potential due to a Central Body's Real Geometry	103
		2.14.3	Real Drag Acceleration Acting on the Actual Satellite	104
		2.14.4	Third-Body Perturbations	105
		2.14.5	Solar Radiation Pressure	106
	2.15	Launch Vehicle Dynamics: Point Mass Model	107
		2.15.1	Systems with Varying Mass	107
		2.15.2	Basic Rocket Thrust Equation	108
	2.16	Applications of the Rocket Equation	109
		2.16.1	Time to Burnout, Velocity, and Altitude in the Boost Phase	109
		2.16.2	Time and Altitude in the Coast Phase	110
		2.16.3	Delta-Vee Solution	110
		2.16.4	Mass-Ratio Decay	110
		2.16.5	Gravity Loss	111
		2.16.6	Specific Impulse	111
	2.17	Effects of Mass Expulsion	111
		2.17.1	Staging and Payloads	112
	2.18	Electric Propulsion	112
		2.18.1	Application to Mission Design	114
	References	115
3	**Space Vehicle Attitude Dynamics and Control**	118
	3.1	Fundamentals of Satellite Attitude Dynamics	118
	3.2	Rigid Body Kinematics and Kinetics	118
		3.2.1	Coordinate Frame Definitions and Transformations	118
		3.2.2	Definition of Frames/ Rotations	118
		3.2.3	The Inertial (i) Frame X–Y–Z	119
		3.2.4	The Local Rotating (r) or Orbiting Frame x–y–z	119
		3.2.5	The Body (b) Frame b_1–b_2–b_3	119
		3.2.6	Defining the Body Frame	120
		3.2.7	Three- and Four-Parameter Attitude Representations	120
	3.3	Spacecraft Attitude Dynamics	121
	3.4	Environmental Disturbances	123
		3.4.1	Gravity Gradient Torques	123
		3.4.2	Aerodynamic Disturbance Torques	125
		3.4.3	Solar Wind and Radiation Pressure	126

- 3.4.4 Thruster Misalignments — 126
  - 3.4.5 Magnetic Disturbance Torques — 126
  - 3.4.6 Control Torques — 129
- 3.5 Numerical Simulation — 129
- 3.6 Spacecraft Stability — 129
  - 3.6.1 Linearized Attitude Dynamic Equation for Spacecraft in Low Earth Orbit — 129
  - 3.6.2 Gravity-Gradient Stabilization — 130
  - 3.6.3 Stability Analysis of the Spacecraft — 131
  - 3.6.4 Influence of Dissipation of Energy on Stability — 133
- 3.7 Introduction and Overview of Spacecraft Attitude Control Concepts — 133
  - 3.7.1 Objectives of Attitude Active Stabilization and Control — 134
  - 3.7.2 Actuators and Thrusters for Spacecraft Attitude Control — 134
  - 3.7.3 Active and Passive Stabilization Techniques — 135
  - 3.7.4 Use of Thrusters on Spinning Satellites — 136
- 3.8 Momentum and Reaction Wheels — 136
  - 3.8.1 Stabilization of Spacecraft — 137
  - 3.8.2 Passive Control with a Gravity-Gradient Boom or a Yo-Yo Device — 139
  - 3.8.3 Reaction Wheel Stabilization — 143
  - 3.8.4 Momentum Wheel and Dual-Spin Stabilization — 145
  - 3.8.5 Momentum Wheel Approximation with MW along Axis 1 — 148
  - 3.8.6 Control Moment Gyroscopes — 149
  - 3.8.7 Example of Control System Based on Reaction Wheels — 149
  - 3.8.8 Quaternion Representation of Attitude — 152
  - 3.8.9 The Relations between the Quaternion Rates and Angular Velocities — 154
  - 3.8.10 The Gravity Gradient Stability Equations in Terms of the Quaternion — 157
- 3.9 Definition of the General Control Problem with CMG Actuation — 158
  - 3.9.1 Nonlinear Attitude Control Laws — 162
  - 3.9.2 Minimum Time Maneuvers — 163
  - 3.9.3 Passive Damping Systems — 163
  - 3.9.4 Spin Rate Damping — 164
- 3.10 Magnetic Actuators — 164
  - 3.10.1 Active Control with Magnetic Actuators — 165
- References — 165

4 Manipulators on Space Platforms: Dynamics and Control — 167
- 4.1 Review of Robot Kinematics — 167
  - 4.1.1 The Total Moment of Momentum and Translational Momentum — 167
  - 4.1.2 The Screw Vector and the Generalized Jacobian Matrix of the Manipulator — 169

	4.2	Fundamentals of Robot Dynamics: The Lagrangian Approach	170
	4.3	Other Approaches to Robot Dynamics Formulation	178
	4.4	Fundamentals of Manipulator Deployment and Control	179
	4.5	Free-Flying Multi-Link Serial Manipulator in Three Dimensions	183
	4.6	Application of the Principles of Momentum Conservation to Satellite-Manipulator Dynamics	185
	4.7	Application of the Lagrangian Approach to Satellite-Manipulator Dynamics	185
	4.8	Gravity-Gradient Forces and Moments on an Orbiting Body	187
		4.8.1 Gravity-Gradient Moment Acting on the Satellite Body and Manipulator Combined	188
	4.9	Application to Satellite-Manipulator Dynamics	189
	4.10	Dynamic Stability of Satellite-Manipulator Dynamics with Gravity-Gradient Forces and Moment	191
	4.11	Three-Axis Control of a Satellite's Attitude with an Onboard Robot Manipulator	196
		4.11.1 Rotation Rate Synchronization Control	196
	References		203
5	**Kinematics, Dynamics, and Control of Mobile Robot Manipulators**		206
	5.1	Kinematics of Wheeled Mobile Manipulators: Non-Holonomic Constraints	206
	5.2	Dynamics of Manipulators on a Moving Base	209
	5.3	Dynamics of Wheeled Mobile Manipulators	209
		5.3.1 Manipulability	211
		5.3.2 Tip Over and Dynamic Stability Issues	212
	5.4	Dynamic Control for Path Tracking by Wheeled Mobile Manipulators	215
	5.5	Decoupled Control of the Mobile Platform and Manipulator	222
	5.6	Motion Planning for Mobile Manipulators	223
	5.7	Non-Holonomic Space Manipulators	224
	References		227
6	**Planetary Rovers and Mobile Robotics**		229
	6.1	Planetary Rovers: Architecture	229
		6.1.1 Vehicle Dynamics and Control	230
		6.1.2 Mission Planning	231
		6.1.3 Propulsion and Locomotion	232
		6.1.4 Planetary Navigation	233
	6.2	Dynamic Modeling of Planetary Rovers	233
		6.2.1 Non-Holonomic Constraints	233
		6.2.2 Vehicle Generalized Forces	235
		6.2.3 Modeling the Suspension System and Limbs	235
		6.2.4 Platform Kinetic and Potential Energies	240

	6.2.5	Assembling the Vehicle's Kinetic and Potential Energies	242
	6.2.6	Deriving the Dynamic Equations of Motion	243
	6.2.7	Considerations of Slip and Traction	243
6.3	Control of Planetary Rovers		248
	6.3.1	Path Following Control: Kinematic Modeling	248
	6.3.2	Estimating Slip	251
	6.3.3	Slip-Compensated Path Following Control Law Synthesis	251
	6.3.4	The Focused D^* Algorithm	254
References			254

7 Navigation and Localization — 257

7.1	Introduction to Navigation		257
	7.1.1	Basic Navigation Activities	257
7.2	Localization, Mapping, and Navigation		258
	7.2.1	Introduction to Localization	259
7.3	Random Processes		264
	7.3.1	Basics of Probability	269
	7.3.2	The Kalman Filter	272
	7.3.3	Probabilistic Methods and Essentials of Bayesian Inference	275
7.4	Probabilistic Representation of Uncertain Motion Using Particles		277
	7.4.1	Monte Carlo Integration, Normalization, and Resampling	277
	7.4.2	The Particle Filter	278
	7.4.3	Application to Rover Localization	282
	7.4.4	Monte Carlo Localization	284
	7.4.5	Probabilistic Localization within a Map, Using Odometry and Range Measurements	285
7.5	Place Recognition and Occupancy Mapping: Advanced Sensing Techniques and Ranging		286
	7.5.1	Place Recognition Using Ranging Signatures: Occupancy Mapping of Free Space and Obstacles	287
7.6	The Extended Kalman Filter		287
	7.6.1	The Unscented Kalman Filter (UKF)	290
7.7	Nonlinear Least Squares, Maximum Likelihood (Ml), Maximum A Posteriori (MAP) Estimation		292
	7.7.1	Nonlinear Least Squares Problems Solution Using Gauss-Newton and Levenberg Marquardt Optimization Algorithms	296
7.8	Simultaneous Localization and Mapping (SLAM)		298
	7.8.1	Introduction to the Essential Principles and Method of SLAM	298
	7.8.2	Multi-Sensor Fusion and SLAM	303
	7.8.3	Large-Scale Map Building via Sub-Maps	304
	7.8.4	Vision-Based SLAM	305
7.9	Localization in Space and Mobile Robotics		305
References			306

8	**Sensing and Estimation of Spacecraft Dynamics**		308
	8.1 Introduction		308
	8.2 Spacecraft Attitude Sensors		308
		8.2.1 The Principle of Operation of Accelerometers and Gyroscopes	308
		8.2.2 Magnetic Field Sensor	311
		8.2.3 Sun Sensors	312
		8.2.4 Earth Horizon Sensors	312
		8.2.5 Star Sensors	313
		8.2.6 Use of Navigation Satellite as a Sensor for Attitude Determination	313
	8.3 Attitude Determination		315
	8.4 Spacecraft Large Attitude Estimation		319
		8.4.1 Attitude Kinematics Process Modeling	320
		8.4.2 Codeless Satellite Navigation Attitude Sensor Model	322
		8.4.3 Application of Nonlinear Kalman Filtering to Attitude Estimation	324
	8.5 Nonlinear State Estimation for Spacecraft Rotation Rate Synchronization with an Orbiting Body		328
		8.5.1 Chaser Spacecraft's Attitude Dynamics	330
		8.5.2 Relative Attitude Dynamics	332
		8.5.3 Nonlinear State Estimation	334
		8.5.4 The Measurements	336
		8.5.5 The Controller Synthesis	338
	8.6 Sensors for Localization		339
	8.7 Sensors for Navigation		341
		8.7.1 Imaging Sensors and Cameras	342
	References		344

Index 349

Preface

This book is about the stability and control of satellites carrying robotic manipulators as well as the stability and control of the manipulators themselves. The book also addresses the issues related to the stability and control of planetary rovers that are used to explore the environment of a planet's surface. The book begins with an introduction to "Autonomous Space Vehicles and Robotics," followed by a chapter on "Space Vehicle Orbit Dynamics." Chapter 3 is entitled "Space Vehicle Attitude Dynamics and Control." The next chapter is about the dynamics of manipulators on board a spacecraft and is entitled "Manipulators on Space Platforms: Dynamics and Control." In particular, the focus of this chapter is on establishing the dynamic models of the robotic manipulators with a finite number of degrees of freedom operating on board a space vehicle. The dynamic stability and control of the coupled manipulator and satellite dynamics are also discussed in this chapter. "Kinematics, Dynamics, and Control of Mobile Robot Manipulators" is discussed in the next chapter, while Chapter 6 is devoted to the dynamics and control of planetary rovers. This chapter deals with mobile robots and is entitled "Planetary Rovers and Mobile Robotics."

The focus of the penultimate chapter is on "Navigation and Localization," and deals with specialized issues related to autonomous navigation and the issue of localization in mobile robotics, including triangulation, trilateration, dead reckoning, localization based on significant environmental characteristics, odometry, statistical and probabilistic approaches, Markov localization, and the use of Kalman filters for localization. Chapter 8 is about "Sensing and Estimation of Spacecraft Dynamics." In particular, the various types of sensors that are typical to the space environment and the problems associated with the estimation of the spacecraft's position and orientation will be discussed here.

I would like to thank my colleagues, in the School of Engineering and Material Science, at Queen Mary University of London, and Professor V. V. Toropov, in particular, for his support in this endeavor. I would also like to express my special thanks to Steven Elliot, Senior Editor, Aeronautical, Biomedical, Chemical, and Mechanical Engineering, Cambridge University Press, New York, for his enthusiastic support for this project.

I would like to thank my wife, Sudha, for her love, understanding, and patience. Her encouragement was a principal factor that provided the motivation to complete the project. I would like to thank my brother Dr. Kosla Vepa, who was always willing to discuss issues related to astronautics and space mechanics. Finally, I must add that my interest in the kinematics of mechanisms and in robotics was inherited from my late father many years ago. To him I owe my deepest debt of gratitude. This book is dedicated to his memory.

Acronyms

AC	Alternating current
AKM	Apogee kick motor
AUV	Autonomous underwater vehicle
BLF	Barrier Lyapunov function
CDM	Code division multiplexing
CG	Center of gravity
CM	Center of mass
CP	Center of (aerodynamic) pressure
DC	Direct current
DOP	Dilution of precision
ECEF	Earth centered earth fixed
ECI	Earth centered inertial
EDM	Enhanced disturbance map
EKF	Extended Kalman filter
ESA	European space agency
FDMA	Frequency division multiple access
GDOP	Geometric dilution of precision
GEO	Geostationary earth orbit
GPE	Gauss planetary equations
GPS	Global positioning system
GTO	Geosynchronous transfer orbit
HCW	Hill Clohessy Wiltshire
HDOP	Horizontal dilution of precision
IAGA	International association of geomagnetism and aeronomy
IGRF	International geomagnetic reference field
IR	Infra-red
ISS	*International space station*
JAXA	Japan aerospace exploration agency
JPL	Jet propulsion laboratory
LADAR	LASER detection and ranging
LASER	Light amplification by stimulated emission of radiation
LEO	Low earth orbit
LERM	Linear equations of relative motion
LIDAR	Light detection and ranging

LORAN	Long range navigation
LPE	Lagrange planetary equations
LQR	Linear-quadratic regulator
LVLH	Local vertical local horizontal
MER	Mars exploration rover
NASA	National aeronautics and space administration
NERM	Nonlinear equations of relative motion
PF	Particle filter
RAAN	Right ascension of the ascending node
RADAR	Radio detection and ranging
ROSCOSMOS	Russian federal space agency
ROV	Remotely operated vehicle
RTG	Radio-isotope thermoelectric generator
SI	Système internationale
SLAM	Simultaneous localization and navigation
SONAR	Sound navigation and ranging
SRP	Solar radiation pressure
SSO	Sun synchronous orbit
SSRMS	Space shuttle remote manipulator system
TDOP	Time dilution of precision
TLE	Two line elements
TVC	Thruster vectoring control
UKF	Unscented Kalman filter
VDOP	Vertical dilution of precision

1 Introduction to Autonomous Space Vehicles and Robotics

1.1 Space Exploration: The Unmanned Spacecraft That Ventured into Space

The dream of exploring the space surrounding the Earth has been around for many hundreds of years. Space exploration needed rockets to go to space, and engineers such as the Russian Konstantin Tsiolkovsky, the German Herman Oberth, and the American Robert H. Goddard were among many who successfully attempted to build rockets for space travel. The dream of space travel was fueled by such science fiction writers as Cyrano de Bergerac (1619–1655), Jules Verne, who published his novel *From the Earth to the Moon* in 1865, and H. G. Wells, whose novel *First Men on the Moon* was published in 1901, and these works inspired scientists such as Robert H. Goddard and Werner von Braun, who led the US effort to send an astronaut to the Moon. On the October 4, 1957 the Soviet Union launched the first autonomous artificial satellite, *Sputnik 1*, illustrated in Figure 1.1, which orbited the Earth in 96.2 minutes in an elliptic orbit at an inclination of 65 degrees.

The satellite was 58 cm in diameter and orbited the Earth at the rate of 8,100 m/s for three weeks before its batteries ran out of power. It was launched on board a Soviet R7 rocket that was just over 29 m long and developed a thrust of 3.9 mega Newtons. It crashed back to Earth six months after its launch. Soon afterward, the Soviet Union launched *Sputnik 2* with a dog on board. Unfortunately, the dog did not survive the flight, although it spent a week in Earth orbit. In 1958 the *Explorer 1* satellite that was responsible for detecting the radiation belts around the Earth was launched by the United States. The first US communications satellite, *Telstar 1*, was launched in July 1962. Over 3,000 satellites have been launched since then for communication and navigation applications, weather observation, space and deep space research, and military applications. The first successful space probe was *Luna 2*, which crashed on the Moon in 1959. After this, several other *Luna* spacecraft visited the Moon, the *Mariner* series of spacecraft visited the planets Venus, Mars, and Mercury, while the *Pioneer 10* and *11* spacecraft visited the planets Jupiter and Saturn, respectively. Early *Pioneer* spacecraft were used to launch probes that landed on the surface of Venus. Figure 1.2 illustrates *Mariner 5*, built as a back-up for *Mariner 4*, the first spacecraft to go to Mars, which eventually went on to probe the planet Venus.

The outer planets Uranus and Neptune were visited by *Voyager 2* in 1986 and 1989, respectively. The asteroid Gaspra was examined by the US spacecraft *Galileo*, while the

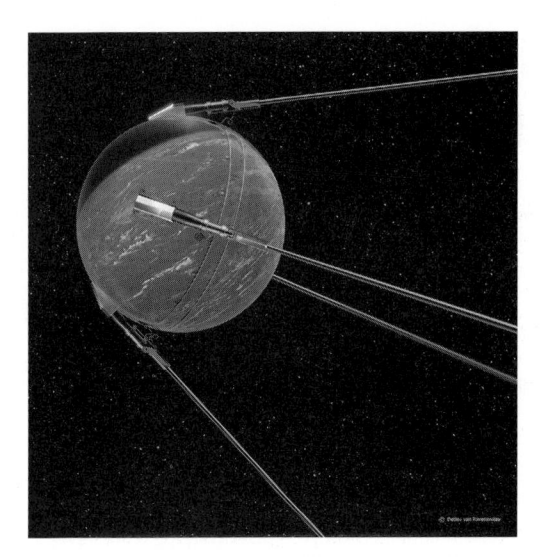

Figure 1.1 *Sputnik 1*. (credit: Detlev van Ravenswaay / Picture Press / Getty Images)

Figure 1.2 The *Mariner 5* spacecraft *en route* to the planet Venus. (credit: Stocktrek Images / Getty Images)

European spacecraft *Giotto* surveyed Halley's comet. Figure 1.3 illustrates NASA's *STARDUST* spacecraft, which was launched in 1999, *en route* to the comet Tempel 1, following a NASA spacecraft's successful fly-by of the comet Hartley 2. NASA's *Mars Atmosphere and Volatile Evolution* (*MAVEN*) satellite, which began orbiting Mars in September 2014, was designed to fly-by the comet C/2013 A1 Siding Spring and is shown in Figure 1.4. The *Near Earth Asteroid Rendezvous* (*NEAR*) satellite mission profile and its journey to the Asteroid 433 EROS, which involved several Deep Space

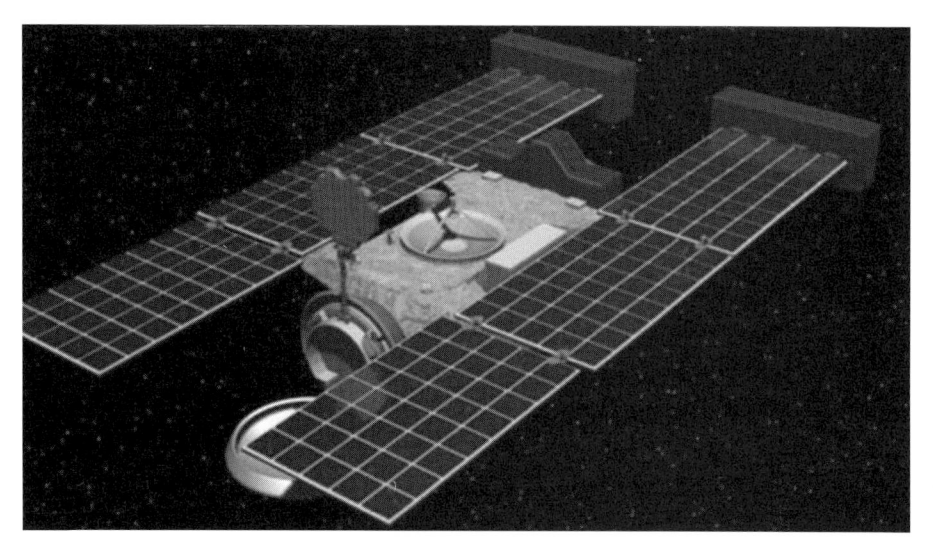

1.1 Space Exploration: The Unmanned Spacecraft that Ventured into Space

Figure 1.3 NASA's *STARDUST* spacecraft *en route* to the comet Tempel 1. (courtesy: NASA)

Figure 1.4 NASA's *Mars Atmosphere and Volatile Evolution* (*MAVEN*) satellite orbiting Mars after a Comet fly-by. (courtesy: NASA)

Maneuvers (DSM) that were executed shortly after the asteroid 253 Mathilde fly-by, on July 3, 1997, is shown in Figure 1.5.

Using a novel slingshot maneuver around the planet Jupiter, the *Ulysses* spacecraft was redirected to the Sun and it passed by the Sun's North Pole in 1995. The first planetary rovers, known as Lunokhods and illustrated in Figure 1.6, were landed on the Moon by the Soviet Union between 1970 and 1973.

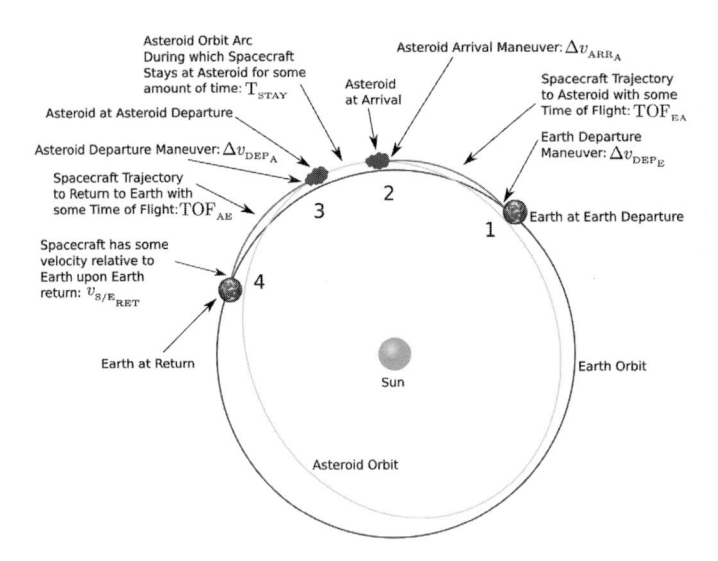

Figure 1.5 The diagram illustrates the parts of a conceptual mission to the asteroid. The outer oval represents Earth's orbit, the inner oval is the asteroid's orbit, and the red arcs are the spacecraft's trajectory to and from the asteroid. (courtesy: NASA, image credit: Brent Barbee)

Figure 1.6 The Lunokhod Moon rover. (courtesy: NASA)

Figure 1.7 shows the space shuttle *Endeavour* docked with the *International Space Station (ISS)* [1].

Although space exploration is extremely expensive, countries like China and India have also designed and launched low-cost missions to the Moon and Mars. Almost all of these spacecraft were autonomous. The *ISS*, led by the United States, involved the use of robotic manipulators in space to assemble several modules over several years to

1.1 Space Exploration: The Unmanned Spacecraft that Ventured into Space

Figure 1.7 Space shuttle *Endeavour* docked with the *International Space Station*. (courtesy: NASA)

Figure 1.8 The *International Space Station*. (credit: Matthias Kulka / Corbis / Getty Images)

make the design and deployment of a permanent artificial orbiting outpost in space, about 400 km above the Earth and moving at the rate of 7.7 km/s, a reality. A space shuttle fitted with a robotic arm was used to shuttle modules from the Earth for assembly and deployment on the *ISS*. The *ISS*, illustrated in Figure 1.8, was assembled using several robotic manipulators and was truly an international effort. The *ISS* and the space shuttle successfully demonstrated the use of tele-operated and autonomous robots in space.

Figure 1.9 A Chinese satellite preparing to dock with a Chinese space station. (credit: AFP / Stringer / Getty Images)

Apart from the US National Aeronautics and Space Administration (NASA), the Russian Federal Space Agency (ROSCOSMOS), and the European Space Agency (ESA), the Japan Aerospace Exploration Agency (JAXA), the China National Space Administration, and the Indian Space Research Organization (ISRO) have also been active in launching satellites. Figure 1.9 illustrates a Chinese satellite preparing to dock with a Chinese space station.

In another development, Arthur C. Clarke predicted in 1945 that satellites could be used for terrestrial communications in an article first published in *Wireless World* magazine in 1945. The first commercial communications satellite, *Intelsat 1*, was launched 20 years later in April 1965. Following the launch of the first communication satellite in 1962, proposals were made to develop radio navigation systems similar in principle to LORAN and DECCA but with satellite transmissions of precise radio navigation signals. This led to the development of the TRANSIT system, involving seven orbiting satellites, where the position of the user was determined from the Doppler shift in the received radio frequency signal. TRANSIT was made available in 1967 and soon after led to the development of the GPS system. After 10 years of development, proposals for establishing the GPS system were approved in the 1970s and the system was made available on a selective basis in the 1980s. Since the 1990s the GPS navigation system has been made available internationally for navigation applications worldwide.

Besides the United States' GPS (with 24 satellites in a constellation, orbiting the Earth at an altitude of 20,200 km), Russia's GLONASS, and the European Union's *Galileo*, the Indian Space Research Organization launched the Indian Regional Navigation Satellite System (IRNSS) satellites, a set of 7 dedicated satellites, in 2016, which form the NavIC (Navigation with Indian Constellation) system for navigation applications. Operating on dual frequency bands using the S and L bands, the NavIC system covers a limited region over the Indian subcontinent. Four of the satellites are

Figure 1.10 A typical IRNSS satellite is shown being assembled in a factory in Bangalore, India. (credit: Pallava Bagla / Corbis News / Getty Images)

geosynchronous, orbiting in pairs so the ground track looks like a figure of eight, north and south of the equator, while the remaining three are geostationary, all orbiting the Earth in a circular orbit at an altitude of 35,787 km above the equator. The Indian space research establishment crossed a major threshold when for the first time a factory owned by the private sector became involved in making a full multi-million dollar navigation satellite. In Figure 1.10, a typical IRNSS satellite is shown being assembled in a factory in Bangalore, India.

The use of the dual frequency band permits corrections to be easily made for transmission times due to atmospheric and tropospheric delays. The deployment of the system demonstrated the feasibility of developing and installing a system that can provide limited navigation coverage over any planet. The China National Space Administration is also in the process of building its own satellite navigation system known as the Beidou Navigation Satellite System.

1.2 Exploring Mars

Mars is one of the most enigmatic planets and a neighbor of our own planet. The manned *Apollo 11* mission to the Moon landed on the lunar surface on July 20,

1969 almost 10 years after the first unmanned lunar landing by a Luna spacecraft. After the successful landing on the Moon, the attention of almost all space researchers turned back to Mars, which had evoked considerable interest from scientists in the early 1900s. It was believed, at that time, that Mars was inhabited and that there were large canals on the surface that were filled with water. This was later considered to be an illusion and probably caused by the reflection of the blood vessels in the human eye in a telescope's eyepiece as astronomers viewed the surface of Mars. However, the more recent discovery of "ghost dunes" by the Mars Reconnaissance Orbiter seems to indicate the presence of extreme environmental conditions (winds and dust storms coupled with extreme temperatures), which seem to be responsible for creating special features on the surface of Mars.

Mars has an equatorial diameter of 6,759 km in comparison to the corresponding equatorial diameter of Earth of 12,756 km, which makes the Earth's diameter almost 1.88 times the diameter of Mars. Mars' orbital period is 687 days, making it also about 1.88 times the orbital period of the Earth (365.25 days). Mars' mean orbital distance from the Sun is 227.94 million km, while the corresponding distance of the Earth is 149.59797 million km, making the distance of Mars about 1.52 times that of Earth or about 1.88 raised to the power of 2/3. The eccentricity of the Martian orbit is a lot greater (0.093) than that of the Earth (0.0167), which makes it 5.57 times that of the Earth. The inclination of Mars' orbit is just less than 2 degrees that of the Earth, while the inclination of the Martian equator is slightly greater than the inclination of Earth's equator to its orbit. Yet the mean density of Mars is only 71.3% of that of Earth, while acceleration due to gravity on the surface of the planet is 3.71 m/s^2, which in comparison to the corresponding value on the surface of the Earth (9.81 m/s^2) is less than 38%. Moreover, Mars has two moons, Phobos and Deimos, orbiting it at distances of 9,370 and 23,460 km, respectively, from its center. Another interesting feature of Mars as seen from Earth, due to its much larger orbital period, is that it appears to briefly move backward as it journeys from one end of the sky to the other.

Following the *Mariner 9* spacecraft's visit to Mars in 1971, two *Viking* spacecraft, *Viking 1* and *Viking 2*, had landed on its surface in 1976 and transmitted pictures of the Martian surface. In 1996, NASA launched *Mars Pathfinder* and *Mars Global Surveyor* followed by *Mars Odyssey* in 2001. It was named in honor of Arthur C. Clarke's novel entitled *2001: A Space Odyssey* and reached Mars in just over six months. The spacecraft returned pictures of the surface of Mars in stunning detail and set the scene for the launch of spacecraft carrying the twin planetary exploration rovers and the planetary science laboratory in the years that followed.

Mars Express Orbiter, a cube-shaped satellite with two sets of solar panels on either side, was launched by the European Space Agency in June 2003 and arrived in the Martian environment in December 2003, after a six-month journey covering the 400 million km distance. *Mars Express Orbiter* is the second longest surviving, continually active spacecraft in orbit around a planet other than Earth. Although it was accompanied by *Beagle 2*, a Mars lander that failed to deploy properly after a successful landing, *Mars Express Orbiter* has now continuously surveyed the structure of the Martian surface for over a decade and in July 2018 it was reported that it had discovered a 12-mile-wide lake containing liquid water beneath the Martian surface, thus indicating the possibility of some life forms existing on the planet.

The *Mars Reconnaissance Orbiter* was launched in 2005, following the launch of the Mars exploration rovers, Spirit and Opportunity, and was fully focused on its mission almost 13 months later. The *Mars Science Laboratory* was launched in 2011, culminating in the landing of the Curiosity rover on the surface of Mars in 2012.

1.3 Robotic Spacecraft for Planetary Landing and Exploration

While NASA's *Mariner 2* was the first spacecraft to reach the planet Venus, it was the Soviet Union's *Venera 3* that made the first landing on the surface of that planet. The first manned spacecraft to land on the surface of the Moon was the Lunar module, which left its mother ship, *Apollo 11*, while it was in a lunar orbit and landed on the lunar surface. NASA plans a return to the lunar surface in the years to come. To realize this project, NASA is launching a satellite, *Orion*, which will fly approximately 100 km above the surface of the Moon, and then use the Moon's gravitational force to move itself into a new deep space orbit. The Orion mission is the first of a series of launches that are designed to explore deep space. Figure 1.11 is an illustration of the *Orion* spacecraft in lunar orbit.

More than 30 missions worldwide have been launched by several countries to explore the Martian environment. Several of these missions have been responsible for taking Martian planetary landers and rovers, such as Pathfinder, Spirit, Opportunity, and Curiosity, and landing them safely on the Martian surface. The latest mission planned as part of the ExoMars program are an *Orbiter* plus an Entry, Descent, and Landing Demonstrator Module, which was launched in 2016. Figure 1.12 is an artist's

Figure 1.11 The *Orion* spacecraft. (courtesy: NASA)

Figure 1.12 A satellite approaching the planet Mars. (credit: Stocktrek Images / Getty Images)

Figure 1.13 The proposed new Mars rover. (courtesy: NASA)

impression of a satellite approaching the planet Mars. Another launch, featuring a new rover, with a launch date in 2020, is also planned. Figure 1.13 is an illustration of the proposed new rover that is to be sent to bring back a rock from the surface of Mars.

1.4 Exploring a Comet

NASA JPL's *Deep Space 1* mission in 1998, on its way to an asteroid and then to a comet, was the first satellite to be powered by an electric propulsion system, using an electrostatic ion thruster that generates thrust by accelerating charged particles or ions by creating electrostatic fields. The European Space Agency tested an electric

Figure 1.14 An artist's impression of the *Rosetta* satellite and the Philae lander approaching the comet. (credit: ESA / Handout / Getty Images News)

propulsion system in space on NASA's *Artemis* spacecraft in 2001. Japan's *Hayabusa* spacecraft used ion electric propulsion in 2003 for an asteroid sample capture and return mission.

Ion electric propulsion is characterized by high specific impulse, which is the momentum added to the spacecraft per unit weight (on Earth) of the propellant (usually Argon, Xenon, or Krypton). Thus, although ion electric propulsion generates low thrust, its endurance is practically very high and can be used for very long periods of time, making it suitable for deep space missions. The variable specific impulse magneto-plasma rockets provide much higher thrusts and can bridge the gap between high thrust chemical propulsion and low thrust electrostatic ion propulsion.

In 2014, history was made with the first ever landing of a lander on a comet. In November of that year the European Space Agency's *Rosetta* spacecraft dispatched a lander named Philae to the surface of the comet 67P/Churyumov-Gerasimenko in 2014, after a 10-year journey of the spacecraft and lander to its vicinity.

The mission revealed in incredible detail the structure of the cometary surface landscape as well as the presence of rocks of relatively large size. Figure 1.14 is an artist's impression of the *Rosetta* mission.

1.5 Grabbing an Asteroid

Three new missions will return samples of asteroids to Earth for future study. JAXA launched its *Hayabusa-2* mission, which will robotically collect samples of asteroid 1999JU3 in 2018 and return them to Earth in 2020. The mission builds on the legacy of

JAXA's earlier *Hayabusa* mission, which explored asteroid Itokawa and returned samples to Earth in 2010.

In 2016 NASA launched its robotic satellite mission, the Origins-Spectral Interpretation-Resource Identification-Security-Regolith Explorer (OSIRIS-Rex), to bring back a sample of an asteroid's soil. The satellite is expected to rendezvous with asteroid 1999 RQ36, also known as "Bennu," in 2019 and bring back samples to the Earth, arriving in 2023. The primary objectives for the mission include finding answers to basic questions about the composition of the very early solar system and the source of organic materials and water that made life possible on Earth. JAXA's *Hayabusa-2* and OSIRIS-Rex missions could help NASA to choose its target for the first-ever ambitious mission to capture and redirect an asteroid. NASA's Asteroid Redirect Mission (ARM) in the 2020s will help NASA test new technologies needed for future human missions to Mars. For the ARM mission, NASA plans to launch a robotic spacecraft to first rendezvous with a near-Earth asteroid. The agency is weighing two concepts for what the spacecraft does next – one would fully capture a small asteroid about 5–10 m in size, using an inflatable mechanism, and the other would retrieve a boulder about 2–5 m (6–15 feet) in size from a much larger asteroid using a robotic arm. The spacecraft will then use the gentle thrust of its solar electric propulsion system and the gravity field of the Earth and Moon to redirect the asteroid into a stable orbit around the Moon, where astronauts will explore it in the mid-2020s, returning to Earth with samples much later. Figure 1.15 illustrates NASA's *Orion* spacecraft approaching the robotic asteroid capture vehicle.

The Near Earth Asteroid Rendezvous – Shoemaker (NEAR Shoemaker), named after planetary scientist Eugene Shoemaker in 1996, was a robotic space probe designed by the Johns Hopkins University Applied Physics Laboratory for NASA

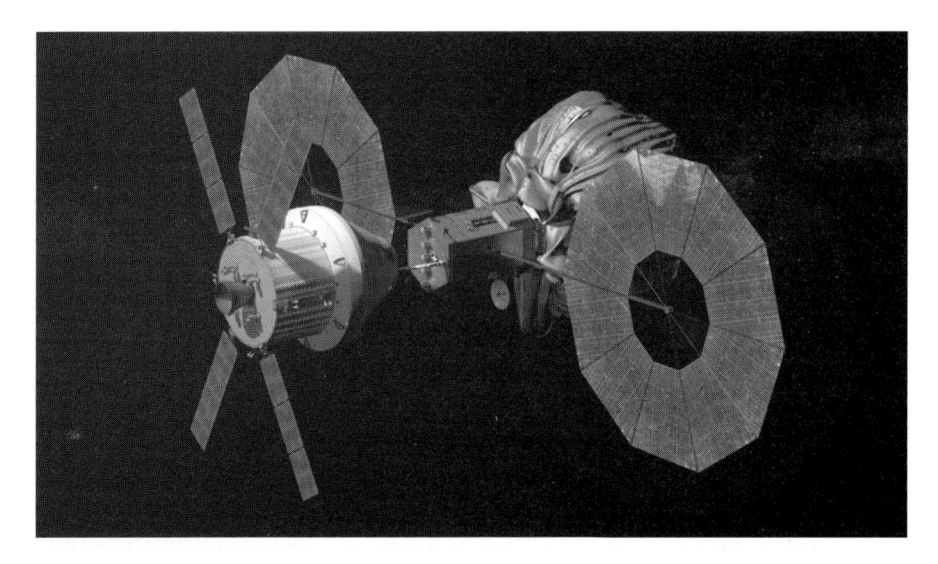

Figure 1.15 Satellite designed for Asteroid capture. (courtesy: NASA)

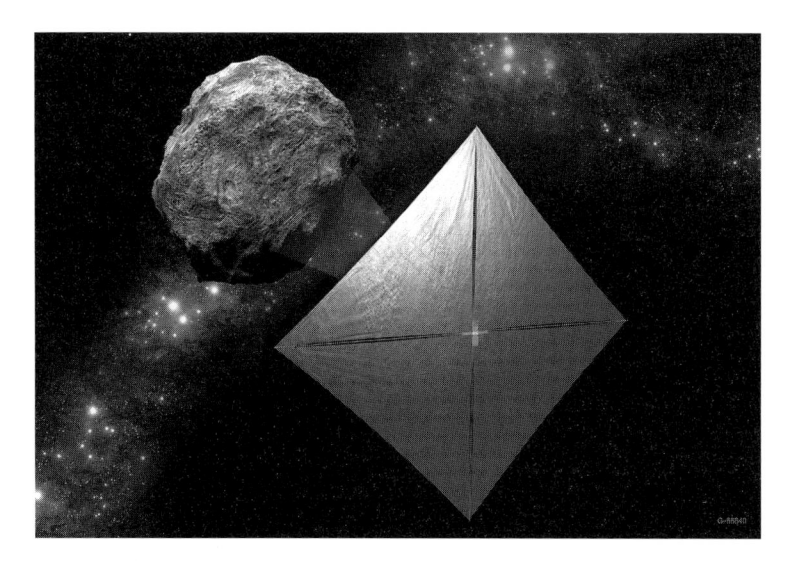

Figure 1.16 NEA Scout CubeSat with its solar sail deployed as it maps a near-Earth asteroid. (courtesy: NASA)

to study near-Earth asteroids. The asteroid 253 Mathilde, which was discovered in 1885, is in the intermediate asteroid belt, and is approximately 50 km in diameter. The first image of asteroid 253 Mathilde was returned by the NEAR spacecraft in 1997. Figure 1.16 illustrates the NEA Scout CubeSat with its solar sail deployed as it maps a near-Earth asteroid.

1.6 Routing Space Debris

There is a growing concern about the amount of debris currently orbiting the Earth, which seems to be particularly concentrated around the geostationary circular orbit above the equator at particular locations along the orbit. According to the European Space Agency's space debris office, there are more than 21,000 objects larger than 10 cm in orbit, while estimates of objects that are larger than 1 cm are in the region of 750,000. It has also been estimated that over 150 million smaller pieces of debris larger than 1 mm and 100 million pieces of debris less than 1 cm in size could be orbiting the Earth. Objects in low Earth orbit naturally decay due to the gradual loss of energy because of atmospheric drag, eventually falling to Earth. They usually get burnt up in the Earth's atmosphere as they gain speed. This method of deorbiting is dependent upon several factors, including size, mass, material composition, and finally altitude. However, objects that are not in low Earth orbit could remain in orbit for as long as 2,000 years and pose a threat to future space missions. On March 13, 2009, when the *ISS* crew failed to anticipate space debris approaching the space station crew were forced into contingency operation. The *ISS* astronauts took cover in a *Soyuz* capsule to reduce the chance of penetration. The debris passed by the station at a safe distance and the crew was able to resume operations [2]. Upon collision with objects in space, the production of debris is only increased and it can be seen that there is indeed a critical density of debris beyond which a serious threat to collision-free space travel is posed. Clearly, there is a need to look at the mitigation of the production of space debris, prediction of the rate at which it is generated, and control of the level of debris that actually exists in space at any given time.

One approach to deorbit a piece of debris is to slow it down significantly so that, as a consequence of the loss of energy, the debris particle will lose altitude and eventually fall to the Earth. The danger of this approach is that an object in space may be mistakenly or deliberately construed as a piece of debris and destroyed, thus possibly sparking a war in space. Yet one method of slowing down an object in space is to fire a LASER in such a direction as to deorbit the object. Claude Phipps's [3, 4] early pioneering concept for space debris removal introduced in 1993 consists of a high-power pulsed, ground-based LASER system used to ablate a fraction of the debris in a specific orientation in order to slow it down. Essentially, this change of velocity would cause the debris to reenter the atmosphere and burn up. Space-based LASERS have also been considered, but a ground-based LASER system was expected to be more advantageous and simpler in operation and maintenance as well as lower on cost. Models have been constructed that allow one to simulate the trajectory of the debris before and after firing a LASER onto its surface. Such a simulation model would provide a useful tool for researchers in the field to investigate the application of LASERS to deorbiting space debris. Clearly, a basic requirement in order to do this is the availability of a catalogue of locations of space debris that is periodically updated. Levit and Marshall [5] have catalogued and recorded the Two Line Element (TLE) orbital data for up to 3 million objects. Levit and Marshall [5] suggest that a high-accuracy catalog based on the publicly available TLEs would be accurate enough to perform studies of changes in the orbital velocities below the cm/s range. Extending the work of Levit and Marshall [5], Mason et al. [6] have catalogued data on debris objects to determine a reasonable area-mass ratio for debris. In Mason et al. [6] the importance of the area-mass ratio of debris objects that must be considered in modeling the orbiting object for LASER ablation is highlighted and a LASER ablation system based on available off-the-shelf LASER technology is proposed. Liedahl [7] and Liedahl et al. [1] have developed the area matrix approach for impulse transfer calculations for ablation of orbiting objects using LASERS and analyzed orbiting bodies of many shapes of space debris, which include cubic, cylindrical, and spherical bodies. While the momentum transfer would depend on the shape of the targeted debris, the analysis has been extended to arbitrarily shaped bodies. From the work of Campbell [8] it is known that LASER interaction with a debris fragment will alter both the frequency and the orientation of the rotation vector. Wang, Zhang, and Wang [9] have performed impulse calculations and characteristic analysis of space debris using pulsed LASER ablation. Based on this calculation, one can estimate the typical impulse requirements for deorbiting an irregular shaped object in orbit around the Earth. A typical simulated trajectory of such an object is shown in Figure 1.17.

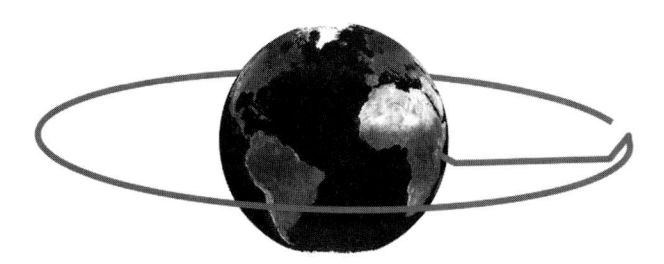

Figure 1.17 Typical example of deorbiting a piece of debris orbiting the Earth.

1.7 Venturing into Deep Space: Spacecraft with Endurance

Voyager 1, which was launched on September 5, 1977, became the only spacecraft to have entered the space beyond the Solar system in 2012, after travelling some 13 billion miles away from the Earth. *Voyager 2*, launched on August 20, 1977, was the only spacecraft to have flown past all four outer planets of the Solar system – Jupiter, Saturn, Uranus, and Neptune – and also visited their moons; Io and Europa orbiting Jupiter, Saturn's moon Titan, and Neptune's moon Triton. It is clear that spacecraft will continue to explore far beyond the solar system in the years to come as several projects have been proposed to explore interstellar space.

The European Space Agency is expected to launch the *JUpiter ICy moons Explorer* or *JUICE*, as the mission has been named, which is intended to explore Jupiter and 3 of its estimated 79 moons. The first three of the four Galilean moons, Europa, Callisto, Ganymede, and Io, are expected to be surveyed by the *JUICE* mission. The orbital spacecraft, which will not be accompanied by a lander, is expected to be the first spacecraft to orbit Ganymede, almost eight years after its launch.

To facilitate deep space missions, NASA has been continuously testing the NASA Evolutionary Xenon Thruster (NEXT) ion engine for about six years – longer than any other space propulsion thruster tested by it. During this period it has consumed less than 10% of the fuel required by a chemically propelled thruster, thus demonstrating its long-term efficiency and its capacity to survive with low fuel consumption. Consequently, ion engines are expected to be the main work horses for future deep space missions.

1.8 Planetary Rovers and Robot Walkers, Hoppers, and Crawlers for Exploration

NASA scientists are testing the new six-wheeled solar-powered rover for future planetary missions, which will carry a host of instruments on board to carry out several missions on Mars. Figure 1.18 is an illustration of the new rover, intended to be a replacement for the rover Curiosity. There is also considerable interest in developing walking robot models that will accompany the planetary rovers. For the development of such walking robot models, simulators have been designed to test the swing and stance

Figure 1.18 The new rover for future planetary missions. (courtesy: NASA)

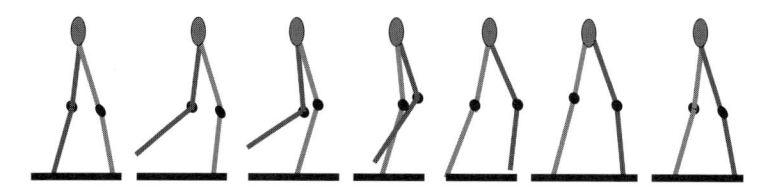

Figure 1.19 Illustration of the swing and stance gait cycle implemented on a robot model.

gait cycle in a remote planetary environment. Figure 1.19 illustrates a typical swing and stance gait cycle, based on Vepa [10].

1.9 Underwater Rovers and Aquanauts

Remotely Operated Vehicles (ROVs) are essentially remote control submarines with cameras that enable one to navigate in the deep waters of the oceans. They can be used for a variety of applications such as off-shore construction, surveillance, underwater inspection, and a host of other applications. Figure 1.20 illustrates the Seahorse-class Autonomous Underwater Vehicle as it is moved into position for launch. In Figure 1.21 is an AUV, designed by Bluefin Robotics for the US Navy, being tested on July 12, 2004, to detect mines and other underwater hazards in waters off Hawaii.

1.10 Humanoid Space Robots and Robonauts

In order to work safely and reliably in the environment around a spacecraft while it is on a space mission, NASA has proposed to design and build a series of humanoid space

1.10 Humanoid Space Robots and Robonauts

Figure 1.20 The Seahorse-class Autonomous Underwater Vehicle is moved into position for launch. (credit: Stocktrek Images / Getty Images)

Figure 1.21 AUV for underwater surveillance. (credit: Marco Garcia / Stringer / Getty Images News)

robots or Robonauts. NASA's Johnson Space Center, in Houston, Texas, produced the first Robonaut in 2000, which was only tested on Earth. Robonaut 2 was deployed on the *ISS* in February 2011. It was designed to work alongside the astronauts, to assist the crew with many of their time-consuming mechanical and repetitive chores, according to IEEE's *Spectrum*. Robonaut 2 was first upgraded with a pair of legs in 2014 to provide it with additional mobility. In 2015 it developed several problems, and in February

Figure 1.22 Robonaut 2 meeting with an astronaut on board the *International Space Station*. (credit: Stocktrek Images / Getty Images)

2018, it was decided by NASA to bring it back to Earth for repairs. Figure 1.22 illustrates the Robonaut 2 performing its tasks on board the *ISS*.

1.11   Robot Arms for Tele-Robotic Servicing

Generally, astronauts perform many chores outside a spacecraft while on a space mission. They could be assisted in this task by several robotic arms onboard the spacecraft. Figure 1.23 illustrates an astronaut anchored to the foot of Canadarm2, the robotic arm assisting the astronauts on the *ISS*. The *ISS* has several robotic arms on board to assist the astronauts. Canadarm2, also known as the Space Station Remote Manipulator System (SSRMS), was ferried to the *ISS* by the space shuttle *Endeavour* on April 19, 2001. It took about five days and a couple of spacewalks by the astronauts on board to install Canadarm2 on to the *ISS*. Since then, the arm has been used extensively outside of the space station, for transporting heavy payloads from one end of the space station to the other. Canadarm2 may be used to grab and dock approaching spacecraft, as and when they arrive at the space station. Flight controllers on the ground in Houston, Texas, can also operate Canadarm2 remotely, usually to assist and to relocate such pieces of equipment as a docking adapter to facilitate two or more spacewalks. The 18-m-long arm is one of three robotic components that now make up the space station's Mobile Servicing System, along with a robotic dexterous "hand" known as Dextre. There is also a base platform known as the Mobile Remote Servicer Base System (MBS), which facilitates the motion of both Canadarm2 and Dextre in the environment in the vicinity of the space station's spinal truss structure by sliding along a prismatic joint along the entire length of the space station.

1.11 Robot Arms for Tele-Robotic Servicing 19

Figure 1.23 An astronaut anchored to the foot of the Canadarm2 robotic arm. (credit: NASA / Handout / Getty Images News)

Figure 1.24 Canadarm2 and Dextre. (credit: Stocktrek Images / Getty Images)

Figure 1.24 shows the Canadarm2 and Dextre in operation, performing their usual chores. Dextre is a remotely operated, dexterous special purpose manipulator weighing 1,500 kg that can be controlled either from inside the space station or from the ground station in Houston, Texas. The manipulator has seven joints in its arms, constituting shoulders, elbows, and wrists, which are configured for maximum reach. One arm holds on to the space station while the other is working. This ensures that the manipulator is

Figure 1.25 The Dextre special purpose dexterous manipulator. (courtesy: NASA)

Figure 1.26 NASA's Double Asteroid Redirection Test (DART) spacecraft. (courtesy: NASA)

balanced and stable and also ensures that there is no collision between the two arms. Its width across its shoulders is about 2.7 m; the length of one arm is 3.5 m and can handle payloads weighing up to 600 kg. The hand is endowed with a force-moment touch sensor so it can grasp an object with the correct controlled force. Dextre is capable of installing or removing small payloads such as batteries. Dextre is shown in Figure 1.25. NASA's *Double Asteroid Redirection Test* (*DART*) spacecraft, which uses the NASA Evolutionary Xenon Thruster – Commercial (NEXT-C) solar electric propulsion system as its primary in-space propulsion system and is expected to be launched in late 2020 or in early 2021, is designed to demonstrate the kinetic impactor technique – a methodology to deliver an impact to the binary near-Earth asteroid (65803) Didymos to shift its orbit and to avoid a potential future collision with it. It is illustrated in Figure 1.26.

Also available on the space station are the European Space Agency's 11-m-long, two-limb European Robotic Arm, which is designed to be used with the Russian segment of the space station. A third arm that is fixed on the Japanese Experiment Module, the Remote Manipulator System (JEM-RMS), uses a similar grapple fixture to Canadarm2. Figure 1.27 shows the extended 2.1-m-long robotic arm of NASA's Mars rover Curiosity. Figure 1.28 shows the Phoenix robotic arm at work. The Phoenix manipulator is a 2.35-m-long arm, with an elbow joint in the middle, allowing the arm to dig about half a meter into the Martian surface and look for water or ice below the planet's surface.

Figure 1.27 The extended 2.1-m-long robotic arm of NASA's Mars rover Curiosity. (courtesy: NASA)

Figure 1.28 NASA's Phoenix robotic arm in action. (courtesy: NASA, JPL)

Figure 1.29 The Strela crane being moved by two astronauts. (courtesy: NASA)

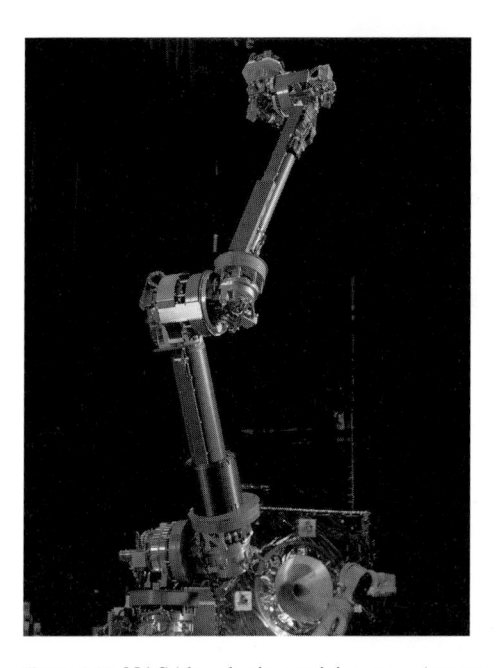

Figure 1.30 NASA's robotic servicing arm. (courtesy: NASA)

Also available on the space station is the Russian-built Strela robotic crane, used to lift and transport payloads in its vicinity. The Strela crane is shown in Figure 1.29 with two astronauts moving it into position.

NASA's robotic servicing arm, which, like a human arm, has seven degrees of freedom, a three-axis shoulder, a pitch actuator at the elbow, and a three-axis spherical wrist as well as a six-axis force torque sensor in the gripper, is shown in Figure 1.30.

Figure 1.31 The "Hedgehog" robot, which gets around by spinning and stopping three internal flywheels using motors and brakes. (credits: NASA, JPL-Caltech / Stanford)

1.12 Tumbling Cubes

As part of its plan to grab an asteroid, NASA is also funding the Tumbling Cubes project, which involves the design, build, and test of robotic tumbling cubes that can move by themselves in a low-gravity environment that may exist on the surface of an asteroid. These cubical objects, known as "Hedgehog" robots, can move essentially by bouncing off the asteroid, which involves jumping up, balancing, and "walking." The cubes contain a set of three flywheels rotating at high speed about three mutually perpendicular axes. The flywheels are used to generate moments acting on the cubes that facilitate the tumbling motion. An artist's impression of the "Hedgehog" robot is shown in Figure 1.31.

1.13 Collaborative Robotic Systems

NASA is also funding the development of an ultra-light robotic system and advanced command and control software that is designed to make use of existing spacecraft equipment and capabilities, including a standard geostationary (GEO) spacecraft platform and processor, to host five or more robotic manipulator arms that will function as a semi-autonomous robotic system for in-space satellite and structures assembly. The NASA Jet Propulsion Laboratory's Planetary Robotics Laboratory, a center for rapid prototyping of advanced robotic systems and the development of algorithms to control them in space, is researching multi-rover coordination for conceptual assembly operations on a planetary surface. A view of Jet Propulsion Laboratory's research test bed is shown in Figure 1.32. This is an example of collaborative robotic systems, albeit on a small scale.

Figure 1.32 A view of the research test bed at the NASA Jet Propulsion Laboratory's Planetary Robotics Laboratory. (courtesy: NASA, JPL)

1.14 The Meaning of Autonomy

Automatic control systems were introduced well before the Second World War, and the study of these systems was pioneered, among others, by the American engineers Harry Nyquist and Herman Bode. Almost always they involve the principle of feedback control and have evolved from the theory of servomechanisms, as they were known in the early days. Currently, there is considerable interest in the subject of autonomous control systems, and often the difference between the newer autonomous control systems and the classical automatic control systems is obscured. It is important to clearly understand the difference and also the uniquely characterizing features of autonomous control systems. First, autonomous systems arise from a synergy of various interacting subsystems. The principle of synergy is that that the sum is greater than the parts. A typical example is the mobile phone, which delivers a range of functions when all its interacting subsystems are assembled so the proper interaction between the subsystems is facilitated. In a classical automatic control system, decision making is done outside the controller. Generally, the controller is under the supervision of the operator in the sense that the set points and plans are set by the operator. There is no possibility of situational or spatial awareness, and it is also generally true that the controller has limited authority. A classical automatic control system tracks a given set point or a plan, an ordered collection of set points. Thus, tracking a desired sequence of set points or a commanded trajectory can be expected of such a system. In the case of an autonomous control system, decision making is within the remit of the controller. The controller is not under the supervision of an operator and it is generally true that the set points and plans are set by itself. To this end it is essential that such a system is capable of spatial awareness and/or situation awareness, which in turn implies monitoring the environment immediately affecting the performance of the control system. Thus, it can be said that an autonomous control system has the additional authority and the responsibility to manage its own "flight" plan, manage all external communications and maintain surveillance, monitor all internal systems for faults and its own "flight" path or trajectory, reconfigure/replan if necessary, manage contingencies, react when there is an obstacle (or conflict), and change its plan, course, or set points. The additional responsibilities generally fall into four categories: mission related issues, spatial and/or situation awareness, "flight" (trajectory) management issues, and systems related issues. At this point, it is important to mention that both automatic control systems and autonomous control systems are expected to deal with uncertainties, feedback architecture is common to both with adaptive or self-tuning control loops, and fault tolerance is also a feature of both. However, only autonomous control systems are expected to deal with changes in mission, (mission replanning), deal with obstacles or conflicts (route replanning), deal with faults both internal and "external" to the system (system reconfiguration), and be able to deal with contingencies (motion replanning). To sum it up, autonomous control systems offer the additional features of replanning and reconfiguration. Unmanned systems are expected to be able to perform autonomously. While almost all terrestrial vehicular systems are continuing to evolve from a manned, manually operated and controlled architecture to unmanned, autonomously controlled systems, space vehicles were naturally conceived as unmanned autonomous systems. Thus, space vehicles could serve as excellent prototypes for the study of the dynamics and control of autonomous vehicles. From a system engineering perspective, the conceptual design and integration of both the systems and components used in unmanned systems, including the locomotion, sensors, and computing systems, are needed to provide inherent autonomy capability (systems for autonomy) and the architectures, algorithms, and methods are needed to enable control and autonomy, including path-tracking control and high-level planning strategies (control for autonomy). A wider class of sensors or transducers and associated software are needed to meet the additional responsibilities (including RADAR, LIDAR, ultra-sonic SONAR, LASER rangers, imaging cameras operating over a wide spectrum, data acquisition, and actuators). To choose between different plans (or algorithms) means maintaining a database or bank of feasible flight plans (or algorithms) or the ability to synthesize them when needed. Synthesis of plans means that mission, route, or motion planning (including fast real time replanning), or all of these together, may be essential. Monitoring tasks may need to be integrated for optimum use of resources (for example, SLAM: Simultaneous localization and Mapping). Thus, there is a need for prototypes of autonomous vehicles that will serve as test beds for new algorithms as well as bench marking autonomous vehicles, which will allow the comparison of alternative algorithms. A hierarchy of vehicles of increasing complexity in terms of the environment they will need to move into is therefore essential. Examples of such cases are planetary rovers, satellites, automobiles, aircraft, and underwater and surface water vehicles.

1.15 Dynamics and Control of Space Vehicles

Space vehicle dynamics is the study of the motion of objects assembled by humans that are placed in an orbit around a planet, star, or comet in space. The subject of space vehicle dynamics is conveniently divided into two parts: Orbital mechanics, where the vehicle is assumed to be very small relative to the size of its orbit and only the motion of its center of mass is of interest; and attitude dynamics, where the motion of the body with reference to its orientation is important. In the first instance the trajectory of motion of the spacecraft or space vehicle's center of mass is the focus. The importance of attitude dynamics arises from the fact that the space vehicle is often required to point in a certain direction. Sometimes a part of the spacecraft is required to point toward the Earth while another part of the spacecraft must point to the Sun.

The study of space vehicle dynamics is also often associated with the need to change the trajectory of motion of its center of mass or the attitude or orientation of the spacecraft. There is a clear need sometimes to control the trajectory of motion of the center of mass of the space vehicle or control the orientation of parts of the spacecraft or both. Thus, one cannot study the dynamics of a spacecraft in isolation but in the context of the application and the orbit and the pointing requirements that the spacecraft must satisfy. When, in addition, robot manipulators or other robotic systems are also present on board a space vehicle, the problems of the dynamics and control of the complete system, including the interactions of the various subsystems with each other, must be viewed and studied from a holistic point of view.

1.16 The Future

In this first chapter a pictorial summary of the fascinating developments in the field of space robotics is presented. It is expected that it will motivate the reader in studying the dynamics and control of autonomous space vehicles and space robotics. One could look to the future to make predictions as to what sort of developments one could expect in this challenging and inspiring field of engineering. The projects that are ongoing and those that are being funded and are currently being pursued provide a glimpse of the kind of developments one could expect. Artificial intelligence is expected to play a significant role in the evolution of future robotic spacecraft and robotic manipulators.

One can expect to see large swarms of humanoid robots working alongside robotic manipulators dealing with a host of difficult tasks associated with space exploration. The coordinated use of robotic manipulators for assembling large structures in space is definitely to be expected in the not too distant future. Tools and techniques for the coordinated control of large swarms of robotic manipulators are currently being researched and developed and are expected to play a key role in pushing the frontiers of space robotics well beyond their current capabilities. The solutions to the emerging problems relating to the coordinated control of space robotic systems will result in the development of newer robotic systems, which will help not only in the exploration of space but also serve in revolutionizing travel and communications on Earth.

In the chapters that follow, the fascinating field of the dynamics and control of robotic spacecraft and space robotics is introduced, which provides the very foundation for all other advanced studies.

References

[1] Liedahl, D. A., Libby, S. B., and Rubenchik, A. (2010) Momentum transfer by laser ablation of irregularly shaped debris, Report Number: LLNL-PROC-423224, Livermore: Lawrence Livermore National Laboratory.

[2] Bergin, C. (2015) Debris from old Russian satellite forced ISS crew into contingency ops. [Online] Available at: www.nasaspaceflight.com/2015/07/debris-russian-satellite-iss-crew-contingency-ops/; Retrieved June 2018.

[3] Phipps C. R. (1994) LISK-BROOM: A laser concept for clearing space junk, in *AIP Conference Proceedings*, 318, *Laser Interaction and Related Plasma Phenomena, 11th International Workshop*, October, 1993, Monterey, CA, Miley, G., ed., New York: American Institute of Physics, 466–468.

[4] Phipps, C. R. et al. (1996) ORION: Clearing near-Earth space debris using a 20 kW 530 nm Earth based repetitively pulsed laser. *Laser and Particle Beams*, 14(1): 1–44.

[5] Levit C., and Marshall, W. (2011) Improved orbit predictions using two-line elements. *Advances in Space Research*, 47(7): 1107–1115.

[6] Mason, J., Stupl J., Marshall, W., and Levit, C. (2011) Orbital debris–debris collision avoidance. *Advances in Space Research*, 48(10): 1643–1655.

[7] Liedahl, D. A., Rubenchik, A., Libby, S. B., Nikolaev, S., and Phipps C. R. (2013) Pulsed laser interactions with space debris: Target shape effects. *Advances in Space Research*, 52 (5): 895–915.

[8] Campbell, J. W. (1996) Project ORION: Orbital debris removal using ground-based sensors and laser, NASA Technical Memorandum 108522.

[9] Wang C., Zhang, Y., and Wang, K. (2016) Impulse calculation and characteristic analysis of space debris by pulsed laser ablation. *Advances in Space Research*, 58, 1854–1863.

[10] Vepa, R. (2016) Bio-inspired modelling and active control of the simulation of human walking dynamics, *CLAWAR (Climbing and Walking Robot) Conference 2016*, September 12–14. In *Advances in Cooperative Robotics*, M. O. Tokhi and G. S. Virk, eds., Singapore: World Scientific, Section 3, pp. 55–62.

2 Space Vehicle Orbit Dynamics

2.1 Orbit Dynamics: An Introduction

All space vehicle flights occur under the influence of the gravitational force of a multitude of planetary objects in the Solar system. In particular, the motion of an artificial Earth satellite is primarily influenced by the gravitational pull of the Earth and the gravitational forces exerted on it by the Moon and the Sun. Given that the Moon is at distance of 384,440 km (semi-major axis), orbits the Earth in a near circular orbit (eccentricity, $e = 0.0549$), and has a mass of only 0.01226 of the Earth's mass, it is conceivable that for artificial Earth satellites orbiting the Earth at a distance of less than approximately 38,444 km the influence of the Moon can be neglected. Furthermore, the Sun is at a distance of $149,599 \times 10^3$ km, which is 400 times further away than the Moon, although it is 332,952 times heavier than the Earth. For an artificial satellite orbiting the Earth at a distance of less than approximately 38,440 km, the influence of the Sun can be assumed to be no different than its influence on the Earth.

2.2 Planetary Motion: The Two-Body Problem

As far as motion around the Earth is concerned, any orbiting artificial satellite and the Earth may be considered in isolation as two interacting independent celestial bodies as long as the artificial satellite is sufficiently close to Earth. This leads to the classical two-body approximation problem that serves as a valuable paradigm for developing the theory of planetary motion. The key question is: How close is sufficiently close? That is a question we shall not seek to answer yet, although it is indeed a fundamental one.

2.2.1 Kepler's Laws

The German astronomer Johann Kepler (1571–1630) formulated three empirical laws of planetary motion based on astronomical data provided to him by the Danish astronomer Tycho Brahe in the late 1590s. The laws were published over a period spanning a decade, at about the same time as Galileo was making his landmark astronomical observations. The laws (see, for example, Deutsch [1], Ball and Osborne [2], Prussing and Conway [3], and Bate [4]) are:

(i) The orbit of each planet is an ellipse with the Sun at one focus;
(ii) The line joining the Sun to the planet sweeps out equal areas in equal lengths of time; and finally,
(iii) The squares of the orbital periods of the planets are proportional to the cubes of their mean distances from the Sun.

2.2.2 Keplerian Motion of Two Bodies

It is to Sir Isaac Newton (1642–1727), the founder of Newtonian mechanics, the theory of gravitation, and differential calculus, that the undisputed honor of being the originator of the mathematical theory of planetary motion falls. When a particle moves in the vicinity of a celestial body, it experiences a force of attraction, which is directed toward a fixed center of attraction, located at the center of mass of the attracting body. The force is mutual in the sense that the attracting body experiences a force directed toward the center of mass of the particle of the same magnitude. This force is given by Newton's law of gravitation as,

$$F = -G\frac{mm_0}{r^2}, \tag{2.1}$$

where m_0 = the mass of the attracting body (fixed), G = the universal gravitational constant, and r = the distance between the centers of masses.

This attracting force F is always in the negative direction of the direction of r. There is no force acting in the direction of the orbit tangent or θ-direction. Hence the force is said to be a central force.

From Newton's Second Law of motion, the force in r-direction is the product of the mass of the particle m, and the acceleration in r-direction.

$$F_r = ma_r. \tag{2.2}$$

The acceleration in r-direction is found to be,

$$a_r = \ddot{r} - r\dot{\theta}^2. \tag{2.3}$$

This force is equal to the central gravitational attraction, thus,

$$-G(mm_0/r^2) = m(\ddot{r} - r\dot{\theta}^2). \tag{2.4}$$

Newton's Law also applies to the θ-direction, i.e.

$$F_\theta = ma_\theta. \tag{2.5}$$

In θ-direction, the angular velocity is equal to,

$$v = r\dot{\theta}. \tag{2.6}$$

The resulting acceleration in the tangential direction is,

$$a_\theta = r\ddot{\theta} + 2\dot{r}\dot{\theta}. \tag{2.7}$$

However, there is no force acting in θ-direction. Hence,

$$F_\theta = ma_\theta = 0 = m(r\ddot{\theta} + 2\dot{r}\dot{\theta}), \tag{2.8}$$

and,

$$0 = m \cdot \frac{1}{r}\frac{d}{dt}(r^2\dot{\theta}). \tag{2.9}$$

Both the mass of the particle and the separated distance are assumed to be nonzero, which results in the product $r^2\dot{\theta}$ being constant. This product is defined to be h, which represents the momentum of the particle per unit mass. Thus, the equation is simply a statement of the principle of conservation of momentum.

In order to obtain the shape of the path followed by m, $r = 1/u$ is substituted.
Therefore,

$$\dot{r} = -\left(\frac{1}{u^2}\right)\dot{u} = -h\left(\frac{du}{d\theta}\right), \quad \ddot{r} = -h^2 u^2 \left(\frac{d^2 u}{d\theta^2}\right). \tag{2.10}$$

Eliminating r from

$$-G\frac{mm_0}{r^2} = m(\ddot{r} - r\dot{\theta}^2), \tag{2.11}$$

the following equation is obtained:

$$-Gm_0 u^2 = -h^2 u^2 (d^2 u/d\theta^2) - \frac{1}{u}(hu^2)^2. \tag{2.12}$$

This is expressed as,

$$\Rightarrow \frac{Gm_0}{h^2} = \frac{d^2 u}{d\theta^2} + u. \tag{2.13}$$

Employing two integration constants, C, δ the solution for u can be expressed as,

$$u = \frac{1}{r} = C\cos(\theta + \delta) + \frac{Gm_0}{h^2}. \tag{2.14}$$

By choosing the x-axis suitably, it is possible to eliminate the phase angle, δ

$$\frac{1}{r} = C\cos\theta + \frac{Gm_0}{h^2}. \tag{2.15}$$

It follows that r has a minimum when $\theta = 0$.

Since a conic section is formed by the locus of a moving point, the ratio e of the distance of the moving point from a focus point to a line (the directrix) is constant.

$$e = \frac{r}{d - r\cos\theta} \Rightarrow \frac{1}{r} = \frac{1}{d}\cos\theta + \frac{1}{ed}. \tag{2.16}$$

By comparing with the equation of the motion for the particle m, the motion is found to be along a conic section with,

2.2 Planetary Motion: The Two-Body Problem

$$d = \frac{1}{C}, \quad ed = \frac{h^2}{Gm_0}, \quad e = \frac{h^2 C}{Gm_0}. \tag{2.17}$$

"e" is the eccentricity defining the shape of the orbit. The eccentricity for a circular orbit is always equal to 0. For a parabolic escape orbit (infinitely elongated ellipse), it is always equal to 1, and it is greater than 1 for a hyperbolic orbit. All elliptic orbits, therefore, have eccentricities somewhere between 0 and 1. Thus since,

$$\frac{1}{r} = \frac{1}{d}\cos\theta + \frac{1}{ed} \Rightarrow r = ed\left(\frac{1}{e\cos\theta + 1}\right). \tag{2.18}$$

Case I: ellipse ($e < 1$)

The distance between the masses of the particle and the attracting body is a minimum when $\theta = 0$ and conversely it is a maximum when $\theta = \pi$. The length of the ellipse is the sum of the minimum and maximum distances. Therefore,

$$2a = \frac{ed}{1+e} + \frac{ed}{1-e} = \frac{2ed}{1-e^2}. \tag{2.19}$$

This equation gives the orbit radius r in terms of the semi-major axis a.

$$\frac{1}{r} = \frac{1 + e\cos\theta}{a(1 - e^2)}. \tag{2.20}$$

It follows that,

$$r = a(1 - e^2)\left(\frac{1}{e\cos\theta + 1}\right), \tag{2.21}$$

which is a mathematical statement of Kepler's First Law as illustrated in Figure 2.1. The angle θ is known as the true anomaly and is denoted by f or ν in the literature. During the short period of time dt, the radius vector of the particle sweeps out an area of,

$$dA = (1/2)r(r\,d\theta). \tag{2.22}$$

This means that the rate at which the area is swept by the radius vector is equal to,

$$\dot{A} = (1/2)r^2\dot{\theta}. \tag{2.23}$$

However, since $r^2\dot{\theta} = h$ is constant, $\dot{A} = h/2$, a constant, and this results in a mathematical statement of Kepler's Second Law, illustrated in Figure 2.2. The period τ for

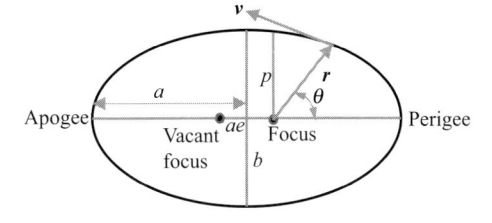

Figure 2.1 Elliptic orbit: Kepler's First Law.

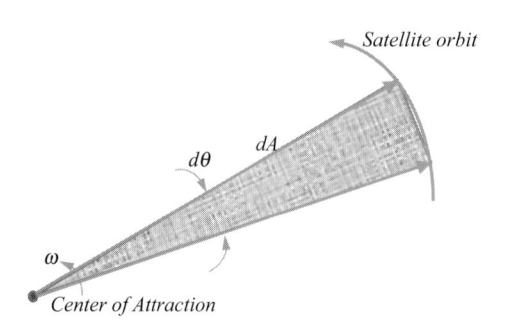

Figure 2.2 Illustration of Kepler's Second Law.

the elliptical orbit is defined by the total area of the ellipse divided by the constant rate at which the area is swept through by the radius vector. So it follows that,

$$\tau = \frac{\pi ab}{\frac{1}{2}r^2\dot\theta} = \frac{2\pi ab}{h}, \qquad (2.24)$$

where

$$b = a\sqrt{1-e^2}. \qquad (2.25)$$

By introducing the product of the mean radius of the central attracting body R and the absolute value of the acceleration due to gravity at the surface of the attracting body g, the period of motion can be expressed as,

$$\tau = \frac{2\pi a^{\frac{3}{2}}}{R\sqrt{g}}, \qquad (2.26)$$

which is Kepler's Third Law.

Case II: parabola ($e = 1$)

Hence one has,

$$\frac{1}{r} = \frac{1}{d}\cos\theta + \frac{1}{ed}. \qquad (2.27)$$

When $e = 1$, this equation of the motion is reduced to,

$$\frac{1}{r} = \frac{1}{d}(1 + \cos\theta). \qquad (2.28)$$

This means that as the angle between the center of the fixed body and the particle becomes π, the radius vector and the dimension become infinite.

Case III: hyperbola ($e > 1$)

Also based on the equation of the motion,

$$\frac{1}{r} = \frac{1}{d}\cos\theta + \frac{1}{ed}. \qquad (2.29)$$

The radial distance, r becomes infinite for the two values of the polar angle θ_1 and $-\theta_1$.

The physically possible motion is between the angles of θ_1 and $-\theta_1$. The angles in the remaining sector have negative values for r; therefore, this cannot exist except for

2.2 Planetary Motion: The Two-Body Problem

repulsive forces. The eccentricity of a hyperbolic orbit is always greater than unity, $e > 1$. The semi-major axis is negative, $a < 0$. Finally, the hyperbolic orbit has a property that no other orbit has. A vehicle traveling the length of the orbit will arrive from some point at infinity, and then fly by through the closest approach point (periapsis), and then leave, going to some point at infinity. The approach direction is from a direction $-v_\infty$ and the departure direction in the direction $+v_\infty$. The angle through which the vehicle turns is called the *turning angle*. This turning angle can be determined from the properties of the hyperbola. Consequent to this behavior, the flight path angle is not periodic but starts at $\pm\pi/2$ and finishes at $\mp\pi/2$.

Elliptic, parabolic, and hyperbolic orbits are compared in Figure 2.3. In Figure 2.4 a typical hyperbolic orbit is shown and compared with an elliptic orbit. Also shown are the focus where the central body is located and the empty or vacant focus, as well as the origin of the reference axes.

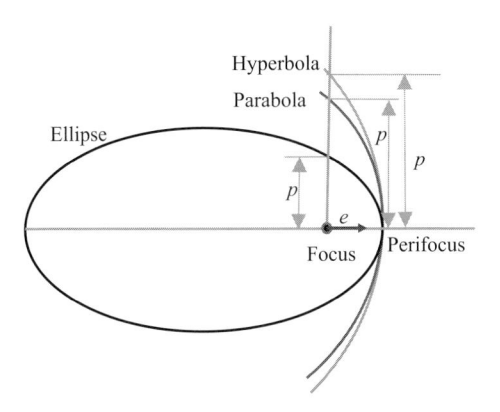

Figure 2.3 Elliptic, parabolic, and hyperbolic orbits.

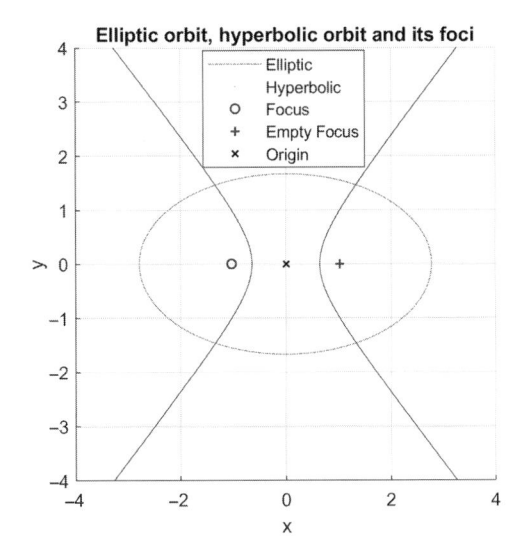

Figure 2.4 Hyperbolic orbit.

2.2.3 Orbital Elements

Seven numbers, known as satellite orbital elements, are required to define a satellite orbit about a planet. This set of seven numbers is called the satellite's "Keplerian" orbital elements, or just elements. These numbers define an ellipse, orient it about the planet, and place the satellite on the ellipse at a particular time. In the Keplerian model, satellites orbit in an ellipse of constant shape and orientation. Uniquely associated with an ellipse are two *foci* and when the two foci coincide, the orbit is circular with a constant radius. The planet is at one focus of the ellipse, not the center (unless the orbit ellipse is actually a perfect circle). The point on the orbit closest to this focus is the *perigee*, while the farthest point is the *apogee*. The minimum separation between the satellite and the planet is said to be at *periapse* and the maximum at *apoapse*. The direction a satellite or other body travels in orbit can be direct, or *prograde*, in which the satellite moves in the same direction as the planet rotates, or *retrograde*, moving in a direction opposite to the planet's rotation.

The primary orbital elements are numbers that: (i) orient the orbital plane in space; (ii) orient the orbital ellipse in the orbital plane; (iii) specify its shape and size; (iv) locate the satellite in the orbital ellipse.

Epoch: A set of orbital elements is a snapshot, at a particular time, of the orbit of a satellite. The epoch is simply a number that specifies the time at which the snapshot was taken.

Orbital inclination (i): The orbit ellipse lies in a plane known as the orbital plane. The orbital plane always goes through the center of the planet, although it may be tilted to any angle relative to the equator. Inclination is the angle between the orbital plane and the equatorial plane. By convention, the inclination is a number between 0 and 180 degrees. Orbits with inclination near 0 degrees are called equatorial orbits, while orbits with an inclination near 90 degrees are called polar. The intersection of the planet's equatorial plane (ecliptic plane) and the orbital plane is a line that is called the line of nodes. *Nodes* are points where an orbit crosses a plane. When an orbiting body crosses the elliptic plane going north, the node is referred to as the *ascending node*, while it is known as the *descending node* when it is southbound.

Right ascension of ascending node (Ω) is the second element that orients the orbital plane in space. Once the orbital inclination is defined, an infinite number of orbital planes are possible. Of the two nodes on the line of nodes, one is the ascending node, where the satellite crosses the equator from south to north. The other is called the descending node, where the satellite crosses the equator going north to south. By convention, we specify the location of the ascending node by the "right ascension of ascending node," which is an angle, measured at the center of the planet from the vernal equinox, a reference point in the sky where right ascension is defined to be zero, to the ascending node. It is an angle measured in the equatorial plane from the vernal equinox.

Argument of perigee (ω): Argument is yet another word for angle. Once the orbital plane is oriented in space, it is essential to orient the orbit ellipse in the orbital plane. This is done by a single angle element known as the argument of perigee.

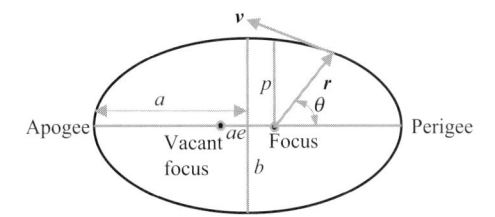

Figure 2.5 Definition of the semi-major axis, the semi-minor axis, the eccentricity, the semi-latus rectum, p, the true anomaly, the apogee, and the perigee.

Eccentricity: In the Keplerian orbit model, the satellite orbit is an ellipse. Eccentricity tells us the "shape" of the ellipse. When $e = 0$, the ellipse is a circle. When e is very near 1, the ellipse is very long and narrow. So far, the orbital elements define the orientation of the orbital plane, the orientation of the orbit ellipse in the orbital plane, and the shape of the orbit ellipse. One still needs to define the "size" of the orbit ellipse.

Mean Motion is usually given in units of revolutions per day. Kepler's third law of orbital motion gives us a precise relationship between the speed of the satellite and its distance from the planet. So by specifying the speed of the satellite or its mean motion, it is possible to define the size of the orbit. Sometimes the *semi-major axis* is specified instead of mean motion. The semi-major axis is one-half the length (measured the long way) of the orbit ellipse, and is directly related to mean motion by a simple equation (Figure 2.5). It now remains to specify exactly where the satellite is on this orbit ellipse at a particular time.

Mean anomaly is simply an angle that marches uniformly in time from 0 to 360 degrees during one revolution. It is defined to be 0 degrees at perigee, and therefore is 180 degrees at apogee. It is related to the *true anomaly* (f or v), which is often employed as an alternate element. It is a term used to describe the locations of various points in an orbit. It is the angular distance of a point in an orbit past the point at periapsis, the point on the orbit referred to as the perigee, measured in degrees at the focus nearer to the perigee, which is also where the planets center is located. For example, a satellite might cross a planet's equator at 30 degrees true anomaly, which defines where the satellite is on this orbit ellipse at a particular time.

There is one important secondary element that is optional: the *drag*. The drag orbital element defines the rate at which mean motion is changing due to drag or other related effects (Figure 2.6).

2.2.4 Two-Body Problem in a Plane: Position and Velocity in an Elliptic Orbit

Consider two planetary bodies, with one orbiting the other. The masses of the two bodies are m_1 and m_2 and the position vector of the second body relative to the first is $\mathbf{r}$. According to Newton's law of gravitation, the force of attraction on body 2 due to the attraction of body 1 is given by,

$$\mathbf{F} = -\frac{Gm_1m_2}{|\mathbf{r}|^2}\left(\frac{\mathbf{r}}{|\mathbf{r}|}\right). \tag{2.30}$$

Space Vehicle Orbit Dynamics

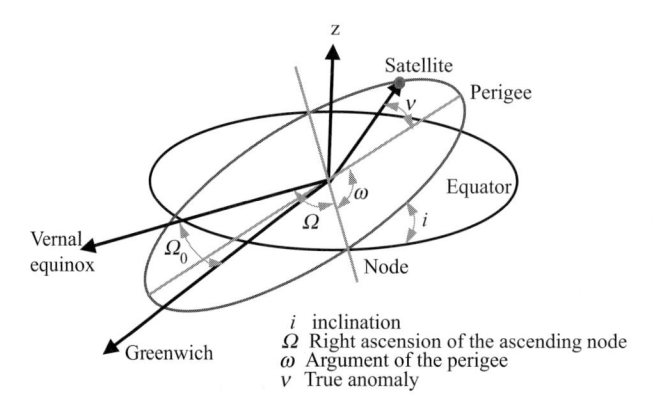

Figure 2.6 Definition of the orbit parameters.

i inclination
Ω Right ascension of the ascending node
ω Argument of the perigee
ν True anomaly

Equation of relative motion: If two masses m_1 and m_2 are separated by a distance r, then the equation of motion due to their mutual gravitation attraction is obtained by inserting Newton's equation for the force due to gravitational attraction into Newton's second and third laws of motion. The forces $\mathbf{F}_1$ and $\mathbf{F}_2$ due to gravitational attraction acting on the masses may be, respectively, expressed as,

$$\mathbf{F}_1 = \frac{Gm_1m_2}{r^2}\left(\frac{\mathbf{r}}{r}\right), \quad \mathbf{F}_2 = -\frac{Gm_1m_2}{r^2}\left(\frac{\mathbf{r}}{r}\right), \quad \mathbf{r} = \mathbf{r}_2 - \mathbf{r}_1. \tag{2.31}$$

In terms of an inertial reference frame (a reference frame completely fixed in space) according to Newton's second law,

$$m_1\ddot{\mathbf{r}}_1 = \mathbf{F}_1, \quad m_2\ddot{\mathbf{r}}_2 = \mathbf{F}_2. \tag{2.32}$$

Hence, the equations of motion for the two masses and the equation of the relative motion, respectively, are,

$$\ddot{\mathbf{r}}_1 = \mathbf{F}_1/m_1, \quad \ddot{\mathbf{r}}_2 = \mathbf{F}_2/m_2, \quad \ddot{\mathbf{r}} = \ddot{\mathbf{r}}_2 - \ddot{\mathbf{r}}_1 = -\left(\frac{G(m_1+m_2)}{r^2}\right)\left(\frac{\mathbf{r}}{r}\right). \tag{2.33}$$

Thus, the equation of the relative motion is,

$$\ddot{\mathbf{r}} + \left(\frac{G(m_1+m_2)}{r^2}\right)\left(\frac{\mathbf{r}}{r}\right) = 0. \tag{2.34}$$

Let, $\mu = G(m_1 + m_2) \approx Gm_1$; note that in *Système International* (SI) units, the units of μ are,

$$\text{Nm}^2/\text{kg} = \text{m}^3/\text{s}^2. \tag{2.35}$$

The equation of relative motion of the two bodies is given by,

$$\ddot{\mathbf{r}} + \frac{G(m_1+m_2)}{|\mathbf{r}|^2}\frac{\mathbf{r}}{|\mathbf{r}|} = \ddot{\mathbf{r}} + \frac{\mu}{|\mathbf{r}|^2}\frac{\mathbf{r}}{|\mathbf{r}|} = \mathbf{0}. \tag{2.36}$$

2.2 Planetary Motion: The Two-Body Problem

This is the equation of *unperturbed* motion due to gravity of mass m_2 with respect to mass m_1.

Position in an elliptic orbit: Consider the cross product $\mathbf{r} \times \dot{\mathbf{r}} = \mathbf{h}$. Differentiating the equation on both sides $\dot{\mathbf{r}} \times \dot{\mathbf{r}} + \mathbf{r} \times \ddot{\mathbf{r}} = \dot{\mathbf{h}} = 0$. But since the cross product of any vector with itself is zero $\dot{\mathbf{r}} \times \dot{\mathbf{r}} = 0$, and it follows that $\mathbf{r} \times \dot{\mathbf{r}} = \mathbf{h}$ provided $\mathbf{r} \times \ddot{\mathbf{r}} = 0$.

However, taking the cross product $\mathbf{r}$ with equation for relative motion,

$$\mathbf{r} \times \ddot{\mathbf{r}} = -\left(\frac{G(m_1 + m_2)}{r^2}\right)\left(\frac{\mathbf{r} \times \mathbf{r}}{r}\right) = 0. \tag{2.37}$$

Hence, $\mathbf{r} \times \dot{\mathbf{r}} = \mathbf{h}$. Thus the equation,

$$\frac{d}{dt}(\mathbf{r} \times \dot{\mathbf{r}}) = \mathbf{0}, \tag{2.38}$$

may be manipulated and integrated to give,

$$\mathbf{r} \times \dot{\mathbf{r}} = \mathbf{h}, \tag{2.39}$$

where $\mathbf{h}$ is a constant. "$\mathbf{h}$" may be interpreted as the moment of momentum vector per unit mass. It is mutually orthogonal to both $\mathbf{r}$ and $\dot{\mathbf{r}}$. Furthermore, given that r is magnitude of the vector $\mathbf{r}$,

$$\mathbf{r} \cdot \mathbf{r} = r^2, \quad \mathbf{r} \cdot \dot{\mathbf{r}} = \frac{1}{2}\frac{d}{dt}(\mathbf{r} \cdot \mathbf{r}) = \frac{1}{2}\frac{d}{dt}r^2 = \dot{r}. \tag{2.40}$$

Moreover, from the equation of relative motion,

$$\ddot{\mathbf{r}} \times \mathbf{h} = -\frac{\mu}{r^3}(\mathbf{r} \times \mathbf{h}) = -\frac{\mu}{r^3}(\mathbf{r} \times (\mathbf{r} \times \dot{\mathbf{r}})) = \frac{\mu}{r^3}(\dot{\mathbf{r}}(\mathbf{r} \cdot \mathbf{r}) - \mathbf{r}(\mathbf{r} \cdot \dot{\mathbf{r}})), \tag{2.41}$$

$$\frac{\mu}{r^3}(\dot{\mathbf{r}}(\mathbf{r} \cdot \mathbf{r}) - \mathbf{r}(\mathbf{r} \cdot \dot{\mathbf{r}})) = \mu\left(\frac{\dot{\mathbf{r}}r}{r^2} - \frac{\mathbf{r}\dot{r}}{r^2}\right) = \mu\frac{d}{dt}\left(\frac{\mathbf{r}}{r}\right). \tag{2.42}$$

Hence, take the cross-product of $\dot{\mathbf{r}}$ for two-body relative motion with $\mathbf{h}$, so the cross product,

$$\dot{\mathbf{r}} \times \mathbf{h} = G(m_1 + m_2)\left(\frac{\mathbf{r}}{|\mathbf{r}|} + \mathbf{e}\right) = \mu\left(\frac{\mathbf{r}}{|\mathbf{r}|} + \mathbf{e}\right), \tag{2.43}$$

where $\mathbf{e}$ is a dimensionless vector constant of integration and for some reference value of the acceleration due to gravity and the magnitude r of the vector $\mathbf{r}$,

$$\mu = G(m_1 + m_2) = g_0 r_0^2. \tag{2.44}$$

First we note that,

$$\dot{\mathbf{r}} = \frac{\mathbf{h}}{h^2} \times \mu\left(\frac{\mathbf{r}}{|\mathbf{r}|} + \mathbf{e}\right). \tag{2.45}$$

But $\dot{\mathbf{r}} \times \mathbf{h}$ is also equal to,

$$\dot{\mathbf{r}} \times \mathbf{h} = (\mathbf{v} \times (\mathbf{r} \times \mathbf{v})) = \dot{\mathbf{r}} \times (\mathbf{r} \times \dot{\mathbf{r}}) = \mathbf{r}(\dot{\mathbf{r}} \cdot \dot{\mathbf{r}}) - \dot{\mathbf{r}}(\dot{\mathbf{r}} \cdot \mathbf{r}) = \mathbf{r}v^2 - \mathbf{v}(\mathbf{v} \cdot \mathbf{r}). \tag{2.46}$$

(Note that for any three vectors, **u**, **v**, and **w**, one has the vector identity, $\mathbf{u} \times (\mathbf{v} \times \mathbf{w}) = \mathbf{v}(\mathbf{u} \cdot \mathbf{w}) - \mathbf{w}(\mathbf{u} \cdot \mathbf{v})$ and $\mathbf{u} \cdot (\mathbf{v} \times \mathbf{w}) = \mathbf{w} \cdot (\mathbf{u} \times \mathbf{v})$.)
Hence,

$$\mathbf{e} = \frac{1}{\mu}\left(\mathbf{r}v^2 - \mathbf{v}(\mathbf{v} \cdot \mathbf{r})\right) - \frac{\mathbf{r}}{|\mathbf{r}|}. \tag{2.47}$$

Furthermore, the pair of equations for **h** and for **e** which are both integrals of the motion and orthogonal to each other, completely characterize the motion of the orbiting body. Moreover, taking the dot product of **e** with **r** gives,

$$\mathbf{r} \cdot \mathbf{e} = \frac{1}{\mu}\left(r^2 v^2 - (\mathbf{v} \cdot \mathbf{r})^2\right) - r, \quad \mu(r + \mathbf{r} \cdot \mathbf{e}) = \left(r^2 v^2 - (\mathbf{v} \cdot \mathbf{r})^2\right) = h^2. \tag{2.48}$$

Hence, the position of the satellite from a focus is,

$$r = \frac{h^2 r}{h^2} = \frac{h^2}{\mu} \times \frac{r}{(r + \mathbf{r} \cdot \mathbf{e})} = \frac{h^2}{\mu}\left(\frac{1}{1 + e \cos v}\right), \tag{2.49}$$

where ae is the distance of the focus from the center.

From the first result obtained in equation 2.49, one may employ the reciprocal of the magnitude of the position vector as the dependent variable, and the solution for the orbit is an *ellipse* that may be expressed in terms of the angle v between the vectors **r** and **e** known as the *true anomaly* and the magnitudes of the vectors **h** and **e**, h, and e, where the latter determines the *eccentricity* of the orbit as:

$$r = \frac{h^2}{G(m_1 + m_2)} \times \frac{1}{(1 + e \cos v)}. \tag{2.50}$$

But from Equations (2.16) and (2.17),

$$r = \frac{e}{C}\left(\frac{1}{1 + e \cos v}\right),$$

where C is constant of integration. Hence,

$$2a = \frac{e/C}{1 + e} + \frac{e/C}{1 - e} = \frac{1}{C} \times \frac{2e}{1 - e^2}, \quad C = \frac{1}{a} \times \frac{e}{1 - e^2}. \tag{2.51}$$

Thus,

$$r = \frac{a(1 - e^2)}{(1 + e \cos v)} = \frac{p}{(1 + e \cos v)}, \tag{2.52}$$

where a is the *semi-major axis* and p is the *semi-latus rectum* of the ellipse.
Since in terms of the *semi-latus rectum p* of the ellipse,

$$h^2 = \mu a(1 - e^2) = \mu p. \tag{2.53}$$

Hence, one also has,

$$p = h^2/\mu, \quad a = p/1 - e^2 = h^2/\mu(1 - e^2). \tag{2.54}$$

***Periapsis* and *apoapsis*:** At the *periapsis* $v = 0$, and $v = \pi$ at the *apoapsis*.
The point of closest approach to the main body or periapsis and the most distant point or *apoapsis* are, respectively, given by,

$$r_p = p/(1 + e) = a(1 - e) \quad \text{and} \quad r_a = p/(1 - e) = a(1 + e). \tag{2.55}$$

Angular rate and orbit period: Since, $h = r^2\dot{\theta} = r^2\dot{v}$, and rearranging, for the time rate of change of the true anomaly, $\dot{v} = h/r^2$. Substituting for r,

$$\dot{v} = (1 + e\cos v)^2 \sqrt{\mu/p^3}. \tag{2.56}$$

Consider the rate at which the area of the ellipse is swept and hence according to Kepler's third law,

$$T = 2\pi ab/h, \quad h^2 = \mu a(1 - e^2), \quad \text{and} \quad b = a\sqrt{1 - e^2}. \tag{2.57}$$

The *orbital period* is given as,

$$T = 2\pi\sqrt{a^3/\mu}. \tag{2.58}$$

If the time of flight is $t - t_0$ and the *mean anomaly M* is defined by,

$$M = (t - t_0)\sqrt{\mu/a^3} = 2\pi(t - t_0)/T = n(t - t_0), \tag{2.59}$$

where n is the mean orbital rate.

2.2.5 Orbital Energy: The Visa-Viva Equation

Consider the total kinetic and potential energy and the total energy of an orbiting body of mass m in a two-body model. The total kinetic energy of a mass is $T = mv^2/2$. The potential energy is obtained from the work done by integrating the gravitational force from r to ∞ and is,

$$V = -\int_r^\infty \frac{m\mu}{r^2} dr = -m\frac{\mu}{r}. \tag{2.60}$$

The total energy is a constant, mC. Hence,

$$\frac{1}{2}m\left(v^2 - \frac{2\mu}{r}\right) = \frac{1}{2}mC, \quad \text{or} \quad \left(v^2 - \frac{2\mu}{r}\right) = C, \quad \frac{1}{r} = \frac{1 + e\cos f}{a(1 - e^2)}. \tag{2.61}$$

To evaluate C the total energy may be evaluated at the periapsis.

$$r_p = \frac{a(1 - e^2)}{(1 + e)} = a(1 - e). \tag{2.62}$$

Since $\mathbf{r} \times \dot{\mathbf{r}} = \mathbf{h}$, at the periapsis where the velocity vector is perpendicular to the position vector $h = v_p r_p$. Thus,

$$v_p = \frac{h}{r_p} = \frac{1}{r_p}\sqrt{\mu p} = \frac{1}{r_p}\sqrt{\mu a(1-e^2)}, \quad v_p^2 = \frac{1}{r_p}\mu(1+e). \tag{2.63}$$

Hence, evaluating the specific total energy constant C,

$$\left(v^2 - \frac{2\mu}{r}\right) = \frac{\mu}{r_p}(1+e-2) = -\frac{\mu}{a}. \tag{2.64}$$

Thus, one obtains the *visa-viva* equation,

$$v^2 = \mu\left(\frac{2}{r} - \frac{1}{a}\right). \tag{2.65}$$

Otherwise observe that the time rate of change of the total specific energy is given by,

$$\frac{dE}{dt} = \gamma_p \cdot \mathbf{v} = \frac{\mu}{2a^2}\frac{da}{dt}, \tag{2.66}$$

where γ_p is the perturbing acceleration in the direction of motion of the orbiting body. Rearranging the visa-viva equation,

$$v^2 = \dot{r}^2 + r^2\dot{\nu}^2 = \dot{r}^2 + \frac{h^2}{r^2} = \mu\left(\frac{2}{r} - \frac{1}{a}\right), \tag{2.67}$$

$$\dot{r} = \pm\frac{1}{r}\sqrt{\mu r\left(2 - \frac{r}{a}\right) - h^2}. \tag{2.68}$$

Circular orbits and the velocity of escape: Consider the case when $e = 0$ and the case when $e = 1$. For a circular orbit $e = 0$,

$$m\frac{v^2}{r} = mg = m\frac{\mu}{r^2}. \tag{2.69}$$

For escape, $e = 1$. Let $a = \infty$ in the energy equation,

$$\left(v^2 - \frac{2\mu}{r}\right) = -\frac{\mu}{a}. \tag{2.64}$$

Thus, the velocity in a circular orbit and the escape velocity are, respectively, given by,

$$V_{circular} = \sqrt{\mu/r}, \quad V_{escape} = \sqrt{2\mu/r}. \tag{2.70}$$

For the velocity at *periapse* (a maximum), let the position be minimum $\left(r_p = a(1-e)\right)$ and *vice-versa*.

Hence, at the apogee and at the perigee the satellite's velocities are,

$$V_{perigee} = \sqrt{\mu(1+e)/a(1-e)} \equiv V_{max}, \quad V_{apogee} = \sqrt{\mu(1-e)/a(1+e)} \equiv V_{min}. \tag{2.71}$$

2.2.6 Position and Time in Elliptic Orbit

We define the position coordinates in the elliptic orbit in terms of the *eccentric anomaly* defined at the center of the ellipse (Figure 2.7) rather than at the focus as,

$$x_c = a\cos E, \quad y_c = b\sin E. \tag{2.72}$$

Hence,

$$r^2 = y_c^2 + (ae - x_c)^2 = b^2 \sin^2 E + (ae - a\cos E)^2. \tag{2.73}$$

It follows that,

$$r^2 = b^2 \sin^2 E + (ae - a\cos E)^2 = a^2(1-e^2)(1-\cos^2 E) + (ae - a\cos E)^2. \tag{2.74}$$

Consequently,

$$r^2/a^2 = (1-e^2)(1-\cos^2 E) + (e-\cos E)^2,$$
$$r^2/a^2 = (1-e^2) - (1-e^2)\cos^2 E + e^2 - 2e\cos E + \cos^2 E, \tag{2.75}$$

which reduces to,

$$r^2/a^2 = 1 + e^2 \cos^2 E - 2e\cos E = (1 - e\cos E)^2. \tag{2.76}$$

Using the fact that $r = a(1 - e\cos E)$ and since,

$$\cos E = \frac{x}{a} = \frac{ae + r\cos v}{a} = e + (1 - e\cos E)\cos v, \quad \sin E = \sqrt{1 - \cos^2 E}. \tag{2.77}$$

Gauss's equations relating the *eccentric anomaly* E to the *true anomaly* v are given by,

$$\cos v = \frac{\cos E - e}{1 - e\cos E}, \quad \sin v = \frac{\sin E\sqrt{1-e^2}}{1 - e\cos E}, \quad \cos E = \frac{\cos v + e}{1 + e\cos v},$$

$$\sin E = \frac{\sin v\sqrt{1-e^2}}{1 + e\cos v}. \tag{2.78}$$

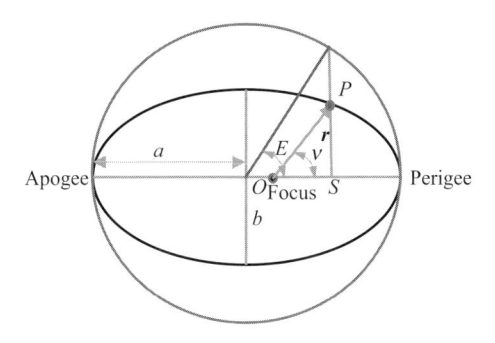

Figure 2.7 Definition of the eccentric anomaly.

Thus,

$$\tan E = \frac{\sin f \sqrt{1-e^2}}{e+\cos f}, \quad \tan\frac{E}{2} = \sqrt{\frac{1-e}{1+e}} \tan\frac{v}{2}. \quad (2.79)$$

Since $r = a(1-e\cos E)$ and $r = p(1+e\cos v)^{-1}$,

$$\dot{r} = ae(\sin E)\dot{E} = re(1+e\cos v)^{-1}\sin v\,\dot{v} = re(1-e^2)^{-1/2}\sin E\,\dot{v}, \quad \dot{v} = h/r^2. \quad (2.80)$$

Hence,

$$r\dot{E} = hb/pa, \quad h^2 = \mu a(1-e^2) = \mu p, \quad b^2 = a^2(1-e^2), \quad a = p/1-e^2. \quad (2.81)$$

Hence,

$$r\dot{E} = hb/pa = \sqrt{\mu/a} = a(1-e\cos E)\dot{E}. \quad (2.82)$$

Thus,

$$\sqrt{\mu/a^3} \equiv n = \dot{M} = (1-e\cos E)\dot{E}. \quad (2.83)$$

Integrating leads to *Kepler's equation* relating the *eccentric anomaly E* to the *mean anomaly M* is obtained as,

$$M = n(t-t_0) = E - e\sin E, \quad (2.84)$$

which relates the time required to travel from the closest approach point or *periapse* through the true anomaly along the orbit or the mean motion of the orbiting body to the eccentric anomaly.

2.2.7 Lambert's Theorem

Lambert's theorem can be considered an alternative form of Kepler's equation. By employing the geometry of the ellipse, Lambert demonstrated for elliptical orbits that the time taken to traverse an arc of the orbit is a function of the semi-major axis a, the sum of the radius vectors at the two extremities of the arc $r_1 + r_2$, and the length of the chord of the arc c. For the position vectors of the satellite at the two extremities of the arc,

$$r_i = a(1 - e\cos E_i), \quad i = 1, 2. \quad (2.85)$$

Hence,

$$r_1 + r_2 = 2a - 2ae\cos\bar{E}\cos\Delta E, \quad \bar{E} = (E_1 + E_2)/2, \quad \Delta E = (E_1 - E_2)/2. \quad (2.86)$$

Also,

$$a(\cos E_2 - \cos E_1) = x_{c2} - x_{c1}, \quad b(\sin E_2 - \sin E_1) = y_{c2} - y_{c1}. \quad (2.87)$$

Thus the chord length satisfies the relation,

$$c^2 = a^2(\cos E_2 - \cos E_1)^2 + b^2(\sin E_2 - \sin E_1)^2 = 4a^2 \sin^2 \Delta E (1 - e^2 \cos^2 \bar{E}). \tag{2.88}$$

Consequently,

$$(r_1 + r_2 \pm c)/a = 2 - 2e\cos\bar{E}\cos\Delta E \pm 2\sin\Delta E \sin\left(\cos^{-1}(e\cos\bar{E})\right), \tag{2.89}$$

It follows that,

$$\frac{r_1 + r_2 \pm c}{a} = 2 - 2\{e\cos\bar{E}\cos\Delta E \mp \sin\Delta E \sin\left(\cos^{-1}(e\cos\bar{E})\right)\}, \tag{2.90}$$

or,

$$\frac{r_1 + r_2 \pm c}{a} = 2 - 2\cos(\pm\Delta E + \cos^{-1}(e\cos\bar{E})), \tag{2.91}$$

and that,

$$2\sin^{-1}\left(\frac{1}{2}\left(\frac{r_1 + r_2 \pm c}{a}\right)^{1/2}\right) = \pm\Delta E + \cos^{-1}(e\cos\bar{E}). \tag{2.92}$$

If we define two angles α and β such that,

$$\frac{\alpha}{2} = \sin^{-1}\left(\frac{1}{2}\left(\frac{r_1 + r_2 + c}{a}\right)^{1/2}\right), \quad \frac{\beta}{2} = \sin^{-1}\left(\frac{1}{2}\left(\frac{r_1 + r_2 - c}{a}\right)^{1/2}\right), \tag{2.93}$$

then it follows that,

$$2\Delta E = \alpha - \beta, \quad e\cos\bar{E} = \cos\left(\frac{\alpha + \beta}{2}\right). \tag{2.94}$$

From Kepler's equations,

$$n(t_1 - t_0) = E_1 - e\sin E_1, \quad n(t_2 - t_0) = E_2 - e\sin E_2, \tag{2.95}$$

$$n(t_1 - t_2) = 2\Delta E - 2e\sin\Delta E \cos\bar{E} = \alpha - \beta - 2\sin\left(\frac{\alpha - \beta}{2}\right)\cos\left(\frac{\alpha + \beta}{2}\right). \tag{2.96}$$

Finally, we have Lambert's theorem, which is,

$$n(t_1 - t_2) = (\alpha - \sin\alpha) - (\beta - \sin\beta). \tag{2.97}$$

2.2.8 Orbit Inclination, Argument of the Ascending Node, Argument of the Perigee, and True Anomaly

Given that the nodal vector, in the direction of the line of nodes and directed toward the ascending node, is defined by $\mathbf{n} = \mathbf{k} \times \mathbf{h}$, and that (**i j k**) are unit vectors representing the directions of the three reference axes of an Earth-centered inertial (ECI) Frame, show that the inclination, the argument of the ascending node, the argument of the perigee, and the true anomaly satisfy the relations,

$$\cos i = \frac{\mathbf{h}\cdot\mathbf{k}}{|\mathbf{h}|}, \quad \cos\Omega = \frac{\mathbf{n}\cdot\mathbf{i}}{|\mathbf{n}|}, \quad \Omega < \pi \text{ when } \mathbf{n}\cdot\mathbf{j} > 0,$$

$$\cos\omega = \frac{\mathbf{n}\cdot\mathbf{e}}{|\mathbf{n}||\mathbf{e}|}, \quad \omega < \pi \text{ when } \mathbf{e}\cdot\mathbf{k} > 0, \quad \text{and} \quad \cos\nu = \frac{\mathbf{e}\cdot\mathbf{r}}{|\mathbf{e}||\mathbf{r}|}. \qquad (2.98)$$

Thus, given that $\mathbf{n} = \mathbf{k} \times \mathbf{h}$, one may summarize the relations between the orbital elements and the vectors $\mathbf{h}$ and $\mathbf{e}$ as,

$$a = |\mathbf{h}|^2/\mu\left(1 - |\mathbf{e}|^2\right), \quad e = |\mathbf{e}|, \quad \cos i = \frac{\mathbf{h}\cdot\mathbf{k}}{|\mathbf{h}|}, \quad \cos\Omega = \frac{\mathbf{n}\cdot\mathbf{i}}{|\mathbf{n}|},$$

$$\cos\omega = \frac{\mathbf{n}\cdot\mathbf{e}}{|\mathbf{n}||\mathbf{e}|}, \quad \cos\nu = \frac{\mathbf{e}\cdot\mathbf{r}}{|\mathbf{e}||\mathbf{r}|}. \qquad (2.99)$$

The relationship between the position of the satellite in the peri-focal coordinate system (the origin of the coordinate system is located at the focus, and two of those axes are in the orbit plane, one of which passes through the perigee) and the inertial reference frame in space is given 3–1–3 Euler angle sequence or Ω–i–ω sequence. Hence, the location of the perigee in the inertial reference frame is given by,

$$[x \ y \ z]^T = \Phi[r_{perigee} \ 0 \ 0]^T, \qquad (2.100)$$

$$\Phi = \begin{bmatrix} \cos\omega\cos\Omega - \sin\omega\sin\Omega\cos i & -\sin\omega\cos\Omega - \cos\omega\sin\Omega\cos i & \sin\Omega\sin i \\ \cos\omega\sin\Omega + \sin\omega\cos\Omega\cos i & -\sin\omega\sin\Omega + \cos\omega\cos\Omega\cos i & -\cos\Omega\sin i \\ \sin\omega\sin i & \cos\omega\sin i & \cos i \end{bmatrix}.$$

$$(2.101)$$

The position and velocity vectors of the satellite can also be defined. The explicit transformations in terms of $\theta = \omega + \nu$ are,

$$\mathbf{r} = r\mathbf{e}_r = r\begin{bmatrix} \cos\theta\cos\Omega - \sin\theta\sin\Omega\cos i \\ \cos\theta\sin\Omega + \sin\theta\cos\Omega\cos i \\ \sin\theta\sin i \end{bmatrix}, \qquad (2.102)$$

$$\dot{\mathbf{r}} = \dot{r}\mathbf{e}_r + r\dot{\mathbf{e}}_r = \dot{r}\mathbf{e}_r + r\dot\theta\frac{d\mathbf{e}_r}{d\theta} \equiv \dot{r}\mathbf{e}_r + r\dot\theta\mathbf{e}_\theta. \qquad (2.103)$$

Hence,

$$\dot{\mathbf{r}} = \dot{r}\begin{bmatrix} \cos\theta\cos\Omega - \sin\theta\sin\Omega\cos i \\ \cos\theta\sin\Omega + \sin\theta\cos\Omega\cos i \\ \sin\theta\sin i \end{bmatrix} + r\dot\theta\begin{bmatrix} -\sin\theta\cos\Omega - \cos\theta\sin\Omega\cos i \\ -\sin\theta\sin\Omega + \cos\theta\cos\Omega\cos i \\ \cos\theta\sin i \end{bmatrix}.$$

$$(2.104)$$

2.2 Planetary Motion: The Two-Body Problem

Consider the equation for motion of an orbiting vehicle relative to large planet given by,

$$\ddot{\mathbf{r}} + \left(\frac{G(m_1 + m_2)}{r^2}\right)\left(\frac{\mathbf{r}}{r}\right) = \ddot{\mathbf{r}} + \frac{\mu \mathbf{r}}{r^3} = 0. \quad (2.105)$$

One may express the position vector in a peri-focal (also known as *dextral*) coordinate system as,

$$\mathbf{r} = \mathbf{r}_P = [r\cos\nu \quad r\sin\nu \quad 0]^T, \quad (2.106)$$

where,

$$r = \frac{a(1-e^2)}{(1+e\cos\nu)} = \frac{p}{(1+e\cos\nu)}. \quad (2.107)$$

Hence,

$$\dot{r} = \frac{pe\sin\nu}{(1+e\cos\nu)^2}\dot{\nu} = \frac{re\sin\nu}{(1+e\cos\nu)}\dot{\nu} = \frac{re\sin\nu}{(1+e\cos\nu)}(1+e\cos\nu)^2\sqrt{\frac{\mu}{p^3}}, \quad (2.108)$$

which is,

$$\dot{r} = e\sin\nu\sqrt{\frac{\mu}{p}}. \quad (2.109)$$

The velocity vector in the peri-focal frame is,

$$\dot{\mathbf{r}}_P = [\dot{r}\cos\nu \quad \dot{r}\sin\nu \quad 0]^T - [r\sin\nu \quad -r\cos\nu \quad 0]^T \dot{\nu}. \quad (2.110)$$

Hence,

$$\dot{\mathbf{r}}_P = \sqrt{\frac{\mu}{p}}[\cos\nu \quad \sin\nu \quad 0]^T e\sin\nu - [\sin\nu \quad -\cos\nu \quad 0]^T(1+e\cos\nu)\sqrt{\frac{\mu}{p}}, \quad (2.111)$$

which is

$$\dot{\mathbf{r}}_P = \sqrt{\frac{\mu}{p}}[-\sin\nu \quad e+\cos\nu \quad 0]. \quad (2.112)$$

To obtain the position and velocity in the inertial frame, we need to transform to the inertial frame. The transformation from the peri-focal to the inertial frame is,

$$\mathbf{T}_{IP} = \begin{bmatrix} \cos\theta\cos\Omega - \sin\theta\sin\Omega\cos i & -\sin\theta\cos\Omega - \cos\theta\sin\Omega\cos i & \sin\Omega\sin i \\ \cos\theta\sin\Omega + \sin\theta\cos\Omega\cos i & -\sin\theta\sin\Omega + \cos\theta\cos\Omega\cos i & -\cos\Omega\sin i \\ \sin\theta\sin i & \cos\theta\sin i & \cos i \end{bmatrix}. \quad (2.113)$$

The position and velocity vectors in the inertial frame are, respectively, given by,

$$\mathbf{r} = \mathbf{T}_{IP}\mathbf{r}_P, \quad \dot{\mathbf{r}} = \mathbf{T}_{IP}\dot{\mathbf{r}}_P. \quad (2.114)$$

The position and velocity vectors in the inertial frame depend only on the orbital elements.

2.2.9 The *f* and *g* Functions

There is important representation of the position and velocity vectors that follows from the definition of the position vector in the peri-focal coordinates and the relations between the true anomaly ν and the eccentric anomaly E. Thus, since,

$$\mathbf{r}_p = r[\cos \nu \quad \sin \nu \quad 0]^T = a[\cos E - e \quad \sin E \sqrt{1-e^2} \quad 0]^T, \tag{2.115}$$

$$\dot{\mathbf{r}}_p = \frac{\sqrt{\mu a}}{r}[-\sin E \quad \cos E \sqrt{1-e^2} \quad 0]. \tag{2.116}$$

Thus, in terms of the unit vectors in the x and y directions of the peri-focal coordinates, $\mathbf{e}_x$ and $\mathbf{e}_y$,

$$\begin{bmatrix} \mathbf{r}_p \\ \dot{\mathbf{r}}_p \end{bmatrix} = \begin{bmatrix} x & y \\ \dot{x} & \dot{y} \end{bmatrix} \begin{bmatrix} \mathbf{e}_x \\ \mathbf{e}_y \end{bmatrix} \equiv \begin{bmatrix} a(\cos E - e) & a\sqrt{1-e^2}\sin E \\ -(\sqrt{\mu a}/r)\sin E & (\sqrt{\mu a}/r)\sqrt{1-e^2}\cos E \end{bmatrix} \begin{bmatrix} \mathbf{e}_x \\ \mathbf{e}_y \end{bmatrix}. \tag{2.117}$$

Thus, at a particular instant of time, $t = t_0$, when $r = r_0$, $\mathbf{r}_p = \mathbf{r}_{p0}$, $\dot{\mathbf{r}}_p = \dot{\mathbf{r}}_{p0}$,

$$\begin{bmatrix} \mathbf{e}_x \\ \mathbf{e}_y \end{bmatrix} = \frac{1}{\sqrt{1-e^2}\sqrt{\mu a}} \begin{bmatrix} (\sqrt{\mu a}/r_0)\sqrt{1-e^2}\cos E_0 & -a\sqrt{1-e^2}\sin E_0 \\ (\sqrt{\mu a}/r_0)\sin E_0 & a(\cos E_0 - e) \end{bmatrix} \begin{bmatrix} \mathbf{r}_{p0} \\ \dot{\mathbf{r}}_{p0} \end{bmatrix}. \tag{2.118}$$

Thus,

$$\mathbf{r}_p = f\,\mathbf{r}_{p0} + g\,\dot{\mathbf{r}}_{p0}, \quad \dot{\mathbf{r}}_p = \dot{f}\,\mathbf{r}_{p0} + \dot{g}\,\dot{\mathbf{r}}_{p0}, \tag{2.119}$$

Where,

$$f = \frac{a}{r_0}\{(\cos E - e)\cos E_0 + \sin E \sin E_0\}, \tag{2.120}$$

$$g = \left(\frac{a^3}{\mu}\right)^{1/2}\{\sin(E - E_0) - e(\sin E - \sin E_0)\}. \tag{2.121}$$

A useful alternate representation is given taking the cross product of $\mathbf{r}_p$ with $\dot{\mathbf{r}}_{p0}$ and $\mathbf{r}_{p0}$, respectively. Thus,

$$f = \frac{\mathbf{r}_p \times \dot{\mathbf{r}}_{p0}}{\mathbf{r}_{p0} \times \dot{\mathbf{r}}_{p0}} = \frac{\mathbf{r}_p \times \dot{\mathbf{r}}_{p0}}{h}, \quad g = \frac{\mathbf{r}_{p0} \times \mathbf{r}_p}{\mathbf{r}_{p0} \times \dot{\mathbf{r}}_{p0}} = \frac{\mathbf{r}_{p0} \times \mathbf{r}_p}{h}. \tag{2.122}$$

Furthermore, since,

$$\mathbf{r}_{p0} \times \dot{\mathbf{r}}_{p0} = \mathbf{r}_p \times \dot{\mathbf{r}}_p = h, \tag{2.123}$$

one has the constraint that,

$$f\dot{g} - \dot{f}g = 1. \tag{2.124}$$

Hence, we may show that,

$$f = \frac{a}{r_0}\{\cos(E - E_0) - e\cos E_0\} = \frac{a\cos(E - E_0)}{r_0} + 1 - \frac{a}{r_0},$$

$$g = t + \left(\frac{a^3}{\mu}\right)^{1/2}\{\sin(E - E_0) - (E - E_0)\},$$

$$\dot{f} = -\frac{\sqrt{\mu a}\sin(E - E_0)}{rr_0}, \quad \dot{g} = \frac{a\cos(E - E_0)}{r} + 1 - \frac{a}{r}. \qquad (2.125)$$

In this form, the f and g functions may be used to determine, $\dot{\mathbf{r}}_p$ and $\dot{\mathbf{r}}_{p0}$ given $\mathbf{r}_p$ and $\mathbf{r}_{p0}$. Hence,

$$\dot{\mathbf{r}}_{p0} = (\mathbf{r}_p - f\,\mathbf{r}_{p0})/g, \quad \dot{\mathbf{r}}_p = (\dot{g}\mathbf{r}_p - \mathbf{r}_{p0})/g. \qquad (2.126)$$

2.3 Types of Orbits

The classification of orbits into different groups makes the study of orbital dynamics relatively easier. One may identify at least six different classes of orbits.

2.3.1 Geosynchronous Earth Orbits

A geosynchronous Earth orbit is a prograde, circular orbit about Earth that has a period of 23 hours 56 minutes 4 seconds. The inclination is generally low and mostly near zero. A spacecraft in geosynchronous orbit appears to remain above Earth at a constant longitude, although it may seem to wander north and south with a narrow footprint on the Earth's surface that is almost symmetric about a fixed longitude.

2.3.2 Geostationary Orbits

A geostationary Earth orbit (GEO) is a geosynchronous orbit with an inclination of either zero, right on the equator, or low enough that the spacecraft can use onboard thrusters to maintain its apparent position above a fixed point on Earth. In a geostationary orbit the spacecraft does not wander about and seems like it has been suspended at a fixed point relative to the Earth. (Any such maneuvering on orbit is a process called *station keeping*.) The orbit can then be called geostationary. This orbit is ideal for certain kinds of communication satellites or meteorological satellites.

2.3.3 Geosynchronous Transfer Orbit

To attain geosynchronous (and also geostationary) Earth orbits, a spacecraft is first launched into an elliptical orbit with an apoapsis altitude in the neighborhood of 37,000 km. This is called a *Geosynchronous Transfer Orbit* (GTO). The spacecraft then circularizes the orbit by turning parallel to the equator at apoapsis and firing its rocket engine. That engine is usually called an *apogee motor*. It serves as a metric for comparison of various launch vehicle capabilities, which is the amount of mass they can lift to GTO.

2.3.4 Polar Orbits

Polar orbits have a fixed 90 degree inclination, useful for spacecraft that carry out mapping or surveillance operations. Since the orbital plane is nominally fixed in inertial space, the planet rotates below a polar orbit, allowing the spacecraft low-altitude access to virtually every point on the surface. The Magellan spacecraft is one example of a spacecraft that used a nearly-polar orbit about Venus. During each periapsis pass, a "swath" of mapping data was taken, and the planet rotated so that "swaths" from consecutive orbits were adjacent to each other. Thus, when the planet rotated once, all 360 degrees longitude had been exposed to the spacecraft.

To achieve a polar orbit of Earth requires more energy, thus more propellant, than does a direct orbit of low inclination. To achieve the latter, the initial launch is normally accomplished near the equator, where the rotational speed of the surface contributes a significant part of the final speed required for orbit. Once in a polar orbit, the satellite will not be able to take advantage of the "free ride" provided by Earth's rotation, and thus the launch vehicle must provide all of the energy for attaining orbital speed.

2.3.5 Walking Orbits

It is possible to choose the parameters of a spacecraft's orbit to take advantage of some or all of the gravitational influences to induce precession, which causes a useful motion of the orbital plane. The result is called a walking orbit or a precessing orbit, since the orbital plane moves slowly with respect to fixed inertial space.

2.3.6 Sun Synchronous Orbits

A sun synchronous orbit is a particular type of walking orbit whose parameters are chosen such that the orbital plane precesses with nearly the same period as the planet's solar orbit period. In such an orbit, the spacecraft crosses periapsis at about the same local time every orbit. This can be useful if instruments on board depend on a certain angle of solar illumination on the surface. *Mars Global Surveyor*'s orbit is a 2 pm Mars Local Time sun-synchronous orbit, chosen to permit well-placed shadows for best viewing. It may not be possible to rely on use of the gravity field alone to exactly maintain the desired synchronous timing, and occasional propulsive maneuvers may be necessary to adjust the orbit.

2.4 Impulsive Orbit Transfer

The whole point about orbit transfers is to be able to change the orbit if a satellite by applying the minimum possible input to the satellite. One approach is to apply an

2.4 Impulsive Orbit Transfer

impulse to the satellite. Thus, the satellite experiences an impulsive change to its velocity vector. Thus, if the initial velocity vector is $\mathbf{v}_i(t)$ and the final desired velocity vector is $\mathbf{v}_f(t)$, one can express the final velocity vector as,

$$|\mathbf{v}_f(t)|^2 = |\mathbf{v}_i(t)|^2 + |\Delta \mathbf{v}(t)|^2 + 2\mathbf{v}_i(t)\Delta\mathbf{v}(t)\cos\beta, \tag{2.127}$$

where β is the angle between the vectors $\mathbf{v}_i(t)$ and $\Delta\mathbf{v}(t)$, where the latter is the impulsive change in the velocity vector applied to the vehicle. Consequently, the increase in the kinetic energy per unit mass is,

$$\Delta E = |\mathbf{v}_f(t)|^2 - |\mathbf{v}_i(t)|^2 = |\Delta\mathbf{v}(t)|^2 + 2\mathbf{v}_i(t)\Delta\mathbf{v}(t)\cos\beta, \tag{2.128}$$

which is a maximum when $\beta = 0$. The maximum increase in the kinetic energy is obtained when $\Delta\mathbf{v}(t)$ is applied in the same direction as $\mathbf{v}_i(t)$ and when $\mathbf{v}_i(t)$ is as high as possible. The Hohmann transfer is based on these principles and involves either co-planar or non-coplanar orbital transfers. Generally, a non-coplanar transfer requires a considerable amount of fuel expenditure and must be carefully evaluated. Approximate computations can often be very misleading.

2.4.1 Co-Planar Hohmann Transfer

Considering a satellite in a standard elliptic orbit, its maximum velocity is when it is at the perigee. Assuming r is maintained constant and given that,

$$v^2 = \mu\left(\frac{2}{r} - \frac{1}{a}\right), \tag{2.65}$$

it follows that,

$$2v\,dv = \mu\frac{da}{a^2}. \tag{2.129}$$

Thus, one obtains for the change in a, due a velocity increment at the perigee, the expression,

$$da = \frac{2a^2 v}{\mu}dv = \frac{2a^2 v_p}{\mu}dv_p. \tag{2.130}$$

The change in the major axis of the ellipse is,

$$2da = \frac{4a^2 v_p}{\mu}dv_p. \tag{2.131}$$

If we now let the satellite orbit along the transfer orbit until it arrives at the apoapse of the transfer orbit and then apply a second impulse, the total change in the major axis of the ellipse is,

$$2da = \frac{(2a)^2 v_p}{\mu} dv_p + \frac{\left(2a + \frac{4a^2 v_p}{\mu} dv_p\right)^2 v_a}{\mu} dv_a = \frac{(2a)^2}{\mu}\left(v_p dv_p + v_a dv_a\right)$$

$$= 2(a_f - a_i). \tag{2.132}$$

The velocities of the satellites in the initial and final orbits, which are both assumed to be in the same plane, are, respectively, given by,

$$v_i^2 = \mu\left(\frac{2}{r} - \frac{1}{a_i}\right), \quad v_f^2 = \mu\left(\frac{2}{r} - \frac{1}{a_f}\right). \tag{2.133}$$

The above expressions are obtained from the visa-viva equation and are therefore independent of the shape of the orbit. To simplify the analysis, we assume that the initial and final orbits are circular. The circular velocities of the satellites in the initial and final orbits are, respectively, given by,

$$v_i^2 = \frac{\mu}{a_i}, \quad v_f^2 = \frac{\mu}{a_f}. \tag{2.134}$$

The transfer orbit is tangent to both circular orbits, first at the perigee of the initial orbit and then at the apogee of the final orbit. Thus, for the transfer orbit, which could be treated as an ellipse with the major axis equal to $a_f + a_i$, we have,

$$v_t^2 = \mu\left(\frac{2}{r} - \frac{2}{a_f + a_i}\right). \tag{2.135}$$

The velocities of the transfer orbit at the perigee and the apogee are, respectively, given by,

$$v_{t,p}^2 = \mu\left(\frac{2}{a_i} - \frac{2}{a_f + a_i}\right), \quad v_{t,a}^2 = \mu\left(\frac{2}{a_f} - \frac{2}{a_f + a_i}\right). \tag{2.136}$$

Thus, the impulses delivered at the perigee and apogee are, respectively, given by,

$$\Delta v_p = v_{t,p} - v_i = \sqrt{\frac{\mu}{a_i}}\left\{\sqrt{2\left(\frac{a_f}{a_f + a_i}\right)} - 1\right\}, \tag{2.137}$$

and

$$\Delta v_a = v_f - v_{t,a} = \sqrt{\frac{\mu}{a_f}}\left\{1 - \sqrt{2\left(\frac{a_i}{a_f + a_i}\right)}\right\}. \tag{2.138}$$

It is possible in principle to also rotate the line of apsides (joining the perigee and the apogee) by applying a single impulse at the point of intersection of initial and final orbits in the radial velocity direction.

2.4.2 Non-Planar Hohmann Transfer

Consider the two-impulse orbit transfer without making any constraining assumptions as far as the orbits are concerned. The initial and final orbit will not be assumed to be in the same plane. It will be assumed that they are inclined to each other by an angle θ. The velocity of the satellite at the perigee of the initial orbit is,

$$v_{i,p} = \sqrt{(\mu/a_i)}\sqrt{(1+e_i)/(1-e_i)}. \qquad (2.139)$$

The velocity of the satellite at the apogee of the final orbit is,

$$v_{f,a} = \sqrt{\mu/a_f}\sqrt{(1-e_f)/(1+e_f)}. \qquad (2.140)$$

For the transfer orbit, the velocity of the satellite at the perigee of the orbit is,

$$v_{t,p} = \sqrt{(\mu/a_t)}\sqrt{(1+e_t)/(1-e_t)}. \qquad (2.141)$$

We have assumed that the semi-major axis and eccentricity of the transfer orbit are a_t and e_t, respectively. For the transfer orbit, the velocity of the satellite at the apogee of the orbit is,

$$v_{t,a} = \sqrt{\mu/a_t}\sqrt{(1-e_t)/(1+e_t)}. \qquad (2.142)$$

The major axis of the transfer orbit is assumed to be the sum of the perigee distance of the initial orbit and the perigee distance of the final orbit. Thus,

$$2a_t = a_f(1+e_f) + a_i(1-e_i). \qquad (2.143)$$

Thus, the eccentricity may be expressed as,

$$e_t = \left(a_f(1+e_f) - a_i(1-e_i)\right)/2a_t. \qquad (2.144)$$

If we assume that the transfer orbit, and the initial and final orbits are co-planar, the total Δv is given by,

$$\Delta v = |v_{t,p} - v_{i,p}| + |v_{f,a} - v_{t,a}|. \qquad (2.145)$$

However, if one assumes that the total plane change required is θ, one can assume that the plane change achieved after the first impulse is α (not to be confused with the angle defined in the section on Lambert's theorem) and the plane change achieved after the second impulse is $\theta - \alpha$. Of course, one is seeking the minimum total Δv and so one could minimize the total Δv with respect to α and determine what proportion of the plane change must be achieved after the first impulse. With the change in planes one has the relations,

$$\Delta v_p^2 = v_{t,p}^2 + v_{i,p}^2 - 2v_{t,p}v_{i,p}\cos\alpha, \qquad (2.146)$$

$$\Delta v_a^2 = v_{f,a}^2 + v_{t,a}^2 - 2v_{f,a}v_{t,a}\cos(\theta - \alpha). \tag{2.147}$$

The total Δv is,

$$\Delta v = |\Delta v_p| + |\Delta v_a|. \tag{2.148}$$

Thus, to minimize Δv and a necessary condition is that,

$$\frac{\partial \Delta v}{\partial \alpha} = 0. \tag{2.149}$$

Following Kamel and Soliman [5], this reduces to the solution of sixth-order polynomial in $x = \cos\alpha$. Kamel and Soliman [5] write the polynomial as,

$$\begin{aligned}&(E_1^2 - E_5^2) + 2(E_1E_2 - E_5E_6)x + (2E_1E_3 - 2E_5E_7 + E_2^2 + E_5^2 - E_6^2)x^2 \\&+ 2(E_1E_4 + E_2E_3 + E_5E_6 - E_6E_7)x^3 + (2E_2E_4 + 2E_5E_7 + E_3^2 + E_6^2 - E_7^2)x^4 \\&+ 2(E_3E_4 + E_6E_7)x^5 + (E_4^2 + E_7^2)x^6 = 0.\end{aligned} \tag{2.150}$$

where the coefficients E_i are defined as,

$$\begin{aligned}E_1 &= v_{i,p}^2 v_{t,p}^2 \left(v_{f,a}^2 + v_{t,a}^2\right) - v_{f,a}^2 v_{t,a}^2 \left(v_{i,p}^2 + v_{t,p}^2\right)\cos^2\theta, \\E_2 &= 2v_{i,p}v_{t,p}v_{f,a}^2 v_{t,a}^2 \cos^2\theta - 2v_{i,p}^2 v_{t,p}^2 v_{f,a} v_{t,a} \cos\theta, \\E_3 &= -v_{i,p}^2 v_{t,p}^2 \left(v_{f,a}^2 + v_{t,a}^2\right) + v_{f,a}^2 v_{t,a}^2 \left(v_{i,p}^2 + v_{t,p}^2\right)(\cos^2\theta - \sin^2\theta), \\E_4 &= 2v_{i,p}^2 v_{t,p}^2 v_{f,a} v_{t,a} \cos\theta - 2v_{i,p}v_{t,p}v_{f,a}^2 v_{t,a}^2 (\cos^2\theta - \sin^2\theta), \\E_5 &= 2v_{i,p}^2 v_{t,p}^2 v_{f,a} v_{t,a} \sin\theta, \\E_6 &= -2\cos\theta\sin\theta v_{f,a}^2 v_{t,a}^2 \left(v_{i,p}^2 + v_{t,p}^2\right), \\E_7 &= 4\cos\theta\sin\theta v_{i,p}v_{t,p}v_{f,a}^2 v_{t,a}^2 - 2\sin\theta v_{i,p}^2 v_{t,p}^2 v_{f,a} v_{t,a}.\end{aligned} \tag{2.151}$$

Once α is determined, the complete plane change is accomplished in two stages to minimize the total Δv required. A part of the plane change given by the angle α is accomplished at the time of the application of the first impulse, while the remaining change given by $\theta - \alpha$ is accomplished during the application of the second impulse. A further discussion of the orbit transfer problem and the application of Hohmann transfers is given in Section 2.12.

There are a number of special cases one could consider. For example, a simple plane change is only necessary to change the direction velocity vector, by providing an impulse at the ascending or descending nodes. In this case, $\Delta v = 2v_{node}\sin(\theta/2)$, where v_{node} is the nodal velocity.

In another example, for a low-inclination orbit, transfer to a zero inclination orbit is obtained by applying the impulse, satisfying the relation given by the law of cosines, $\Delta v^2 = v_i^2 + v_f^2 - 2v_iv_f\cos(\theta)$.

2.5 Preliminary Orbit Determination

We have already considered the case of determining the orbital elements of a Keplerian orbit, given the position and velocity of the satellite at any point in the orbit. Although any satellite orbiting a celestial body does not in reality execute a perfectly Keplerian orbit, the Keplerian orbit can be used as a first approximation for subsequent orbit estimation and tracking. For this reason, it is important how to obtain the Keplerian elements of an orbiting body, given various sets of data. Typically, these could be (a) Two position vectors of the satellite and the corresponding times, (b) Three position vectors of the satellites and the time corresponding to any one of these vectors, (c) two sets of observations at two different times of the distance (range) of the satellite from three distinct points on the surface of Earth, or (d) observations at one time of the distance and speed (range and range rate) of the satellite from three distinct points on the surface of Earth. There are other cases such as Gauss's problem (see, for example, Deutsch [1]) which involves the determination of the satellite's orbit from three measurements of the direction of the orbiting body from the surface of the Earth. We shall now consider all of the above typical cases.

2.5.1 Two Position Vectors of the Satellite

Let the position vectors of the satellites in Earth-centered, Earth-fixed (ECEF) coordinates be:

$$\mathbf{r}_l = r_{l1}\mathbf{i} + r_{l2}\mathbf{j} + r_{l3}\mathbf{k}, \quad l = 1, 2. \tag{2.152}$$

The unit normal to the orbit plane is then given by,

$$\mathbf{h}_n = (\mathbf{r}_1 \times \mathbf{r}_2)/|\mathbf{r}_1 \times \mathbf{r}_2|. \tag{2.153}$$

The inclination angle is obtained from the inverse cosine of $\mathbf{h}_n \cdot \mathbf{k}$ and the vector representing the line of nodes from $\mathbf{n} = \mathbf{h}_n \times \mathbf{k}$. The argument of the ascending node is then obtained from its cosine and sine, which are, respectively, given by $\mathbf{n} \cdot \mathbf{i}$ and $\mathbf{n} \cdot \mathbf{j}$. The semi-major axis a is obtained by applying Lambert's theorem,

$$n(t_1 - t_2) = (\alpha - \sin\alpha) - (\beta - \sin\beta),$$

$$\frac{\alpha}{2} = \sin^{-1}\left(\frac{1}{2}\left(\frac{|\mathbf{r}_1| + |\mathbf{r}_2| + c}{a}\right)^{1/2}\right), \quad \frac{\beta}{2} = \sin^{-1}\left(\frac{1}{2}\left(\frac{|\mathbf{r}_1| + |\mathbf{r}_2| - c}{a}\right)^{1/2}\right). \tag{2.154}$$

And,

$$c^2 = |\mathbf{r}_1|^2 + |\mathbf{r}_2|^2 - 2\mathbf{r}_1 \cdot \mathbf{r}_2. \tag{2.155}$$

The solution for the semi-major axis a is obtained numerically by methods similar to those used to solve Kepler's equation. To solve

$$n(t_1 - t_2) - (\alpha - \sin\alpha) + (\beta - \sin\beta) = \varepsilon = 0, \tag{2.156}$$

approximately, it is written as,

$$n(t_1 - t_2) \approx (\alpha^3/3) - (\beta^3/3), \qquad (2.157)$$

with,

$$\alpha = \left(\frac{|\mathbf{r}_1| + |\mathbf{r}_2| + c}{a}\right)^{1/2}, \quad \beta = \left(\frac{|\mathbf{r}_1| + |\mathbf{r}_2| - c}{a}\right)^{1/2}. \qquad (2.158)$$

Thus,

$$a^{3/2} \approx \frac{1}{3} \frac{(|\mathbf{r}_1| + |\mathbf{r}_2| + c)^{3/2} - (|\mathbf{r}_1| + |\mathbf{r}_2| - c)^{3/2}}{n(t_1 - t_2)}. \qquad (2.159)$$

Once an approximate solution is obtained for the semi-major axis a, it is improved iteratively by minimizing ε^2. The arguments of the position vectors from the line nodes are then found as,

$$\cos(\omega + v_i) \equiv \cos u_i = \mathbf{r}_i \cdot \mathbf{n}/|\mathbf{r}_i|. \qquad (2.160)$$

But,

$$\left(a(1 - e^2)/|\mathbf{r}_k|\right) - 1 = e \cos v_k = e \cos \omega \cos u_k + e \sin \omega \sin u_k, \quad k = 1, 2. \qquad (2.161)$$

Thus,

$$\begin{bmatrix} \cos u_1 & \sin u_1 \\ \cos u_2 & \sin u_2 \end{bmatrix} \begin{bmatrix} e \cos \omega \\ e \sin \omega \end{bmatrix} = \begin{bmatrix} \frac{a(1-e^2)}{|\mathbf{r}_1|} - 1 & \frac{a(1-e^2)}{|\mathbf{r}_2|} - 1 \end{bmatrix}^T. \qquad (2.162)$$

It follows that,

$$\begin{bmatrix} e \cos \omega \\ e \sin \omega \end{bmatrix} = \begin{bmatrix} \cos u_1 & \sin u_1 \\ \cos u_2 & \sin u_2 \end{bmatrix}^{-1} \begin{bmatrix} \frac{a(1-e^2)}{|\mathbf{r}_1|} - 1 & \frac{a(1-e^2)}{|\mathbf{r}_2|} - 1 \end{bmatrix}^T, \qquad (2.163)$$

$$e^2 = [e \cos \omega \quad e \sin \omega][e \cos \omega \quad e \sin \omega]^T. \qquad (2.164)$$

The above are iteratively solved for the eccentricity e and the argument of the perigee ω. The time of passage at the perigee t_0 is then obtained from either,

$$n(t_1 - t_0) = E_1 - e \sin E_1 \text{ or } n(t_2 - t_0) = E_2 - e \sin E_2. \qquad (2.165)$$

2.5.2 Three Position Vectors of the Satellite

Let the position vectors of the satellites in ECEF coordinates be,

$$\mathbf{r}_l = r_{l1}\mathbf{i} + r_{l2}\mathbf{j} + r_{l3}\mathbf{k}, \quad l = 1, 2, 3. \qquad (2.166)$$

It is important to check that the vectors are co-planar. Thus, $(\mathbf{r}_1 \times \mathbf{r}_2) \cdot \mathbf{r}_3 = 0$.
The unit normal to the orbit plane is then given by,

$$\mathbf{h}_n = (\mathbf{r}_1 \times \mathbf{r}_2)/|\mathbf{r}_1 \times \mathbf{r}_2|. \qquad (2.167)$$

2.5 Preliminary Orbit Determination

The inclination angle is obtained from the inverse cosine of $\mathbf{h}_n \cdot \mathbf{k}$ and the vector representing the line of nodes from $\mathbf{n} = \mathbf{h}_n \times \mathbf{k}$. The argument of the ascending node is then obtained from its cosine and sine, which are, respectively, given by $\mathbf{n} \cdot \mathbf{i}$ and $\mathbf{n} \cdot \mathbf{j}$. The arguments of the position vectors from the line nodes are then found as,

$$\cos(\omega + v_i) \equiv \cos u_i = \mathbf{r}_i \cdot \mathbf{n}/|\mathbf{r}_i|. \tag{2.168}$$

But,

$$\left(a(1-e^2)/|\mathbf{r}_k|\right) - 1 = e\cos v_k = e\cos\omega\cos u_k + e\sin\omega\sin u_k, \quad k=1,2,3. \tag{2.169}$$

Thus,

$$\begin{bmatrix} \cos u_1 & \sin u_1 & |\mathbf{r}_1|^{-1} \\ \cos u_2 & \sin u_2 & |\mathbf{r}_2|^{-1} \\ \cos u_3 & \sin u_3 & |\mathbf{r}_3|^{-1} \end{bmatrix} \begin{bmatrix} e\cos\omega \\ e\sin\omega \\ a(1-e^2) \end{bmatrix} = -\begin{bmatrix} 1 \\ 1 \\ 1 \end{bmatrix}, \tag{2.170}$$

which is solved for the eccentricity e, the argument of the perigee ω, and the semi-major axis a. The time of passage at the perigee t_0 is then obtained from,

$$n(t_1 - t_0) = E_1 - e\sin E_1. \tag{2.171}$$

2.5.3 Two Sets of Observations of the Range at Three Locations

If it is assumed that one has three radars located at three distinct sites, with each capable of measuring the spacecraft's distance from the site in ECEF coordinates. Then the position vector of the satellite at the time $t = t_k$,

$$\mathbf{r}(t_k) = \mathbf{T}_{IE}\boldsymbol{\rho}_j + \mathbf{R}_{kj}, \tag{2.172}$$

where the position vector of the site $\boldsymbol{\rho}_j$ is defined in terms of geocentric latitude ϕ_j, geocentric longitude λ_j, and the distance of the site from the Earth's center ρ_{0j}, as,

$$\boldsymbol{\rho}_j = \rho_{0j}\left(\cos\phi_j\cos\lambda_j\mathbf{e}_x + \cos\phi_j\sin\lambda_j\mathbf{e}_y + \sin\phi_j\mathbf{e}_z\right). \tag{2.173}$$

The transformation $\mathbf{T}_{IE}$ is the transformation relating a vector in the ECEF frame to the ECI frame. The vector $\mathbf{R}_{kj}$ is the position vector of the satellite at $t = t_k$ from the site j. Writing $\mathbf{r}_{ej}(t_k) = \mathbf{T}_{IE}\boldsymbol{\rho}_j$, one obtains,

$$\mathbf{R}_{kj} = \mathbf{r}(t_k) - \mathbf{r}_{ej}(t_k). \tag{2.174}$$

It follows that,

$$\mathbf{R}_{kj} \cdot \mathbf{R}_{kj} = \left(\mathbf{r}(t_k) - \mathbf{r}_{ej}(t_k)\right) \cdot \left(\mathbf{r}(t_k) - \mathbf{r}_{ej}(t_k)\right) = R_{kj}^2, \tag{2.175}$$

and

$$R_{kj}^2 = |\mathbf{r}(t_k)|^2 + |\mathbf{r}_{ej}(t_k)|^2 - 2\mathbf{r}(t_k) \cdot \mathbf{r}_{ej}(t_k). \tag{2.176}$$

Differencing across two different sites, and rearranging,

$$2\mathbf{r}(t_k) \cdot \left\{\mathbf{r}_{ej}(t_k) - \mathbf{r}_{em}(t_k)\right\} = |\mathbf{r}_{ej}(t_k)|^2 - |\mathbf{r}_{em}(t_k)|^2 - R_{kj}^2 + R_{km}^2. \tag{2.177}$$

For three sites and at two time instants, the above are six linear equations for the six components of the satellite's position vector at the two time instants, which can be solved by successive elimination.

2.5.4 Range and Range Rates Measured at Three Locations

When the range and range rate are measured at one time instant at three sites, we consider the relation,

$$R_{kj}^2 = (\mathbf{r}(t_k) - \mathbf{r}_{ej}(t_k)) \cdot (\mathbf{r}(t_k) - \mathbf{r}_{ej}(t_k)), \qquad (2.178)$$

which may be solved in principle for the components of $\mathbf{r}(t_k)$ by unconstrained function minimization.

Differentiating the above relation, one obtains,

$$\dot{R}_{kj} R_{kj} = \dot{\mathbf{r}}(t_k) \cdot (\mathbf{r}(t_k) - \mathbf{r}_{ej}(t_k)) - \dot{\mathbf{r}}_{ej}(t_k) \cdot (\mathbf{r}(t_k) - \mathbf{r}_{ej}(t_k)), \qquad (2.179)$$

which is then solved for the components of $\dot{\mathbf{r}}(t_k)$.

2.6 Lambert's Problem

One of the fundamental problems of astrodynamics is the transfer of a spacecraft from a point P_1 to a point P_2 in a specified time interval. Lambert's problem is to determine the orbit connecting the two points in space. This has been extensively discussed by Prussing and Conway [3] and Hu [6].

Let F_c be the focus where the central attracting body is located and is the center of attraction. A possible solution to the problem is the ellipse shown in Figure 2.8. Also in the diagram, the vacant or empty focus is denoted by F_v. The triangle $P_1 F_c P_2$ is known as the *space triangle* and the chord connecting the points P_1 and P_2 is denoted by c. If one considers different solutions with slightly smaller semi-major axes, the points of intersection of the two circles would trace a locus that would resemble a hyperbola. A fundamental property of ellipses is that the $F_c P_1 + P_1 F_v = 2a$, where a is the semi-major axis. Similarly, $F_c P_2 + P_2 F_v = 2a$. It follows that that the vacant focus must be located at distances $2a - r_1$ from point P_1 and $2a - r_2$ from P_2. If circles of radii

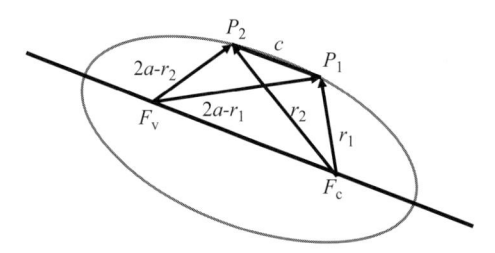

Figure 2.8 Transfer orbit geometry and property of the space triangle.

2.6 Lambert's Problem

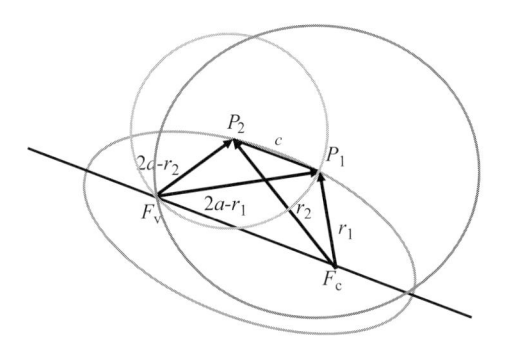

Figure 2.9 Vacant foci solutions.

$2a - r_1$ and $2a - r_2$ are drawn centered at P_1 and P_2, respectively, they would intersect at two points, one of which is the vacant focus F_v shown in Figure 2.9.

When the vacant focus lies on the chord joining the points P_1 and P_2, the two points of intersection would be the same and the semi-major axis for this case would represent the smallest value of $a = a_{\min}$ that results in a feasible solution. Thus, the case when $a = a_{\min}$ represents a minimum energy solution and all ellipses with lower values of the semi-major axes are infeasible. From the chord of the *space triangle*, when $a = a_{\min}$, $2a - r_1 + 2a - r_2 = c$. Hence, $4a_{\min} = c + r_1 + r_2 = 2s$, where s is the semi-perimeter of the *space triangle*. Furthermore, since the radius vector along the orbit satisfies the equation,

$$r + \mathbf{e} \cdot \mathbf{r} = p, \tag{2.180}$$

it follows that,

$$\mathbf{e} \cdot (\mathbf{r}_2 - \mathbf{r}_1) = -(r_2 - r_1). \tag{2.181}$$

Further, since $\mathbf{r}_2 = \mathbf{r}_1 + \mathbf{c}$, $(\mathbf{r}_2 - \mathbf{r}_1)/c$ is a unit vector along $\mathbf{c}$ and it follows that,

$$e_{\min} = (r_2 - r_1)/c, \tag{2.182}$$

is the minimum eccentricity of the transfer ellipse corresponding to $a = a_{\min}$. Moreover,

$$P_2 F_v = s - r_2, \quad P_1 F_v = s - r_1, \quad F_c F_v = 2a_{\min} e_{\min}, \quad \text{and} \quad P_2 F_c = r_2. \tag{2.183}$$

Hence, it follows that,

$$(2a_{\min} e_{\min})^2 = ((s - r_2) \sin \alpha)^2 + (r_2 - (s - r_2) \cos \alpha)^2, \tag{2.184}$$

which is also equal to,

$$(2a_{\min} e_{\min})^2 = s^2 - 2r_2(s - r_2)(1 + \cos \alpha). \tag{2.185}$$

From Figure 2.10,

$$\cos \alpha = \frac{r_2^2 - r_1^2 + c^2}{2r_2 c} = \frac{4s^2 - 4sr_1 - 2r_2 c}{2r_2 c} = \frac{2s(s - r_1)}{r_2 c} - 1. \tag{2.186}$$

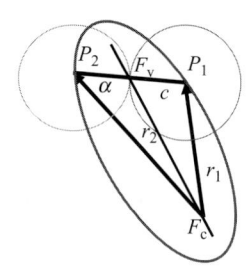

Figure 2.10 The minimum energy orbit.

$$(2a_{\min}e_{\min})^2 = s^2 - 4\frac{s}{c}(s - r_1)(s - r_2). \tag{2.187}$$

But,

$$p_{\min} = a_{\min}(1 - e_{\min}^2), \quad 4a_{\min}^2 e_{\min}^2 = 4a_{\min}^2 - 4a_{\min}p_{\min} = s^2 - 2sp_{\min}. \tag{2.188}$$

It follows that,

$$p_{\min} = \frac{2}{c}(s - r_1)(s - r_2) \quad \text{and} \quad e_{\min} = \sqrt{1 - \frac{2p_{\min}}{s}}. \tag{2.189}$$

We can now invoke Lambert's theorem from Section 2.2.7 to show that for any transfer ellipse,

$$n(t_1 - t_2) = (\alpha - \sin\alpha) - (\beta - \sin\beta), \tag{2.190}$$

where

$$\frac{\alpha}{2} = \sin^{-1}\left(\frac{1}{2}\left(\frac{r_1 + r_2 + c}{a}\right)^{1/2}\right), \quad \frac{\beta}{2} = \sin^{-1}\left(\frac{1}{2}\left(\frac{r_1 + r_2 - c}{a}\right)^{1/2}\right). \tag{2.191}$$

For the minimum energy ellipse this reduces to,

$$n(t_1 - t_2) = \pi - (\beta_{\min} - \sin\beta_{\min}). \tag{2.192}$$

We can also invoke the f and g functions from Section 2.2.9 to show that for the minimum energy ellipse,

$$\mathbf{v}_1 = \frac{\sqrt{\mu p_{\min}}}{r_1 r_2 \sin(\Delta \nu)}\left\{\mathbf{r}_2 - \left(1 - \frac{r_2}{p_{\min}}(1 - \cos(\Delta \nu))\right)\mathbf{r}_1\right\}. \tag{2.193}$$

2.7 Third Body and Other Orbit Perturbations

Planets are not perfectly spherical, and they do not have evenly distributed surface mass. The Earth, for example, is flattened at the poles and the equator is not a perfect circle. The effect of Earth flattening and equatorial bulge is a steady drift of the orbit plane as well as a drift within the orbit toward two fixed points along the equator. Also, they do not exist in a gravity "vacuum."

Other bodies, such as the Moon, the Sun, and its other natural satellites, the other planets, as well as the other planetary satellites, contribute their gravitational influences to a spacecraft in orbit about a planet. There are also a host of other perturbing forces, such as atmospheric drag, forces and moments due to thermal variations, radiation effects, and electro-magnetic effects that influence a satellite's orbit.

2.7.1 Circular Restricted Three-Body Problem

Joseph Louis Lagrange (1736–1813) showed that three bodies can occupy positions at the apexes of an equilateral triangle that rotates in its plane and still remain in relative equilibrium.

Consider a spacecraft, the second body, orbiting a primary body, denoted by the subscript E and influenced also by a third body, denoted by the subscript M. The equation of motion of the spacecraft orbiting the primary body and subjected to an inverse-square gravitational force field, in the vicinity of the third body is,

$$\ddot{\mathbf{r}} + \frac{G(m_E + m_s)}{|\mathbf{r}|^3}\mathbf{r} = -\nabla_\mathbf{r} R_{1,2}, \quad R_{1,2} = -Gm_M \left(\frac{1}{r_{1,M}} - \frac{\mathbf{r}\cdot\mathbf{r}_M}{|\mathbf{r}_M|^3}\right), \tag{2.194}$$

where $\mathbf{r}_M$ is the distance of M from E, $r_{1,M} = \sqrt{(\mathbf{r}-\mathbf{r}_M)^2}$ is the distance of the spacecraft from the third planet, and the m's are the masses of the primary body, the spacecraft, and the third body, and G is the universal gravitational constant.

The main assumptions of the circular restricted three-body problem, on which the analysis is based, are:

(i) the gravitational forces of the primary and the third body are the only forces taken into account;
(ii) the spacecraft's mass is neglected;
(iii) the eccentricity of the orbit of M around E is neglected, so that the motion of the primary and third bodies is considered to be circular about their common center of mass, the barycenter.

Let μ be the ratio of the mass of M to the mass of E, and x, y, z the nondimensional position components of the spacecraft in each axis of the rotating coordinate system, rotating about the z axis, with the mean speed of motion of the two body system. The mean motion of the barycenter of the primary body and the third body is given by:

$$n = \sqrt{\frac{G(m_E + m_M)}{d^3}}, \tag{2.195}$$

where $m_E + m_M$ is the total mass of the primary and third bodies, and d is the distance between the two. All linear displacements are scaled so $d = 1$. Define $\tau = nt$ as the dimensionless time, and convert all time derivatives so that they are done with respect to τ, and,

$$() = \frac{d}{d\tau} = \frac{1}{n}\frac{d}{dt}. \tag{2.196}$$

The nonlinear governing equations of the spacecraft's motion are,

$$\ddot{x} - 2\dot{y} - x = -\left\{\frac{\mu[x - (1-\mu)]}{\rho_{MS}^3} + \frac{(1-\mu)[x+\mu]}{\rho_{ES}^3}\right\} + \bar{U}_x, \tag{2.197}$$

$$\ddot{y} + 2\dot{x} - y = -y\left\{\frac{\mu}{\rho_{MS}^3} + \frac{(1-\mu)}{\rho_{ES}^3}\right\} + \bar{U}_y, \tag{2.198}$$

$$\ddot{z} = -z\left\{\frac{\mu}{\rho_{MS}^3} + \frac{(1-\mu)}{\rho_{ES}^3}\right\} + \bar{U}_z, \tag{2.199}$$

where $\rho_{MS} = \sqrt{[x-(1-\mu)]^2 + y^2 + z^2}$, $\rho_{ES} = \sqrt{[x+\mu]^2 + y^2 + z^2}$ and $\bar{U}_x, \bar{U}_y$, and $\bar{U}_z$ are the normalized, externally applied, or disturbance acceleration components acting on the spacecraft.

Consider a system with two large bodies such as the Moon orbiting the Earth (or the Earth orbiting the Sun). The barycenter is the location of the center of mass of the two body system, the Earth and the Moon in this case. The third body, such as a spacecraft or an asteroid, might occupy any of five points of equilibrium, which are known as the *Lagrange points*. A Lagrange point is a point in space between the Earth and the Moon where gravitational attractions of these two bodies are in balance with one another. For a space vehicle launched into a Lagrange point, the orbit remains motionless with respect to both the gravitational fields. It is possible for an object to be in a stable orbit around this point, such as in a "HALO" orbit. There are five Lagrange points: translunar (L_1), cislunar (L_2), trans-earth (L_3), and triangular (L_4, L_5) equilibrium points. In line with the two large bodies are the L_1, L_2, and L_3 points. The leading apex of the triangle is L_4; the trailing apex is L_5. These last two are also called *Trojan points*. The nondimensional coordinates of the location of these points are shown in Table 2.1. The Lagrange points for the Earth–Moon case are illustrated in Figure 2.11.

HALO orbits are a family of simple periodic orbits of the circular restricted three-body problem. These orbits remain in the vicinity of a collinear Lagrangian point and are among the most useful orbits for many types of mission. They can be used to monitor potentially damaging particles travelling toward Earth from the Sun (Figure 2.12).

So one can get an idea of the relative distances and certain characteristic distances in space, they are listed in Table 2.2.

Table 2.1 Lagrangian equilibrium point relative locations (The Earth–Moon distance is assumed to be unity)

Equilibrium point	X	Y
Translunar, L_1	1.155682	0
Cislunar, L_2	0.836915	0
Trans-Earth, L_3	−1.005063	0
Triangular, L_4, L_5	0.487849	±0.866025

2.7 Third Body and Other Orbit Perturbations

Table 2.2 Characteristic distances in space

Characteristic	Distance	Characteristic	Distance
Troposphere	0–6 miles	Exosphere	300–600 miles
Stratosphere	6–50	Remote sensing	500
Ionosphere	50–300	Nav. satellites	6,000–13,000
Shuttle Orbit	150	Geo stat. sat	22,000
Manned Spacecraft	90–300	Lunar orbit	238,850

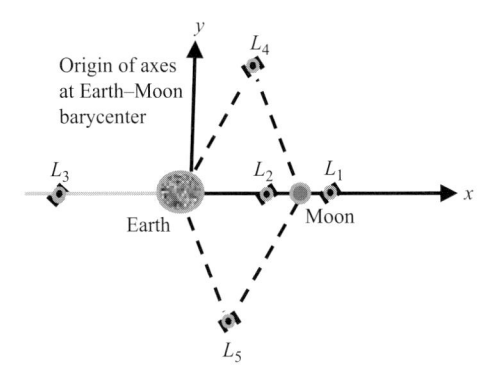

Figure 2.11 The Lagrange points: Note that L_1, L_2, and L_3 are unstable, while L_4 and L_5 are stable.

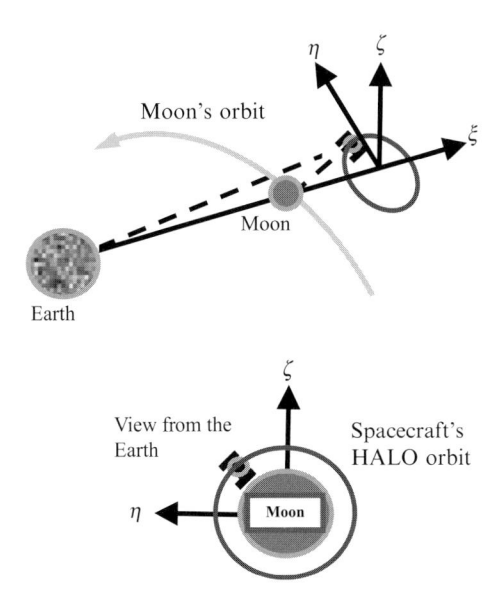

Figure 2.12 A typical HALO orbit.

2.8 Lagrange Planetary Equations

Basic dynamics of a satellite in an Earth orbit are described by Newton's second law, $F = Ma$, where the force F is composed of gravitational forces, radiation pressure (drag is negligible for a GPS satellite, for example), and thruster firings (not directly modelled); M is the mass of the satellite; and a the total acceleration of the satellite relative to an inertially fixed reference frame, fixed in space. Basic orbit behavior is given by the Keplerian orbital dynamics equation. The relevant gravitational constant is,

$$GM_e = \mu = 3{,}986{,}006 \times 10^8 \text{ m}^3/\text{s}^2. \tag{2.200}$$

The Keplerian orbit is the basic analytical solution to the central force problem. The basic central force potential (energy) may be expressed as, $U_0 = \mu/r$, where r is the distance of the satellite from the Earth's center.

For Earth satellites, these are elliptical orbits, where the mean motion n in terms of period P is given by,

$$n = \frac{2\pi}{P} = \frac{1}{a}\sqrt{\frac{\mu}{a}}. \tag{2.201}$$

The semi-minor axis is represented by b. The mean anomaly and the semi-latus rectum are given by,

$$M = n(t - t_{perigee}), \quad p = b^2/a = a(1 - e^2), \quad h = \sqrt{\mu \times p}, \tag{2.202}$$

where, for a typical GPS satellite orbit, $a \approx 26{,}400$ km corresponding to an almost 12–sidereal-hour period, with an inclination of 55.5 degrees and an eccentricity near 0 (largest 0.02). In total, there are six orbital planes with four to five satellites per plane on average.

Although the central force is the main force acting on a satellite, there are other significant perturbations. The effects of these perturbations on the orbital elements are governed by the Lagrange planetary equations (LPE), which have expressions for rates of change of orbital elements as a function of all the disturbing potentials not including the principal central force potential. The disturbing potential V is defined as,

$$V = U - \frac{GM}{r} = U - \frac{\mu}{r}. \tag{2.203}$$

It contains the non-spherical gravitational contributions as well as due to other disturbing perturbations to the potential. The nonspherical gravitational contribution arises from the nonspherical mass or density distribution of the body. The net effect of this nonuniformity is the creation of a disturbing force on the orbiting body, which in turn causes a change in the orbital elements over time.

Physically, the orbit in the inertial configuration space is always tangential to an "instantaneous" ellipse (or hyperbola), defined by the "instantaneous" values of the time-varying orbital elements. Thus, the perturbed physical trajectory would coincide with the Keplerian orbit that the body would follow if the perturbing force was switched off instantaneously. This instantaneous elliptic orbit is called an *osculating* orbit, and the corresponding orbital elements are called osculating orbital elements.

The LPE describe the effects of a disturbing function on an orbiting body. The orbital elements are found by numerically integrating the LPE, which are:

$$\frac{da}{dt} = \frac{2}{na} \times \frac{\partial V}{\partial M}, \quad \frac{de}{dt} = \frac{\sqrt{1-e^2}}{na^2 e} \times \left(\sqrt{1-e^2} \times \frac{\partial V}{\partial M} - \frac{\partial V}{\partial \omega} \right), \qquad (2.204)$$

$$\frac{d\omega}{dt} = \frac{1}{na^2} \times \left(\frac{\sqrt{1-e^2}}{e} \times \frac{\partial V}{\partial e} - \frac{1}{\sqrt{1-e^2}} \times \frac{\cos i}{\sin i} \times \frac{\partial V}{\partial i} \right), \qquad (2.205)$$

$$\frac{di}{dt} = \frac{1}{na^2} \frac{1}{\sqrt{1-e^2}} \frac{1}{\sin i} \left(\cos i \frac{\partial V}{\partial \omega} - \frac{\partial V}{\partial \Omega} \right), \qquad (2.206)$$

$$\frac{d\Omega}{dt} = \frac{1}{na^2} \frac{1}{\sqrt{1-e^2}} \frac{1}{\sin i} \frac{\partial V}{\partial i}, \qquad (2.207)$$

$$\frac{dM}{dt} = n - \frac{1}{na^2} \left(\frac{1-e^2}{e} \frac{\partial V}{\partial e} + 2a \frac{\partial V}{\partial a} \right). \qquad (2.208)$$

They may be expressed in matrix form as,

$$\frac{d}{dt}[a \quad e \quad i \quad \omega \quad \Omega \quad M]^T = [0 \quad 0 \quad 0 \quad 0 \quad 0 \quad n]^T$$

$$+ \frac{1}{na^2} \begin{bmatrix} 0 & 0 & 0 & 0 & 0 & 2a \\ 0 & 0 & 0 & -\gamma/e & 0 & \gamma^2/e \\ 0 & 0 & 0 & 1/\gamma \tan i & -1/\gamma \sin i & 0 \\ 0 & \gamma/e & -1/\gamma \tan i & 0 & 0 & 0 \\ 0 & 0 & 1/\gamma \sin i & 0 & 0 & 0 \\ -2a & -\gamma^2/e & 0 & 0 & 0 & 0 \end{bmatrix} \partial V/\partial \begin{bmatrix} a \\ e \\ i \\ \omega \\ \Omega \\ M \end{bmatrix},$$

(2.209)

with, $\gamma = \sqrt{1-e^2}$. The 6 × 6 coefficient matrix is known as the *Poisson* matrix and is skew-symmetric. The orbital elements on the right-hand side of the equations are the osculating orbital elements. The LPE, written in terms of osculating orbital elements, assume the compact form,

$$\frac{d\boldsymbol{\alpha}}{dt} = \mathbf{L}^{-1} \left[\frac{\partial V}{\partial \boldsymbol{\alpha}} \right]^T, \qquad (2.210)$$

where $\boldsymbol{\alpha}^T = [a \quad e \quad i \quad \omega \quad \Omega \quad M]$. The matrix $\mathbf{L}$ is the skew-symmetric Lagrange matrix, whose entries are defined in terms of the *Lagrange brackets*,

$$\mathbf{L}_{ij} = [\alpha_i, \quad \alpha_j] = \left[\frac{\partial \mathbf{r}}{\partial \alpha_i} \right]^T \left[\frac{\partial \mathbf{v}}{\partial \alpha_j} \right] - \left[\frac{\partial \mathbf{r}}{\partial \alpha_j} \right]^T \left[\frac{\partial \mathbf{v}}{\partial \alpha_i} \right]. \qquad (2.211)$$

To evaluate the Lagrange brackets, one must evaluate the derivatives of $\mathbf{r}$ and $\mathbf{v}$ with respect to the orbital elements. There are in fact only three nonzero derivatives of

r and **v**, each with respect to the orbital elements. The most important property of the matrix **L** is that it is not an explicit function of time, and so it can be evaluated at any point in the orbit and the same numerical values will result. Thus, the Lagrange brackets are easily evaluated and the matrix **L** inverted to obtain the full set of LPE. The last of the equation in the LPE is expressed in terms of the rate of change of the *mean anomaly* rather than in terms of the rate of change of the *true anomaly*, as most often it is the mean motion of the orbiting body, averaged over several orbital periods, that is of primary interest. An alternative set are given by,

$$\frac{da}{dt} = \frac{2}{na} \times \frac{\partial V}{\partial M_0}, \quad \frac{de}{dt} = \frac{b}{na^3 e}\left(\frac{b}{a}\frac{\partial V}{\partial M_0} - \frac{\partial V}{\partial \omega}\right),$$

$$\frac{di}{dt} = \frac{1}{nab}\frac{1}{\sin i}\left(\cos i \frac{\partial V}{\partial \omega} - \frac{\partial V}{\partial \Omega}\right), \quad \frac{d\Omega}{dt} = \frac{1}{nab}\frac{1}{\sin i}\frac{\partial V}{\partial i},$$

$$\frac{d\omega}{dt} = \frac{b}{na^3 e}\frac{\partial V}{\partial e} - \frac{\cos i}{nab \sin i}\frac{\partial V}{\partial i}, \quad \frac{dM_0}{dt} = -\frac{1}{na^2}\left(\frac{b^2}{a^2 e}\frac{\partial V}{\partial e} + 2a\frac{\partial V}{\partial a}\right), \quad (2.212)$$

where $M_0 = nt_{perigee} \equiv nt_0$.

The LPE are the result of a direct coordinate transformation of the equations of motion from the three Cartesian displacement coordinates and the three Cartesian velocity coordinates to the six classical orbital elements. The LPE are employed in orbital calculations because they enable direct integration and calculation of the classical orbital elements. The advantage in this is that one could employ relatively large time steps in the numerical integration. The use of the mean anomaly facilitates this. There are several other applications of the LPE, such as the deflection of a spacecraft's trajectory in the vicinity of an asteroid or comet.

The gravitational potential field originating from a satellite's planetary host is the source of the main external force affecting an orbiting satellite. The Earth is not uniformly spherical, as it is an oblate spheroid, nor is the mass distribution homogeneous and uniform. The associated nonspherical gravitational field may be expressed in terms of Legendre and associated Legendre polynomials, the distance r of the satellite from the Earth's center, the longitude L of the satellite, measured from the Greenwich meridian and positive eastward, the latitude l of the satellite taken to be positive toward north, and the equatorial Earth radius $R_e = 6,378.16$ km as (see for example, Deutsch [1]),

$$U = \frac{\mu}{r} - \frac{\mu}{r}$$

$$\times \left[\sum_{n=2}^{\infty}\left(\frac{R_e}{r}\right)^n J_n \times P_n(\sin l) - \sum_{n=2}^{\infty}\sum_{m=1}^{\infty}\left(\frac{R_e}{r}\right)^n J_{nm} \times P_{nm}(\sin l) \cos(m(L - L_{nm}))\right],$$

$$(2.213)$$

where,

$$P_n(x) = \frac{1}{2^n n!} \times \frac{d^n}{dx^n}(x^2 - 1)^n, \quad P_{nm}(x) = (1 - x^2)^{\frac{m}{2}} \times \frac{d^m}{dx^m} P_n(x). \quad (2.214)$$

J_n, J_{nm}, and L_{mn} are constants characterizing the Earth's mass distribution. J_n are the zonal harmonics related to the Earth's oblateness and the later constants are associated with the tesseral harmonics related to the ellipticity of the equator that results in a 150 m difference between the Earth's major and minor axes.

The principal perturbation to orbital elements in an Earth satellite's orbit is due to the Earth's flattening, given by J_2 in the expression for the total potential energy possessed by a satellite by virtue of the Earth's gravitational field. The effect of the J_2 perturbation can be computed from the LPE. For the Earth, $J_2 = 1.08284 \times 10^{-3}$, $J_3 = -2.56 \times 10^{-6}$, $J_4 = -1.58 \times 10^{-6}$, and J_2 is at least about 1,000 times larger, in magnitude, than all other J_n, for $n > 4$. For the Earth, $J_{21} = 0$, $J_{22} = 1.86 \times 10^{-6}$, $J_{31} = 2.1061 \times 10^{-6}$, and all other J_n are at most equal to 1.0×10^{-6}.

Assuming that only the J_2 term influences the orbital elements significantly, only the orbital elements Ω, ω, and n are affected, and this results in a secular perturbation, the node of the orbit precesses, the argument of perigee rotates around the orbit plane, and the satellite moves with a slightly different mean motion.

2.8.1 Geostationary Satellites

The disadvantage of employing the standard orbital elements as the dependent variables in the equations is the inability to address certain types of orbits, such as equatorial orbits and circular orbits, as some of the elements may not be defined and could be extraneous. The Earth's gravitational potential may be expressed as,

$$U = \frac{\mu}{r} - \frac{\mu}{r}\left(\frac{R_e}{r}\right)^2 \left[J_2\left(1 - \frac{3}{2}\cos^2\delta\right) - 3J_{22}\cos^2\delta \cos(2(L - L_{22}))\right], \quad (2.215)$$

with $L_{22} = -15°$. As a geostationary satellite orbits the Earth along the equator, the latitude δ from the equatorial plane is always zero, and since $R_e/r \ll 1$, the Earth's gravitational potential may be simplified.

One could obtain the corresponding LPE for a geostationary satellite. (For a geostationary orbit $a = 42,164.5$ km, the orbit velocity is $v = 3,074.7$ m/s and the rotation period of the Earth is $P = 1,436.0683$ minutes.)

2.9 Gauss' Planetary Equations: Force Perturbations

Rather than employ the gradients of the disturbing potential function, the planetary equations may be directly expressed in terms of the generalized disturbing forces and the true anomaly f, as in the Gauss planetary equations (GPE). The generalized disturbing forces are assumed to be resolved in the direction of the position vector, normal to it but in the orbital plane and with the third force being mutually perpendicular to the first two, with the triad forming a right-handed system of vectors. If these axes are denoted as $(g_1 = r, g_2, g_3)$, the corresponding generalized (specific) forces are defined as,

$$F_r = \frac{\partial V}{\partial g_1} = \frac{\partial V}{\partial x}\frac{x}{r} + \frac{\partial V}{\partial y}\frac{y}{r} + \frac{\partial V}{\partial z}\frac{z}{r}, \quad F_s = \frac{\partial V}{\partial g_2} \quad \text{and} \quad F_n = \frac{\partial V}{\partial g_3}. \tag{2.216}$$

Thus, it can be shown that,

$$\frac{\partial V}{\partial a} = \frac{r}{a} F_r, \quad \frac{\partial V}{\partial \omega} = rF_s, \quad \frac{\partial V}{\partial i} = F_n r \sin(\omega + v),$$

$$\frac{\partial V}{\partial e} = a\left(-F_r \cos v + F_s\left(1 + \frac{r}{p}\right)\sin v\right), \quad \frac{\partial V}{\partial \Omega} = -F_n r \cos(\omega + v) \sin i - rF_s \cos i,$$

$$\frac{\partial V}{\partial M} = \frac{a}{\sqrt{1-e^2}}\left(F_r e \sin v + F_s \frac{p}{r}\right). \tag{2.217}$$

The planetary equations may be expressed as,

$$\frac{da}{dt} = \frac{2}{na}\frac{\partial V}{\partial M} = \frac{2}{na} \times \frac{a}{\sqrt{1-e^2}}\left(F_r e \sin v + F_s \frac{p}{r}\right),$$

$$\frac{de}{dt} = \frac{\sqrt{1-e^2}}{na^2 e}\left(\sqrt{1-e^2}\frac{\partial V}{\partial M} - \frac{\partial V}{\partial \omega}\right) = \frac{\sqrt{1-e^2}}{na^2 e} a\left(F_r e \sin v + F_s\left(\frac{p}{r} - \frac{r}{a}\right)\right),$$

$$\frac{d\omega}{dt} = \frac{1}{na^2} \times \left(\frac{\sqrt{1-e^2}}{e} \times \frac{\partial V}{\partial e} - \frac{1}{\sqrt{1-e^2}} \times \frac{\cos i}{\sin i} \times \frac{\partial V}{\partial i}\right)$$

$$= -\frac{\cos v}{nae}\sqrt{1-e^2}F_r + \frac{p}{eh}\left[\sin v\left(1 + \frac{1}{1+e\cos v}\right)\right]F_s - \frac{r}{na^2\sqrt{1-e^2}}\frac{\sin \theta}{\sin i}F_n \cos i$$

or

$$\frac{d\omega}{dt} = \frac{1}{na} \times \frac{\sqrt{1-e^2}}{e}\left(-F_r \cos v + F_s\left(1 + \frac{r}{p}\right)\sin v\right) - \cos i \frac{d\Omega}{dt}$$

$$\frac{di}{dt} = \frac{1}{na^2}\frac{1}{\sqrt{1-e^2}}\frac{1}{\sin i}\left(\cos i \frac{\partial V}{\partial \omega} - \frac{\partial V}{\partial \Omega}\right) = \frac{1}{na^2}\frac{1}{\sqrt{1-e^2}}F_n r \cos(\omega + v),$$

$$\frac{d\Omega}{dt} = \frac{1}{na^2}\frac{1}{\sqrt{1-e^2}}\frac{1}{\sin i}\frac{\partial V}{\partial i} = \frac{1}{na^2}\frac{1}{\sqrt{1-e^2}}\frac{1}{\sin i}F_n r \sin(\omega + v),$$

and

$$\frac{dM}{dt} = n - \frac{2}{na}\left(\frac{1-e^2}{2ae}\frac{\partial V}{\partial e} + \frac{\partial V}{\partial a}\right)$$

$$= n - \frac{1}{na}\left(\frac{2r}{a} - \frac{1-e^2}{e}\cos v\right)F_r - \frac{\sin v}{na}\left(\frac{1-e^2}{e}\right)\left(1 + \frac{r}{p}\right)F_s, \tag{2.218}$$

where the true anomaly is related to the mean anomaly via the eccentric anomaly and Kepler's equation.

(Note: Sometimes, rather than employ the argument of the peri-center ω and the mean anomaly it may be more convenient to employ the *true orbital longitude* $l = \Omega + \omega + v$ and the *mean orbital longitude* $\lambda = \Omega + \omega + M$ as the dependent variables. An alternative form of these equations, with the true anomaly f as the independent variable and in a slightly different notation, is given by Fortescue and Stark [7]. At other times, the orbital element Ω is replaced by the element $\bar{\lambda} = \Omega - \nu$.)

One could employ the relations,

$$p = a(1 - e^2), \qquad (2.219)$$

$$h^2 = \mu p = \mu\, a(1 - e^2) = a^4 n^2 (1 - e^2) \quad \text{or} \quad h = a^2 n \sqrt{1 - e^2}, \qquad (2.220)$$

where $n = \sqrt{\mu/a^3}$, to obtain the following alternate form of the planetary equations or the GPE as,

$$\frac{da}{dt} = \frac{2a^2}{h}[F_r e \sin \nu + (1 + e \cos \nu)F_s] = \frac{2e \sin \nu}{n\sqrt{1-e^2}} F_r + \frac{2a\sqrt{1-e^2}}{nr} F_s$$

$$\frac{de}{dt} = \frac{\sqrt{1-e^2} \sin \nu}{na} F_r + \frac{\sqrt{1-e^2}}{nea^2}\left[\frac{a^2(1-e^2)}{r} - r\right]F_s,$$

$$\frac{di}{dt} = \frac{r \cos \theta}{na^2 \sqrt{1-e^2}} F_n = \frac{r \cos \theta}{h} F_n, \quad \frac{d\Omega}{dt} = \frac{r}{na^2 \sqrt{1-e^2}} \frac{\sin \theta}{\sin i} F_n = \frac{r \sin \theta}{h \sin i} F_n,$$

$$\frac{d\omega}{dt} + \frac{d\Omega}{dt} \cos i = -\frac{\sqrt{1-e^2}}{nae}\left[F_r \cos \nu - \left(1 + \frac{1}{1 + e \cos \nu}\right)F_s \sin \nu\right].$$

$$\frac{dM}{dt} = n - \frac{2r}{na^2} F_r - \sqrt{1-e^2}\left(\frac{d\omega}{dt} + \frac{d\Omega}{dt} \cos i\right), \theta = \omega + \nu. \qquad (2.221)$$

Hence, one may invert these relations and obtain expressions for the triple F_r, F_s, and F_n in terms of the rates of change of the orbital elements. Since these are the forces per unit mass (specific forces), the above equations are in terms of perturbation acceleration components. The second and third equations are singular when the inclination is zero. The last three equations may also be expressed as,

$$\frac{de}{dt} = \frac{p \sin \nu}{h} F_r + \frac{1}{h}[(p + r)\cos f + re]F_s,$$

$$\frac{d\omega}{dt} + \frac{d\Omega}{dt} \cos i = -\frac{1}{he}[F_r p \cos \nu - (p + r)F_s \sin \nu],$$

$$\frac{d\nu}{dt} = \frac{h}{r^2} + \frac{1}{he}[F_r p \cos \nu - (p + r)F_s \sin \nu] = \frac{h}{r^2} - \left(\frac{d\omega}{dt} + \frac{d\Omega}{dt} \cos i\right).$$

$$(2.222)$$

The last two equations are singular for a circular orbit.

2.9.1 Effect of Atmospheric Drag

The influence of atmospheric drag results in one of the most significant perturbations to a satellite in low Earth orbit, especially below 400–600 km. Taking into account the rotation of the upper atmosphere with the Earth, the drag acceleration on a satellite is,

$$d\mathbf{F}_{drag} = F_r \mathbf{e}_r + F_s \mathbf{e}_s + F_n \mathbf{e}_n = -\frac{1}{2} C_D \rho A v_{rel} \mathbf{v}_{rel}, \qquad (2.223)$$

where C_D is the drag coefficient, A is the projected satellite area, ρ is the atmospheric density, which is assumed to be locally constant, although it is generally not,

$$\mathbf{v}_{rel} = \mathbf{v} - \omega_{ua}\mathbf{e}_z \times r\mathbf{e}_r, \tag{2.224}$$

and ω_{ua} is the rotational angular velocity of the Earth's upper atmosphere. (C_D is the drag coefficient, which depends to very large extent on the shape and surface of the satellite. For a sphere it is less than 2.2 and for a cylinder it is about 3. The drag coefficient C_D is not as trivial to evaluate as it may seem. Since the density is very low at the altitudes of satellite orbits, even low Earth orbits, the ordinary continuum-flow theory of conventional aerodynamics does not apply and the appropriate regime is that of free-molecule flow.) At an altitude of 450 km, the density is approximately given by, $\rho = 1.585 \times 10^{-12}$ kg/m^3.

Assuming a circular orbit and neglecting terms of the order of $(\omega_{ua}/n)^2$, it follows that,

$$v_{rel} = na(1 - (\omega_{ua}/n)\cos i), \quad F_r = 0,$$
$$F_s\mathbf{e}_s + F_n\mathbf{e}_n = -\frac{1}{2}C_D\rho A a^2 n^2((1 - (\omega_{ua}/n))\cos i)\mathbf{e}_s + (\omega_{ua}/n)\sin i \cos\theta\,\mathbf{e}_n). \tag{2.225}$$

One could then obtain the rate of change per unit time, averaged over several orbits with respect to the variable $\theta = \omega + f$, the *argument of the latitude* of the: (a) orbital energy, where $d\varepsilon/dt = \mathbf{v}\cdot d\mathbf{F}$; (b) scalar moment of momentum, h where $dh/dt = rT$; (c) semimajor axis; (d) inclination and, (e) longitude of the ascending node Ω.

2.9.2 Space Shuttle in a Low Earth Orbit

It is not always essential to consider the LPE in orbital flight. Consider, for example, a space vehicle such as the space shuttle in a low Earth orbit. If we restrict the motion to a plane, the equations of motion for planar atmospheric flight, based on a point mass model (Ashley [8], Vinh [9]), and ignoring aerodynamic heating effects in the atmosphere, are:

$$\frac{dr}{dt} = V\sin\gamma, \quad m\frac{dV}{dt} = -D - mg\sin\gamma = -\frac{\rho SC_D V^2}{2} - mg\sin\gamma, \tag{2.226}$$

$$mV\frac{d\gamma}{dt} = -L - m\left(g - \frac{V^2}{r}\right)\cos\gamma = -\frac{\rho SC_L V^2}{2} - m\left(g - \frac{V^2}{r}\right)\cos\gamma, \tag{2.227}$$

where γ is the flight path angle, V the vehicle velocity, r the distance of the vehicle from the Earth center, L the aerodynamic lift, D the aerodynamic drag, S, the vehicle surface area, and ρ, the atmospheric density,

$$g = \mu/r^2 \quad \text{and} \quad C_D = C_{D_0} + KC_L^2. \tag{2.228}$$

Nominally, the space shuttle is at 6,400 km from the Earth's center, $C_{D_0} = 0.2$, $K = 1.67$, $C_L = 2C_D$, and the mass to surface area ratio, $m/S = 300$ kg/m^2, an initial

2.9 Gauss' Planetary Equations: Force Perturbations 69

entry flight path angle of $-6°$. The atmosphere may be assumed to be modeled according to the US Standard Atmosphere Model, 1976.

(Note that the atmosphere may be assumed to be within a distance of 6,498 km, and above this distance the density may be assumed to be zero. The atmospheric density may be approximated by employing a look-up table in the range of 6,378 km $\leq r \leq$ 6,498 km. Any other equivalent approximation scheme is also acceptable. Also note that the above planar model is only approximate. Since the Earth's atmosphere rotates almost synchronously with the Earth, i.e. about the Earth's axis of rotation, there is a component of the drag force normal to the orbit plane that tends to change the inclination of the orbit, thus making it non-planar. To analyze such orbits, in general, one needs to append the equations for the rates of change of the *azimuth angle*, *longitude*, and *latitude* to the set presented here. The dependent variables then form an alternate set of orbital elements.)

2.9.3 Lunar Orbits

Another interesting orbital motion is the motion of a typical lunar orbiter, orbiting the Moon. When considering a Lunar orbiter, it is important to consider appropriate Lunar gravitation models. The gravitational potential field originating from the Moon is the main source of the external disturbing forces affecting a Lunar orbiter. The Lunar gravitational model is usually expressed in a slightly different but equivalent form as its Earth counterpart. It is usually expressed in the form,

$$U = \frac{\mu}{r} + \frac{\mu}{r}\left[\sum_{n=2}^{\infty}\sum_{m=0}^{\infty}\left(\frac{R}{r}\right)^n P_{nm}(\sin l)(C_{nm}\cos(mL) + S_{nm}\sin(mL))\right], \quad (2.229)$$

where C_{nm} and S_{nm} are constants characterizing the Moon's mass distribution. $C_{n0} = -J_n$ are the zonal harmonics related to the Moon's oblateness that describe the axially symmetric potential independent of the longitude. C_{nm} and S_{nm} with $n = m \neq 0$ are constants associated with the sectorial harmonics (zero potential along meridians of longitude), while in the case of $n \neq m$, $m \neq 0$, and $n \neq 0$, they are constants associated with the tesseral harmonics (zero potentials along parallels of latitude). The numerical values in the coefficients of the Ferrari model are tabulated in Table 2.3.

(i) Note that the coefficient C_{20} only affects the *nodal longitude* and the *argument of the perigee* and that the LPE, averaged over several orbits with respect to the variable $\theta = \omega + f$, the *argument of the latitude*, reduce to the pair,

$$\frac{d\omega}{dt} = -\frac{3n}{4}\frac{1}{(1-e^2)^2}\frac{R^2}{a^2}C_{20}(4 - 5\sin^2 i), \quad \frac{d\Omega}{dt} = \frac{3n}{2}\frac{1}{(1-e^2)^2}\frac{R^2}{a^2}C_{20}\cos i,$$

and

$$\frac{dM}{dt} = n - \frac{2r}{na^2}\hat{F}_r - \sqrt{1-e^2}\left(\frac{d\omega}{dt} + \frac{d\Omega}{dt}\cos i\right), \quad (2.230)$$

Table 2.3 Numerical values (un-normalized) of coefficients C_{nm} and S_{nm} in Ferrari model (the shaded areas represent the most important coefficients; all others may be set to zero)

Degree n	Order m	$C_{nm} \times 10^5$	$S_{nm} \times 10^5$	Degree n	Order m	$C_{nm} \times 10^5$	$S_{nm} \times 10^5$
2	0	−29.215	0	4	2	−0.1440	−0.2884
2	1	−0.01014	0	4	3	−0.0085	−0.0789
2	2	2.2304	0.00173	4	4	−0.01549	0.00564
3	0	−1.2126	0	5	0	−4.46	0
3	1	3.071	0.56107	5	1	−0.326	0.673
3	2	0.48884	0.1687	5	2	0.1556	−0.0522
3	3	0.1436	−0.033435	5	3	−0.0148	0.0127
4	0	0.015	0	5	4	0.00598	0.00456
4	1	−0.718	0.295	5	5	0.00122	0.00137

where $\hat{F}_r$ may be approximated when $e \ll 1$ as,

$$\hat{F}_r = \frac{3n^2}{2r} \frac{a^2}{(1-e^2)^3} \frac{R^2}{a^2} C_{20} \left[\left(1 - \frac{3}{2}\sin^2 i\right)(1 + 3e^2) + \frac{9}{4}e^2 \sin^2 i \cos 2\Omega_s \right], \quad (2.231)$$

where $\Omega_s = \Omega - \omega_m t$ is the orbital selenographic node longitude and ω_m is the local Moon rotation rate. Hence, the critical inclination angles when $\dot{\omega} = 0$ may be determined.

(ii) In the case of almost circular orbits, the coefficients C_{30} and C_{50} significantly influence the *eccentricity*, although they do also effect the *inclination*, the *nodal longitude*, and the *argument of the perigee*.

(iii) The coefficient C_{22} only effects the *inclination*, the *nodal longitude*, and the *argument of the perigee*, and the LPE, averaged over several orbits with respect to the variable $\theta = \omega + \nu$, the *argument of the latitude*, reduce to,

$$\frac{di}{dt} = 3n \frac{1}{(1-e^2)^2} \frac{R^2}{a^2} C_{22} \sin i \sin 2(\Omega - \omega_e t),$$

$$\frac{d\Omega}{dt} = 3n \frac{1}{(1-e^2)^2} \frac{R^2}{a^2} C_{22} \cos i \cos 2(\Omega - \omega_e t),$$

$$\frac{d\omega}{dt} = -\frac{3n}{2} \frac{1}{(1-e^2)^2} \frac{R^2}{a^2} C_{22} (2 - 5\sin^2 i) \cos 2(\Omega - \omega_e t),$$

$$\frac{dM}{dt} = n - \frac{2r}{na^2} \hat{F}_r - \sqrt{1-e^2} \left(\frac{d\omega}{dt} + \frac{d\Omega}{dt} \cos i \right), \quad (2.232)$$

where $\hat{F}_r$ may be approximated when $e \ll 1$ as,

$$\hat{F}_r = -\frac{9n^2 R^2 C_{22}}{2r(1-e^2)^3}$$
$$\times \left[\sin^2 i \cos 2\Omega_s (1 + 3e^2) - \frac{3}{2}e^2 \left(\cos i \sin 2\Omega_s \sin 2\omega + (2 - \sin^2 i) \cos 2\Omega_s \cos 2\omega \right) \right], \quad (2.233)$$

where $\Omega_s = \Omega - \omega_m t$ is the orbital selenographic node longitude and ω_m is the local Moon rotation rate.

(iv) The coefficient C_{31} only affects the *eccentricity*, the *inclination*, the *nodal longitude*, and the *argument of the perigee*.

2.9.4 Third-Body Perturbation and Orbital Elements in Earth Orbit

Consider the third-body perturbations (Sun or Moon) of a typical Earth satellite and show that there are only two elements that are influenced significantly and that the equations governing the averaged element dynamics can be expressed as,

$$d\omega/dt = \bar{K}_1\left(4 - 5\sin^2 i\right), \quad d\Omega/dt = -\bar{K}_2 \cos i, \qquad (2.234)$$

where $\bar{K}_1$ and $\bar{K}_2$ are constants. Hence, the argument of the perigee and the right ascension of ascending node may be expressed in terms of the orbital period T in days/orbit, as,

$$\omega = K_1 T\left(4 - 5\sin^2 i\right), \quad \Omega = -K_2 T \cos i. \qquad (2.235)$$

The values of the constants are tabulated in Table 2.4, for information.

2.10 Spacecraft Relative Motion

One of the most important problems in space dynamics is the definition of the relative equations of motion of an orbiting body relative to a nominal or reference satellite in a circular or an elliptic orbit. In the following sections, we shall first develop the nonlinear equations of motion of an orbiting body relative to a nominal or reference satellite in a circular orbit in the absence of perturbation forces, and then consider the influence of perturbations in general. Following this, we will extend the equations of motion to the case of the satellite in elliptic motion. Finally, we will consider the change in the independent coordinates of the equations of motion and express them in terms of the true anomaly of the nominal or reference satellite.

2.10.1 Hill-Clohessy-Wiltshire Equations

Consider a nominal satellite in a circular orbit. In the satellite's local vertical local horizontal (LVLH) frame, the position vector $\mathbf{r}_d$ of another orbiting body is given by the following,

$$\mathbf{r}_d = \mathbf{r} + \vec{\rho} = (r+x)i + yi + zk, \quad \mathbf{r}_d \cdot \mathbf{r}_d = (r+x)^2 + y^2 + z^2, \qquad (2.236)$$

Table 2.4 Constants K_1 and K_2 for perturbations due to the Sun and the Moon

Sun	$K_1 = 0.00077$	$K_2 = 0.00154$	Moon	$K_1 = 0.00169$	$K_2 = 0.00338$

where **r** is the position vector of the satellite, r is the magnitude of **r**, and $\vec{\rho} = xi + yj + zk$ is the position vector of the other orbiting body relative to the satellite. The angular velocity and acceleration of the LVLH frame, which is moving with the satellite, are given by $\omega = \dot{v}k = (h/r^2)k$ and $\alpha = \ddot{v}k = -(2\dot{r}/r)\omega = -(2\dot{r}/r)(h/r^2)k$, respectively, where v is the true anomaly and $h = |\mathbf{r} \times \dot{\mathbf{r}}|$ is the magnitude of the orbital angular momentum. From kinematics, the equation of motion for the other orbiting body in the satellite's LVLH frame is given by the following,

$$\ddot{\mathbf{r}}_d = \ddot{\mathbf{r}} + \ddot{\rho} + 2\omega \times \dot{\rho} + \omega \times (\omega \times \rho) + \alpha \times \rho. \tag{2.237}$$

Thus, in component form we have,

$$\ddot{\mathbf{r}}_d = (\ddot{r} + \ddot{x} - 2\dot{v}\dot{y} - \dot{v}^2 x - \ddot{v}y)i + (\ddot{y} + 2\dot{x}\dot{v} - \dot{v}^2 y + \ddot{v}x)j + \ddot{z}k. \tag{2.238}$$

Using Kepler's two-body equations of motion for the satellite and for the other orbiting body, both orbiting the Earth, $\ddot{r} = -\mu_e/r^2$ and $\ddot{\mathbf{r}}_d = -\mu_e \mathbf{r}_d/r_d^3$, respectively, the equations of motion are expressed as,

$$-\mu_e \mathbf{r}_d/r_d^3 = (-\mu_e/r^2 + \ddot{x} - 2\dot{v}\dot{y} - \dot{v}^2 x - \ddot{v}y)i + (\ddot{y} + 2\dot{x}\dot{v} - \dot{v}^2 y + \ddot{v}x)j + \ddot{z}k. \tag{2.239}$$

In component form we have,

$$\ddot{x} - 2\dot{v}\dot{y} - \dot{v}^2 x - \ddot{v}y = (\mu_e/r^2) - \mu_e(r+x)/r_d^3, \tag{2.240}$$

$$\ddot{y} + 2\dot{x}\dot{v} - \dot{v}^2 y + \ddot{v}x = -\mu_e y/r_d^3, \tag{2.241}$$

$$\ddot{z} = -\mu_e z/r_d^3. \tag{2.242}$$

But,

$$r_d^{-3} = r^{-3}\left(1 + 2\frac{x}{r} + \frac{x^2 + y^2 + z^2}{r^2}\right)^{-3/2}. \tag{2.243}$$

Hence, the equations for the other orbiting body's relative position coordinates are,

$$\ddot{x} - 2\dot{v}\dot{y} - \dot{v}^2 x - \ddot{v}y = \frac{\mu_e r}{r^3} - \frac{\mu_e(r+x)}{r_d^3}$$

$$= \frac{\mu_e r}{r^3}\left(1 - \left(1 + \frac{x}{r}\right)\left(1 + 2\frac{x}{r} + \frac{x^2 + y^2 + z^2}{r^2}\right)^{-3/2}\right), \tag{2.244}$$

$$\ddot{y} + 2\dot{x}\dot{v} - \dot{v}^2 y + \ddot{v}x = -\frac{\mu_e y}{r^3}\left(1 + 2\frac{x}{r} + \frac{x^2 + y^2 + z^2}{r^2}\right)^{-3/2}, \tag{2.245}$$

$$\ddot{z} = -\frac{\mu_e z}{r^3}\left(1 + 2\frac{x}{r} + \frac{x^2 + y^2 + z^2}{r^2}\right)^{-3/2}. \tag{2.246}$$

2.10 Spacecraft Relative Motion

Linearizing Equations (2.244)–(2.246),

$$\ddot{x} - 2\dot{\nu}\dot{y} - \dot{\nu}^2 x - \ddot{\nu} y = 2\mu_e x/r^3, \tag{2.247}$$

$$\ddot{y} + 2\dot{x}\dot{\nu} - \dot{\nu}^2 y + \ddot{\nu} x = -\mu_e y/r^3, \tag{2.248}$$

$$\ddot{z} = -\mu_e z/r^3. \tag{2.249}$$

These equations are called the linearized equations of relative motion (LERM).

Thus, the nonlinear equations of relative motion (NERM) can be expressed as,

$$\ddot{x} - 2\dot{\nu}\dot{y} - \left(\dot{\nu}^2 + 2\frac{\mu_e}{r^3}\right)x - \ddot{\nu} y$$
$$= \frac{\mu_e r}{r^3}\left(1 - 2\frac{x}{r} - \left(1 + \frac{x}{r}\right)\left(1 + 2\frac{x}{r} + \frac{x^2 + y^2 + z^2}{r^2}\right)^{-3/2}\right) \tag{2.250}$$

$$\ddot{y} + 2\dot{x}\dot{\nu} - \left(\dot{\nu}^2 - \frac{\mu_e}{r^3}\right)y + \ddot{\nu} x = \frac{\mu_e y}{r^3}\left(1 - \left(1 + 2\frac{x}{r} + \frac{x^2 + y^2 + z^2}{r^2}\right)^{-3/2}\right), \tag{2.251}$$

$$\ddot{z} + \frac{\mu_e z}{r^3} = \frac{\mu_e z}{r^3}\left(1 - \left(1 + 2\frac{x}{r} + \frac{x^2 + y^2 + z^2}{r^2}\right)^{-3/2}\right). \tag{2.252}$$

To obtain the Hill-Clohessy-Wiltshire (HCW) equations, we let $\dot{\nu} - n = \ddot{\nu} = 0$, $\dot{\nu}^2 = \mu_e/r^3 = n^2$, where n represents the mean rate of motion of the satellite or the mean motion. To write Equations (2.250)–(2.252) in state space form one defines the velocity components in the LVLH frame by,

$$\begin{bmatrix} \Delta v_x \\ \Delta v_y \\ \Delta v_z \end{bmatrix} = \begin{bmatrix} \dot{x} \\ \dot{y} \\ \dot{z} \end{bmatrix} + \boldsymbol{\omega} \times \begin{bmatrix} x \\ y \\ z \end{bmatrix} = \begin{bmatrix} \dot{x} \\ \dot{y} \\ \dot{z} \end{bmatrix} + \begin{bmatrix} 0 & -\dot{\nu} & 0 \\ \dot{\nu} & 0 & 0 \\ 0 & 0 & 0 \end{bmatrix} \begin{bmatrix} x \\ y \\ z \end{bmatrix}. \tag{2.253}$$

Thus,

$$\begin{bmatrix} \Delta \dot{v}_x \\ \Delta \dot{v}_y \\ \Delta \dot{v}_z \end{bmatrix} + \begin{bmatrix} 0 & -\dot{\nu} & 0 \\ \dot{\nu} & 0 & 0 \\ 0 & 0 & 0 \end{bmatrix} \begin{bmatrix} \Delta v_x \\ \Delta v_y \\ \Delta v_z \end{bmatrix} = \begin{bmatrix} \ddot{x} \\ \ddot{y} \\ \ddot{z} \end{bmatrix} + 2\begin{bmatrix} 0 & -\dot{\nu} & 0 \\ \dot{\nu} & 0 & 0 \\ 0 & 0 & 0 \end{bmatrix} \begin{bmatrix} \dot{x} \\ \dot{y} \\ \dot{z} \end{bmatrix} - \begin{bmatrix} \dot{\nu}^2 & \ddot{\theta} & 0 \\ -\ddot{\nu} & \dot{\nu}^2 & 0 \\ 0 & 0 & 0 \end{bmatrix} \begin{bmatrix} x \\ y \\ z \end{bmatrix}. \tag{2.254}$$

Hence, one may write Equations (2.253)–(2.254) in state space form as,

$$\begin{bmatrix} \dot{x} \\ \dot{y} \\ \dot{z} \end{bmatrix} = \begin{bmatrix} \Delta v_x \\ \Delta v_y \\ \Delta v_z \end{bmatrix} - \begin{bmatrix} 0 & -\dot{\nu} & 0 \\ \dot{\nu} & 0 & 0 \\ 0 & 0 & 0 \end{bmatrix} \begin{bmatrix} x \\ y \\ z \end{bmatrix}, \tag{2.255}$$

$$\begin{bmatrix} \Delta \dot{v}_x \\ \Delta \dot{v}_y \\ \Delta \dot{v}_z \end{bmatrix} = -\begin{bmatrix} 0 & -\dot{\nu} & 0 \\ \dot{\nu} & 0 & 0 \\ 0 & 0 & 0 \end{bmatrix} \begin{bmatrix} \Delta v_x \\ \Delta v_y \\ \Delta v_z \end{bmatrix} + \frac{\mu_e}{r^3}\begin{bmatrix} 2x \\ -y \\ -z \end{bmatrix} + \frac{\mu_e}{r^3}\begin{bmatrix} rf_1 \\ yf_2 \\ zf_3 \end{bmatrix}. \tag{2.256}$$

In state-space form, the LERM are given by the equation,

$$\dot{\mathbf{x}} = \mathbf{A}(t)\mathbf{x}, \tag{2.257}$$

where,

$$\mathbf{x} = \begin{bmatrix} x & y & z & \Delta v_x & \Delta v_y & \Delta v_z \end{bmatrix}^T, \tag{2.258}$$

$$\mathbf{A}(t) = \begin{bmatrix} \begin{bmatrix} 0 & -\dot{v} & 0 \\ -\dot{v} & 0 & 0 \\ 0 & 0 & 0 \end{bmatrix} & \begin{bmatrix} 1 & 0 & 0 \\ 0 & 1 & 0 \\ 0 & 0 & 1 \end{bmatrix} \\ \frac{\mu_e}{r^3}\begin{bmatrix} 2 & 0 & 0 \\ 0 & -1 & 0 \\ 0 & 0 & -1 \end{bmatrix} & -\begin{bmatrix} 0 & -\dot{v} & 0 \\ \dot{v} & 0 & 0 \\ 0 & 0 & 0 \end{bmatrix} \end{bmatrix}. \tag{2.259}$$

The matrix $\mathbf{A}(t)$ is time varying, as r, $\dot{v}$, and $\ddot{v}$ vary with time. Thus, $\mathbf{A}(t)$ is periodic with the period equal to $T = 2\pi/n$. Thus, in terms of the mean rate of motion n, the NERM are,

$$\ddot{x} - 2\dot{y}n - 3n^2 x = 2\dot{y}(\dot{v} - n) + \left(\dot{v}^2 + 2\frac{\mu_e}{r^3} - 3n^2\right)x + \ddot{v}y$$

$$+ \frac{\mu_e r}{r^3}\left(1 - 2\frac{x}{r} - \left(1 + \frac{x}{r}\right)\left(1 + 2\frac{x}{r} + \frac{x^2 + y^2 + z^2}{r^2}\right)^{-3/2}\right), \tag{2.260}$$

$$\ddot{y} + 2\dot{x}n = 2\dot{x}(n - \dot{v}) + \left(\dot{v}^2 - \frac{\mu_e}{r^3}\right)y - \ddot{v}x + \frac{\mu_e y}{r^3}\left(1 - \left(1 + 2\frac{x}{r} + \frac{x^2 + y^2 + z^2}{r^2}\right)^{-\frac{3}{2}}\right) \tag{2.261}$$

$$\ddot{z} + n^2 z = \left(n^2 - \frac{\mu_e}{r^3}\right)z + \frac{\mu_e z}{r^3}\left(1 - \left(1 + 2\frac{x}{r} + \frac{x^2 + y^2 + z^2}{r^2}\right)^{-3/2}\right). \tag{2.262}$$

The HCW equations are obtained by setting the right-hand sides of Equations (2.260)–(2.262) to zero.

2.10.2 Linear Orbit Theory with Perturbations

For a spacecraft orbiting a primary body, in terms of a general gravitational field, the equations of motion are,

$$\ddot{\mathbf{r}} = \mathbf{g}(\mathbf{r}) + \mathbf{U}(\mathbf{r}), \tag{2.263}$$

where the last term represents the externally applied or disturbance acceleration components acting on the spacecraft and $\mathbf{r}(t)$ represents the position vector of the orbit vehicle in a reference frame fixed at the barycenter of the two-mass system, the primary body and the spacecraft. Let $\mathbf{r}^*(t)$ be a known reference orbit and the vector difference of the spacecraft's position and the reference position, evaluated at some time t, the relative position vector $\delta\,\mathbf{r}(t)$ is:

$$\delta\,\mathbf{r}(t) = \mathbf{r}(t) - \mathbf{r}^*(t). \tag{2.264}$$

(i) The governing equation of motion may be linearized and $\delta\,\mathbf{r}(t)$ satisfies the equation,

$$\delta\,\ddot{\mathbf{r}}(t) = \frac{\partial}{\partial\,\mathbf{r}^*}\mathbf{g}(\mathbf{r}^*)\delta\,\mathbf{r}(t) + \mathbf{U}(\mathbf{r}) = G(\mathbf{r}^*)\delta\,\mathbf{r}(t) + \mathbf{U}(\mathbf{r}). \tag{2.265}$$

In the case of the inverse-square gravitational force field,

$$G(\mathbf{r}) = \left(\mu/|\mathbf{r}|^5\right)\left(3\mathbf{r}\mathbf{r}^T - |\mathbf{r}|^2\mathbf{I}_{3\times 3}\right), \tag{2.266}$$

where $\mathbf{I}_{3\times 3}$ is the 3×3 identity matrix.

(ii) Define $\tau = nt$ as the dimensionless time, where n is the mean motion of the barycenter of the primary body and the second body, and convert all time derivatives so that they are done with respect to τ, and

$$(\dot{\,}) = d/d\tau = d/(ndt), \quad n = \sqrt{\mu/(\mathbf{r}^*)^3}. \tag{2.267}$$

Consider a reference frame that rotates with the reference position vector $\mathbf{r}^*(t)$, with the speed of the mean motion of the two-body system rotating about an axis normal to the plane of the orbit, and let x, y, z be the nondimensional position components of the spacecraft in each axis of the rotating coordinate system.

The spacecraft's motion is governed by the HCW equations,

$$\ddot{x} - 2\dot{y} - x = 2x + \bar{U}_x, \quad \ddot{y} + 2\dot{x} - y = -y + \bar{U}_y, \quad \ddot{z} = -z + \bar{U}_z, \tag{2.268}$$

where $\bar{U}_x$, $\bar{U}_y$, and $\bar{U}_z$ are the normalized, externally applied, or disturbance acceleration components acting on the spacecraft.

2.10.3 Nonlinear Equations of Relative Motion with Perturbations

Consider a vector of perturbations acting on the nominal satellite and given by, $\mathbf{f}_S$. The equations of motion of the nominal satellite are given by,

$$\ddot{\mathbf{r}} = -\mu_e\left(\mathbf{r}/|\mathbf{r}|^3\right) + \mathbf{f}_S = -\mu_e\left(\mathbf{r}/r^3\right) + \mathbf{f}_S. \tag{2.269}$$

For the other orbiting body, with perturbations given by $\mathbf{f}_B$, the corresponding equations are,

$$\ddot{\mathbf{r}}_b = -\mu_e\left(\mathbf{r}_b/|\mathbf{r}_b|^3\right) + \mathbf{f}_B = -\mu_e\left(\mathbf{r}_b/r_b^3\right) + \mathbf{f}_B, \tag{2.270}$$

where $\mathbf{r}_b$ is the vector position of the other orbiting body in ECI coordinates. Hence,

$$-\mu_e(\mathbf{r}_b/r_b^3) + \mathbf{f}_B - \mathbf{f}_S$$
$$= (-\mu_e/r^2 + \ddot{x} - 2\dot{\nu}\dot{y} - \dot{\nu}^2 x - \ddot{\nu}y)\mathbf{i} + (\ddot{y} + 2\dot{x}\dot{\nu} - \dot{\nu}^2 y + \ddot{\nu}x)\mathbf{j} + \ddot{z}\mathbf{k}. \quad (2.271)$$

In the ECI frame, let $\tilde{\mathbf{f}}_{BS} = \mathbf{f}_B - \mathbf{f}_S$. Then it follows that,

$$\ddot{x} - 2\dot{\nu}\dot{y} - \dot{\nu}^2 x - \ddot{\nu}y = (\mu_e/r^2) - \mu_e(r+x)/r_d^3 + \tilde{\mathbf{f}}_{BS} \cdot \hat{\mathbf{x}}, \quad (2.272)$$

$$\ddot{y} + 2\dot{x}\dot{\nu} - \dot{\nu}^2 y + \ddot{\nu}x = -\mu_e y/r_d^3 + \tilde{\mathbf{f}}_{BS} \cdot \hat{\mathbf{y}}, \quad (2.273)$$

$$\ddot{z} = -\mu_e z/r_d^3 + \tilde{\mathbf{f}}_{BS} \cdot \hat{\mathbf{z}}, \quad (2.274)$$

where the unit vectors $\hat{\mathbf{x}}, \hat{\mathbf{y}}, \hat{\mathbf{z}}$ are defined as, $\hat{\mathbf{x}} = \mathbf{r}/r, \hat{\mathbf{y}} = \dot{\mathbf{r}}/|\dot{\mathbf{r}}|, \hat{\mathbf{z}} = \hat{\mathbf{x}} \times \hat{\mathbf{y}}$. Thus, the NERM can be expressed as,

$$\ddot{x} - 2\dot{y}\dot{\nu} - \left(\dot{\nu}^2 + 2\frac{\mu_e}{r^3}\right)x - \ddot{\nu}y,$$
$$= \frac{\mu_e r}{r^3}\left(1 - 2\frac{x}{r} - \left(1 + \frac{x}{r}\right)\left(1 + 2\frac{x}{r} + \frac{x^2 + y^2 + z^2}{r^2}\right)^{-3/2}\right) + \tilde{\mathbf{f}}_{BS} \cdot \hat{\mathbf{x}} \quad (2.275)$$

$$\ddot{y} + 2\dot{x}\dot{\nu} - \left(\dot{\nu}^2 - \frac{\mu_e}{r^3}\right)y + \ddot{\nu}x = \frac{\mu_e y}{r^3}\left(1 - \left(1 + 2\frac{x}{r} + \frac{x^2 + y^2 + z^2}{r^2}\right)^{-3/2}\right) + \tilde{\mathbf{f}}_{BS} \cdot \hat{\mathbf{y}}, \quad (2.276)$$

$$\ddot{z} + \frac{\mu_e z}{r^3} = \frac{\mu_e z}{r^3}\left(1 - \left(1 + 2\frac{x}{r} + \frac{x^2 + y^2 + z^2}{r^2}\right)^{-3/2}\right) + \tilde{\mathbf{f}}_{BS} \cdot \hat{\mathbf{z}}. \quad (2.277)$$

Using the binomial series expansion we may express,

$$f_2(x, y, z, r) \equiv 1 - (1 + \delta)^{-3/2}$$
$$\cong 3\frac{x}{r} - \frac{15}{2}\frac{x^2}{r^2}\left(1 - \frac{7x}{3r}\right) + \frac{3x^2 + y^2 + z^2}{2}\frac{1}{r^2}\left(1 - 5\frac{x}{r}\right), \quad (2.278)$$

$$f_1(x, y, z, r) \equiv 1 - 2\frac{x}{r} - \left(1 + \frac{x}{r}\right)(1 + \delta)^{-3/2}$$
$$\cong 1 - 2\frac{x}{r} - \left(1 + \frac{x}{r}\right)\left(1 - 3\frac{x}{r} + \frac{15 x^2}{2 r^2}\left(1 - \frac{7x}{3r}\right) - \frac{3x^2 + y^2 + z^2}{2}\frac{1}{r^2}\left(1 - 5\frac{x}{r}\right)\right)$$
$$= \frac{x^2}{r^2}\left(\frac{15}{2}\left(1 - \frac{7x}{3r}\right)\left(1 + \frac{x}{r}\right) - 3\right) - \frac{3x^2 + y^2 + z^2}{2}\frac{1}{r^2}\left(1 - 5\frac{x}{r}\right)\left(1 + \frac{x}{r}\right), \quad (2.279)$$

where,

$$\delta = 2\frac{x}{r} + \frac{x^2 + y^2 + z^2}{r^2}. \tag{2.280}$$

The approximations for $f_i(x, y, z, r)$ can be used for $\delta \leq 0.001$ and the exact expressions otherwise.

2.10.4 Nonlinear Equations of Relative Motion with Reference to an Elliptic Orbit

The orbital equations of an orbiting body are now defined in the LVLH frame, which is assumed to be attached to a nominal satellite in an elliptic orbit. The position vector and radial distance of an orbiting body are defined, respectively, in the LVLH frame as,

$$\vec{r}_s = (r+x)\hat{x} + y\hat{y} + z\hat{z}, \quad r_s = |\vec{r}_s| = \sqrt{(r+x)^2 + y^2 + z^2}, \tag{2.281}$$

where $\hat{x}$, $\hat{y}$, $\hat{z}$, are the unit vectors in the LVLH frame. The time rate of change of the position vector and the acceleration vector are, respectively, given by,

$$\dot{\vec{r}}_s = (\dot{r} + \dot{x} - \omega_z y)\hat{x} + (\dot{y} + r\omega_z + x\omega_z - z\omega_x)\hat{y} + (\dot{z} + \omega_x y)\hat{z}, \tag{2.282}$$

$$\ddot{\vec{r}}_s = (\ddot{r} + \ddot{x} - \omega_z \dot{y} - \dot{\omega}_z y)\hat{x} + (\ddot{y} + (\dot{r} + \dot{x})\omega_z - \dot{z}\omega_x$$
$$+ (r+x)\dot{\omega}_z - z\dot{\omega}_x)\hat{y} + (\ddot{z} + \omega_x \dot{y} + \dot{\omega}_x y)\hat{z} + (\dot{r} + \dot{x} - \omega_z y)\omega_z \hat{y}$$
$$+ (\dot{y} + r\omega_z + x\omega_z - z\omega_x)(\omega_x \hat{z} - \omega_z \hat{x}) - (\dot{z} + \omega_x y)\omega_x \hat{y}. \tag{2.283}$$

In these equations, ω_x, ω_y, and ω_z are the angular velocity components of the LVLH frame. The acceleration vector may be expressed in matrix-vector form as,

$$\ddot{\vec{r}}_s = \begin{bmatrix} \ddot{r} - r\omega_z^2 \\ 2\dot{r}\omega_z + r\dot{\omega}_z \\ r\omega_z \omega_x \end{bmatrix} + \begin{bmatrix} \ddot{x} \\ \ddot{y} \\ \ddot{z} \end{bmatrix} + \dot{\boldsymbol{\omega}} \times \begin{bmatrix} x \\ y \\ z \end{bmatrix} + 2\boldsymbol{\omega} \times \begin{bmatrix} \dot{x} \\ \dot{y} \\ \dot{z} \end{bmatrix} + \boldsymbol{\omega} \times \boldsymbol{\omega} \times \begin{bmatrix} x \\ y \\ z \end{bmatrix}. \tag{2.284}$$

Eliminating the position and velocity of the reference orbit, using Newton's law of gravitation, and assuming there are no other perturbation forces,

$$\ddot{\vec{r}}_s = \begin{bmatrix} \ddot{x} \\ \ddot{y} \\ \ddot{z} \end{bmatrix} + \dot{\boldsymbol{\omega}} \times \begin{bmatrix} x \\ y \\ z \end{bmatrix} + 2\boldsymbol{\omega} \times \begin{bmatrix} \dot{x} \\ \dot{y} \\ \dot{z} \end{bmatrix} + \boldsymbol{\omega} \times \boldsymbol{\omega} \times \begin{bmatrix} x \\ y \\ z \end{bmatrix} - \frac{\mu}{r^2} \begin{bmatrix} 1 \\ 0 \\ 0 \end{bmatrix}. \tag{2.285}$$

Define the perturbation velocity states Δv_x, Δv_y, Δv_z,

$$\begin{bmatrix} \Delta v_x \\ \Delta v_y \\ \Delta v_z \end{bmatrix} = \begin{bmatrix} \dot{x} \\ \dot{y} \\ \dot{z} \end{bmatrix} + \boldsymbol{\omega} \times \begin{bmatrix} x \\ y \\ z \end{bmatrix}. \tag{2.286}$$

Hence,

$$\begin{bmatrix} \Delta\dot{v}_x \\ \Delta\dot{v}_y \\ \Delta\dot{v}_z \end{bmatrix} = \begin{bmatrix} \ddot{x} \\ \ddot{y} \\ \ddot{z} \end{bmatrix} + \dot{\omega} \times \begin{bmatrix} x \\ y \\ z \end{bmatrix} + \omega \times \begin{bmatrix} \dot{x} \\ \dot{y} \\ \dot{z} \end{bmatrix}. \qquad (2.287)$$

Thus,

$$\ddot{\vec{r}}_s = \begin{bmatrix} \Delta\dot{v}_x \\ \Delta\dot{v}_y \\ \Delta\dot{v}_z \end{bmatrix} + \omega \times \begin{bmatrix} \dot{x} \\ \dot{y} \\ \dot{z} \end{bmatrix} + \omega \times \omega \times \begin{bmatrix} x \\ y \\ z \end{bmatrix} - \frac{\mu}{r^2}\begin{bmatrix} 1 \\ 0 \\ 0 \end{bmatrix}. \qquad (2.288)$$

For any orbiting body, assuming no other perturbation forces other than the gravitational force of the central spherical body, one may also write the equations for the rate of change of its position vector as,

$$\ddot{\vec{r}}_s = -\frac{\mu \vec{r}_s}{r_s^2 r_s} = -\frac{\mu_s}{r_s^2}\left(\frac{(r+x)\hat{x}}{r_s} + \frac{y\hat{y}}{r_s} + \frac{z\hat{z}}{r_s}\right). \qquad (2.289)$$

It is expressed in matrix-vector form as,

$$\ddot{\vec{r}}_s = -(\mu_s/r_s^3)[r+x \quad y \quad z]^T. \qquad (2.290)$$

From Equations (2.287) and (2.285) it follows that,

$$\ddot{\vec{r}}_s = \begin{bmatrix} \Delta\dot{v}_x \\ \Delta\dot{v}_y \\ \Delta\dot{v}_z \end{bmatrix} + \omega \times \begin{bmatrix} \dot{x} \\ \dot{y} \\ \dot{z} \end{bmatrix} + \omega \times \omega \times \begin{bmatrix} x \\ y \\ z \end{bmatrix} - \frac{\mu}{r^2}\begin{bmatrix} 1 \\ 0 \\ 0 \end{bmatrix} = -\frac{\mu_s}{r_s^3}\begin{bmatrix} r+x \\ y \\ z \end{bmatrix}. \qquad (2.291)$$

Using Equation (2.286),

$$\begin{bmatrix} \Delta\dot{v}_x \\ \Delta\dot{v}_y \\ \Delta\dot{v}_z \end{bmatrix} + \omega \times \begin{bmatrix} \Delta v_x \\ \Delta v_y \\ \Delta v_z \end{bmatrix} = -\left[\frac{\mu_s}{r_s^3}\begin{bmatrix} r+x \\ y \\ z \end{bmatrix} - \frac{\mu}{r^2}\begin{bmatrix} 1 \\ 0 \\ 0 \end{bmatrix}\right]. \qquad (2.292)$$

Hence, the equations of motion of a body relative to the LVLH frame are expressed as,

$$\begin{bmatrix} \Delta\dot{v}_x \\ \Delta\dot{v}_y \\ \Delta\dot{v}_z \end{bmatrix} + \omega \times \begin{bmatrix} \Delta v_x \\ \Delta v_y \\ \Delta v_z \end{bmatrix} = -\left[\frac{\mu_s}{r_s^3}\begin{bmatrix} r+x \\ y \\ z \end{bmatrix} - \frac{\mu}{r^2}\begin{bmatrix} 1 \\ 0 \\ 0 \end{bmatrix}\right], \quad \begin{bmatrix} \Delta v_x \\ \Delta v_y \\ \Delta v_z \end{bmatrix} = \begin{bmatrix} \dot{x} \\ \dot{y} \\ \dot{z} \end{bmatrix} + \omega \times \begin{bmatrix} x \\ y \\ z \end{bmatrix}. \qquad (2.293)$$

Equation (2.293) may also be expressed as,

$$\begin{bmatrix} \Delta\dot{v}_x \\ \Delta\dot{v}_y \\ \Delta\dot{v}_z \end{bmatrix} = -\left[\frac{\mu_s}{r_s^3}\begin{bmatrix} r+x \\ y \\ z \end{bmatrix} - \frac{\mu}{r^2}\begin{bmatrix} 1 \\ 0 \\ 0 \end{bmatrix}\right] - \begin{bmatrix} 0 & \Delta v_z & -\Delta v_y \\ -\Delta v_z & 0 & \Delta v_x \\ \Delta v_y & -\Delta v_x & 0 \end{bmatrix}\begin{bmatrix} \omega_x \\ \omega_y \\ \omega_z \end{bmatrix}, \qquad (2.294)$$

2.10 Spacecraft Relative Motion

$$\begin{bmatrix} \dot{x} \\ \dot{y} \\ \dot{z} \end{bmatrix} = \begin{bmatrix} \Delta v_x \\ \Delta v_y \\ \Delta v_z \end{bmatrix} - \begin{bmatrix} 0 & z & -y \\ -z & 0 & x \\ y & -x & 0 \end{bmatrix} \begin{bmatrix} \omega_x \\ \omega_y \\ \omega_z \end{bmatrix}. \quad (2.295)$$

It may be noted that,

$$r_s = |\vec{r}_s| = r\sqrt{1 + 2\frac{x}{r} + \frac{x^2 + y^2 + z^2}{r^2}} \equiv r\sqrt{1+\delta}, \quad \delta = 2\frac{x}{r} + \frac{x^2 + y^2 + z^2}{r^2}. \quad (2.296)$$

For a 3–1–3 Euler angle sequence, one may express the angular velocity vector or the rotation rate of the LVLH frame in terms orbital inclination i, the orbit's right ascension of the ascending node Ω, and the rate of change of the satellite's longitude θ as,

$$\begin{bmatrix} \omega_x \\ \omega_y \\ \omega_z \end{bmatrix} = \boldsymbol{\omega} = \begin{bmatrix} i\cos\theta + \dot{\Omega}\sin i \sin\theta \\ -i\sin\theta + \dot{\Omega}\sin i \cos\theta \\ \dot{\theta} + \dot{\Omega}\cos i \end{bmatrix} = \begin{bmatrix} \sin i \sin\theta & \cos\theta & 0 \\ 0 & 0 & 0 \\ \cos i & 0 & 1 \end{bmatrix} \begin{bmatrix} \dot{\Omega} \\ \dot{i} \\ \dot{\theta} \end{bmatrix}, \quad (2.297)$$

where the planetary equations of Gauss have been used to show that $\omega_y = 0$. Furthermore, from Gauss' planetary equations for the rate of change of Ω, i, and θ one can show that,

$$\begin{bmatrix} \dot{\Omega} \\ \dot{i} \\ \dot{\theta} \end{bmatrix} = \eta\zeta\left(\frac{a}{r}\right)^2 \begin{bmatrix} 0 \\ 0 \\ 1 \end{bmatrix} + \frac{r}{\eta\zeta a^2}\begin{bmatrix} \sin\theta/\sin i \\ \cos\theta \\ -\sin\theta/\tan i \end{bmatrix} F_z, \quad (2.298)$$

where F_z is the perturbation force acting on the nominal satellite normal to its orbit plane. Eliminating the $\boldsymbol{\omega}$ vector using Equation (2.298) and assuming that $\mu_s = \mu_e$, the equations of motion of a body or satellite relative to the LVLH frame are the final state space equations for the relative motion of the satellite in the LVLH frame. They are given by,

$$\begin{bmatrix} \Delta\dot{v}_x \\ \Delta\dot{v}_y \\ \Delta\dot{v}_z \end{bmatrix} = \eta\zeta\left(\frac{a}{r}\right)^2 \begin{bmatrix} \Delta v_y \\ -\Delta v_x \\ 0 \end{bmatrix} + \begin{bmatrix} \Delta\tilde{v}_x \\ \Delta\tilde{v}_y \\ \Delta\tilde{v}_z \end{bmatrix}$$

$$- \begin{bmatrix} -\Delta v_y \cos i & 0 & -\Delta v_y \\ -\Delta v_z \sin i \sin\theta + \Delta v_x \cos i & -\Delta v_z \cos\theta & \Delta v_x \\ \Delta v_y \sin i \sin\theta & \Delta v_y \cos\theta & 0 \end{bmatrix} \begin{bmatrix} \dot{\Omega} \\ \dot{i} \\ \Delta\dot{\theta} \end{bmatrix}, \quad (2.299)$$

$$\begin{bmatrix} \dot{x} \\ \dot{y} \\ \dot{z} \end{bmatrix} = \begin{bmatrix} \Delta v_x \\ \Delta v_y \\ \Delta v_z \end{bmatrix} + \eta\zeta\left(\frac{a}{r}\right)^2 \begin{bmatrix} y \\ -x \\ 0 \end{bmatrix}$$

$$- \begin{bmatrix} -y\cos i & 0 & -y \\ -z\sin i \sin\theta + x\cos i & -z\cos\theta & x \\ y\sin i \sin\theta & y\cos\theta & 0 \end{bmatrix} \begin{bmatrix} \dot{\Omega} \\ \dot{i} \\ \Delta\dot{\theta} \end{bmatrix}, \quad (2.300)$$

where,

$$\Delta\dot\theta = \dot\theta - \eta\zeta(a/r)^2, \tag{2.301}$$

$$\begin{bmatrix}\dot\Omega\\\dot i\\\Delta\dot\theta\end{bmatrix} = \frac{r}{\eta\zeta a^2}\begin{bmatrix}\sin\theta/\sin i\\\cos\theta\\-\sin\theta/\tan i\end{bmatrix}F_{z,r},\quad \begin{bmatrix}\Delta\tilde v_x\\\Delta\tilde v_y\\\Delta\tilde v_z\end{bmatrix} = \frac{\mu_e}{r^3}\begin{bmatrix}2x\\-y\\-z\end{bmatrix} + \frac{\mu_e}{r^3}\begin{bmatrix}rf_1\\yf_2\\zf_3\end{bmatrix}, \tag{2.302}$$

with $f_1 = 1 - 2x/r - (1+\delta)^{-3/2}(1+x/r)$, and $f_2 = 1 - (1+\delta)^{-3/2}$. The rates of change of Ω, i, and $\Delta\theta$ may now be eliminated. The resulting equations are,

$$\begin{bmatrix}\Delta\dot v_x\\\Delta\dot v_y\\\Delta\dot v_z\end{bmatrix} = -\frac{1}{\eta\zeta a^2 r^2}\begin{bmatrix}0 & (\eta\zeta a^2)^2 & 0\\-(\eta\zeta a^2)^2 & 0 & -r^3 F_{z,r}\\0 & r^3 F_{z,r} & 0\end{bmatrix}\begin{bmatrix}\Delta v_x\\\Delta v_y\\\Delta v_z\end{bmatrix}$$

$$+ \frac{\mu_e}{r^3}\begin{bmatrix}2x\\-y\\-z\end{bmatrix} + \frac{\mu_e}{r^3}\begin{bmatrix}rf_1\\yf_2\\zf_3\end{bmatrix}, \tag{2.303}$$

$$\begin{bmatrix}\dot x\\\dot y\\\dot z\end{bmatrix} = \begin{bmatrix}\Delta v_x\\\Delta v_y\\\Delta v_z\end{bmatrix} - \frac{1}{\eta\zeta a^2 r^2}\begin{bmatrix}0 & (\eta\zeta a^2)^2 & 0\\-(\eta\zeta a^2)^2 & 0 & -r^3 F_{z,r}\\0 & r^3 F_{z,r} & 0\end{bmatrix}\begin{bmatrix}x\\y\\z\end{bmatrix}. \tag{2.304}$$

Equations (2.303)–(2.304) are cast in the same form as Equations (2.255)–(2.256). Comparing Equations (2.255)–(2.256) with Equations (2.303)–(2.304) it is observed that Equations (2.303)–(2.304) reduce to Equations (2.255)–(2.256) when $\dot\theta = \eta\zeta(a/r)^2$ and $F_{z,r} = 0$. When all perturbation forces acting on both the nominal satellite in elliptic orbit and the orbiting bodies are included, Equations (2.255) and (2.303) are, respectively, given by,

$$\begin{bmatrix}\Delta\dot v_x\\\Delta\dot v_y\\\Delta\dot v_z\end{bmatrix} = -\begin{bmatrix}0 & -\dot\theta & 0\\\dot\theta & 0 & 0\\0 & 0 & 0\end{bmatrix}\begin{bmatrix}\Delta v_x\\\Delta v_y\\\Delta v_z\end{bmatrix} + \frac{\mu_e}{r^3}\begin{bmatrix}2x\\-y\\-z\end{bmatrix} + \frac{\mu_e}{r^3}\begin{bmatrix}rf_1\\yf_2\\zf_3\end{bmatrix} + \begin{bmatrix}\Delta F_x\\\Delta F_y\\\Delta F_z\end{bmatrix}, \tag{2.305}$$

$$t\begin{bmatrix}\Delta\dot v_x\\\Delta\dot v_y\\\Delta\dot v_z\end{bmatrix} = -\frac{1}{\eta\zeta a^2 r^2}\begin{bmatrix}0 & (\eta\zeta a^2)^2 & 0\\-(\eta\zeta a^2)^2 & 0 & -r^3 F_{z,r}\\0 & r^3 F_{z,r} & 0\end{bmatrix}\begin{bmatrix}\Delta v_x\\\Delta v_y\\\Delta v_z\end{bmatrix}$$

$$+ \frac{\mu_e}{r^3}\begin{bmatrix}2x\\-y\\-z\end{bmatrix} + \frac{\mu_e}{r^3}\begin{bmatrix}rf_1\\yf_2\\zf_3\end{bmatrix} + \begin{bmatrix}\Delta F_x\\\Delta F_y\\\Delta F_z\end{bmatrix}, \tag{2.306}$$

with,

$$\begin{bmatrix}\Delta F_x\\\Delta F_y\\\Delta F_z\end{bmatrix} = \begin{bmatrix}F_{x,p}\\F_{y,p}\\F_{z,p}\end{bmatrix} - \begin{bmatrix}F_{x,r}\\F_{y,r}\\F_{z,r}\end{bmatrix}, \tag{2.307}$$

where $F_{x,p}$, $F_{y,p}$, and $F_{z,p}$ are the perturbation forces acting on the orbiting body, and $F_{x,r}$, $F_{y,r}$, and $F_{z,r}$ are the corresponding perturbation forces acting on the nominal satellite in the reference orbit. Equations (2.304), (2.306), and (2.307) are the most general equations of relative motion. Comparing these with Equations (2.256), (2.305), and (2.307), it is clear that a circular reference orbit may be used whenever it is known a priori that $F_{z,r} = 0$. Yet it is also important, when a circular reference orbit is used, that $\dot{\theta}$ is appropriately chosen.

2.10.5 The Extended Nonlinear Tschauner-Hempel Equations

The Tschauner-Hempel equations are obtained by using the true anomaly $f = \nu$ as the independent variable. Moreover, Tschauner and Hempel [10] also applied a coordinate scaling prior to transforming the equations. The scaled state variable $\tilde{x}$ is related to the original state by the relation $\tilde{x} = (1 + e\cos f)x$. One of the major applications of these equations is in the control of the orbits of secondary satellites relative to a primary or reference satellite. Further,

$$d/df = d/\dot{f}\,dt = (r^2/h)d/dt, \quad r = p/(1 + e\cos f), \tag{2.308}$$

where p is the semi-latus rectum and $h = \sqrt{\mu p}$. It follows that,

$$\frac{d\tilde{x}}{df} = -ex\sin f + (1 + e\cos f)\frac{dx}{df}, \quad \frac{dx}{df} = \frac{r^2}{h}\frac{dx}{dt} = \frac{p^2}{h(1 + e\cos f)^2}\frac{dx}{dt}. \tag{2.309}$$

Thus,

$$\frac{d\tilde{x}}{df} = -ex\sin f + \frac{p^2}{h(1 + e\cos f)}\frac{dx}{dt}, \tag{2.310}$$

It follows that,

$$\begin{bmatrix}\tilde{x}\\\tilde{y}\\\tilde{z}\end{bmatrix} = (1 + e\cos f)\begin{bmatrix}x\\y\\z\end{bmatrix}, \quad \frac{d}{df}\begin{bmatrix}\tilde{x}\\\tilde{y}\\\tilde{z}\end{bmatrix} = -e\sin f\begin{bmatrix}x\\y\\z\end{bmatrix} + \frac{p^2}{h(1 + e\cos f)}\frac{d}{dt}\begin{bmatrix}x\\y\\z\end{bmatrix}. \tag{2.311}$$

The inverse transformations are,

$$\begin{bmatrix}x\\y\\z\end{bmatrix} = \frac{1}{(1 + e\cos f)}\begin{bmatrix}\tilde{x}\\\tilde{y}\\\tilde{z}\end{bmatrix}, \quad \frac{d}{df}\begin{bmatrix}x\\y\\z\end{bmatrix} = \frac{p^2}{h(1 + e\cos f)^2}\frac{d}{dt}\begin{bmatrix}x\\y\\z\end{bmatrix}. \tag{2.312}$$

Thus,

$$\frac{(1 + e\cos f)}{\dot{f}}\frac{d}{dt}\begin{bmatrix}x\\y\\z\end{bmatrix} = \frac{d}{df}\begin{bmatrix}\tilde{x}\\\tilde{y}\\\tilde{z}\end{bmatrix} + \frac{e\sin f}{(1 + e\cos f)}\begin{bmatrix}\tilde{x}\\\tilde{y}\\\tilde{z}\end{bmatrix}. \tag{2.313}$$

Consequently, we obtain,

$$\frac{d}{df}\begin{bmatrix}x\\y\\z\end{bmatrix} = \frac{e\sin f}{(1 + e\cos f)^2}\begin{bmatrix}\tilde{x}\\\tilde{y}\\\tilde{z}\end{bmatrix} + \frac{1}{(1 + e\cos f)}\frac{d}{df}\begin{bmatrix}\tilde{x}\\\tilde{y}\\\tilde{z}\end{bmatrix}. \tag{2.314}$$

Thus, the transformation relating the state vector takes the form $\tilde{\mathbf{x}} = \mathbf{T}\mathbf{x}$,

$$\tilde{\mathbf{x}} = [\tilde{x} \ \ \tilde{y} \ \ \tilde{z} \ \ \tilde{x}' \ \ \tilde{y}' \ \ \tilde{z}']^T, \quad \mathbf{x} = [x \ \ y \ \ z \ \ \dot{x} \ \ \dot{y} \ \ \dot{z}]^T, \tag{2.315}$$

$$\mathbf{T} = \begin{bmatrix} (1+e\cos f)\mathbf{I} & \mathbf{0} \\ -(e\sin f)\mathbf{I} & \dfrac{p^2}{h(1+e\cos f)}\mathbf{I} \end{bmatrix}, \quad \mathbf{T}^{-1} = \begin{bmatrix} \dfrac{1}{(1+e\cos f)}\mathbf{I} & \mathbf{0} \\ \dfrac{h(e\sin f)}{p^2}\mathbf{I} & \dfrac{h(1+e\cos f)}{p^2}\mathbf{I} \end{bmatrix}. \tag{2.316}$$

Furthermore,

$$\frac{d^2}{df^2}\begin{bmatrix} \tilde{x} \\ \tilde{y} \\ \tilde{z} \end{bmatrix} = -e\cos f \begin{bmatrix} x \\ y \\ z \end{bmatrix} + \frac{p^2}{h(1+e\cos f)}\frac{d}{dt}\frac{d}{df}\begin{bmatrix} x \\ y \\ z \end{bmatrix}. \tag{2.317}$$

Hence,

$$\frac{d^2}{df^2}\begin{bmatrix} \tilde{x} \\ \tilde{y} \\ \tilde{z} \end{bmatrix} = -e\cos f \begin{bmatrix} x \\ y \\ z \end{bmatrix} + \frac{1+e\cos f}{\dot{f}^2}\frac{d^2}{dt^2}\begin{bmatrix} x \\ y \\ z \end{bmatrix}. \tag{2.318}$$

It follows that,

$$\frac{d^2}{df^2}\begin{bmatrix} \tilde{x} \\ \tilde{y} \\ \tilde{z} \end{bmatrix} = -\begin{bmatrix} \tilde{x} \\ \tilde{y} \\ \tilde{z} \end{bmatrix} + \frac{1}{1+e\cos f}\begin{bmatrix} \tilde{x} \\ \tilde{y} \\ \tilde{z} \end{bmatrix} + \frac{1+e\cos f}{\dot{f}^2}\frac{d^2}{dt^2}\begin{bmatrix} x \\ y \\ z \end{bmatrix}. \tag{2.319}$$

From the preceding sections, the NERM may be expressed compactly as,

$$\frac{d^2}{dt^2}\begin{bmatrix} x \\ y \\ z \end{bmatrix} = -\dot{f}^2 \mathbf{K}_{lerm} \begin{bmatrix} x \\ y \\ z \end{bmatrix} - \dot{f}\mathbf{G}_{lerm}\frac{d}{dt}\begin{bmatrix} x \\ y \\ z \end{bmatrix} + \begin{bmatrix} \tilde{a}_{nx} \\ \tilde{a}_{ny} \\ \tilde{a}_{nz} \end{bmatrix}, \tag{2.320}$$

where,

$$\mathbf{K}_{lerm} = \frac{1}{\dot{f}^2}\begin{bmatrix} -\dot{f}^2 - 2\dfrac{\mu_e}{r^3} & -\ddot{f} & 0 \\ \ddot{f} & -\dot{f}^2 + \dfrac{\mu_e}{r^3} & 0 \\ 0 & 0 & \dfrac{\mu_e}{r^3} \end{bmatrix}, \quad \mathbf{G}_{lerm} = \begin{bmatrix} 0 & -2 & 0 \\ 2 & 0 & 0 \\ 0 & 0 & 0 \end{bmatrix},$$

$$\begin{bmatrix} \tilde{a}_{nx} \\ \tilde{a}_{ny} \\ \tilde{a}_{nz} \end{bmatrix} = \begin{bmatrix} a_{px} - \left(\mu_e(r+x)/|\mathbf{r}+\vec{\rho}|^3\right) - (\mu_e/r^2)(1-2x/r) \\ a_{py} - \left(\mu_e y/|\mathbf{r}+\vec{\rho}|^3\right) + (\mu_e/r^2)(y/r) \\ a_{pz} - \left(\mu_e z/|\mathbf{r}+\vec{\rho}|^3\right) + (\mu_e/r^2)(z/r) \end{bmatrix},$$

and,

$$\begin{bmatrix} a_{px} \\ a_{py} \\ a_{pz} \end{bmatrix} = \begin{bmatrix} \tilde{\mathbf{f}}_{BS} \cdot \hat{\mathbf{x}} \\ \tilde{\mathbf{f}}_{BS} \cdot \hat{\mathbf{y}} \\ \tilde{\mathbf{f}}_{BS} \cdot \hat{\mathbf{z}} \end{bmatrix} + \mathbf{u}, \tag{2.321}$$

2.10 Spacecraft Relative Motion

where **u** is a vector of control accelerations applied to an orbiting body relative to the nominal or reference satellite. Noting that $p/r = (1 + e\cos f)$, from Equation (2.320) we have,

$$\frac{d^2}{df^2}\begin{bmatrix}\tilde{x}\\\tilde{y}\\\tilde{z}\end{bmatrix} = -\begin{bmatrix}\tilde{x}\\\tilde{y}\\\tilde{z}\end{bmatrix} + \frac{r}{p}\begin{bmatrix}\tilde{x}\\\tilde{y}\\\tilde{z}\end{bmatrix} - \mathbf{K}_{lerm}\begin{bmatrix}\tilde{x}\\\tilde{y}\\\tilde{z}\end{bmatrix} - \mathbf{G}_{lerm}\left\{\frac{d}{df}\begin{bmatrix}\tilde{x}\\\tilde{y}\\\tilde{z}\end{bmatrix} + \frac{re\sin f}{p}\begin{bmatrix}\tilde{x}\\\tilde{y}\\\tilde{z}\end{bmatrix}\right\}$$

$$+\frac{r}{p\dot{f}^2}\begin{bmatrix}\tilde{a}_{nx}\\\tilde{a}_{ny}\\\tilde{a}_{nz}\end{bmatrix}. \qquad (2.322)$$

Thus,

$$\frac{d^2}{df^2}\begin{bmatrix}\tilde{x}\\\tilde{y}\\\tilde{z}\end{bmatrix} = -(\mathbf{K}_{lerm} + \mathbf{I})\begin{bmatrix}\tilde{x}\\\tilde{y}\\\tilde{z}\end{bmatrix} - \frac{re\sin f}{p}\mathbf{G}_{lerm}\begin{bmatrix}\tilde{x}\\\tilde{y}\\\tilde{z}\end{bmatrix} + \frac{\mathbf{I}r}{p}\begin{bmatrix}\tilde{x}\\\tilde{y}\\\tilde{z}\end{bmatrix} - \mathbf{G}_{lerm}\frac{d}{df}\begin{bmatrix}\tilde{x}\\\tilde{y}\\\tilde{z}\end{bmatrix}$$

$$+\frac{r}{p\dot{f}^2}\begin{bmatrix}\tilde{a}_{nx}\\\tilde{a}_{ny}\\\tilde{a}_{nz}\end{bmatrix}. \qquad (2.323)$$

But,

$$\mathbf{K}_{lerm} + \mathbf{I} + \frac{r(e\sin f)}{p}\mathbf{G}_{lerm} - \frac{r}{p}\mathbf{I}$$

$$= \left(\frac{1}{\dot{f}} - 1\right)\frac{2re\sin f}{p}\begin{bmatrix}0 & 2 & 0\\-2 & 0 & 0\\0 & 0 & 0\end{bmatrix} + \frac{r}{p}\begin{bmatrix}-2-1 & 0 & 0\\0 & 1-1 & 0\\0 & 0 & 1+1+e\cos f\end{bmatrix}. \qquad (2.324)$$

Hence,

$$\mathbf{K}_{lerm} + \mathbf{I} + \frac{(e\sin f)}{(1+e\cos f)}\mathbf{G}_{lerm} - \frac{1}{1+e\cos f}\mathbf{I}$$

$$= \left(\frac{p^2}{h(1+e\cos f)^2} - 1\right)\frac{2e\sin f}{(1+e\cos f)}\begin{bmatrix}0 & 2 & 0\\-2 & 0 & 0\\0 & 0 & 0\end{bmatrix} + \frac{1}{(1+e\cos f)}\begin{bmatrix}-3 & 0 & 0\\0 & 0 & 0\\0 & 0 & 2+e\cos f\end{bmatrix}. \qquad (2.325)$$

Hence, if we let,

$$\tilde{\mathbf{K}}_{lerm} = \frac{1}{(1+e\cos f)} \left\{ \begin{bmatrix} -3 & 0 & 0 \\ 0 & 0 & 0 \\ 0 & 0 & 2+e\cos f \end{bmatrix} + 2e\sin f \left(\frac{p^2}{h(1+e\cos f)^2} - 1 \right) \begin{bmatrix} 0 & 2 & 0 \\ -2 & 0 & 0 \\ 0 & 0 & 0 \end{bmatrix} \right\},$$

$$\tilde{\mathbf{G}}_{lerm} = \begin{bmatrix} 0 & -2 & 0 \\ 2 & 0 & 0 \\ 0 & 0 & 0 \end{bmatrix}, \tag{2.326}$$

the nonlinear controlled Tschauner-Hempel (TH) equations, with control inputs $\tilde{\mathbf{u}}(\tilde{\mathbf{x}})$, derived in Vepa [11], are given by,

$$\frac{d^2}{df^2}\begin{bmatrix}\tilde{x}\\\tilde{y}\\\tilde{z}\end{bmatrix} = -\tilde{\mathbf{K}}_{lerm}\begin{bmatrix}\tilde{x}\\\tilde{y}\\\tilde{z}\end{bmatrix} - \tilde{\mathbf{G}}_{lerm}\frac{d}{df}\begin{bmatrix}\tilde{x}\\\tilde{y}\\\tilde{z}\end{bmatrix} + \frac{p^4}{h^2(1+e\cos f)^3}\begin{bmatrix}\tilde{a}_{nx}(\tilde{x},\tilde{y},\tilde{z})\\\tilde{a}_{ny}(\tilde{x},\tilde{y},\tilde{z})\\\tilde{a}_{nz}(\tilde{x},\tilde{y},\tilde{z})\end{bmatrix} + \tilde{\mathbf{u}}(\tilde{\mathbf{x}}),$$
(2.327)

where $\tilde{a}_{n*}(\tilde{x},\tilde{y},\tilde{z})$ are the perturbation accelerations expressed in terms of the scaled coordinates. Given a full state control input vector in the scaled TH frame as,

$$\tilde{\mathbf{u}}(\tilde{\mathbf{x}}) = -\tilde{\mathbf{K}}_f \tilde{\mathbf{x}} = -\tilde{\mathbf{K}}_f \begin{bmatrix} \tilde{x} & \tilde{y} & \tilde{z} & \frac{d\tilde{x}}{df} & \frac{d\tilde{y}}{df} & \frac{d\tilde{z}}{df} \end{bmatrix}^T, \tag{2.328}$$

since,

$$\tilde{\mathbf{x}} = \mathbf{T}\mathbf{x}, \quad \mathbf{T} = \begin{bmatrix} (1+e\cos f)\mathbf{I} & \mathbf{0} \\ -(e\sin f)\mathbf{I} & \frac{p^2}{h(1+e\cos f)}\mathbf{I} \end{bmatrix}, \tag{2.329}$$

the physical control input vector is,

$$\mathbf{u}(\mathbf{x}) = \frac{h^2(1+e\cos f)^3}{p^4}\tilde{\mathbf{u}}(\mathbf{Tx}) = -\frac{h^2(1+e\cos f)^3}{p^4}\tilde{\mathbf{K}}_f \mathbf{Tx}$$

$$= -\frac{h^2(1+e\cos f)^3}{p^4}\tilde{\mathbf{K}}_f \mathbf{T}\begin{bmatrix} x & y & z & \frac{dx}{df} & \frac{dy}{df} & \frac{dz}{df}\end{bmatrix}^T. \tag{2.330}$$

Thus, in terms of the original state variables, the physical control input vector is,

$$\mathbf{u}(\mathbf{x}) = -\mathbf{K}_f \mathbf{x} = -\mathbf{K}_f \begin{bmatrix} x & y & z & \frac{dx}{df} & \frac{dy}{df} & \frac{dz}{df} \end{bmatrix}^T, \tag{2.331}$$

with

$$\mathbf{K}_f = -\frac{h^2(1+e\cos f)^3}{p^4}\tilde{\mathbf{K}}_f \mathbf{T}, \tag{2.332}$$

or

$$\mathbf{K}_f = -\frac{h^2(1+e\cos f)^3}{p^4}\tilde{\mathbf{K}}_f \begin{bmatrix} (1+e\cos f)\mathbf{I} & 0 \\ -(e\sin f)\mathbf{I} & \frac{p^2}{h(1+e\cos f)}\mathbf{I} \end{bmatrix}. \quad (2.333)$$

2.11 Orbit Control

The motion of spacecraft along an orbit can be considered to be either absolute or relative motion. The absolute (or inertial) motion of a spacecraft is governed by the LPE, while its relative motion, relative to reference circular orbit, is governed by the HCW equations. One of the applications of the LPE is to control the orbit traversed by a spacecraft. However, the classical set of orbital elements and the corresponding LPE cannot be used for predicting the near circular orbits or orbits at zero inclination. Thus we need to define an alternative set of orbital elements and transform the LPE accordingly. However, we shall first define the basic equations defining the position vector of the spacecraft. The equation of relative motion of the two bodies under the action of a central gravitational force field and disturbing or control acceleration $\mathbf{d}$ is given by,

$$\ddot{\mathbf{r}} + \frac{G(m_1+m_2)}{|\mathbf{r}|^3}\mathbf{r} = \mathbf{d}. \quad (2.334)$$

The corresponding equations for dynamics of $\mathbf{h}$ and $\mathbf{e}$, which are no longer constant, are,

$$\frac{d\mathbf{h}}{dt} = \mathbf{r} \times \mathbf{d}, \quad (2.335)$$

$$\mu\frac{d\mathbf{e}}{dt} = (\mathbf{v} \times \mathbf{r}) \times \mathbf{d} + \mathbf{v} \times (\mathbf{r} \times \mathbf{d}), \quad (2.336)$$

where $\mathbf{h} = \mathbf{r} \times \mathbf{v}$ and $\mu \mathbf{e} = (rv^2 - \mathbf{v}(\mathbf{v}\cdot\mathbf{r})) - \mu\mathbf{r}/|\mathbf{r}|$.

Thus, the scalar components of $\mathbf{h}$ and $\mathbf{e}$ satisfy,

$$\frac{d}{dt}\begin{bmatrix} h_r \\ h_s \\ h_n \end{bmatrix} = \begin{bmatrix} 0 & -r_n & r_s \\ r_n & 0 & -r_r \\ -r_s & r_r & 0 \end{bmatrix}\begin{bmatrix} F_r \\ F_s \\ F_n \end{bmatrix} = \begin{bmatrix} F_n r_s - F_s r_n \\ F_r r_n - F_n r_r \\ F_s r_r - F_r r_s \end{bmatrix}, \quad (2.337)$$

$$\mu\frac{d}{dt}\begin{bmatrix} e_r \\ e_s \\ e_n \end{bmatrix} = \begin{bmatrix} 0 & -r_s v_r + r_r v_s & r_r v_n - r_n v_r \\ r_s v_r - r_r v_s & 0 & -r_n v_s + r_s v_n \\ -r_r v_n + r_n v_r & r_n v_s - r_s v_n & 0 \end{bmatrix}\begin{bmatrix} F_r \\ F_s \\ F_n \end{bmatrix} + \begin{bmatrix} 0 & -v_n & v_s \\ v_n & 0 & -v_r \\ -v_s & v_r & 0 \end{bmatrix}\begin{bmatrix} 0 & -r_n & r_s \\ r_n & 0 & -r_r \\ -r_s & r_r & 0 \end{bmatrix}\begin{bmatrix} F_r \\ F_s \\ F_n \end{bmatrix}, \quad (2.338)$$

where the radial component F_r of the disturbing or control acceleration is assumed to be parallel to the position vector, the transverse component F_s, to be tangential to the orbit and in the orbit plane, and the orbit normal component F_n to be normal to the orbit plane. The subscripts r, s, and n refer to components in the directions parallel to the position vector, tangential to the orbit and in the orbit plane, and normal to the orbit plane. The six governing equations in terms of the scalar components of **h** and **e** may be transformed to the LPE or GPE for the six orbital elements.

2.11.1 Delaunay Elements

The Delaunay elements are a set of canonical, physically meaningful action and angle variables that are commonly used to describe satellite orbits. The classical Keplerian element, the argument of perigee, becomes singular when the eccentricity is zero. As eccentricity approaches zero, the line of apsides also becomes indeterminate. Several physically meaningful orbital elements, including the Delaunay elements, are also singular for zero inclination and/or zero eccentricity. The Delaunay element set consists of three conjugate action and angle pairs. By convention, lowercase letters are used to represent the angles and uppercase letters are used to represent the conjugate actions. The representation gives the canonical actions used in Hamilton's equations of motion. The action elements are L, which is related to the two-body orbital energy, G, which is the magnitude of the orbital angular momentum, and H, which is the z component of the orbital angular momentum. These three elements are expressed in terms of distance squared, divided by time, where distance is measured in standard units and time is measured in seconds. The angles are l, which is the mean anomaly, g, which is the argument of perigee, and h is the right ascension of the ascending node.

2.11.2 Non-Singular Element Sets

To define the orbit in the presence of the disturbing or control acceleration components, one may define from the classical elements a set of non-singular elements, such as the semi-major axis a, the mean longitude of the satellite $\lambda = \Omega + \omega + M = \varpi + M$, with $\varpi = \Omega + \omega$, which is the longitude of the peri-center, $\xi = e\cos\varpi$, $\eta = e\sin\varpi$, $P = \sin(i/2)\cos\Omega$, and $Q = \sin(i/2)\sin\Omega$ for $i \neq \pi$ and $e < 1$.

Hence, the inverse relations are,

$$e = \sqrt{\xi^2 + \eta^2}, \quad i = 2\sin^{-1}\left(\sqrt{P^2 + Q^2}\right), \quad \Omega = \tan^{-1}(Q/P), \quad \varpi = \tan^{-1}(\eta/\xi)$$

$$\omega = \varpi - \Omega = \tan^{-1}(\eta/\xi) - \Omega \quad \text{and} \quad M = \lambda - \varpi. \tag{2.339}$$

As $\gamma = \sqrt{1 - e^2}$ and V is the potential of the disturbing or control forces, the complete set of equations defining the evolution of the elements are,

$$\frac{da}{dt} = \frac{2}{na}\frac{\partial V}{\partial \lambda}, \tag{2.340}$$

$$\frac{d\xi}{dt} = -\frac{\gamma}{na^2(\gamma+1)}\left(\xi\frac{\partial V}{\partial \lambda}\right) - \frac{\gamma}{na^2}\frac{\partial V}{\partial \eta} - \frac{\eta}{2na^2\gamma}\left(P\frac{\partial V}{\partial P} + Q\frac{\partial V}{\partial Q}\right), \tag{2.341}$$

$$\frac{d\eta}{dt} = -\frac{\gamma}{na^2(\gamma+1)}\left(\eta\frac{\partial V}{\partial \lambda}\right) + \frac{\gamma}{na^2}\frac{\partial V}{\partial \xi} + \frac{\xi}{2na^2\gamma}\left(P\frac{\partial V}{\partial P} + Q\frac{\partial V}{\partial Q}\right), \tag{2.342}$$

$$\frac{dP}{dt} = -\frac{1}{2na^2\gamma}\left(P\frac{\partial V}{\partial \lambda}\right) - \frac{1}{4na^2\gamma}\frac{\partial V}{\partial Q} + \frac{P}{2na^2\gamma}\left(\eta\frac{\partial V}{\partial \xi} - \xi\frac{\partial V}{\partial \eta}\right), \quad (2.343)$$

$$\frac{dQ}{dt} = -\frac{1}{2na^2\gamma}\left(Q\frac{\partial V}{\partial \lambda}\right) + \frac{1}{4na^2\gamma}\frac{\partial V}{\partial P} + \frac{Q}{2na^2\gamma}\left(\eta\frac{\partial V}{\partial \xi} - \xi\frac{\partial V}{\partial \eta}\right), \quad (2.344)$$

$$\frac{d\lambda}{dt} = n - \frac{2}{na}\frac{\partial V}{\partial a} + \frac{\gamma}{2na^2}\left(\xi\frac{\partial V}{\partial \xi} + \eta\frac{\partial V}{\partial \eta}\right) + \frac{1}{2na^2\gamma}\left(P\frac{\partial V}{\partial P} + Q\frac{\partial V}{\partial Q}\right). \quad (2.345)$$

The equations may also be expressed in matrix form, but the coefficient matrix is no longer skew-symmetric,

$$\frac{d}{dt}\begin{bmatrix} a \\ \xi \\ \eta \\ P \\ Q \\ \lambda \end{bmatrix} = \begin{bmatrix} 0 \\ 0 \\ 0 \\ 0 \\ 0 \\ n \end{bmatrix} + \frac{1}{2na^2\gamma}\begin{bmatrix} 0 & 0 & 0 & 0 & 0 & 4a\gamma \\ 0 & 0 & -2\gamma^2 & -P\eta & -Q\eta & -2\xi\gamma^2/(\gamma+1) \\ 0 & 2\gamma^2 & 0 & P\xi & Q\xi & -2\eta\gamma^2/(\gamma+1) \\ 0 & P\eta & -P\xi & 0 & -1/2 & -P \\ 0 & Q\eta & -Q\xi & 1/2 & 0 & -Q \\ -4a\gamma & \gamma^2\xi & \gamma^2\eta & P & Q & 0 \end{bmatrix}\partial V/\partial\begin{bmatrix} a \\ \xi \\ \eta \\ P \\ Q \\ \lambda \end{bmatrix}. \quad (2.346)$$

These equations can also be expressed in terms of the perturbing acceleration components.

2.11.3 Equinoctial Elements

The six governing equations of a satellite's orbital motion in terms of the scalar components of the angular momentum vector **h** and the eccentricity vector **e** may be transformed to the GPE in terms of a set of non-singular elements, which are the semi-major axis a, the true longitude of the satellite $\lambda = \varpi + M$ with the longitude of the peri-center $\varpi = \Omega + \omega$, $\xi = e\cos\varpi$, $\eta = e\sin\varpi$, $h_x = \tan(i/2)\cos\Omega$, and $h_y = \tan(i/2)\sin\Omega$. The inverse relations are,

$$e = \sqrt{\xi^2 + \eta^2}, \quad i = 2\tan^{-1}\left(\sqrt{h_x^2 + h_y^2}\right), \quad \Omega = \tan^{-1}(h_y/h_x), \quad \varpi = \tan^{-1}(\eta/\xi)$$

$$\omega = \varpi - \Omega = \tan^{-1}(\eta/\xi) - \Omega \quad \text{and} \quad M = \lambda - \varpi. \quad (2.347)$$

Sometimes the true longitude $l = \Omega + \omega + f = \varpi + f$ is substituted by the mean longitude λ and the true anomaly is $f = l - \varpi$. These are usually referred to as the *equinoctial elements*. The equinoctial coordinates use the center of the Earth as the origin and the plane of the satellite's orbit as the reference plane. Orbits are generally classified as retrograde, which means they have their singularities at an inclination of 0 degrees, or posigrade, which means they have their singularities at an inclination of 180 degrees. The advantage of this element set is that singularities are absent in *posigrade*, near equatorial orbits. When the inclination of an orbit is zero, with respect to the

Keplerian elements, the right ascension of the ascending node (RAAN) is undefined. It is numerically unstable as the inclination approaches zero, as the line of nodes becomes indeterminate. Thus, in the case of several satellite constellations, the equinoctial elements are used during the orbit estimation process. Apart from the semi-major axis, which is half the length of the major axis of the orbital ellipse, and the mean motion, which is the average angular rate of the satellite based on two-body motion, the equinoctial elements are defined from a set of numbers that describe the shape of the satellite's orbit and the position of perigee, and specify a satellite's position within its orbit at epoch by the sum of the classical RAAN, argument of the perigee, and the mean anomaly. Equations similar to those obtained for the previous set may be obtained for the time rates of change of this alternative set of non-singular elements. In terms of the radial, transverse, and orbit normal control acceleration components F_r, F_s, F_n, the equations are,

$$\frac{da}{dt} = \frac{2a^2}{h}\left[F_r(\xi \sin l - \eta \cos l) + \frac{p}{r}F_s\right], \tag{2.348}$$

$$\frac{d\xi}{dt} = \frac{r}{h}\left[-\frac{p}{r}F_r \sin l + \left(\xi + \left(1 + \frac{p}{r}\right)\cos l\right)F_s - \eta\left(h_y \cos l - h_x \sin l\right)F_n\right], \tag{2.349}$$

$$\frac{d\eta}{dt} = \frac{r}{h}\left[-\frac{p}{r}F_r \cos l + \left(\eta + \left(1 + \frac{p}{r}\right)\sin l\right)F_s - \xi\left(h_y \cos l - h_x \sin l\right)F_n\right], \tag{2.350}$$

$$\frac{dh_y}{dt} = \frac{r}{2h}\left(1 + h_x^2 + h_y^2\right)F_n \sin l, \quad \frac{dh_x}{dt} = \frac{r}{2h}\left(1 + h_x^2 + h_y^2\right)F_n \cos l, \tag{2.351}$$

$$\frac{d\lambda}{dt} = n - \frac{r}{h}\left(\frac{a}{a+b}\right)\left(\frac{p}{r}\right)(\eta \sin l + \xi \cos l) + \frac{2b}{a}\right)F_r$$
$$- \frac{ar}{(a+b)h}\left(1 + \frac{p}{r}\right)(\eta \cos l - \xi \sin l)F_s - \frac{r}{h}\left(h_y \cos l - h_x \sin l\right)F_n. \tag{2.352}$$

The disadvantage of using the mean anomaly is that one must necessarily use Kepler's equation to evaluate the true anomaly. Thus, rather than using h_x, h_y, and λ, one may use the orbital elements with the orbit plane quaternion replacing the Euler angles in the 3–1–3 sequence: Ω–i–θ, where $\theta = \omega + f$, the satellite's longitude.

2.11.4 Orbital Elements with the Orbit Plane Quaternion Replacing the Euler Angles in the 3–1–3 Sequence

A very important non-singular element set is obtained with the orbit plane quaternion replacing the Euler angles in the 3–1–3 sequence: Ω–i–θ. The orbit quaternion parameters are used for computation of attitude relative to the orbiting frame to avoid the singularity problem that arises when Euler angles are used to represent the attitude. A quaternion representation is a four-part complex number that can describe the sequential three-axes rotations as a single rotation about a direction or axis. The quaternion parameters are related to the rotation angle by the relation,

2.11 Orbit Control

$$[q_1 \quad q_2 \quad q_3]^T = \mathbf{n}\sin(\theta_r/2), \quad q_4 = \cos(\theta_r/2), \tag{2.353}$$

where the angle θ_r is the total, single rotation angle about a direction defined by the unit vector $\mathbf{n}$. The quaternion is then defined by the four parameters,

$$\mathbf{q} = [q_1 \quad q_2 \quad q_3 \quad q_4]^T. \tag{2.354}$$

Quaternions also satisfy the constraint,

$$q_4^2 + q_1^2 + q_2^2 + q_3^2 = \mathbf{q}^T\mathbf{q} = 1. \tag{2.355}$$

For the 3–1–3 Euler angle sequence Ω, i, $\theta = \omega + f$, the body axes are related to a succession of frames obtained by rotating the inertial frame about the sequence of angles,

$$\begin{bmatrix} x_b \\ y_b \\ z_b \end{bmatrix} = \begin{bmatrix} c\theta & s\theta & 0 \\ -s\theta & c\theta & 0 \\ 0 & 0 & 1 \end{bmatrix} \begin{bmatrix} x'' \\ y'' \\ z'' \end{bmatrix}, \quad \begin{bmatrix} x'' \\ y'' \\ z'' \end{bmatrix} = \begin{bmatrix} 1 & 0 & 0 \\ 0 & ci & si \\ 0 & -si & ci \end{bmatrix} \begin{bmatrix} x' \\ y' \\ z' \end{bmatrix}, \tag{2.356}$$

$$\begin{bmatrix} x' \\ y' \\ z' \end{bmatrix} = \begin{bmatrix} c\Omega & s\Omega & 0 \\ -s\Omega & c\Omega & 0 \\ 0 & 0 & 1 \end{bmatrix} \begin{bmatrix} x \\ y \\ z \end{bmatrix}. \tag{2.357}$$

In Equations (2.356) and (2.357), the prefixes "c" and "s" stand for the *cosine* and *sine* of the angle that follows the prefix. Eliminating the intermediate frames, the body axes are related to the inertial axes by,

$$\begin{bmatrix} x_b \\ y_b \\ z_b \end{bmatrix} = \begin{bmatrix} c\theta & s\theta & 0 \\ -s\theta & c\theta & 0 \\ 0 & 0 & 1 \end{bmatrix} \begin{bmatrix} 1 & 0 & 0 \\ 0 & ci & si \\ 0 & -si & ci \end{bmatrix} \begin{bmatrix} c\Omega & s\Omega & 0 \\ -s\Omega & c\Omega & 0 \\ 0 & 0 & 1 \end{bmatrix} \begin{bmatrix} x \\ y \\ z \end{bmatrix}. \tag{2.358}$$

Thus,

$$\begin{bmatrix} x_b \\ y_b \\ z_b \end{bmatrix} = \begin{bmatrix} -s\Omega ci s\theta + c\Omega c\theta & c\Omega ci s\theta + s\Omega c\theta & si s\theta \\ -s\Omega ci c\theta - c\Omega s\theta & c\Omega ci c\theta - s\Omega s\theta & si c\theta \\ s\Omega si & -c\Omega si & ci \end{bmatrix} \begin{bmatrix} x \\ y \\ z \end{bmatrix}. \tag{2.359}$$

Comparing with the transformation relating the body axes to the inertial axes in terms of the components of a quaternion,

$$\mathbf{R}_{IB}(\mathbf{q}) = \begin{bmatrix} q_4^2 + q_1^2 - q_2^2 - q_3^2 & 2(q_1q_2 - q_3q_4) & 2(q_1q_3 + q_2q_4) \\ 2(q_1q_2 + q_3q_4) & q_4^2 - q_1^2 + q_2^2 - q_3^2 & 2(q_2q_3 - q_1q_4) \\ 2(q_1q_3 - q_2q_4) & 2(q_2q_3 + q_1q_4) & q_4^2 - q_1^2 - q_2^2 + q_3^2 \end{bmatrix}, \tag{2.360}$$

one may write,

$$q_4^2 - q_1^2 - q_2^2 + q_3^2 = \cos i, \quad 2\sqrt{(q_2q_3 + q_1q_4)^2 + (q_1q_3 - q_2q_4)^2} = \sin i, \tag{2.361}$$

and,

$$\Omega = \tan^{-1}(2(q_1q_3 - q_2q_4), -2(q_2q_3 + q_1q_4)), \qquad (2.362)$$

$$\theta = \tan^{-1}(2(q_1q_3 + q_2q_4), 2(q_2q_3 - q_1q_4)). \qquad (2.363)$$

Thus,

$$i = \tan^{-1}\left(2\sqrt{(q_2q_3 + q_1q_4)^2 + (q_1q_3 - q_2q_4)^2}, q_4^2 - q_1^2 - q_2^2 + q_3^2\right). \qquad (2.364)$$

The inverse relations are,

$$q_1 = \frac{1}{4q_4}(R_{23} - R_{32}) = \pm\frac{1}{2}\sqrt{1 + R_{11} - R_{22} - R_{33}}, \qquad (2.365)$$

$$q_2 = \frac{1}{4q_1}(R_{12} + R_{21}) = \frac{1}{4q_4}(R_{23} - R_{32}), \qquad (2.366)$$

$$q_3 = \frac{1}{4q_4}(R_{12} - R_{21}) = \frac{1}{4q_1}(R_{31} + R_{13}), \qquad (2.367)$$

$$q_4 = \frac{1}{4q_1}(R_{23} - R_{32}) = \pm\frac{1}{2}\sqrt{1 + R_{11} + R_{22} + R_{33}}, \qquad (2.368)$$

where $\mathbf{R}_{BI} = \{R_{ij}\}$.

The orbit plane quaternion rates are related to the orbit plane's angular velocity components by,

$$\frac{d\mathbf{q}}{dt} = \frac{1}{2}\Omega_\omega(\omega)\mathbf{q}, \quad \mathbf{q} = \begin{bmatrix} q_1 \\ q_2 \\ q_3 \\ q_4 \end{bmatrix}, \quad \Omega_\omega(\omega) = \begin{bmatrix} 0 & r & -q & p \\ -r & 0 & p & q \\ q & -p & 0 & r \\ -p & -q & -r & 0 \end{bmatrix}, \qquad (2.369)$$

$$\frac{d}{dt}\begin{bmatrix} q_1 \\ q_2 \\ q_3 \\ q_4 \end{bmatrix} = \frac{1}{2}\Omega_\omega(\omega)\begin{bmatrix} q_1 \\ q_2 \\ q_3 \\ q_4 \end{bmatrix}, \qquad (2.370)$$

$$\frac{d}{dt}\begin{bmatrix} q_1 \\ q_2 \\ q_3 \\ q_4 \end{bmatrix} = \frac{1}{2}\begin{bmatrix} q_4 & -q_3 & q_2 \\ q_3 & q_4 & -q_1 \\ -q_2 & q_1 & q_4 \\ -q_1 & -q_2 & -q_3 \end{bmatrix}\begin{bmatrix} p \\ q \\ r \end{bmatrix}. \qquad (2.371)$$

This relation may be compactly expressed as,

$$\frac{d}{dt}\begin{bmatrix} q_1 \\ q_2 \\ q_3 \\ q_4 \end{bmatrix} = \frac{1}{2}\begin{bmatrix} q_4\mathbf{I} + \mathbf{S}(\mathbf{q}^3) \\ -\mathbf{q}^3 \end{bmatrix}\begin{bmatrix} p \\ q \\ r \end{bmatrix}, \quad \mathbf{S}(\varepsilon) = -\mathbf{S}(\varepsilon)^T = \begin{bmatrix} 0 & -q_3 & q_2 \\ q_3 & 0 & -q_1 \\ -q_2 & q_1 & 0 \end{bmatrix}. \qquad (2.372)$$

2.11 Orbit Control

For the 3–1–3 Euler angle sequence,

$$\begin{bmatrix} p \\ q \\ r \end{bmatrix} = \omega = \begin{bmatrix} \dot{i}\cos\theta + \dot{\Omega}\sin i \sin\theta \\ -\dot{i}\sin\theta + \dot{\Omega}\sin i \cos\theta \\ \dot{\theta} + \dot{\Omega}\cos i \end{bmatrix}. \tag{2.373}$$

2.11.5 Gauss Planetary Equations in Terms of Orbit Quaternion Parameters

From Gauss' classical planetary equations, $\dot{\Omega}\sin i \cos\theta = \dot{i}\sin\theta$. Thus, it follows that,

$$\begin{bmatrix} p \\ q \\ r \end{bmatrix} = \omega = \begin{bmatrix} \dot{i}\cos\theta + \dot{\Omega}\sin i \sin\theta \\ -\dot{i}\sin\theta + \dot{\Omega}\sin i \cos\theta \\ \dot{\theta} + \dot{\Omega}\cos i \end{bmatrix} = \begin{bmatrix} \sin i \sin\theta & \cos\theta & 0 \\ 0 & 0 & 0 \\ \cos i & 0 & 1 \end{bmatrix} \begin{bmatrix} \dot{\Omega} \\ \dot{i} \\ \dot{\theta} \end{bmatrix}. \tag{2.374}$$

Also from Gauss' classical planetary equations, it follows that,

$$\begin{bmatrix} \dot{\Omega} \\ \dot{i} \\ \dot{\theta} \end{bmatrix} = n\sqrt{1-e^2}\left(\frac{a}{r}\right)^2 \begin{bmatrix} 0 \\ 0 \\ 1 \end{bmatrix} + \frac{r}{na^2\sqrt{1-e^2}} \begin{bmatrix} \sin\theta/\sin i \\ \cos\theta \\ -\sin\theta/\tan i \end{bmatrix} F_z, \tag{2.375}$$

$$\begin{bmatrix} p \\ q \\ r \end{bmatrix} = n\sqrt{1-e^2}\left(\frac{a}{r}\right)^2 \begin{bmatrix} 0 \\ 0 \\ 1 \end{bmatrix} + \frac{r}{n\sqrt{1-e^2}a^2} \begin{bmatrix} \sin i \sin\theta & \cos\theta & 0 \\ 0 & 0 & 0 \\ \cos i & 0 & 1 \end{bmatrix} \begin{bmatrix} \sin\theta/\sin i \\ \cos\theta \\ -\sin\theta/\tan i \end{bmatrix} F_z. \tag{2.376}$$

Thus,

$$\begin{bmatrix} p \\ q \\ r \end{bmatrix} = n\sqrt{1-e^2}\left(\frac{a}{r}\right)^2 \begin{bmatrix} 0 \\ 0 \\ 1 \end{bmatrix} + \frac{r}{n\sqrt{1-e^2}a^2} \begin{bmatrix} 1 \\ 0 \\ 0 \end{bmatrix} F_z. \tag{2.377}$$

Since, from Equation (2.372),

$$\frac{d}{dt}\mathbf{q} = \frac{1}{2}\begin{bmatrix} q_4\mathbf{I} + \mathbf{S}(\mathbf{q}^3) \\ -\mathbf{q}^3 \end{bmatrix}\begin{bmatrix} p \\ q \\ r \end{bmatrix}, \tag{2.378}$$

it follows that,

$$\frac{d}{dt}\mathbf{q} = \frac{1}{2}\begin{bmatrix} q_4\mathbf{I} + \mathbf{S}(\mathbf{q}^3) \\ -\mathbf{q}^3 \end{bmatrix}\left\{ n\sqrt{1-e^2}\left(\frac{a}{r}\right)^2 \begin{bmatrix} 0 \\ 0 \\ 1 \end{bmatrix} + \frac{r}{n\sqrt{1-e^2}a^2}\begin{bmatrix} 1 \\ 0 \\ 0 \end{bmatrix} F_z \right\}. \tag{2.379}$$

The complete set of seven equations of a satellite's orbital motion in terms of the extended equinoctial elements are,

$$\frac{da}{dt} = \frac{2a^2}{h}\left[F_x(\xi \sin l - \eta \cos l) + \frac{p}{r}F_y \right], \tag{2.380}$$

$$\frac{d\xi}{dt} = \frac{r}{h}\left[-\frac{p}{r}F_x \sin l + \left(\xi + \left(1+\frac{p}{r}\right)\cos l\right)F_y - \eta(h_y \cos l - h_x \sin l)F_z\right], \quad (2.381)$$

$$\frac{d\eta}{dt} = \frac{r}{h}\left[-\frac{p}{r}F_x \cos l + \left(\eta + \left(1+\frac{p}{r}\right)\sin l\right)F_y - \xi(h_y \cos l - h_x \sin l)F_z\right], \quad (2.382)$$

$$\frac{d}{dt}\mathbf{q} = \frac{1}{2}\begin{bmatrix}q_4\mathbf{I} + \mathbf{S}(\mathbf{q}^3)\\-\mathbf{q}^3\end{bmatrix}\left\{n\sqrt{1-e^2}\left(\frac{a}{r}\right)^2\begin{bmatrix}0\\0\\1\end{bmatrix} + \frac{r}{n\sqrt{1-e^2}a^2}\begin{bmatrix}1\\0\\0\end{bmatrix}F_z\right\}. \quad (2.383)$$

The parameters l, h_x, and h_y in the complete set of seven equations of a satellite's orbital motion are, $l = \Omega + \omega + f = \varpi + f = \omega + \theta$, $h_x = \tan(i/2)\cos\Omega$, and $h_y = \tan(i/2)\sin\Omega$. The inverse parametric relations are,

$$i = 2\tan^{-1}\left(\sqrt{h_x^2 + h_y^2}\right), \quad \Omega = \tan^{-1}(h_y/h_x), \quad i = \tan^{-1}\left(2\sqrt{h_x^2 + h_y^2}, 1 - h_x^2 - h_y^2\right). \quad (2.384)$$

Thus, the classical orbital elements and ephemeris data recovered from seven extended equinoctial elements are,

$$\Omega = \tan^{-1}(2(q_1q_3 - q_2q_4), -2(q_2q_3 + q_1q_4)),$$

$$\theta = \tan^{-1}(2(q_1q_3 + q_2q_4), 2(q_2q_3 - q_1q_4)),$$

$$i = \tan^{-1}\left(2\sqrt{(q_2q_3 + q_1q_4)^2 + (q_1q_3 - q_2q_4)^2}, 1 - 2q_1^2 - 2q_2^2\right), \quad e = \sqrt{\xi^2 + \eta^2},$$

$$\varpi = \tan^{-1}(\eta/\xi), \quad \omega = \varpi - \Omega = \tan^{-1}(\eta/\xi) - \Omega, \quad f = l - \varpi = \theta - \omega,$$

$$r = a(1-e^2)/(1 + e\cos f), \quad h = a^2 n\sqrt{1-e^2}, \quad n = \sqrt{\mu/a^3}, \quad p = h^2/\mu,$$

$$a = p/1 - e^2, \quad \text{and} \quad p = a(1-e^2) = r(1 + e\cos f). \quad (2.385)$$

It is observed that both the eccentricity and the inclination are not present in the denominators of the right-hand sides of the equations, which are non-singular as long as $0 \leq i < 180°$. Sometimes the semi-latus rectum is used in place of the semi-major axis as one of the seven primary elements.

2.11.6 Other Nonclassical Elements

Kustaanheimo-Stiefel (K-S) Elements

Like the Levi-Civita transformation for planar orbits, the Kustaanheimo-Stiefel transformation provides a method of regularization and also has the property of linearizing the equations of motion of the two-body problem. It was developed by Kustaanheimo and Stiefel [12] and led to the development of a pair of quaternion-like elements (eight in all) to describe the orbit in the case of the two-body problem as discussed by Stiefel and Scheifele [13]. Each of the two quaternion-like element sets satisfies a normalization constraint, and consequently there are only six independent elements. By the use of these elements, the differential equations for the evolution of the elements representing the orbit were regularized so they behave as a four-dimensional harmonic oscillator, thus eliminating any secular behavior among the elements.

Synchronous Elements
These are a set of orbital elements, similar to the equinoctial elements, which are particularly useful in describing satellites in near geostationary orbits. These elements use the drift rate, where the semi-major axis rate is scaled by the radius of the geostationary orbit (see, for example, Li [14]).

Geo-Synchronous Elements
Geo-synchronous orbits are different geo-stationary orbits that must necessarily be circular. In the case of geo-synchronous orbits, both the eccentricity and inclination are nonzero. A number of possible combinations exist for such circumstances, which have been discussed by Gazzino [15].

2.12 Orbit Maneuvers

Orbital maneuvers may be classified in terms of the number of impulses they require and whether or not they require a plane change. Single-impulse orbit changes can achieve a number of maneuvers such as,

(i) a maneuver at the apoapsis can push the periapsis further by changing a and e;
(ii) an orbit rotation maneuver within a plane to place the periapsis at a particular latitude requires a change in ω;
(iii) a maneuver at the periapse can change the orbital period by changing a and e;
(iv) a maneuver at the line of nodes to change the inclination requires a plane change.

However, two-impulse maneuvers such as the co-planar Hohmann transfer, which is a minimum-fuel optimal transfer (although not a global minimum energy transfer), can be energy saving. In some cases, the fuel required for two-impulse transfers can be as low as 50% of the fuel required for a corresponding single-impulse transfer.

Consider the co-planar transfer of a spacecraft in a circular orbit of radius $\mathbf{r}_1$ to another circular orbit of radius $\mathbf{r}_2$. Assuming no external forces during the transfer, the transfer orbit is also an ellipse. The first impulse $\Delta \mathbf{v}_1$ is applied to place the spacecraft on the transfer orbit, and the second $\Delta \mathbf{v}_2$ is meant to re-circularize the orbit at $\mathbf{r}_2$. The periapsis of the transfer orbit cannot lie outside the inner orbit, and the apoapsis must lie outside or be tangent to the outer orbit. Consequently, the transfer orbit satisfies the conditions,

$$\mathbf{r}_p = a(1-e) \leq \mathbf{r}_1, \quad \mathbf{r}_a = a(1+e) \geq \mathbf{r}_2. \qquad (2.386)$$

Before the application of $\Delta \mathbf{v}_1$, the spacecraft's velocity is,

$$|\mathbf{v}_1^-| = \sqrt{\mu/r_1}. \qquad (2.387)$$

Immediately after the application of $\Delta \mathbf{v}_1$ it is,

$$|\mathbf{v}_1^+| = \sqrt{2\mu\{(1/r_1) - 1/(r_a + r_p)\}}. \tag{2.388}$$

Before the application of $\Delta \mathbf{v}_2$, the spacecraft's velocity is,

$$|\mathbf{v}_2^-| = \sqrt{2\mu\{(1/r_2) - 1/(r_a + r_p)\}}. \tag{2.389}$$

Immediately after the application of $\Delta \mathbf{v}_2$ it is,

$$|\mathbf{v}_2^+| = \sqrt{\mu/r_2}. \tag{2.390}$$

The angle between the two vectors $\mathbf{v}_1^-$ and $\mathbf{v}_1^+$ is given by considering the angular momentum,

$$h = r_1 |\mathbf{v}_1^+| \cos\gamma_1 = \sqrt{2\mu r_p r_a/(r_p + r_a)}. \tag{2.391}$$

Then,

$$\Delta \mathbf{v}_1 = \sqrt{|\mathbf{v}_1^-|^2 + |\mathbf{v}_1^+|^2 - 2|\mathbf{v}_1^-||\mathbf{v}_1^+|\cos\gamma_1}. \tag{2.392}$$

For the second impulse, we obtain by a similar argument that,

$$\Delta \mathbf{v}_2 = \sqrt{|\mathbf{v}_2^-|^2 + |\mathbf{v}_2^+|^2 - 2|\mathbf{v}_2^-||\mathbf{v}_2^+|\cos\gamma_2}. \tag{2.393}$$

The Hohmann transfer corresponds to $\gamma_1 = \gamma_2 = 0$, results in a minimum $\Delta \mathbf{v}$, and corresponds to a minimum fuel transfer.

Apart from the co-planar and non-planar Hohmann transfers considered in Sections 2.4.1 and 2.4.2, there are also bi-parabolic and bi-elliptic transfers, as well as several other transfers that are not considered here.

2.12.1 Feedback Control Laws for Low-Thrust Transfers Based on the GPE

Alternatives to multi-impulse transfer are continuous low-thrust transfers, where either a tangential (tangential to the orbit) or a circumferential (orthogonal to the radius vector) acceleration is applied over a finite time interval. Such maneuvers are used mainly for escape to interplanetary trajectories. In both cases, the trajectory is a spiral, although the former transfer is generally faster but uses slightly more energy than the latter. Practical considerations generally dictate the choice of transfer used in practice.

The feedback control of orbital transfers between two elliptic Keplerian orbits is indeed feasible. To examine the feasibility of feedback control, the GPE may also be expressed as,

$$\frac{da}{dt} = u_1 = \frac{2a^2}{rh}[F_r re \sin f + pF_s], \tag{2.394}$$

2.12 Orbit Maneuvers

$$\frac{de}{dt} = u_2 = \frac{p \sin f}{h} F_r + \frac{1}{h}[(p+r)\cos f + re]F_s, \qquad (2.395)$$

$$\frac{d\omega}{dt} = u_3 = -\frac{\cos f}{nae}\sqrt{1-e^2}F_r + \frac{p}{eh}\left[\sin f\left(1 + \frac{1}{1+e\cos f}\right)\right]F_s$$
$$- \frac{r}{na^2\sqrt{1-e^2}} \frac{\sin \theta}{\sin i} F_n \cos i, \qquad (2.396)$$

$$\frac{di}{dt} = u_4 = \frac{1}{na^2}\frac{1}{\sqrt{1-e^2}} F_n r \cos(\omega + f), \qquad (2.397)$$

$$\frac{d\Omega}{dt} = u_5 = \frac{1}{na^2}\frac{1}{\sqrt{1-e^2}}\frac{1}{\sin i} F_n r \sin(\omega + f), \qquad (2.398)$$

and,

$$\frac{dM}{dt} = u_6 = n - \frac{1}{na}\left(\frac{2r}{a} - \frac{1-e^2}{e}\cos f\right)F_r - \frac{\sin f}{na}\left(\frac{1-e^2}{e}\right)\left(1 + \frac{r}{p}\right)F_s, \qquad (2.399)$$

where $h = a^2 n \sqrt{1-e^2}$. From the first two GPE we have,

$$F_r e \sin f + F_s \frac{p}{r} = \frac{\sqrt{1-e^2}}{a}\frac{na}{2}\frac{da}{dt}; \quad F_r e \sin f + F_s\left(\frac{p}{r} - \frac{r}{a}\right) = \frac{na^2 e}{a\sqrt{1-e^2}}\frac{de}{dt}. \qquad (2.400)$$

Hence,

$$F_s = \frac{\sqrt{1-e^2}}{r}\frac{na}{2}\frac{da}{dt} - \frac{na^2 e}{r\sqrt{1-e^2}}\frac{de}{dt}, \quad F_r = \frac{1}{e\sin f}\left(\frac{\sqrt{1-e^2}}{a}\frac{na}{2}\frac{da}{dt} - F_s\frac{p}{r}\right). \qquad (2.401)$$

Further, since,

$$F_r \cos f - \left(1 + \frac{1}{1+e\cos f}\right)F_s \sin f = -\frac{nae}{\sqrt{1-e^2}}\left(\frac{d\omega}{dt} + \frac{d\Omega}{dt}\cos i\right), \qquad (2.402)$$

it also follows that,

$$F_s\left\{\left(\frac{p}{er} - \frac{r}{ea}\right)\cos f + \left(1 + \frac{1}{1+e\cos f}\right)\sin^2 f\right\}$$
$$= \frac{na^2}{a\sqrt{1-e^2}}\frac{de}{dt}\cos f + \frac{nae}{\sqrt{1-e^2}}\left(\frac{d\omega}{dt} + \frac{d\Omega}{dt}\cos i\right)\sin f, \qquad (2.403)$$

$$F_r \cos f = -\frac{nae}{\sqrt{1-e^2}}\left(\frac{d\omega}{dt} + \frac{d\Omega}{dt}\cos i\right) + \left(1 + \frac{1}{1+e\cos f}\right)F_s \sin f, \qquad (2.404)$$

$$F_n = \frac{h}{r\cos\theta}\frac{di}{dt}. \qquad (2.405)$$

Since a force or an impulse vector has only three components, only three of the six orbital elements can be controlled independently. Even then it is not always possible seek any three combinations of the orbital elements, as some combinations may be impossible to achieve. If $F_n = 0$, i and Ω cannot be altered. In the case of planar manoeuvers, $F_n = 0$. Orbit changes or maneuvers that do not change the direction of angular momentum but change only a, e, ω, and/or t_0 (the time of perigee passage) are in-plane or co-planar maneuvers. The Δv for such a manoeuver must therefore lie within the orbit plane.

(i) Case I $F_r = 0$, $F_s \neq 0$, $F_n \neq 0$

In this case, only a tangential force or impulse is applied. From the GPE,

$$u_1 = \frac{2a^2}{h}\frac{p}{r}F_s \equiv u_{1d}, \tag{2.406}$$

$$u_2 = \frac{1}{h}[(p+r)\cos f + re]F_s \equiv u_{2d}, \tag{2.407}$$

$$u_3 = \frac{p}{eh}\left[\sin f\left(1 + \frac{1}{1 + e\cos f}\right)\right]F_s - \frac{r}{na^2\sqrt{1-e^2}}\frac{\sin\theta}{\sin i}F_n\cos i \equiv u_{3d}, \tag{2.408}$$

$$u_4 = \frac{1}{na^2}\frac{1}{\sqrt{1-e^2}}F_n r \cos(\omega + f) \equiv u_{4d}, \tag{2.409}$$

$$u_5 = \frac{1}{na^2}\frac{1}{\sqrt{1-e^2}}\frac{1}{\sin i}F_n r \sin(\omega + f) \equiv u_{5d}, \tag{2.410}$$

and

$$n - u_6 = \frac{\sin f}{na}\left(\frac{1-e^2}{e}\right)\left(1 + \frac{r}{p}\right)F_s \equiv n - u_{6d}. \tag{2.411}$$

Thus, one could specify either u_{1d}, u_{2d}, u_{6d}, or u_{4d}, u_{5d}. Any of these could be replaced by u_{3d}. Hence, one of the following applies, and,

$$u_1 = K_1(a_d - a) \equiv u_{1d}, \tag{2.412}$$

$$u_2 = K_2(e_d - e) \equiv u_{2d}, \tag{2.413}$$

$$u_6 = n - K_1(M_{0d} - M_0) \equiv u_{6d}, \tag{2.414}$$

Or,

$$u_4 = K_4(i_d - i) \equiv u_{4d}, \tag{2.415}$$

$$u_5 = K_5(\Omega_d - \Omega) \equiv u_{5d}. \tag{2.416}$$

The complete set of the strictly positive control gains are determined by the use of an appropriate optimal control synthesis method or by Lyapunov's method. Thus, F_s and F_n are determined as two of the orbital elements. The remaining four elements are then determined from the GPE.

2.12 Orbit Maneuvers

(ii) *Case II* $F_r \neq 0$, $F_s \neq 0$, $F_n \neq 0$

Consider the case when u_1, u_2, and u_3 are specified. For example, consider the feedback laws,

$$\begin{bmatrix} u_1 \\ u_2 \\ u_3 \end{bmatrix} = \begin{bmatrix} K_1(a_d - a) \\ K_2(e_d - e) \\ K_3(\omega_d - \omega) \end{bmatrix} \equiv \begin{bmatrix} u_{1d} \\ u_{2d} \\ u_{3d} \end{bmatrix}. \tag{2.417}$$

From the first two GPE,

$$\begin{bmatrix} u_{1d} \\ u_{2d} \end{bmatrix} = \frac{1}{h} \begin{bmatrix} 2a^2 e \sin f & 2a^2 p/r \\ p \sin f & (p+r)\cos f + re \end{bmatrix} \begin{bmatrix} F_r \\ F_s \end{bmatrix}. \tag{2.418}$$

Hence,

$$\begin{bmatrix} F_r \\ F_s \end{bmatrix} = \frac{1}{\Delta} \begin{bmatrix} (p+r)\cos f + re & -2a^2 p/r \\ -p \sin f & 2a^2 e \sin f \end{bmatrix} \begin{bmatrix} u_{1d} \\ u_{2d} \end{bmatrix}, \tag{2.419}$$

with

$$\Delta = \frac{2a^2 r \sin f((p/r + 1)e \cos f + e^2 - p^2/r^2)}{h^2}$$

From the equation for u_3,

$$F_n = \frac{h \tan i}{r \sin \theta} \left\{ \frac{p}{eh} \sin f \left(1 + \frac{1}{1 + e \cos f}\right) F_s - u_{3d} - \frac{\cos f}{nae}\sqrt{1 - e^2} F_r \right\}. \tag{2.420}$$

The other three orbital elements are then found.

Consider the case when u_1 and u_6 are specified. The feedback control laws take the form,

$$\begin{bmatrix} u_1 \\ u_6 \end{bmatrix} = \begin{bmatrix} K_1(a_d - a) \\ n - K_1(M_{0d} - M_0) \end{bmatrix} \equiv \begin{bmatrix} u_{1d} \\ u_{6d} \end{bmatrix}. \tag{2.421}$$

Thus,

$$\begin{bmatrix} u_{1d} \\ n - u_{6d} \end{bmatrix} = \frac{1}{h} \begin{bmatrix} 2a^2 e \sin f & 2a^2 p/r \\ \frac{2rh}{na^2} - \frac{1 - e^2}{nae} h \cos f & \frac{h \sin f}{na}\left(\frac{1 - e^2}{e}\right)\left(1 + \frac{r}{p}\right) \end{bmatrix} \begin{bmatrix} F_r \\ F_s \end{bmatrix}. \tag{2.422}$$

Hence, for this case,

$$\begin{bmatrix} F_r \\ F_s \end{bmatrix} = h \begin{bmatrix} 2a^2 e \sin f & 2a^2 p/r \\ \frac{2rh}{na^2} - \frac{1 - e^2}{nae} h \cos f & \frac{h \sin f}{na}\left(\frac{1 - e^2}{e}\right)\left(1 + \frac{r}{p}\right) \end{bmatrix}^{-1} \begin{bmatrix} u_{1d} \\ n - u_{6d} \end{bmatrix}. \tag{2.423}$$

Furthermore, if u_4 or u_5 are specified,

$$u_4 = K_4(i_d - i) \equiv u_{4d} \quad \text{or} \quad u_5 = K_5(\Omega_d - \Omega) \equiv u_{5d}. \tag{2.424}$$

Hence,

$$F_n = \frac{\sqrt{1-e^2}}{r\cos(\omega+f)} na^2 u_{4d} \quad \text{or} \quad F_n = \frac{\sin i \sqrt{1-e^2}}{r\sin(\omega+f)} na^2 u_{5d}. \qquad (2.425)$$

For most space missions it is necessary to conserve fuel, and for this reason one approach is to base the control laws on the optimal control theory outlined in Bryson and Ho [16], which has been derived from the calculus of variations. The general approach is to minimize a performance index or cost function, such as the magnitude of the total applied thrust subject to the trajectory, satisfying the differential equations of motion and a set of terminal state constraints. The problem is solved by introducing adjoint variables (or Lagrange variables) and numerically solving a Two-Point Boundary Value Problem. The solution is usually obtained numerically, as outlined by Conway [17]. Sometimes the problem may be simplified by time averaging the equations for a, e, ω, i, and Ω, which are slowly varying functions of the fast variable, the true anomaly f, to obtain an expression for the impulsive thrust or for the continuous thrust orbit transfers. One could also use the *primer vector* formulation of Lawden [18], where Lawden [18] replaces one of the adjoint vectors, which always satisfies the adjoint differential equation arising from the relationship between the position and velocity vectors of the spacecraft, by the *primer vector*, which not only simplifies the adjoint equations but also is zero in the terminal state. Moreover, the direction of the control thrust is parallel to the primer vector, thus facilitating the choice of a bang-bang control law in certain simplified cases.

2.12.2 Feedback Control Laws with Constraints on the Control Accelerations

Often it is necessary to impose additional constraints on the control accelerations while defining the feedback control laws. For example, when going into a transfer orbit that takes the spacecraft from an outer orbit to an inner orbit, it is often necessary to maintain the semi-major axis. On the other hand, when going into a transfer orbit that takes the spacecraft from an inner orbit to an outer orbit, it is often necessary to maintain the semi-minor axis. The overall manoeuver may involve going from an initial orbit to a final orbit while maintaining the eccentricity of the orbit at a constant value. In the first case, $da/dt = 0$, while in the second case, $db/dt = 0$, and in the third case, $de/dt = 0$.

Thus, in the first case,

$$\frac{da}{dt} = u_1 = \frac{2a^2}{rh}[F_r re\sin f + pF_s] = 0 \Rightarrow F_s = -F_r re\sin f/p. \qquad (2.426)$$

In the second case, consider the fact, $b^2 = (1-e^2)a^2$. Hence, one obtains the pair of equations,

$$\frac{db}{dt} = \sqrt{1-e^2}\frac{da}{dt} - \frac{ae}{\sqrt{1-e^2}}\frac{de}{dt} = 0. \qquad (2.427\text{a})$$

2.12 Orbit Maneuvers

$$\left(2a\sqrt{1-e^2} - \frac{p}{\sqrt{1-e^2}}\right) F_r ae \sin f$$
$$+ \left[\left(p+r+2ea^2\sqrt{1-e^2}\right)\cos f + re + 2a^2\sqrt{1-e^2}\right] F_s = 0. \qquad (2.427b)$$

In the third case,

$$\frac{de}{dt} = u_2 = \frac{p \sin f}{h} F_r + \frac{1}{h}[(p+r)\cos f + re] F_s \Rightarrow F_s = -\frac{p \sin f}{(p+r)\cos f + re} F_r. \qquad (2.428)$$

Once the constraints on the control accelerations are applied, for planar changes in the orbit, there is just a single input that must be specified.

One approach to applying the optimal control methodology is based on Edelbaum's solution [19]. Edelbaum's approach is to replace the Gauss planetary equations by averaged equations over a single circular orbit. Apart from assuming a circular orbit over which the orbital quantities are averaged, Edelbaum also assumes constant thrusting to obtain the optimum minimum time solution to the orbit transfer problem, over the timeframe considered. The problem involves the transfer of the vehicle from one given circular orbit to another given circular orbit in minimum time. Several improvements to Edelbaum's model have been implemented, and these have been summarized by Cerf [20]. At a given time, the orbital plane is defined in the Earth's inertial reference frame by the averaged inclination I and the averaged right ascension of the ascending node (RAAN). The averaged inclination I is the angle of the orbital plane with the Earth's equatorial plane. The averaged RAAN is denoted by the angle Ω between the X axis of the Earth's inertial reference frame and the direction of the ascending node (node crossed with a northward motion). The circular orbit shape is defined equivalently by the radius "a" or by the velocity V. The upper case variables V, I, and Ω denote the averaged quantities. In Edelbaum's model, it is assumed that the thrust direction is normal to the radius vector and that it makes a constant angle β with the orbital plane during one period, with a sign change at the anti-nodes. This results in a null RAAN change over one period while maximizing the inclination change for the current value of the out of plane angle β. In the model defined here, following Cerf [20], Edelbaum's dynamics is extended by taking into account the first zonal term (denoted J_2) due to the Earth's oblateness. This perturbation causes no secular change in the semi-major axis, the eccentricity, or the inclination. The averaged orbit remains circular, with radius "a" and inclination I. However, a RAAN precession rate is induced, depending on the orbit radius "a" and the inclination I.

The dynamics model consists of three first-order state equations representing the evolution of the averaged orbital variables (V, I, Ω), defined by,

$$\dot{V} = -f \cos \beta,$$
$$\dot{I} = -2f \sin \beta / \pi V,$$
$$\dot{\Omega} = -(3/2) J_2 \sqrt{\mu} R_E^2 a^{-7/2} \cos I = -k(\mu/a)^{-7/2} \cos I = -kV^7 \cos I, \qquad (2.429)$$

where $R_E = 6,378,137$ m is Earth's equatorial radius, $\mu = 3.9860005 \times 10^{14}$ m^3/s^2 is the gravitational constant of the Earth, $J_2 = 1.08266$ is the first zonal term due to the Earth's flattening, and $k = 3J_2R_E^2/(2\mu^3) = 1.0425 \times 10^{-33}$ s^6/m^7. The control variables are the acceleration level f and its direction β over each period. The acceleration can take one of two values, either zero or a maximum value $f_{\max}$. The minimum time solution is obtained by the optimal control methodology. Edelbaum's strategy in defining the transfer trajectory is to split it into a first propelled transfer from the initial orbit toward a drift orbit, a coast phase on the drift orbit, and a second propelled transfer from the drift orbit toward the final orbit. Fuel consumption is directly proportional to the durations of the two propelled transfers.

The simplifying approach consists of using Edelbaum's analytical minimum time solution to define the two propelled transfers. The problem reduces to optimizing the drift orbit parameters that should be defined in order to minimize the cost while meeting the final RAAN constraint. This simplified transfer strategy is suboptimal, unlike the optimal control solution, although it is quite close to it as long as the propelled duration remains small compared to the drift duration. Moreover, the RAAN change is performed during the drift phase. This assumption is the underlying basis for the formulation of Edelbaum's split transfer strategy. Edelbaum's approach is particularly suitable for changing a satellite's altitude and inclination. The details are not presented here.

For a satellite in a near circular orbit, using Newton's law, $\dot{v} = f$, where f is the applied thrust in the tangential direction, and the corresponding visa-viva equation for the conservation of energy, $v^2 = \mu/a$, one can obtain equations for the orbit-raising rate of change of the radius $\dot{a}$, the Δv required for raising the orbit from one height to another, the time of flight for the orbit-raising operation, and the fuel required over the timeframe.

2.13   Interception and Rendezvous

The process of interception and rendezvous, followed by docking and berthing, requires the deployment of control forces to execute a series of orbital maneuvers that bring a chaser vehicle into the vicinity of, to eventually make contact with, the target vehicle. At the end of the launch phase, the chaser spacecraft is generally placed in an orbit near the target and in the same orbital plane as the target. After the chaser satellite is in the chaser orbit, its relative phase must be adjusted so it is either just behind or just ahead of the target. The height of the apogee/perigee must be changed so that the two orbits have the same size or semi-major axis. To raise the apogee, a tangential thrust must be provided at the perigee, while to raise the perigee, a tangential thrust must be provided at the perigee. When necessary, the inclination and argument of the ascending node of the chaser orbit are also corrected. To change the inclination, a thrust normal to the orbit plane must be applied. Moreover, this correction must be applied at the ascending or descending node. A typical example of an orbit transfer involving an inclination change is illustrated in Figure 2.13. For a change in the argument of the ascending node of the chaser orbit, the correction normal to the orbit plane must be applied at a point that is

2.13 Interception and Rendezvous

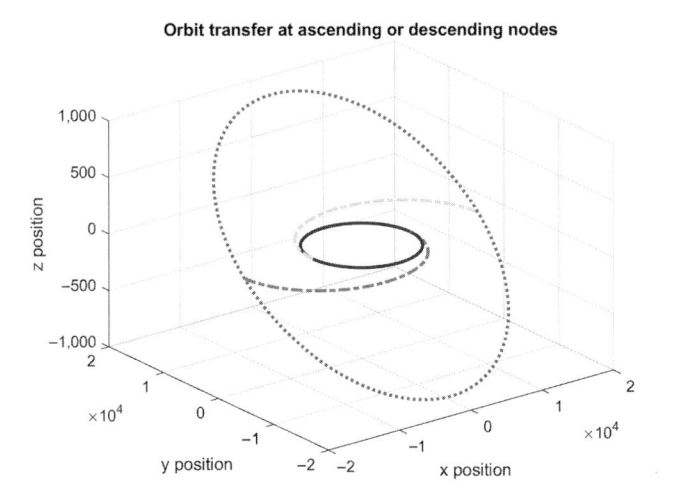

Figure 2.13 Orbit transfer involving a change in inclination, with the thrust deployed at the ascending or descending nodes.

midway between the ascending and descending nodes and on the orbit. Once in the vicinity of the target, the axes of rotation must be aligned, and the rotation speeds must be synchronized before rendezvous can be attempted. Thus, the approach trajectory of the chaser must be within limits of the position, the velocity, and angular velocity of the target vehicle before the chaser's motion is synchronized with the target and rendezvous and docking are completed. Before docking is initiated, the relative velocity of the chaser with respect to the target is asymptotically reduced to zero to facilitate a rendezvous with minimum impact. While phasing the chaser spacecraft with respect to the target spacecraft, the navigation is based on absolute requirements in a navigation frame.

After the initiation of rendezvous, navigation is based on motion requirements relative to the target spacecraft. This requires the relative measurement of range and direction. Thus, at the end of phasing the chase spacecraft relative to the target, the open loop maneuvers must bring the chaser within the range of the relative navigation sensor to facilitate the acquisition of the target, which is usually within 100 m. The motion of the target satellite relative to the chaser satellite satisfies the equations of relative orbital motion that were introduced in Section 2.10. All filters for relative position measurement are mechanized and implemented with these relative motion equations. The problem has been dealt with by Fehse [21], who has devoted a whole book to the subject of automated rendezvous and docking of spacecraft.

For interception, the determination of the interception and rendezvous transfer orbits requires that both the departure and arrival times are specified. Typically, the initial position $\mathbf{r}_1$ in the orbit is specified at a time t_1 and the final position $\mathbf{r}_2$ at time t_2. The problem now reduces to solving for the semi-major axis a, that meets the time constraint imposed by the solution to Lambert's problem in Section 2.6. A numerical, iterative method is adopted to solve this problem by choosing an initial estimate of the semi-latus rectum p until an acceptable solution for a is found.

2.14 Advanced Orbit Perturbations

Gravitational perturbations due to the geometry of the central body can be classified into two groups: (i) zonal harmonics associated with a perfect oblate spheroid model, and (ii) zonal, sectoral, and tesseral harmonics associated with a real model of the central body.

2.14.1 Gravitational Potential of a Perfect Oblate Spheroid Model of the Central Body

Three primary perturbation forces, due to third-body gravitation, solar radiation pressure, and aerodynamic drag, apart from the central body gravitational perturbations, should be considered in the definition of the orbit of a real satellite, particularly when considering Earth-orbiting satellites. Moreover, perturbations could be probabilistic in nature.

Generally, one is permitted to ignore certain perturbation accelerations on the grounds of smallness of their magnitudes, and these perturbations that are not included or ignored in the analysis may be treated as random, band-limited, or colored process noise sources. When the bandwidth of the noise spectrum is sufficiently wide, the sources could be modeled as white noise sources. For the reference orbit, one could consider the zonal harmonic model for the gravitation perturbation model of the central body, which is assumed to be a perfect oblate spheroid and is given in terms of Legendre functions P_k of the latitude ϕ by,

$$U = \frac{GM}{r}\left\{1 - \sum_{k=2}^{\infty} J_k \left(\frac{R_{eq}}{r}\right)^k P_k(\sin\phi)\right\}, \qquad (2.430)$$

where R_{eq} is the equatorial radius of the central body.

For this type of model, which is axially symmetric, the rotation of the celestial central body does not affect the accelerations of the orbiting body. The series is truncated to a finite series and the perturbation acceleration on the orbiting body is estimated using the equations developed in Eyer [22].

The perturbation accelerations on the orbiting body are expressed compactly in the form,

$$\vec{a}_{J_{2n}} = -\frac{J_{2n}}{\beta_{2n}} \frac{\mu}{R^2} \left(\frac{R_{eq}}{R}\right)^{2n} \begin{bmatrix} \bar{x} \sum_{k=2n,\, 2n-2\cdots}^{0} C_{k,2n}^{x} \bar{z}^k \\ \bar{y} \sum_{k=2n,\, 2n-2\cdots}^{0} C_{k,2n}^{x} \bar{z}^k \\ \bar{z} \sum_{k=2n,\, 2n-2\cdots}^{0} C_{k,2n}^{z} \bar{z}^k \end{bmatrix}, \qquad (2.431)$$

Table 2.5 The coefficients (a) β_k, (b) $C^x_{k,m}$, and (c) $C^z_{k,m}$

(a) β_k

k	2	3	4	5	6
β_k	2	2	8	8	16

(b) $C^x_{k,m}$

$m = j+1 \to$ $k = 2*i - 2 \downarrow 0$	2	3	4	5	6
	3	45	−15	−105	−35
2	−15	−105	210	630	945
4			−315	−693	−3,465
6					3,003

(c) $C^z_{k,m}$

$m = j+1 \to$ $k = 2*i - 2 \downarrow 0$	2	3	4	5	6
	9	−3	−70	15	245
2	−15	30	350	−351	2,205
4		−35	−315	945	4,851
6				−693	−3,003

$$\vec{a}_{J_{2n+1}} = -\frac{J_{2n+1}}{\beta_{2n+1}} \frac{\mu}{R^2} \left(\frac{R_{eq}}{R}\right)^{2n+1} \begin{bmatrix} \bar{x}\bar{z} \sum_{k=2n, 2n-2\cdots}^{0} C^x_{k, 2n+1} \bar{z}^k \\ \bar{y}\bar{z} \sum_{k=2n, 2n-2\cdots}^{0} C^x_{k, 2n+1} \bar{z}^k \\ \bar{z} \sum_{k=2n+2, 2n\cdots}^{0} C^z_{k, 2n+1} \bar{z}^k \end{bmatrix}, \quad n = 1, 2, 3..., \quad (2.432)$$

where $\bar{x} = x/R, \bar{y} = y/R, \bar{z} = z/R, R = \sqrt{x^2 + y^2 + z^2}$, and in this section, the variables $[x \ y \ z]$ are the position coordinates of the satellite in a body's centered reference frame, with the z axis aligned with the body's north-south axis normal to the equatorial plane. The nonzero coefficients β_k, $C^x_{k,m}$, and $C^z_{k,m}$ are listed in Table 2.5.

The formulae in Table 2.5 are used to estimate oblateness effects of the planet on the nominal reference orbiting body. These are generally small if the orbiting body is relatively far away from the planet, and, in these cases, they are ignored.

2.14.2 Gravitational Potential due to a Central Body's Real Geometry

For the perturbations acting on the nominal satellite, the gravitational potential model of the real geometry of the central body, if known, may also be considered. The potential is assumed to be given by all of the zonal, sectoral, and tesseral spherical harmonics in terms of associate Legendre functions P_{nm} by,

$$U = \frac{\mu}{r} - \frac{\mu}{r}\left[\sum_{n=2}^{\infty}\sum_{m=0}^{\infty}\left(\frac{R_m}{r}\right)^n J_{nm} \times P_{nm}(\sin\phi)\cos(m(\lambda - \lambda_{nm}))\right], \quad (2.433)$$

where in a Cartesian body-centered coordinate frame fixed in the central body,

$$r = \sqrt{x^2 + y^2 + z^2}, \quad \sin(\phi) = z/r, \quad \cos(\phi) = \sqrt{x^2 + y^2}/r,$$

$$\sin(\lambda) = y/\sqrt{x^2 + y^2},$$

$$\cos(\lambda) = x/\sqrt{x^2 + y^2}, \quad R_m = R_{eq}, \quad \mu = G(m_2 + m_1), \quad (2.434)$$

And,

$$[x \quad y \quad z]^T = r[\cos\phi\cos\lambda \quad \cos\phi\sin\lambda \quad \sin\phi]^T. \quad (2.435)$$

The expression for the perturbation potential model V is truncated for Earth-orbiting satellites and expressed as,

$$V = \frac{\mu}{r}\sum_{l=2}^{8}\left(\frac{R_m}{r}\right)^l \sum_{m=0}^{8-l} P_{lm}(\sin(\phi))\{C_{lm}\cos m\lambda + S_{lm}\sin m\lambda\}. \quad (2.436)$$

The perturbation acceleration vector,

$$\vec{a} = \frac{\partial V}{\partial r}u_r + \frac{1}{r\cos\phi}\frac{\partial V}{\partial \lambda}u_\lambda + \frac{1}{r}\frac{\partial V}{\partial \phi}u_\phi, \quad (2.437)$$

where,

$$\begin{bmatrix}u_r \\ u_\lambda \\ u_\phi\end{bmatrix} = \begin{bmatrix}\cos\phi\cos\lambda & \cos\phi\sin\lambda & \sin\phi \\ -\sin\lambda & \cos\lambda & 0 \\ -\sin\phi\cos\lambda & -\sin\phi\sin\lambda & \cos\phi\end{bmatrix}\frac{1}{r}\begin{bmatrix}\vec{x} \\ \vec{y} \\ \vec{z}\end{bmatrix}. \quad (2.438)$$

The gradients of the perturbing potential are given by,

$$\left[\frac{\partial V}{\partial r} \quad \frac{1}{r\cos\phi}\frac{\partial V}{\partial \lambda} \quad \frac{1}{r}\frac{\partial V}{\partial \phi}\right]^T = \frac{\mu}{r^2} \times \sum_{l=2}^{\infty}\left(\frac{R_m}{r}\right)^l \sum_{m=0}^{l}\begin{bmatrix}-(l+1)P_{lm}(\sin\phi)C_{lm} \\ \frac{m}{\cos\phi}P_{lm}(\sin\phi)S_{lm} \\ P_{lm}'(\sin\phi)C_{lm}\cos\phi\end{bmatrix}\cos m\lambda$$

$$+ \begin{bmatrix}-(l+1)P_{lm}(\sin\phi)S_{lm} \\ -\frac{m}{\cos\phi}P_{lm}(\sin\phi)C_{lm} \\ P_{lm}'(\sin\phi)S_{lm}\cos\phi\end{bmatrix}\sin m\lambda. \quad (2.439)$$

2.14.3 Real Drag Acceleration Acting on the Actual Satellite

A model of the drag acceleration that includes wind speed variations in the atmosphere (in the inertial reference frame, ECI) is used for the relative velocity in the expression for the drag acceleration $\vec{a}_{drag}$. The nonconservative drag acceleration is,

2.14 Advanced Orbit Perturbations

$$\vec{a}_{drag} = -\frac{1}{2}\rho\frac{C_D A}{m_S}|\mathbf{v}_{rel}|^2 \frac{\mathbf{v}_{rel}}{|\mathbf{v}_{rel}|}, \qquad (2.440)$$

where A is the projected cross-sectional area perpendicular to the velocity direction. For evaluating the nominal drag acceleration, the relative velocity of the satellite with reference to the planetary atmosphere is approximated as,

$$\mathbf{v}_{rel} = \vec{v}_s - \vec{\omega}_e \times \vec{r}_s + \vec{v}_{wind, LVLH}. \qquad (2.441)$$

In terms of the unit vectors of the LVLH frame,

$$\vec{\omega}_e = \omega_e \hat{Z} = \omega_e(\sin\theta \sin i \hat{x} + \cos\theta \sin i \hat{y} + \cos i \hat{z}), \qquad (2.442)$$

with $\omega_e = 7.2921 \times 10^{-5}$ rad/s. It follows that,

$$\vec{v}_s - \vec{\omega}_e \times \vec{r}_s = v_{x,r}\hat{x} + \frac{h}{r}\hat{y} - \vec{\omega}_e \times \vec{r} = v_{x,r}\hat{x} + \left(\frac{h}{r} - \omega_e r \cos i\right)\hat{y} + \omega_e r \cos\theta \sin i \hat{z}. \qquad (2.443)$$

The wind vector is initially obtained in the ECEF coordinate frame and transformed to the LVLH frame. In the ECEF frame,

$$\vec{v}_{wind, ECEF} = v_w [\cos\alpha_w \sin\delta_w \quad \sin\alpha_w \sin\delta_w \quad \cos\delta_w]^T. \qquad (2.444)$$

Furthermore, $v_w = v_w(\vec{r}_s)$ is an estimate of the neutral wind speed at the satellite's position, which is obtained from a standard wind model, α_w is the wind vector's azimuth angle, and δ_w is the wind vector's declination as measured from the ECEF coordinate frame. The transformation from the ECI to the ECEF frame is assumed to be,

$$\begin{bmatrix} x_{ef} \\ y_{ef} \\ z_{ef} \end{bmatrix} = \begin{bmatrix} \cos\omega_e(t-t_0) & \sin\omega_e(t-t_0) & 0 \\ -\sin\omega_e(t-t_0) & \cos\omega_e(t-t_0) & 0 \\ 0 & 0 & 1 \end{bmatrix} \begin{bmatrix} x_i \\ y_i \\ z_i \end{bmatrix}. \qquad (2.445)$$

The transformation from the ECI to the LVLH frame is easily obtained and will not be given here.

2.14.4 Third-Body Perturbations

In the case of a satellite in Earth orbit under the influence of the Moon, the two primary masses move in two-body motion about each other, and the satellite is significantly affected by both masses. If one lets $\mu = G(m_\oplus + m_S) \approx Gm_\oplus$, then in the case of J third bodies, one may derive the following relationship for the satellite's acceleration,

$$\ddot{\vec{r}}_{\oplus \to S} = -\mu \frac{\vec{r}_{\oplus \to S}}{r_{\oplus \to S}^3} + \sum_{j=1}^{J} Gm_{B3j}\left(\frac{\vec{r}_{\oplus \to B3j} - \vec{r}_{\oplus \to S}}{\left|\vec{r}_{\oplus \to B3j} - \vec{r}_{\oplus \to S}\right|^3} - \frac{\vec{r}_{\oplus \to B3j}}{r_{\oplus \to B3j}^3}\right), \qquad (2.446)$$

where the subscripts $\oplus$, $B3j$, and S refer to the Earth, the third body, and the satellite, respectively, and the subscript $\oplus \to S$ indicates a vector that originates at the Earth and ends at the satellite. The first term in Equation (2.446) is the familiar two-body equation of motion; the second term arises from the third-body perturbations. Equation (2.446) is expressed as,

$$\ddot{\mathbf{r}} = -\mu \frac{\mathbf{r}}{|\mathbf{r}|^3} + \mathbf{f}_{3S}\left(\vec{r}_{\oplus \to B3j}, \vec{r}_{\oplus \to S}\right) = -\mu \frac{\mathbf{r}}{r^3} + \mathbf{f}_{3S}\left(\vec{r}_{\oplus \to B3j}, \vec{r}_{\oplus \to S}\right), \qquad (2.447)$$

with $\mathbf{r} = \vec{r}_{\oplus \to S}$. Thus the third-body perturbation force vector is given by,

$$\begin{bmatrix} F_{x,p}^{3rd} \\ F_{y,p}^{3rd} \\ F_{z,p}^{3rd} \end{bmatrix} = \mathbf{f}_{3S}\left(\vec{r}_{\oplus \to B3j}, \vec{r}_{\oplus \to S}\right) = \sum_{j=1}^{J} Gm_{B3j} \left(\frac{\vec{r}_{\oplus \to B3j} - \vec{r}_{\oplus \to S}}{\left|\vec{r}_{\oplus \to B3j} - \vec{r}_{\oplus \to S}\right|^3} - \frac{\vec{r}_{\oplus \to B3j}}{r_{\oplus \to B3j}^3} \right). \qquad (2.448)$$

2.14.5 Solar Radiation Pressure

Solar radiation pressure (SRP) is one of the remaining dominant disturbing forces acting on a satellite in Earth orbit.

SRP is a force acting on the satellite's surface that is caused by the impact of sunlight on the satellite's surface. The acceleration of the satellite due to the force of solar radiation is proportional to the effective satellite surface area, the reflectivity of the surface, and the solar flux; it is inversely proportional to the velocity of light. Acceleration resulting from SRP is,

$$\vec{a}_{srp} = \frac{\vec{f}_{srp}}{m_{sat}} = -\gamma C_r \frac{|\vec{r}_{sun}|^2}{|\vec{r}_{sat} - \vec{r}_{sun}|^2} \frac{\vec{r}_{sat} - \vec{r}_{sun}}{|\vec{r}_{sat} - \vec{r}_{sun}|} \frac{A}{m_{sat}} \frac{\Phi}{c}, \qquad (2.449)$$

γ is the shadow factor, $\gamma = 1$ for complete sunlight, $\gamma = 0$ when the satellite is within the umbra, and $0 < \gamma < 1$ when it is within the penumbra. C_r is the coefficient of surface reflectivity, $\vec{r}_{sun}$ is the geocentric vector position of the Sun, $\vec{r}_{sat}$ is the geocentric vector position of the satellite, A is the satellite cross-sectional area normal to the Sun vector, m_{sat} is the mass of the satellite, and c is the speed of light in vacuum. The solar flux in the vicinity of the Earth's surface Φ and the velocity of light c may be assumed to be 1,372.5398 W/m² and 299,792,458 m/s, respectively, so $\Phi/c =$ 4.5783e−06 N/m². In general, for a surface element, C_r is a function of coefficient of diffuse reflectivity δ, the coefficient of specular reflectivity ρ, and the angle β between the direction of the Sun $\mathbf{s}$ and the surface normal $\mathbf{n}$. Thus, integrating over the entire surface of the satellite, one may express C_r as,

$$C_r = \frac{1}{A} \int_A \left((1-\rho)\mathbf{s} + 2\left(\frac{\delta}{3} + \rho \cos\beta\right)\mathbf{n} \right) \mathbf{n} \cdot d\mathbf{A}. \qquad (2.450)$$

Thus, C_r depends on the outer geometry of the satellite, its rate of spin, its orbital rate, and the orbital rate of the Sun. While a refined method of SRP estimation is discussed by Wetterer et al. [23], Kubo-oka and Sengoku's [24] method of estimating the magnitude of the SRP is adopted here. If the satellite is assumed to be a sphere, based on the so-called cannonball model, the radiation pressure coefficient can be expressed as in Kubo-oka and Sengoku [24], $C_r = 1 + 4\delta/9$, where δ is the coefficient of diffuse reflectivity, which is assumed to be about four times the coefficient of specular reflectivity ρ of the surface of the satellite. If the surface is made up of solar panels, $\delta = 0.168$ and $C_r = 1.0747$. If the satellite body is covered by gold foil, $\delta = 0.736$ and $C_r = 1.3271$ (Kubo-oka and Sengoku [24]. If the satellite is assumed to be a sphere, the projected surface area of the satellite body is expressed as, $A = \left(36\pi V_{sat}^2\right)^{1/3}/4 = 1.209 V_{sat}^{2/3}$. Additionally, the satellite may be powered by solar sails, which are always pointing to the Sun. Thus, if the total area of the sails is A_{sails}, the total projected area may be expressed as, $A = 1.209 V_{sat}^{2/3} + A_{sails}$. Similar formulae can be obtained for satellites of different geometries.

The radiation around the vicinity of the Earth includes the direct solar radiation from the Sun, the albedo, and the thermal or infrared radiation emitted by the surface of the Earth. Solar radiation that reflects from a planet is referred to as the albedo. The effect of the pressure due to the albedo on a satellite is much more difficult to model than the SRP. It depends on the orbital position vector relative to the central body, the Earth, and on the relative position between the satellite and the Sun, which in turn determines the so-called view factors. The combined secondary radiation pressure is generally less than 10% of the primary SRP, with albedo contributing about 70% of that. Equations similar to Equation (2.249), with the appropriate reflectivity coefficients and radiation flux per unit velocity of the radiation, are used to determine these secondary acceleration effects. Thus, the total secondary SRP-related acceleration is expressed as,

$$\vec{a}_{ssrp} = \frac{\vec{f}_{ssrp}}{m_{sat}} = -\frac{(r_{earth\_radius})^2}{|\vec{r}_{sat}|^2} \frac{\vec{r}_{sat}}{|\vec{r}_{sat}|} \frac{A}{m_{sat}} \left(C_{tr} \frac{\Phi_{tr}}{c} + 0.07 F \times C_r \frac{\Phi}{c} \right), \qquad (2.451)$$

where the second term in the brackets is due to the albedo, C_{tr} is the reflectivity coefficient for thermal radiation that could be assumed to be the same as C_r, Φ_{tr} is the thermal radiation flux at the Earth's surface, and F is the view factor that depends on the incident angle of the radiation on the satellite. For satellites orbiting the Earth, the dynamic or process uncertainties are assumed to be determined by the primary and secondary SRP effects, as well as the primary gravitation effects related to bodies and influences such as tidal motions that were not accounted for. It is in fact the SRP forces that dominate, as many of the gravitation influences were accounted for. The magnitudes of these uncertainties may be estimated based on the work of Scheeres, Williams, and Miller [25] using standard methods of primary and secondary SRP estimation for satellites.

2.15 Launch Vehicle Dynamics: Point Mass Model

2.15.1 Systems with Varying Mass

We now develop the equations of linear motion of a system whose mass varies with time. The motion of a space vehicle often involves flight with variable mass, as illustrated in Figure 2.14.

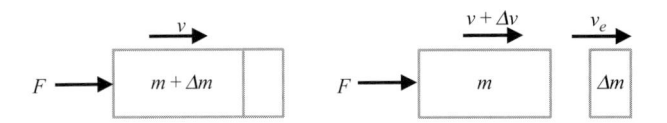

Figure 2.14 Change in momentum in a variable mass system.

Referring to Figure 2.14, the initial momentum and the final momentum to first order are,

$$\mathbf{p} = (m + \Delta m)\mathbf{v} \quad \text{and} \quad \mathbf{p} + \Delta \mathbf{p} = m(\mathbf{v} + \Delta \mathbf{v}) + \Delta m \mathbf{v}_e. \tag{2.452}$$

Hence, the thrust acting on the space vehicle in the same direction as the force is,

$$\mathbf{T} = -(\mathbf{v}_e - \mathbf{v})dm/dt. \tag{2.453}$$

Since,

$$\mathbf{c} = (\mathbf{v}_e - \mathbf{v}), \tag{2.454}$$

is the relative velocity of Δm relative to $(m + \Delta m)$, we may express the thrust as,

$$\mathbf{T} = -\mathbf{c} dm/dt. \tag{2.455}$$

2.15.2 Basic Rocket Thrust Equation

One may now adapt the thrust equation developed in the previous section to a rocket subjected to an aerodynamic drag and force due to the average static pressure across the nozzle exit plane. The forces acting on the rocket are shown in Figure 2.15. It is assumed that the rocket is a closed-body type, wherein the fuel is burnt within a combustor and expelled to the atmosphere via a nozzle. Applying Newton's law, the equation of vertical motion is,

$$m\dot{v} = T + pA - D - mg, \quad T = -cdm/dt, \tag{2.456}$$

where D is the total drag and c is the nozzle exit velocity. Note that the rate of change of mass due to the combustion of the fuel is always negative, and this implies that the thrust is always positive.

Often it is convenient to absorb the thrust increment due to the static pressure acting across the nozzle exit area in the expression for the thrust and define an equivalent nozzle exit velocity such that,

$$c_e(dm/dt) = c(dm/dt) - pA. \tag{2.457}$$

Recognizing that the rate of change of mass is negative and letting,

$$m_f = -dm/dt, \quad m\dot{v} = m_f c_e - D - mg. \tag{2.458}$$

Since the last two terms represent the sum of the external forces F_{ext}, the equation of rocket motion is,

$$m\dot{v} = m_f c_e + F_{ext}. \tag{2.459}$$

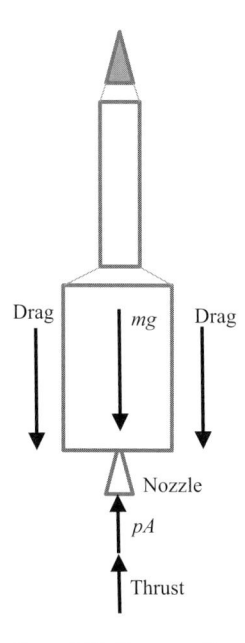

Figure 2.15 Forces acting on a rocket.

2.16 Applications of the Rocket Equation

2.16.1 Time to Burnout, Velocity, and Altitude in the Boost Phase

Combining the total thrust and weight, the rocket equation is expressed as,

$$m\dot{v} = T' - D. \tag{2.460}$$

Expressing the drag as,

$$D = kv^2, \quad T' = kv_T^2, \quad m\dot{v} = k\left(v_T^2 - v^2\right). \tag{2.461}$$

Thus, the time to burnout is,

$$t = \frac{m}{k} \int_0^{v_t} \frac{dv}{\left(v_T^2 - v^2\right)} = \frac{m}{2kv_T} \ln \frac{v_T + v_t}{v_T - v_t}. \tag{2.462}$$

The terminal velocity at burnout follows as,

$$v_t = \frac{1 - \exp\left(-\dfrac{2kv_T t}{m}\right)}{1 + \exp\left(-\dfrac{2kv_T t}{m}\right)} v_T. \tag{2.463}$$

Similarly, the height at burnout is,

$$h_B = \frac{m}{k} \int_0^{v_t} \frac{v\, dv}{(v_T^2 - v^2)} = \frac{m}{2k} \ln \frac{v_T^2}{v_T^2 - v_t^2}. \qquad (2.464)$$

2.16.2 Time and Altitude in the Coast Phase

In the coast phase, $T = 0$ and it can be shown that,

$$h_C = \frac{m_C}{2k} \ln \frac{(m_C g/k) + v_t^2}{m_C g/k}. \qquad (2.465)$$

The time of coast from burnout to when the velocity is zero (apogee) is,

$$t_C = \frac{\sqrt{m_C g/k}}{g} \int_{v_t}^{0} \frac{dv}{((m_C g/k) + v^2)} = \frac{m_C}{kg} \arctan\left(\frac{v_t}{\sqrt{m_C g/k}}\right). \qquad (2.466)$$

2.16.3 Delta-Vee Solution

When $F_{ext} \equiv 0$, a solution for the equation of motion may be derived. In this case, the equation of motion may be expressed as,

$$m \frac{dv}{dt} = -\frac{dm}{dt} c_e \quad \text{or as,} \quad dv = -\frac{dm}{m} c_e. \qquad (2.467)$$

Assuming that the relative exhaust velocity c_e is constant, which is true for a number of high thrust producing chemical rocket engines, and integrating both sides,

$$\Delta v = v - v_0 = c_e \ln\left(\frac{m_0}{m}\right). \qquad (2.468)$$

The quantity Δv, given the initial rocket velocity v_0, is known as the *delta-vee* or the characteristic velocity increment associated with a particular manoeuver characterized by the three parameters, the nozzle exit velocity c_e, the initial mass m_0, and the final mass m.

2.16.4 Mass-Ratio Decay

The solution for the mass of the propellant consumed during the delta-vee maneuver is,

$$\frac{\Delta m}{m_0} = \frac{m - m_0}{m_0} = 1 - \exp\left(-\frac{\Delta v}{c_e}\right). \qquad (2.469)$$

2.16.5 Gravity Loss

In the preceding analysis it was assumed that $F_{ext} \equiv 0$. When F_{ext} is directly proportional to the mass of the rocket, and can be expressed as $F_{ext} = -mg_e$, where g_e is a constant and is the equivalent acceleration due to gravity, the delta-vee solution must be corrected and expressed as,

$$\Delta v = v - v_0 = c_e \ln\left(\frac{m_0}{m}\right) - g_e t. \tag{2.470}$$

The reduction due to the last term is the gravity loss and represents the decrease in the velocity change in the presence of gravity.

2.16.6 Specific Impulse

The total mechanical impulse imparted to the rocket per unit weight is given by the ratio of the impulse due to the flow of exhaust gasses across the nozzle exit and the weight of an equivalent mass of fuel and is,

$$I_{mech} = \frac{c_e \dot{m} \Delta t}{g \dot{m} \Delta t} = \frac{c_e dm}{g dm} = \frac{c_e}{g}. \tag{2.471}$$

The specific impulse is defined as the maximum possible value of the mechanical impulse per unit weight of the fuel and is,

$$I_{sp} = \frac{(c_e)_{max}}{g} = \frac{c_{e-max}}{g}. \tag{2.472}$$

The seconds are the units of the specific impulse, and it is a characteristic of the fuel employed in the rocket to generate the thrust. It represents the maximum possible thrust that the rocket engine can deliver with that particular fuel. For the *Saturn V*'s main engine, it is about 265 s and about 450 s for the space shuttle's main engine. For a modern high-thrust chemical rocket, it is about 500 s, and 850–2,000 s or more for a thermonuclear rocket. For a typical V-2 type WWII rocket engine, it is about 200 s. A reasonable assumption for most current chemical rocket launchers is about 300 s. Thus, a rocket engine using fuel with a high specific impulse needs less fuel than one that uses fuel with a lower specific impulse.

2.17 Effects of Mass Expulsion

Mass expulsion has a number of effects on orbit and attitude dynamics. While the effects on orbit dynamics are directly related to the reduction in mass, the effects on attitude dynamics are due to the changes in the moments of inertia. Furthermore, the effect of mass expulsion on a moving and rotating object is to introduce Coriolis–type forces and gyroscopic torques that can seriously affect the motion of the vehicle.

2.17.1 Staging and Payloads

We observe that the initial mass of the rocket m_0 is the sum of three contributions: the mass of the payload m_p, the expended mass of the propellant and oxidizer or fuel m_e, and the mass of the remaining structure m_s. Thus, ignoring the gravity loss, the maximum velocity increment over the flight of a single-stage rocket, assuming that the payload is not jettisoned during the flight, is given by:

$$\Delta v_m = v_f - v_0 = c_e \ln\left(\frac{m_p + m_s + m_e}{m_p + m_s}\right) = c_e \ln\left(1 + \frac{m_e}{m_p + m_s}\right). \tag{2.473}$$

It is seen that this expression can be made as large as possible by reducing the mass of the structure m_s. It is also true that Δv_m can be made as large as possible by jettisoning much of the mass that contributes to the mass of the structure during the flight. This is the principle of *multi-staging*. Typically, the rocket's flight progresses in two or three stages. As an example, consider a three-stage rocket. The mass of the payload for each stage is the total mass of the remaining stages. Thus, the total Δv_m for a three-stage rocket can be expressed in terms of the fuel and structural masses of each stage and the payload mass as,

$$\Delta v_m = c_e \ln(s),$$

$$s = \left(1 + \frac{m_{e1}}{m_p + m_{e2} + m_{e3} + m_s}\right)\left(1 + \frac{m_{e2}}{m_p + m_{e3} + m_{s3} + m_{s2}}\right)\left(1 + \frac{m_{e3}}{m_p + m_{s3}}\right), \tag{2.474}$$

where m_s is the total mass of the structure in all of the stages. It can be shown after some analysis that this is larger than the corresponding expression for a single-stage rocket. In fact, it is possible to determine the optimal structural and fuel masses for each stage in order to maximize Δv_m.

2.18 Electric Propulsion

In the case of electric propulsion, one may assume that the power plant power supplied is given by $P_s = m_s h_s$, where h_s is the specific power of the power plant. Here we have made the assumption that the entire structural mass is equal to the propulsion system mass. The power P_s is related to the actual power converted into useful power by the jet as $P_T = \eta P_s$, where η is the overall mechanical power conversion efficiency. Assuming a uniform burn rate, it is related to the propellant mass m_e, the relative exhaust velocity c_e, and the burn time t_b by the relation,

$$P_T = \frac{1}{2}\left(\frac{m_e}{t_b}\right) c_e^2. \tag{2.475}$$

Hence, it follows that,

$$m_s = m_e \frac{c_e^2}{2\eta t_b h_s} = m_e \frac{c_e^2}{v_c^2}, \quad v_c^2 = 2\eta t_b h_s. \tag{2.476}$$

Thus,

$$m_0 = m_s + m_e + m_p = m_e \left(\frac{v_c^2 + c_e^2}{v_c^2}\right) + m_p, \qquad (2.477)$$

and,

$$m_0 - m_p = m_e \left(\frac{v_c^2 + c_e^2}{v_c^2}\right). \qquad (2.478)$$

Hence, we may write,

$$m_e = (m_0 - m_p)\left(\frac{v_c^2}{v_c^2 + c_e^2}\right), \quad m_s = (m_0 - m_p)\left(\frac{c_e^2}{v_c^2 + c_e^2}\right), \quad \frac{m_s}{m_0} = \left(1 - \frac{m_p}{m_0}\right)\left(\frac{c_e^2}{v_c^2 + c_e^2}\right),$$

$$\frac{m_s}{m_0} + \frac{m_p}{m_0} = \left(\frac{c_e^2}{v_c^2 + c_e^2}\right) + \left(\frac{m_p}{m_0}\right)\left(1 - \frac{c_e^2}{v_c^2 + c_e^2}\right). \qquad (2.479)$$

Substituting in the rocket equation, and eliminating m_s, one obtains an estimate of the Δv_m that could be obtained from electric propulsion for a given specific power output h_s, burn time t_b, and overall mechanical power conversion efficiency η, payload to overall mass ratio, and relative exhaust velocity c_e as,

$$\Delta v_m = c_e \ln\left(\frac{m_0}{m_p + m_s}\right) = c_e \ln\left(\frac{1 + \frac{v_c^2}{c_e^2}}{1 + \frac{m_p}{m_0}\frac{v_c^2}{c_e^2}}\right). \qquad (2.480)$$

An alternative form of this equation is,

$$\Delta v_m = c_e \ln\left(\frac{m_0}{m_p + m_s}\right) = c_e \ln\left(\frac{1 + \frac{c_e^2}{v_c^2}}{\frac{m_p}{m_0} + \frac{c_e^2}{v_c^2}}\right). \qquad (2.481)$$

Thus,

$$1 + \frac{c_e^2}{v_c^2} = \exp\left(\frac{\Delta v_m}{c_e}\right)\left(\frac{m_p}{m_0} + \frac{c_e^2}{v_c^2}\right). \qquad (2.482)$$

Rearranging, and solving for the payload mass ratio is expressed as,

$$\frac{m_p}{m_0} = 1 - \left(1 - \exp\left(-\frac{\Delta v_m}{c_e}\right)\right) - \frac{c_e^2}{v_c^2}\left(1 - \exp\left(-\frac{\Delta v_m}{c_e}\right)\right). \qquad (2.483)$$

This expression assumes that the initial mass of the system is made up of the structural mass (m_s), the mass of the expended propellant (m_e), and the mass of the payload (m_p), and that the entire structural mass is equal to the propulsion system mass. The structural mass m_s may be considered to include the mass of the fuel tanks (m_t) as well as the propulsion system mass (m_w). To modify this expression for the payload mass ratio $\mu_p = m_p/m_0$, define the specific mass ratio of the power system mass $\alpha = 1/h_s = m_w/P_{input}$ as the ratio of the propulsion system mass to the power input to the thrusters, and $t_b = m_e/\dot{m}$, which is the thrusting time. But, $T/P_{input} = 2\eta/c_e$ and $T = \dot{m}c_e$. Thus,

$$m_w = \alpha c_e T/2\eta = \alpha c_e^2 m_e/2\eta t_b. \tag{2.484}$$

Note that,

$$P_T = P_{out} = \frac{1}{2}\dot{m}c_e^2 = \frac{1}{2}\left(\frac{m_e}{t_b}\right)c_e^2. \tag{2.485}$$

Recall that $m_0 = m_s + m_e + m_p$. Hence, one has,

$$\frac{m_p}{m_0} = 1 - \frac{m_s}{m_0} - \frac{m_e}{m_0}. \tag{2.486}$$

Since the structural mass is the sum of the propulsion system mass and the mass of the tanks, $m_s = m_w + m_t$, it follows that,

$$\frac{m_p}{m_0} = 1 - \frac{m_t}{m_0} - \frac{m_w}{m_0} - \frac{m_e}{m_0}. \tag{2.487}$$

Thus,

$$\mu_p = 1 - \mu_t - \mu_w - \mu_e. \tag{2.488}$$

Assuming that $\mu_t = K_t \mu_e$, $\mu_w = \alpha c_e^2 \mu_e/2\eta t_b$, $\mu_e = 1 - \exp(-\Delta v_m/c_e)$, $v_c^2 = 2\eta t_b/\alpha$, it follows that,

$$\mu_p = \frac{m_p}{m_0} = \exp\left(-\frac{\Delta v_m}{c_e}\right) - \left(\frac{K_t}{\alpha v_c^2} + \frac{c_e^2}{v_c^2}\right)\left(1 - \exp\left(-\frac{\Delta v_m}{c_e}\right)\right). \tag{2.489}$$

Consequently, Equation (2.481) is modified as,

$$\frac{\Delta v_m}{v_c} = \frac{c_e}{v_c} \ln\left\{\frac{1 + \dfrac{K_t}{\alpha v_c^2} + \dfrac{c_e^2}{v_c^2}}{\dfrac{m_p}{m_0} + \dfrac{K_t}{\alpha v_c^2} + \dfrac{c_e^2}{v_c^2}}\right\}. \tag{2.490}$$

The modification has the effect of effectively increasing the exhaust velocity c_e or the specific impulse.

2.18.1 Application to Mission Design

In designing for a specific mission, the first objective is to design a set of feasible orbit transfer and propulsion system combinations for a given mission requirement. The steps involved are:

(i) Compute the maximum Δv available.

(ii) Compute the Δv required for all the orbit transfers. This may involve a number of orbit transfer strategies consisting of Hohmann transfers, elliptical and hyperbolic transfers, Hohmann transfer segments, and spiral orbit transfers. Typically, one could use Equations (2.148) and (2.150) to estimate the required Δv for each transfer. For electrically propelled spacecraft, expressions for the thrust and the relative velocity of the exhaust may be obtained from Goebel and Katz [26]. For low-thrust electric propulsion, the required Δv may be approximated, as suggested by Oleson [27], who provides an analytical method for predicting an electric propulsion system's performance, optimal specific impulse, optimal payload ratios, and the corresponding vehicle parameters. This method uses an analytical expression given by Edelbaum [19], which can be used to determine the performance of electric propulsion orbit transfer. Oleson [27] also gives analytical expressions for the efficiency (η) of ion, Magneto-plasma dynamic, and arcjet propulsion thrusters. Edelbaum [19] gives an analytical expression that approximates the Δv required by a low-thrust vehicle performing an orbital transfer and a plane change near the Earth suitable for electric propulsion applications. This expression is,

$$\Delta v^2 = v_0^2 + v_f^2 - 2v_0 v_f \cos\left(\frac{\pi}{2}\Delta i\right), \quad (2.491)$$

where $v_0 = \sqrt{\mu/r_0}$ is the circular velocity of the original orbit, $v_f = \sqrt{\mu/r_f}$ is the circular velocity of the final orbit, and Δi is the desired change in the inclination. From the rocket equation it also follows that,

$$m_i = m_f \exp\left(\frac{\Delta v}{c_e}\right), \quad (2.492)$$

(iii) Compute the propulsion system mass for chemical (liquid or solid) and electric propulsion systems. For an electric propulsion system, the above simple analysis may be used. A comparable analysis for chemical propulsion is much more complex and tedious. The mass of fuel is then estimated as,

$$m_{fuel} = m_i - m_f = m_f\left(\exp\left(\frac{\Delta v}{c_e}\right) - 1\right). \quad (2.493)$$

One may then compute the total mass of the system, the transfer time, and the thrust required, and would also have a candidate propulsion system for the mission.

References

[1] Deutsch, R. (1963) *Orbital Dynamics of Space Vehicles*, Prentice Hall: Engelwood Cliffs, NJ.

[2] Ball, K. J. and Osborne, G. F. (1967) *Space Vehicle Dynamics*, Oxford: Clarendon Press.

[3] Prussing, J. E. and Conway, B. A. (2012) *Orbital Mechanics*, 2nd edn, New York: Oxford University Press.
[4] Bate, R. (2015) *Fundamentals of Astrodynamics*, 2nd edn, New York: Dover Publications.
[5] Kamel, O. M. and Soliman, A. S. (2011) On the generalized Hohmann transfer with plane change using energy concepts, part I. *Mechanics and Mechanical Engineering*, 15(2): 183–191.
[6] Hu, W. (2015) *Fundamental Spacecraft Dynamics and Control*, New York: Wiley.
[7] Fortescue, P. and Stark, J. (1991) *Spacecraft Systems Engineering*, chapter 4, Chichester: John Wiley & Sons.
[8] Ashley, H. (1974) *Engineering Analysis of Flight Vehicles*, chapter 11, Reading, MA: Addison-Wesley Publishing Co.
[9] Vinh, N. X. (1981) *Optimal Trajectories in Atmospheric Flight*, New York: Elsevier Scientific Publishing Co.
[10] Tschauner, J. and Hempel, P. (1965) Rendezvous with a target describing an elliptical trajectory (In German). *Astronautica Acta*, 11(2): 104–109.
[11] Vepa, R. (2018) Application of the nonlinear Tschauner-Hempel equations to satellite relative position estimation and control. *Journal of Navigation*, 71(1): 44–64.
[12] Kustaanheimo, P. and Stiefel, E. L. (1965) Perturbation theory of Kepler motion based on spinor regularization. *Journal fur die Reine und Angewandte Mathematik*, 218: 204–219.
[13] Stiefel, E. L. and Scheifele, G. (1971) *Linear and Regular Celestial Mechanics*, New York: Springer.
[14] Li, H. (2014) *Geostationary Satellites Collocation*, Berlin/Heidelberg: Springer.
[15] Gazzino, C. (2017) Dynamics of a geostationary satellite, [Research Report] Rapport LAAS n° 17432, LAAS-CNRS, hal-01644934v2.
[16] Bryson, A. E. and Ho, Y. C. (1975) *Applied Optimal Control*, Washington, DC: Hemisphere Publishing Co.
[17] Conway, B. A. (Ed.) (2010) *Spacecraft Trajectory Optimisation*, New York: Cambridge University Press.
[18] Lawden, D. F. (1963) *Optimal Trajectories for Space Navigation*, London: Butterworths.
[19] Edelbaum, T. N. (1961) Propulsion requirements for controllable satellites. *ARS Journal*, 31: 1079–1089.
[20] Cerf, M. (2018) Low-thrust transfer between circular orbits using natural precession, AIRBUS Defence & Space, 7 Les Mureaux, France, https://arxiv.org/pdf/1503.00644, Retrieved July, 2018.
[21] Fehse, W. (2003) *Automated Rendezvous and Docking of Spacecraft*, Cambridge Aerospace Series, New York: Cambridge University Press.
[22] Eyer, J. K. (2009) A Dynamics and Control Algorithm for Low Earth Orbit Precision Formation Flying Satellites, Doctor of Philosophy Thesis, Department of Aerospace Science and Engineering, University of Toronto, Toronto.
[23] Wetterer, C. J., Linares, R., Crassidis, J. L., Kelecy, T. M., Ziebart, M. K., Jah, M. K., and Cefola, P. J. (2014) Refining space object radiation pressure modeling with bidirectional reflectance distribution functions. *Journal of Guidance, Control, and Dynamics*, 37(1): 185–196.
[24] Kubo-oka, T. and Sengoku, A. (1999) Solar radiation pressure model for the relay satellite of SELENE. *Earth Planets Space*, 51: 979–986.

[25] Scheeres, D. J., Williams, B. G., and Miller, J. K. (2000) Evaluation of the dynamic environment of an asteroid applications to 433 Eros. *Journal of Guidance, Control, and Dynamics*, 23(3): 466–475.

[26] Goebel, D. M. and Katz, I. (2008) *Fundamentals of Electric Propulsion: Ion and Hall Thrusters*, Pasadena, CA: Jet Propulsion Laboratory (JPL), Volume 1, pp. 6–7, 22, 26, 148, 189, 192, 243–244, 325, 327.

[27] Oleson, S. R. (1993) An analytical optimization of electric propulsion orbit transfers, NASA Contractor Report, NASA CR 191129, NASA Lewis Research Center.

3 Space Vehicle Attitude Dynamics and Control

3.1 Fundamentals of Satellite Attitude Dynamics

The determination and control of the orientation of satellites is a problem of fundamental importance. In particular, it is always essential that the satellite's orientation is stable with respect to either an orbiting reference frame or a reference frame fixed in space. Thus, it is necessary to study the stability of the attitude dynamics of a satellite, which primarily depends on its dynamics. This chapter will focus on the dynamics of the attitude of a satellite and its control. The problem of the determination of a satellite's attitude from relevant measurements will be discussed in Chapter 8, as it involves complex and nonlinear methods of state estimation from noise measurements made from observations of the satellite's motion. First, an introduction to satellite attitude dynamics is presented. For a more advanced analysis of satellite attitude dynamics the reader is referred to the more advanced texts on the subject by Kane, Likins, and Levinson [1] and Rimrott [2].

3.2 Rigid Body Kinematics and Kinetics

3.2.1 Coordinate Frame Definitions and Transformations

The first step to developing the dynamics of an orbiting space vehicle or spacecraft such as a satellite is to define a set of coordinates and coordinate frames in which the model will be developed. Transformations relating these sets of coordinates to each other must also be defined at the outset. A minimum of three coordinate frames is essential to model the dynamics of the space vehicle: an inertial frame, a local rotating or orbiting frame, and a body-fixed (also rotating) frame.

3.2.2 Definition of Frames/Rotations

In the case of a spacecraft that is pointing toward the center of the Earth (nadir-pointing spacecraft), it is traditional to define the vehicle's orientation relative to the local-level coordinate system, which follows the spacecraft around its orbit. The local-level system's $+z$ axis points toward the nadir, and its y axis is perpendicular to both the nadir vector and the instantaneous orbital velocity vector as measured with respect to an

Earth-centered, inertially fixed reference frame. The $+y$ axis points toward negative orbit normal. The x axis, defined by the right-hand rule, points approximately along the velocity vector. The other important reference frame is spacecraft-fixed. When the nadir-pointing spacecraft has the desired attitude, this body-fixed reference frame is aligned with the local-level reference frame.

3.2.3 The Inertial (*i*) Frame *X–Y–Z*

The first frame defined is an inertial frame, labeled *X–Y–Z* and hereafter called simply the inertial frame. The origin is located at the center of the Earth with the *X* axis defined to point in the direction of the Vernal Equinox, *Z* pointing north, and *Y* completing the right-handed frame. This frame is used primarily to calculate the latitude and longitude of the spacecraft's center of mass as it moves along its orbit. The process by which the latitude and longitude are calculated is described later.

3.2.4 The Local Rotating (*r*) or Orbiting Frame *x–y–z*

Next, a rotating frame, labeled *x–y–z* and hereafter referred to as simply the rotating frame, is defined with its origin centered on the spacecraft's center of mass. The *x* axis points east in the plane of the orbit, *y* points south perpendicular to the orbital plane, and *z* points toward the center of the Earth (i.e. it is nadir-pointing). Although this frame is defined with its origin at the center of mass and rotates around the orbital plane with the spacecraft, it is not fixed in the body of the satellite. Therefore, the *z* axis will always be pointing toward the center of the Earth. The inertial and rotating frames are graphically depicted in Figure 3.1.

3.2.5 The Body (*b*) Frame *b1–b2–b3*

A third frame, labeled b_1–b_2–b_3 and hereafter called the body-fixed frame or simply a body frame, is also needed. The origin of this frame is centered on the spacecraft's center of mass and is defined such that b_1 points along the principal or geometric axis of the orbiting vehicle, while b_2 and b_3 point along suitable chosen directions mutually perpendicular to each other as well as to the first axis.

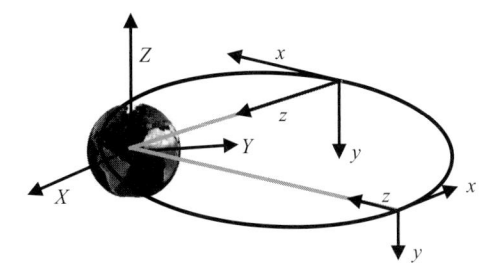

Figure 3.1 Definition of the inertial and orbital frames.

3.2.6 Defining the Body Frame

One of the primary goals of the simulation is to generate a time history of the orientation of the body frame with respect to the rotating frame. The body frame is defined such that it is fixed in the body of the spacecraft and thus will be used to determine the orientation of the spacecraft with respect to the rotating frame. Since the z axis is always nadir-pointing, the b_3 axis is offset from the nadir and it can easily be determined. In order to relate the rotating frame to the body frame, an Euler angle sequence of rotations such as a 1–2–3 body Euler rotation sequence is performed. Any other sequence may also be chosen, and the choice of the sequence depends to a large extent on the nature of the attitude control problem. Euler equations and Euler angles constitute a classical method for attitude description of a rigid body in space, with respect to an inertial coordinate system. They permit the formal parameterization of a rotation matrix that may be employed to transform a set of vector quantities in the body frame to the local-level base frame and vice-versa.

3.2.7 Three- and Four-Parameter Attitude Representations

Representation of attitude is not as straightforward as position representation. There are three rotational degrees of freedom, hence the attitude can be represented by three parameters. However, a three-parameter representation is singular for some attitude. To avoid singularity, more parameters are needed, but then there is redundancy in the representation, and it must be subjected to constraints.

Initially, the body frame can be assumed to be aligned with the rotating frame, that is, b_1 aligned along the x-axis, b_2 along the y-axis, and b_3 along the z-axis. First, the body frame is rolled ϕ degrees about the x axis and the resulting frame is labeled as the $x'y'z'$ frame. Next, the resulting intermediate frame is pitched θ degrees about the y' axis and finally yawed ψ degrees about the z'' axis. These three rotations, ϕ, θ, and ψ, are a set of three-parameter representations of the body attitude relative to the rotating frame.

Combining the results of these three rotations leads to a direction cosine matrix allowing the transformation of any vector from the rotating frame to the body frame. The matrix becomes,

$$\begin{bmatrix} b_1 \\ b_2 \\ b_3 \end{bmatrix} = \mathbf{T}_{BR} \times \begin{bmatrix} x \\ y \\ z \end{bmatrix}, \qquad (3.1)$$

where the transformation $\mathbf{T}_{BR}$ for the 1–2–3 orbiting to body frame Euler rotation sequence is,

$$\mathbf{T}_{BR} = \begin{bmatrix} \cos\psi & \sin\psi & 0 \\ -\sin\psi & \cos\psi & 0 \\ 0 & 0 & 1 \end{bmatrix} \begin{bmatrix} \cos\theta & 0 & -\sin\theta \\ 0 & 1 & 0 \\ \sin\theta & 0 & \cos\theta \end{bmatrix} \begin{bmatrix} 1 & 0 & 0 \\ 0 & \cos\phi & \sin\phi \\ 0 & -\sin\phi & \cos\phi \end{bmatrix}. \qquad (3.2)$$

Hence,

$$\begin{bmatrix} b_1 \\ b_2 \\ b_3 \end{bmatrix} = \begin{bmatrix} c\theta\, c\psi & s\phi\, s\theta\, c\psi + c\phi\, s\psi & -c\phi\, s\theta\, c\psi + s\phi\, s\psi \\ -c\theta\, s\psi & -s\phi\, s\theta\, s\psi + c\phi\, c\psi & c\phi\, s\theta\, s\psi + s\phi\, c\psi \\ s\theta & -s\phi\, c\theta & c\phi\, c\theta \end{bmatrix} \begin{bmatrix} x \\ y \\ z \end{bmatrix}, \quad (3.3)$$

where $c\theta = \cos\theta$, $c\phi = \cos\phi$, $c\psi = \cos\psi$, $s\theta = \sin\theta$, $s\phi = \sin\phi$, and $s\psi = \sin\psi$.

When the angles are assumed to be small,

$$\mathbf{T}_{BR} = \begin{bmatrix} 1 & \psi & -\theta \\ -\psi & 1 & \phi \\ \theta & -\phi & 1 \end{bmatrix}, \quad \begin{bmatrix} b_1 \\ b_2 \\ b_3 \end{bmatrix} = \begin{bmatrix} 1 & \psi & -\theta \\ -\psi & 1 & \phi \\ \theta & -\phi & 1 \end{bmatrix} \begin{bmatrix} x \\ y \\ z \end{bmatrix}. \quad (3.4)$$

Another alternative is to employ four-parameter attitude representations. Deviations of the reference frame's attitude from that of the local-level base frame are parameterized by the attitude quaternion **q**. The orthonormal transformation matrix from local-level coordinates to the spacecraft coordinates is a function of **q**.

3.3 Spacecraft Attitude Dynamics

The vector form of the equations of motion of a rigid body in rotation is given by,

$$\frac{d\mathbf{h}}{dt} = \mathbf{M}. \quad (3.5)$$

Equation (3.5) refers to a fixed frame of reference. Equation (3.5) is a mathematical statement of an important dynamical law when **h** is identified as the moment of momentum of a rigid body and **M** is the moment of the externally applied external forces,

The time rate of change of moment of momentum of a rigid body rotating about any axis is equal to the moment of the applied external forces about the same axis.

Since the rate of change of the moment of momentum may be regarded as the velocity of the end of the vector **h**, it follows that,

The velocity of the end of the moment of momentum vector is equal to the moment of the external forces.

The angular velocity of a rigid body B with reference to a fixed frame F in terms of a right-handed set of mutually perpendicular unit vectors, $\mathbf{n}_1$, $\mathbf{n}_2$, and $\mathbf{n}_3$ fixed in B along the three principal axes is given by,

$$^F\boldsymbol{\omega}_B = p_B \mathbf{n}_1 + q_B \mathbf{n}_2 + r_B \mathbf{n}_3. \quad (3.6)$$

If a vector $\mathbf{h} = {}^F\mathbf{h}_B$ is resolved along the principal axes of the body, then,

$$^F\mathbf{h}_B = I_1 p_B \mathbf{n}_1 + I_2 q_B \mathbf{n}_2 + I_3 r_B \mathbf{n}_3, \quad (3.7)$$

where I_1, I_2, and I_3 are the principal moments of inertia and are constants, the time derivative of $^F\mathbf{h}_B$ in F can be employed to express the equations of motion of a rigid body in rotation as,

$$\frac{d^F\mathbf{h}_B}{dt} + \begin{bmatrix} 0 & -r_B & q_B \\ r_B & 0 & -p_B \\ -q_B & p_B & 0 \end{bmatrix} {}^F\mathbf{h}_B = \mathbf{M}_B. \tag{3.8}$$

The moment of the externally applied torques is defined by the vector,

$$\mathbf{M}_B = M_1\mathbf{n}_1 + M_2\mathbf{n}_2 + M_3\mathbf{n}_3. \tag{3.9}$$

The scalar equations of motion are,

$$I_1\dot{p}_B + (I_3 - I_2)q_B r_B = M_1, \tag{3.10}$$

$$I_2\dot{q}_B + (I_1 - I_3)r_B p_B = M_2, \tag{3.11}$$

$$I_3\dot{r}_B + (I_2 - I_1)p_B q_B = M_3. \tag{3.12}$$

Considering the attitudinal motion of a space vehicle, the angular velocity of the body $^F\boldsymbol{\omega}_B$ may be expressed in terms of the angular velocity vector of the orbital frame R and the angular velocity vector of the body relative to the orbital frame is,

$$^F\boldsymbol{\omega}_B = {}^F\boldsymbol{\omega}_R + {}^R\boldsymbol{\omega}_B. \tag{3.13}$$

It is assumed that, when resolved in the body axes,

$$^R\boldsymbol{\omega}_B = \omega_1\mathbf{n}_1 + \omega_2\mathbf{n}_2 + \omega_3\mathbf{n}_3. \tag{3.14}$$

It is also assumed that the angular velocity of the orbital frame orbiting with the spacecraft, when resolved in the orbital frame, is given by,

$$^F\boldsymbol{\omega}_B = \omega_x\mathbf{i} + \omega_y\mathbf{j} + \omega_z\mathbf{k} = -n\mathbf{j}. \tag{3.15}$$

The body angular velocity relative to the orbital frame in terms of the Euler angle rates is,

$$^R\boldsymbol{\omega}_B = \dot{\phi}\,\mathbf{i} + \dot{\theta}\,\mathbf{j}'' + \dot{\psi}\,\mathbf{n}_3, \tag{3.16}$$

where $\mathbf{j}''$ is a unit vector in the y'' direction. Resolving the body angular velocity relative to the orbital frame, in terms of the body axes, it follows that,

$$\omega_1 = \dot{\theta}\sin\psi + \dot{\phi}\cos\theta\cos\psi, \quad \omega_2 = \dot{\theta}\cos\psi - \dot{\phi}\cos\theta\sin\psi,$$
$$\omega_3 = \dot{\phi}\sin\theta + \dot{\psi}. \tag{3.17}$$

In terms of the body axes, the angular velocity of the orbital frame is,

$$^F\boldsymbol{\omega}_R = -nT_{BR}[0\ 1\ 0]^T. \tag{3.18}$$

Hence it follows that,

$$\begin{bmatrix} p_B \\ q_B \\ r_B \end{bmatrix} = \begin{bmatrix} \dot{\theta} \sin \psi + \dot{\phi} \cos \theta \cos \psi \\ \dot{\theta} \cos \psi - \dot{\phi} \cos \theta \sin \psi \\ \dot{\phi} \sin \theta + \dot{\psi} \end{bmatrix} - nT_{BR} \begin{bmatrix} 0 \\ 1 \\ 0 \end{bmatrix}. \qquad (3.19)$$

These quantities may be substituted into the scalar equations of motion to obtain the governing equations for large motion attitude dynamics.

3.4 Environmental Disturbances

The total torque t acting on the satellite body is made up from several sources. This includes a number of disturbances from various sources. The major disturbance effects are,

Gravity Gradient: This results in a "Tidal" force-like disturbance due to $1/r^2$ gravitational field variation for long, extended bodies (e.g. large satellites, space shuttle, tethered vehicles) and can destabilize a spacecraft.

Aerodynamic Drag: Aerodynamic drag results in a "Weathervane" effect due to an offset between the center of mass (CM) and the center of the aerodynamic pressure (CP). It is only a factor in low earth orbits when there is a significant atmosphere.

Magnetic Torques: These are induced by the residual magnetic moment within the Earth's magnetosphere. The spacecraft acts effectively as a magnetic dipole.

Solar Radiation: Torques are induced by the offset of the solar CP from the CM. They can be compensated by reaction wheels or differential reflectivity.

Mass Expulsion: Torques are induced by leaks or jettisoned objects or combusted fuel.

Internal: The satellite itself will generate different torques due to the presence of onboard rotating equipment (machinery, wheels, cryocoolers, pumps, etc.). The deployment of both antennae and boom will produce great torques, but they will be short lived. Although there is no net effect, there is internal momentum exchange, which affects the attitude.

The effect of the environmental torques is two-fold: (i) they dissipate energy and thus reduce the useful life of the satellite and (ii) they contribute to the coupling between the orbital and attitude dynamics.

3.4.1 Gravity Gradient Torques

The gravity field from the Earth makes a most important contribution. Although not the largest, it has a major influence on the satellite's stability. It acts on the mass distribution of the spacecraft, and the net result is a torque acting on it. Although small in magnitude, it can have a significant effect on the stability of the satellite, particularly in non-geostationary satellites. Assuming that the Earth is spherically symmetric (point mass), it can be shown that the gravitational effects acting on the spacecraft contribute to two forces acting on it; one at its mass center and the other away from the mass center. While the former force *does not* contribute to the torque on the spacecraft, the latter force does and is known as the gravity gradient torque.

A mass element *dm* of a satellite is attracted to the Earth with a mass m_e by force determined by Newton's law of gravitation, given by,

$$d\mathbf{F} = -\left(Gm_e dm/|\mathbf{r}|^3\right) \times \mathbf{r}. \tag{3.20}$$

Assuming that the body axes may be transformed by a sequence of rotations, first about the roll axes, then about the pitch and finally about the yaw axis, and aligned with the orbit axes, **r** may be expressed as,

$$\mathbf{r} = \mathbf{a} + \boldsymbol{\rho} = a(\cos\phi\mathbf{i} + \sin\phi\cos\theta\mathbf{j} - \sin\phi\sin\theta\mathbf{k}) + (x\mathbf{i} + y\mathbf{j} + z\mathbf{k}), \tag{3.21}$$

where $\boldsymbol{\rho}$ is the position vector of the mass element relative to the satellite's center of mass and **a** is the position vector of the satellite's center of mass relative to the center of the Earth. The moment of the force,

$$d\mathbf{M} = \boldsymbol{\rho} \times d\mathbf{F}, \tag{3.22}$$

may be expressed as,

$$\begin{aligned} d\mathbf{M} = &\left(Gm_e a dm/|\mathbf{r}|^3\right) \\ &\times [(y\sin\phi\sin\theta + z\sin\phi\cos\theta)\mathbf{i} - (x\sin\phi\sin\theta + z\cos\phi)\mathbf{j} \\ &- (x\sin\varphi\cos\theta - y\cos\phi)\mathbf{k}]. \end{aligned} \tag{3.23}$$

The quantity, $1/|\mathbf{r}|^3$ may be expressed as,

$$\frac{1}{|\mathbf{r}|^3} \approx \frac{1}{a^3}\left(1 - \frac{3}{a}(\mathbf{a}\cdot\boldsymbol{\rho})\right) = \frac{1}{a^3}\left(1 - \frac{3}{a}(x\cos\phi + y\sin\phi\cos\theta - z\sin\phi\sin\theta)\right). \tag{3.24}$$

Assuming that the body axes may be transformed by a sequence of rotations, first about the roll axes, then about the pitch and finally about the yaw axis, and aligned with the orbit axes, and that the Earth is spherical, the gravity force on the space vehicle results in torque components about the body axes equal to,

$$M_x = \frac{3Gm_e}{R_0^3}\frac{(I_{yy} - I_{zz})}{2}\sin 2\theta \sin^2\phi, \quad M_y = \frac{3Gm_e}{R_0^3}\frac{(I_{zz} - I_{xx})}{2}\sin 2\phi \sin\theta,$$
$$M_z = \frac{3Gm_e}{R_0^3}\frac{(I_{yy} - I_{xx})}{2}\sin 2\phi \cos\theta, \tag{3.25}$$

where R_0 is the distance of the Earth's center of mass to the space vehicle's center of mass. The body axes may be assumed to be the principal axes of the space vehicle. Thus, in general, gravity gradient torque on the spacecraft may be expressed as,

$$\mathbf{g}^B = 3\omega_0^2 \mathbf{c}_3 \times \mathbf{I}^B \cdot \mathbf{c}_3, \tag{3.26}$$

where c_3 is the direction cosine from the rotation matrix that transforms vectors from the orbital to the body frame, i.e. a unit vector in the **r** direction given by, $\mathbf{c}_3 = \mathbf{r}/|\mathbf{r}|$ and ω_0 is the angular velocity of the satellite in orbit.

3.4.2 Aerodynamic Disturbance Torques

For satellites in low earth orbits, the atmosphere is still present and the atmospheric drag will be nonzero. Aerodynamic forces and torques arise through the transfer of momentum between atmospheric particles and the spacecraft's exposed or "wetted" surfaces. The atmospheric drag is a force creating an acceleration a_D in the opposite direction of the satellite's velocity. The magnitude of this acceleration depends on the density of the atmosphere ρ, the cross section area S, and mass m of the satellite and of course the magnitude of the velocity V, and is given by,

$$a_D = (1/2)\rho V^2 C_D S/m, \qquad (3.27)$$

where C_D is the drag coefficient. The drag coefficient is further discussed in Wertz and Larson [3], and a suggested approximation is $C_D \approx 2.2$. The atmospheric drag is a breaking force and hence dissipates energy from the satellite in orbit. The net effect of the drag force, which is the result of a distributed pressure, is a torque acting on the spacecraft.

In the upper reaches of the atmosphere, the flow is considerably rarefied, and it is then possible to make some simplifying assumptions to model the flow. The molecular mean free path is the average distance traveled by a molecule before encountering another. The atmospheric density is small enough, and one may assume that the molecular mean free path is large compared with the dimensions of a typical spacecraft. The incoming flow of molecules colliding with the spacecraft's outer surfaces may treated as independent particles and separately from the flow of molecules moving away from those surfaces. This simplified model of the aerodynamic flow is referred to as *free molecular flow*. It may also be assumed that the speed of the spacecraft relative to the atmosphere is much larger than the mean thermal speed of the atmospheric molecules. Thus, the aerodynamic flow about the spacecraft is said to be hyperthermal.

In considering the momentum of the molecules after collision with the spacecraft one needs to also consider whether or not the collisions are elastic. In the former case, the molecules are said to be specularly reflected, while in the case of completely inelastic collisions, they are diffusely absorbed or accommodated to the surface, in the sense that the molecules leave the surface after a short time interval with a random velocity and in a random direction. As accommodation could take place differently in both the normal and tangential directions, the aerodynamic forces and moments are estimated in terms of two empirically determined accommodation coefficients. For an excellent discussion of the aerodynamic torques acting on a spacecraft, and in fact of all the environmental torques, the reader is referred to P. C. Hughes's classic textbook, Hughes [4].

3.4.3 Solar Wind and Radiation Pressure

The largest disturbance torque is the pressure from solar radiation. The sun radiates a vast amount of particles known as solar wind and electromagnetic radiation. Electromagnetic radiation exerts a pressure on the spacecraft in the same way as atmospheric molecules. The photons carrying the energy are known to have a certain momentum and this momentum may be transferred to the spacecraft's surface just as with incident molecules, resulting in a pressure distribution over the spacecraft's outer surfaces known as solar radiation pressure. Both solar wind and solar pressure will establish a torque on the satellite. While the Sun is the primary source of radiation, reflected and emitted radiation from the Earth are secondary sources of radiation that also contribute to the radiation-induced torques. Specifically, the solar radiation pressure force is computed theoretically as,

$$\mathbf{F}_s = g\mathbf{r}_g, \quad g = G_1 dS/m|\mathbf{R}|^2, \tag{3.28}$$

where $\mathbf{r}_g$ is the unit vector from the Sun to the spacecraft, g is the magnitude of the acceleration acting on the spacecraft, $G_1 = 1 \times 10^8$ kg km³/s²/m², m is the spacecraft mass, dS is the projected area of a surface element, and $\mathbf{R}$ is the distance from the Sun to a small elemental surface area on the spacecraft's surface. This force can be expressed in terms of a potential force acting on the spacecraft,

$$\mathbf{R}_s = \mathbf{F}_s \cdot \mathbf{r}, \tag{3.29}$$

where $\mathbf{r}$ is the position vector of the spacecraft's center of mass with respect to the small elemental area.

In practice, the perturbation acceleration caused by solar radiation pressure acting over a surface area dS has a magnitude of

$$da_R = -4.5 \times 10^{-6} \times (1 + \beta) dS/m, \tag{3.30}$$

where β is a reflection factor between 1 (total reflection) and -1 (total transmission) and is 0 for total reception (black body). The perturbations due to solar radiation are in the same magnitude as atmospheric drag perturbations for altitudes at 800 km, and less for lower orbits (Wertz and Larson [3]). The net effect of the solar radiation pressure force that results in a distributed pressure acting over the spacecraft is a torque acting on the spacecraft.

3.4.4 Thruster Misalignments

These are responsible for significant amounts of disturbance torques, as it often becomes necessary to align the direction of the thrusters dynamically. Consequently, there are always net misalignments that result in residual torques acting on the satellite.

3.4.5 Magnetic Disturbance Torques

The Earth's magnetic field interacts with a spacecraft's magnetic moment, resulting in a magnetic torque acting on the spacecraft. The strength of the magnetic field varies as the

3.4 Environmental Disturbances

inverse cube of the distance from the Earth's center, so its influence is more significant for spacecraft in near-Earth orbits than in high altitude orbits. Although the torque due to the Earth's magnetic field is considerably weaker than the gravitational and aerodynamic torques, it is particularly significant when electro-magnetic torque coils are used on board a spacecraft to deliberately control its attitude. The magnetic torque acting on the satellite is the cross product of the satellite's magnetic field and the Earth's magnetic field. The satellite's magnetic field consists of the field produced by the control torquers, electromagnets, and other magnetic disturbances in the satellite. All electric components on board might produce electromagnetic fields interacting with the Earth's magnetic field in the same way as the control torquers. The other magnetic disturbances are caused by permanent magnets in the spacecraft and are sometimes ignored when they are much smaller than the controlled magnetic field.

In order to model the interaction of the magnets within the spacecraft with the Earth's magnetic field, a reliable model for the Earth's magnetic field is needed. The Earth's magnetic field could be represented by a finite series of spherical harmonics and associated Legendre polynomials, weighted by a set of estimated coefficients. The standard field model most commonly used is the International Geomagnetic Reference Field (IGRF) model. The magnetic field varies strongly over the Earth's surface. On the surface of the earth, the field varies from being horizontal and of magnitude about 30,000 nT near the equator, to vertical and about 60,000 nT near the poles. The internal geomagnetic field also varies in time, on a timescale of months and longer, in an unpredictable manner. The International Geomagnetic Reference Field (IGRF) is an attempt by the International Association of Geomagnetism and Aeronomy (IAGA) to provide a model acceptable to a variety of users. It is meant to give a reasonable approximation, near and above the Earth's surface, to that part of the Earth's magnetic field that has its origin in the Earth's core. At any one time, the IGRF specifies the numerical coefficients of a truncated spherical harmonic series. At present, the truncation is at $n = 13$, so there are 195 coefficients. The IGRF model is specified every five years, for epochs 2000.0, 2005.0, ..., which is currently valid to 2020.0 (IGRF 2012). Sometimes, a first-degree truncation that constitutes a simpler series expansion known as the Dipole Model and models the Earth's magnetic field as a magnetic dipole is also used.

The IGRF model comprises a set of spherical harmonic coefficients called Gauss coefficients, g_n^m and h_n^m, in a truncated series expansion of the geomagnetic potential function of internal origin,

$$V = R_e \left[\sum_{n=1}^{N} \sum_{m=0}^{M} \left(\frac{R_e}{r}\right)^{n+1} P_n^m(\cos\theta)\left(g_n^m \cos(m\phi) + h_n^m \sin(m\phi)\right)\right], \qquad (3.31)$$

where R_e is the mean radius of the Earth (6,371.2 km), and r, ϕ, θ are the geocentric spherical coordinates. r is the distance from the center of the Earth, ϕ is the longitude eastward from Greenwich, and θ is the co-latitude equal to 90 degrees minus the latitude. The functions $P_n^m(\cos\theta)$ are Schmidt quasi-normalized associated Legendre functions of degree n and order m, where $n \geq 1$ and $m \leq n$. The maximum spherical harmonic degree of the expansion is N. As the magnetic field revolves with the Earth, the magnetic field B from an IGRF model is in the Earth-centered Earth frame and must be transformed to the rotating orbital frame.

This model includes the main or core field without external sources, such as the interaction of the field with the solar wind. The field models are valid for altitudes up to 30,000 km and for the years 1945–2020. Two sets of magnetic field data with different resolutions are usually obtained. One is a fine grid and the other is a coarse grid, both compiled for intervals of a 500 km change in altitude. Simulations can be run using the coarse grid for the purpose of debugging the code. Once the bugs are fixed, and one obtains trustworthy results, the grid definition can be increased to the fine grid for more accurate results.

For the coarse grid model, a magnetic dipole approximation of the Earth's magnetic field may be adopted. When coupled with the assumptions of no Earth rotation and no orbit precession, the magnetic dipole approximation yields the following periodic model for the magnetic field vector (see, for example, Wertz (ed.) [5]), given by,

$$\mathbf{B}_{md} = \frac{\mu_{md}}{|\mathbf{r}_{eci}|^3}\left(3\left(\frac{\mathbf{r}_{eci}}{|\mathbf{r}_{eci}|}\cdot \mathbf{m}_e\right)\frac{\mathbf{r}_{eci}}{|\mathbf{r}_{eci}|} - \mathbf{m}_e\right), \quad \mathbf{m}_e = \begin{bmatrix} \sin\theta_m \cos(\omega_e t + \alpha_0) \\ \sin\theta_m \sin(\omega_e t + \alpha_0) \\ \cos\theta_m \end{bmatrix}, \quad (3.32)$$

where μ_{md} is the total dipole strength, $\mathbf{r}_{eci}$ is the spacecraft's position vector in the Earth-centered inertial frame, and $\mathbf{m}_e$ is the vector of the direction cosines of the Earth's magnetic field direction expressed in the Earth-centered inertial frame. It is approximated in terms of three parameters, the magnetic dipole's co-elevation θ_m, the Earth's average rotation rate ω_e, and the right ascension of the dipole at time $t = 0$, α_0. For the year 2010, $\mu_{md} = 7.746 \times 10^{15}$ Wbm and for the year 2000, $\mu_{md} = 7.9 \times 10^{15}$ Wbm. For the year 2010, $\theta_m = 170.0°$. The magnitude of the total field at the pole is twice as strong as at the equator.

Time is measured from $t = 0$ at the ascending-node crossing of the magnetic equator. The field's dipole strength μ_{md} is completely specified by the coefficients g_1^0, g_1^1, and h_1^1. A simple expression for $\mathbf{B}_{md}$ in an Earth-fixed frame is,

$$\mathbf{B}_{md} = \begin{bmatrix} B_r \\ B_\lambda \\ B_\phi \end{bmatrix} = \frac{\mu_{md}}{|\mathbf{r}_{eci}|^3}\begin{bmatrix} -2\sin\lambda \\ \cos\lambda \\ 0 \end{bmatrix}, \quad (3.33)$$

where λ is the magnetic latitude ($\lambda = 0$ at the magnetic north pole and $\lambda = 90°$ at the magnetic equator).

In order to obtain the field vector for longitude and latitude values that lie within the grid points, bilinear interpolation is commonly employed. The core magnetic field of the Earth is illustrated in Figure 3.2.

The moment due to the Earth's magnetic field can be modeled as, $\boldsymbol{\tau} = \mathbf{m} \times \mathbf{B}$,

where $\boldsymbol{\tau}$ is the resulting torque applied to the spacecraft, $\mathbf{m}$ is the magnetic dipole moment of the spacecraft, and $\mathbf{B}$ is the local magnetic field vector of Earth's magnetic field. The total magnetic dipole moment of the magnets on the spacecraft has to be estimated first both in magnitude and direction.

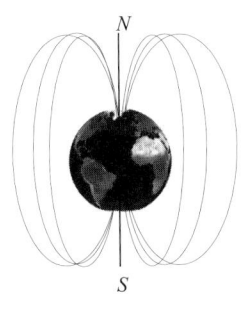

Figure 3.2 The Earth's core magnetic field.

3.4.6 Control Torques

Spacecraft rocket engines and thrusters are employed to provide both orbit and attitude control. Thrust generation is purely by mass expulsion, which is nonenvironmental, and thrusters may be employed both for attitude and orbit control. Thrusters may be employed independently to control both translational and rotational motion. A pair of thrusters fired "symmetrically" relative to the spacecraft's center of mass would result in pure translational motion, while thrusters fired anti-symmetrically result in pure rotational motion.

3.5 Numerical Simulation

MATLAB or SIMULINK may be chosen as the tool to numerically integrate the EOMs described in the previous section, as in the case of aircraft dynamics simulation. When an Euler angle sequence, such as the 1–2–3 sequence, is used to describe the spacecraft's attitude, the simulation encounters a singularity when the pitch offset (θ) of the body frame with respect to the rotating frame passes through 90 degrees. This is most likely to occur near the equator. Therefore, the simulation is limited to runs of half an orbit. However, for the purposes described previously, this is enough due to the symmetry in the Earth's magnetic field. Motion over a half, or even a quarter, of an orbit can be extrapolated across the entire orbit. Alternatively, continuation methods may be employed by explicitly deriving the equations of motion in the vicinity of $\theta = 90°$; i.e. as θ passes through θ.

A preferred alternative is to employ the four-parameter attitude representations discussed earlier in place of the Euler angle sequence to describe the spacecraft's attitude.

3.6 Spacecraft Stability

3.6.1 Linearized Attitude Dynamic Equation for Spacecraft in Low Earth Orbit

We now consider the linearized dynamic equation for small attitude disturbances of a Earth-pointing satellite orbiting the Earth in a circular orbit. The equations of motion are,

$$I_1\dot{p}_B + (I_3 - I_2)q_B r_B = M_1, \quad (3.34)$$

$$I_2\dot{q}_B + (I_1 - I_3)r_B p_B = M_2, \quad (3.35)$$

$$I_3\dot{r}_B + (I_2 - I_1)p_B q_B = M_3. \quad (3.36)$$

Using the (3–2–1) sequence from the inertial to body axes, so,

$$\mathbf{T}_{BR} = \begin{bmatrix} 1 & 0 & 0 \\ 0 & \cos\phi & \sin\phi \\ 0 & -\sin\phi & \cos\phi \end{bmatrix} \begin{bmatrix} \cos\theta & 0 & -\sin\theta \\ 0 & 1 & 0 \\ \sin\theta & 0 & \cos\theta \end{bmatrix} \begin{bmatrix} \cos\psi & \sin\psi & 0 \\ -\sin\psi & \cos\psi & 0 \\ 0 & 0 & 1 \end{bmatrix}, \quad (3.37)$$

the angular velocity components are,

$$\begin{bmatrix} p_B \\ q_B \\ r_B \end{bmatrix} = \begin{bmatrix} 1 & 0 & -\sin\theta \\ 0 & \cos\phi & \sin\phi\cos\theta \\ 0 & -\sin\phi & \cos\phi\cos\theta \end{bmatrix} \begin{bmatrix} \dot\phi \\ \dot\theta \\ \dot\psi \end{bmatrix} - n\mathbf{T}_{BR} \begin{bmatrix} 0 \\ 1 \\ 0 \end{bmatrix}, \quad (3.38)$$

which upon linearization, reduces to,

$$\begin{bmatrix} p_B \\ q_B \\ r_B \end{bmatrix} = \begin{bmatrix} \dot\phi \\ \dot\theta \\ \dot\psi \end{bmatrix} - n\begin{bmatrix} \psi \\ 1 \\ -\phi \end{bmatrix} \quad \text{and} \quad \begin{bmatrix} \dot p_B \\ \dot q_B \\ \dot r_B \end{bmatrix} = \begin{bmatrix} \ddot\phi \\ \ddot\theta \\ \ddot\psi \end{bmatrix} - n\begin{bmatrix} \dot\psi \\ 0 \\ -\dot\phi \end{bmatrix}. \quad (3.39)$$

Thus

$$\dot p_B = \ddot\phi - n\dot\psi, \quad \dot q_B = \ddot\theta, \quad \dot r_B = \ddot\psi + n\dot\phi, \quad p_B = \dot\phi - n\psi, \quad q_B = \dot\theta - n, \quad r_B = \dot\psi + n\phi, \quad (3.40)$$

and

$$q_B r_B = -n\dot\psi - n^2\phi, \quad r_B p_B \approx 0, \quad p_B q_B = -n\dot\phi + n^2\psi. \quad (3.41)$$

Hence,

$$I_1(\ddot\phi - n\dot\psi) - n(I_3 - I_2)\dot\psi - n^2(I_3 - I_2)\phi = M_1, \quad I_2\ddot\theta = M_2,$$
$$I_3(\ddot\psi + n\dot\phi) - (I_2 - I_1)n\dot\phi + (I_2 - I_1)n^2\psi = M_3. \quad (3.42)$$

The equations may be expressed as,

$$I_1\ddot\phi - n(I_1 + I_3 - I_2)\dot\psi + n^2(I_2 - I_3)\phi = M_1, \quad I_2\ddot\theta = M_2,$$
$$I_3\ddot\psi + n(I_1 + I_3 - I_2)\dot\phi + n^2(I_2 - I_1)\psi = M_3. \quad (3.43)$$

3.6.2 Gravity-Gradient Stabilization

As far as the disturbing torques are concerned, we assume that only the gravity-gradient torques are significant. Thus, with the vector $\mathbf{c}_3$ given by the last column of the linearized transformation matrix,

$$\mathbf{T}_{BR} = \begin{bmatrix} 1 & \psi & -\theta \\ -\psi & 1 & \phi \\ \theta & -\phi & 1 \end{bmatrix}, \quad \text{i.e.} \quad \mathbf{c}_3 = \begin{bmatrix} -\theta \\ \phi \\ 1 \end{bmatrix}, \quad (3.44)$$

and with,

$$\mathbf{I}^B = \begin{bmatrix} I_1 & 0 & 0 \\ 0 & I_2 & 0 \\ 0 & 0 & I_3 \end{bmatrix}, \quad (3.45)$$

the gravity-gradient torque is given by,

$$\mathbf{M}_{gg} = \mathbf{g}^B = 3n^2[(I_3 - I_2)c_{32}c_{33} \quad (I_1 - I_3)c_{33}c_{31} \quad (I_2 - I_1)c_{31}c_{22}]^T, \quad (3.46)$$

where c_{13}, c_{23}, c_{33} are the components of the direction cosine vector $\mathbf{c}_3$. It is linearized and approximated as,

$$\mathbf{M}_{gg} = \mathbf{g}^B \approx 3n^2[(I_3 - I_2)\phi \quad (I_3 - I_1)\theta \quad 0]^T. \quad (3.47)$$

The equations may be expressed as,

$$I_1\ddot{\phi} - n(I_1 + I_3 - I_2)\dot{\psi} + n^2(I_2 - I_3)\phi = 3n^2(I_3 - I_2)\phi,$$

$$I_2\ddot{\theta} = 3n^2(I_3 - I_1)\theta,$$

$$I_3\ddot{\psi} + n(I_1 + I_3 - I_2)\dot{\phi} + n^2(I_2 - I_1)\psi = 0.$$

We may now investigate the stability of the spacecraft to see if the moments due to the gravity-gradients enhance the restoring torque acting on the spacecraft.

3.6.3 Stability Analysis of the Spacecraft

A primary objective of stabilization is to ensure that the spacecraft does not suffer from the problems of *librations* (sustained vibrations) or *nutation* (sustained periodic rotations). Nutation and libration may be damped by either passive means of by active dampers. These problems are also often avoided by meeting the conditions for stability.

We observe that the second of the three equations in Equation (3.48) for the pitch attitude is given by,

$$I_2\ddot{\theta} + 3n^2(I_1 - I_3)\theta = 0. \quad (3.49)$$

The stability of the pitch attitude response is guaranteed when, $I_1 > I_3$. The remaining two equations are,

$$I_1\ddot{\phi} - n(I_1 + I_3 - I_2)\dot{\psi} + 4n^2(I_2 - I_3)\phi = 0,$$
$$I_3\ddot{\psi} + n(I_1 + I_3 - I_2)\dot{\phi} + n^2(I_2 - I_1)\psi = 0. \quad (3.50)$$

To express the equations in a compact way we introduce two nondimensional inertia parameters, $k_1 = (I_2 - I_3)/I_1$ and $k_3 = (I_2 - I_1)/I_3$. In terms of these parameters, which are known as the *Smelt parameters*, the condition for pitch stability is $k_1 > k_3$ ($I_1 > I_3$). The equations then take the form,

$$\ddot{\phi} - n(1 - k_1)\dot{\psi} + 4n^2 k_1 \phi = 0, \quad \ddot{\psi} + n(1 - k_3)\dot{\phi} + n^2 k_3 \psi = 0. \quad (3.51)$$

These equations have the same form as the governing dynamical equations of a pair of gyroscopically coupled two-degree-of-freedom vibrating systems. The characteristic stability polynomial is given by,

$$\det \left| \lambda^2 \begin{bmatrix} 1 & 0 \\ 0 & 1 \end{bmatrix} + \lambda n \begin{bmatrix} 0 & -(1-k_1) \\ (1-k_3) & 0 \end{bmatrix} + n^2 \begin{bmatrix} 4k_1 & 0 \\ 0 & k_3 \end{bmatrix} \right| = 0. \quad (3.52)$$

The determinant may be expanded to give the stability quartic,

$$(\lambda/n)^4 + (1 + 3k_1 + k_1 k_3)(\lambda/n)^2 + 4k_1 k_3 = 0. \quad (3.53)$$

The conditions for stability are,

$$(1 + 3k_1 + k_1 k_3) > 0, \quad 4k_1 k_3 > 0, \quad \text{and} \quad (1 + 3k_1 + k_1 k_3)^2 > 4(4k_1 k_3). \quad (3.54)$$

The full set of conditions for stability may be expressed as,

$$k_1 > k_3, \quad k_1 k_3 > 0, \quad \|1 + 3k_1 + k_1 k_3\| > 4\sqrt{k_1 k_3} > 0. \quad (3.55)$$

These stability conditions were first discovered by DeBra and Delp [6] and are known as the DeBra-Delp stability conditions. In particular, they were able to show that the system was dynamically stable even when it was statically unstable ($k_1 < 0$ and $k_3 < 0$). They also provided a unique graphical representation of the stability conditions, shown in Figure 3.3.

There are two regions of stability in this representation: the Lagrangian region, with $k_1 > k_3 > 0$, resulting in $I_2 > I_1 > I_3$, and the DeBra-Delp region, where the DeBra-Delp conditions are met. When the stability conditions are met, the gravity-gradient torques act as a set of restoring torques, thereby further stabilizing the spacecraft. Several of the Applications Technology Satellite (ATS, 2, 4, and 5) series of spacecraft

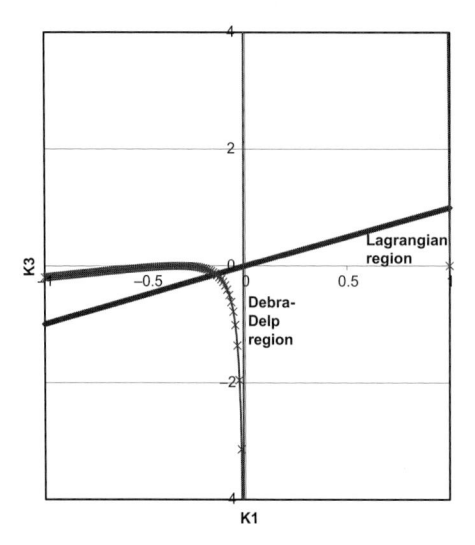

Figure 3.3 Stability diagram: $k_1 > k_3$, (Lagrangian condition for static stability, region below the straight line), $\|1 + 3k_1 + k_1 k_3\| > 4\sqrt{k_1 k_3} > 0$. (DeBra-Delp condition).

were gravity stabilized. One method of influencing the gravity gradient torque is to employ a deployable and extendable boom with a tip mass. By controlling the extension of the boom, one could effectively control the gravity-gradient torque.

3.6.4 Influence of Dissipation of Energy on Stability

Primarily, a spacecraft is an energy-conserving system with relatively little or no dissipation of energy. Contrary to the popular belief that damping in general and energy dissipation can be stabilizing in the case of a lightly stable dynamic system, the effect of any form of energy dissipation on a spacecraft could be disastrous. This could be demonstrated by adopting a suitable mathematical model. The situation is quite similar to the torque-free dynamics of a top, which tends to lose its upright and stable equilibrium attitude due to the frictional effects with the surface over which it is rotating.

Physically, the spacecraft's motion is facilitated by a balance of forces and the conservation of the total kinetic and potential energy. Any form of dissipation and the loss of energy from the total energy would result in the spacecraft's inability to either maintain its orbit, orbital speed, or attitude. The consequences of such energy dissipation could therefore lead not only to the spacecraft behaving erratically but also result in its total loss.

3.7 Introduction and Overview of Spacecraft Attitude Control Concepts

One of the major torques experienced by a spacecraft that is not caused by the environment but is deliberately applied is the spacecraft's attitude torque. All spacecraft have some form of attitude and orbit control system, which is expressly designed for the purpose of maintaining the spacecraft in some particular orientation and flying along a desired orbit. Typically, for example, an Earth observation satellite is usually oriented so that one axis always points at the Earth, while another points along the positive orbit normal, and the body frame must continuously be aligned with the orbital frame. On the other hand, an astronomical satellite may have to be controlled so one axis of the spacecraft is pointed in some inertially fixed direction, possibly toward the Sun or some other star in the celestial sky, while the satellite itself traverses a desired orbit. Considering the Earth observation satellite, the attitude control task is to ensure that the axes system in the spacecraft is always aligned with the orbital frame, and this means that the satellite is asymptotically stable in the presence of small disturbances that might cause the orientation to change. Thus, the spacecraft must be designed so that it naturally aligns itself as desired in its equilibrium state with the orbital frame. If this is not naturally so, a passive or active stabilization system is employed to make this happen.

An attitude and orbit control system consists of a complement of sensors that measure a sufficient number of the spacecraft's attitude or orbit parameters, so the attitude or orbit may be effectively estimated as required; and one or more computers to determine estimates of the attitude or orbital element deviations from the required values and compute, in accordance with an algorithm known as a control law, corrective commands to the thrusters or other onboard actuators employed to apply the control forces and torques to the spacecraft. For example, in the case of an attitude controller that attempts to bring the body frame into coincidence with the orbital frame, the error signal to the controller is the angular displacement vector of the body frame relative to the orbital frame. This angular displacement error vector is generally assumed to be small when specifying the control law, and is known as the attitude error vector. When the attitude error vector is designated by the symbol ϕ, and is a column vector containing the components of the attitude error about the three body axes, the control law defining the corrective commands to the thrusters typically takes the form, $\mathbf{u} = -\left(\mathbf{k}_p \phi + \mathbf{k}_v \dot{\phi}\right)$, where $\dot{\phi}$ is the attitude rate, $\mathbf{k}_p$ and $\mathbf{k}_v$ are fixed constants known as the control gains, and the negative sign ensures negative feedback, to ensure asymptotic stability in that the attitude error vector ϕ continually tends to a zero vector. The attitude computer on board the spacecraft is responsible for computing the corrective commands to the thrusters based on this control law.

3.7.1 Objectives of Attitude Active Stabilization and Control

There are many objectives in introducing some form of attitude control in a satellite. Some of the main objectives are,

(i)   Increase damping in least stable mode and attenuate the effect of disturbance torques;
(ii) Attitude stabilization reference to planet/Sun or star;
(iii) Maneuver control: perform a desired maneuver in a controlled and stable manner;
(iv) Inertial guidance: Inertial guidance systems often require attitude-stabilized platforms in space;
(v)   Attitude Acquisition: Pointing maneuvers require attitude acquisition.

3.7.2 Actuators and Thrusters for Spacecraft Attitude Control

Passive attitude control utilizes environmental torques to maintain spacecraft orientation. Gravity-gradient booms and solar sails are common examples of passive attitude control devices. The most common sources of torque for active attitude control are cold and hot mass and ion expulsion jets, electromagnets, momentum and reaction wheels, and control moment gyroscopes.

All jets or thrusters produce thrust by expelling mass or ions in the opposite direction. Mass expulsion jets can be hot gas jets, when energy is derived from a chemical reaction, or cold gas jets when energy is derived from the latent heat of a phase change. Ionic emissions are related either by thermionic emission or by electro-static discharge. Ions in the form of plasma can be redirected by magnetic fields and expelled through a nozzle to create a thrust.

Reaction and momentum wheels, and control moment gyros are mechanical devices used for the storage and regulation of angular momentum. Normally, a flywheel is a rotating mechanical wheel used to store momentum. A momentum wheel is just a flywheel operating at a biased momentum state. It is also capable of storing momentum about its rotation axis, which is usually fixed in the vehicle. Momentum wheels provide for a bias to the momentum vector by nominally spinning at a particular rate, and consequently provide gyroscopic stiffness. A reaction wheel is also a flywheel operating at zero momentum bias. A control moment gyro uses a single- or double-gimbaled wheel spinning at a constant rate. The gimbaled rings facilitate the alteration of the direction of the flywheel momentum vector in the spacecraft body. All of these inertial devices are generally used to dynamically alter the spin rate and attitude about the wheel axis. A complete momentum wheel assembly consists of a flywheel and other components, including bearings to reduce friction, speed-controlled torque motors, tachometers used for feedback control of speed, and a built-in control electronic board. Torque motors are used to transfer the momentum between the wheel and the spacecraft's body. Reaction wheels are similar to momentum wheels, except their spin rate can be varied or reversed as required so as to create a reaction torque on the satellite's body.

Magnetic torquers are coils of uniform wire, wound on a core of nonmagnetic material, to avoid nonlinearity and hysteresis. When a voltage is applied across the coil winding, a current is passed, which in turn creates a magnetic field. The strength and direction of the field depends on the magnitude and direction of the current in the windings, the number of turns of wire, and the total area enclosed by the coil. Torques are generated when the field produced by the torquer's coil interacts with the Earth's magnetic field. The torques are employed to control the attitude of the spacecraft.

3.7.3 Active and Passive Stabilization Techniques

There are a number of stabilization and control techniques, and the most important among these are also summarized in Table 3.1. The design of these control systems could be performed by one of the many control system synthesis techniques.

The conditions of stability of equilibrium referred to earlier can be influenced significantly by active feedback. Active stabilization systems generally employ a host of control torque generating systems that are either inertial or non-inertial and are broadly classified in Table 3.1. Gas-jet thrusters that are generally operated by a pulsing technique such as pulse width modulation are particularly suitable for proportional feedback control, while integral and proportional-integral control could be quite easily implemented with reaction wheels.

Table 3.1 Passive/active methods of stabilization

Passive	Passive	Active	Active
Applied torque	Inertial	Applied torque	Inertial
Solar sail	Spin stabilization	Gas thrusters	Momentum wheel/bias
Solar radiation	Gravity-gradient	E-magnetic torquers	Reaction wheel
Permanent magnet	Nutation dynamics	Ion, Electro-static	Control moment gyro
Special purpose	Special purpose	Special purpose	Special purpose

Furthermore, it ought to be recognized that the overall stabilization system design for a spacecraft has several modes of operation, such as ascent mode, attitude acquisition mode, long-term beam pointing mode, station-keeping mode, and attitude re-acquisition mode, as well as one of several back-up modes. During the periods of thrust application, the spacecraft's attitude is generally maintained along a desired trajectory by thrust vector control (TVC), which is actuated by appropriate servo systems. The TVC system is expected to react promptly to the maximum starting transient, accounting for the thrust misalignment about the center of gravity to maintain the attitude within prescribed limits of the desired program. The control of the transient is an important consideration in the choice of hardware. The residual angular velocity at thrust cut-off provides the initial condition for the design of the attitude control system during coasting. During this period, the sensor characteristics and disturbance torques play a significant role in the control system design. The nature of the mode and the accuracy requirements are the main drivers in the design of the attitude stabilization system. There are indeed a number of other methods of passive stabilization and these are briefly classified in Table 3.1.

3.7.4 Use of Thrusters on Spinning Satellites

When a spacecraft is launched and placed into an orbit, it is initially not spinning. To set it spinning about a particular axis, thrusters on board the spacecraft are fired. Thrusters could also be used to subsequently de-spin a section of the spacecraft or to set the overall attitude of the spacecraft by delivering a set of attitude re-orientation torques. Once the initial attitude of the satellite is set, it is maintained in this position by an attitude-stabilizing feedback controller.

3.8 Momentum and Reaction Wheels

A main difference between a momentum wheel and a reaction wheel is that the reaction wheel can operate with zero momentum bias, while a momentum wheel operates about a fixed momentum bias. A momentum wheel operates on the principle that by adding or removing momentum from a flywheel, a torque is applied to a single axis of the spacecraft. Thus, a momentum wheel is used in dual-spin Earth-orbiting spacecraft. Such a spacecraft has two sections with different spin rates. One section is usually de-spun using a momentum wheel, while the other acts as a flywheel. In such a spin-stabilized spacecraft, the momentum wheel has its axis of rotation along the same axis as the stabilized spin axis of the spacecraft. However, it spins in a direction opposite to the direction of spin of the spacecraft at nearly the same angular momentum. Thus, the de-spun section of the spacecraft's body has a net total momentum equal to zero. The direction of the spin axis coincides with the pitch axis of the satellite, which in turn is parallel to the orbit normal. (The roll axis is tangential to the orbit, while the yaw axis points to the Earth). Reaction wheels are used to execute an attitude re-orientation manoeuver by applying torques to the spacecraft's body to rotate it about a given axis.

The angular momentum vector is inertially fixed, while only the attitude angles change as the angular momentum vector components change in the body-fixed coordinates. Reaction wheels are specified by their total momentum capacity and the maximum torque delivered. Reaction and momentum wheels are generally both driven by high performance brushless DC motors.

3.8.1 Stabilization of Spacecraft

The two important configurations for wheel and spin satellite stabilization are body-stabilized and spin-stabilized. Body-stabilized satellites are simpler in concept but require complex additional hardware. They are usually stationary in the orbit frame and may be zero momentum systems controlled by reaction wheels. Furthermore, momentum wheels are generally much larger in comparison with reaction wheels.

Typically a horizon (IR- or RF-type) sensor is employed to determine the pitch and roll with reference to the Earth. A yaw sensor (gyro or star tracker) is also employed. At least three wheels are needed to control the rotation about three axes, and additional wheels may be employed for redundancy. In a passive system, the wheels are spun at a constant speed. Passive stabilization involves no feedback. Gas-jet thrusters, which generate thrust by expelling mono-propellants such as Nitrogen $(I_{sp} \approx 65\text{ s})$, stored as a cold gas, or hydrogen peroxide $(I_{sp} \approx 150\text{ s})$, may also be employed for the purpose of dumping the momentum acquired by the wheels. These spacecraft often suffer from the problem of nutation, while the problem of libration is associated with gravity-gradient stabilized spacecraft. Active nutation stabilization is often the preferred solution and could also involve the use of control moment gyros (CMGs). Control moment gyros are also essentially rotating wheels, each with a constant angular momentum, with the proviso that the directions of these momentum vectors can be controlled, by virtue of the gimballed mechanism. Yet, reaction wheels are the most popular method of control, with a proportional-integral type feedback loop to control the speed of the reaction wheels. Lead-lag compensation of some form is included in the forward path of the loop. A typical block diagram of a spacecraft control system employing CMGs as the principal actuators and magnetic torquers for the purpose of momentum dumping is illustrated in Figure 3.4.

A spin-stabilized satellite of the type shown in Figure 3.5 has a significant section that rotates at a rapid rate. Because the spin rate is significantly different from the orbital rate, the gravity-gradient torques are significantly different from when the satellite is in synchronous orbit. It is assumed that the center of mass is on the spin axis, that the orbit is circular, $I_3 = I_1$, and that the spin axis is normal to the orbit plane. The equations of motion then take the form,

$$I_1\ddot{\phi} - (n+\omega_s)(I_1+I_3-I_2)\dot{\psi} + (n+\omega_s)^2(I_2-I_3)\phi = 3n^2(I_3-I_2)\phi$$
$$I_2\ddot{\theta} = 3n^2(I_3-I_1)\theta,$$
$$I_3\ddot{\psi} + (n+\omega_s)(I_1+I_3-I_2)\dot{\phi} + (n+\omega_s)^2(I_2-I_1)\psi = 0, \qquad (3.56)$$

where ω_s is the spin rate. When $I_3 \neq I_1$ the linear equations now have time-varying coefficients or parameters, as they are functions of $\cos(nt+\omega_s t)$ and $\sin(nt+\omega_s t)$,

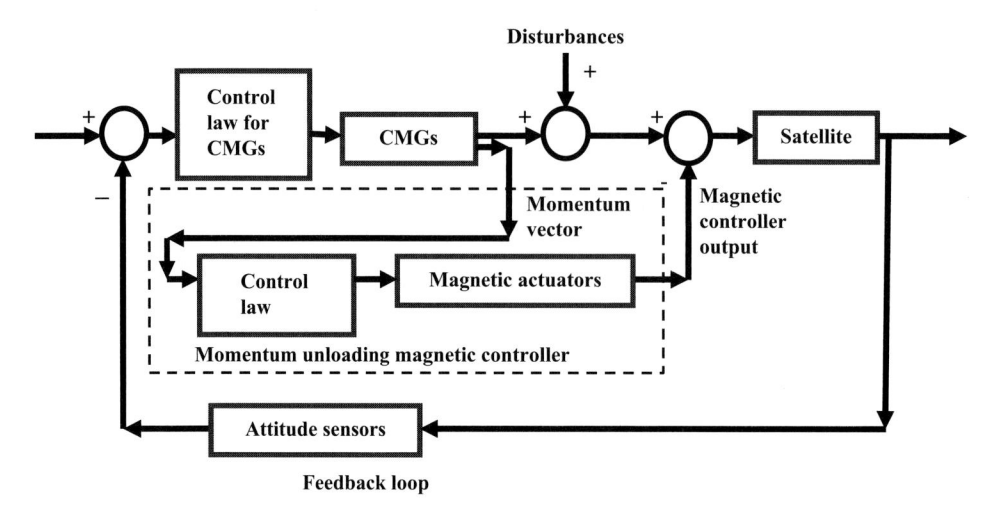

Figure 3.4 Attitude and momentum feedback and control law loops.

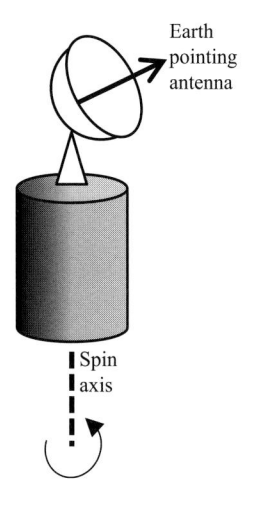

Figure 3.5 Example of spin-stabilized satellite.

with $n \neq \omega_s$ as in the synchronous case. The analysis of stability is no longer as simple as in the case of systems with constant coefficients. In some situations, however, as in this case, it is possible to transform the equations into a set of equations with constant coefficients and the analysis of stability is similar to the case of a satellite in synchronous orbit. Conditions of stability similar to the DeBra-Delp conditions but involving an additional parameter, the ratio of the spin rate to the orbit, are obtained. The conditions of stability of orbital equilibrium associated with Wilson could also be generalized to the case when the spin axis is inclined to the orbit plane. These conditions of stability of equilibrium are associated with Likins and Pringle. Like the DeBra-Delp conditions, the influence of a small amount of dissipation could be catastrophic.

Small simple satellites may be single bodies that rotate about the axis with a maximum moment of inertia. They may have an omnidirectional or toroidal antenna that is de-spun mechanically or electrically. In this case, antenna direction is controlled independently by an antenna pointing control system associated with a gimbal-mounted pointing mechanism. The control method is a simple programmed sequence of control inputs with no feedback. Although unconditionally stable, these spacecraft also often suffer from the problem of nutation. Nutations are damped by spinning the spacecraft about the axis of maximum principal moment of inertia, and they often occur when a space vehicle does not spin about such an axis. It is also possible to employ a host of eddy current, moving inertia, viscous, and magnetic dampers to act as nutation (or libration) dampers. The inclusion of a nutation damper, however, does alter the conditions of passive stability slightly when compared with the undamped case, indicating that the stability analysis with the damper must be performed to confirm the system performance.

3.8.2 Passive Control with a Gravity-Gradient Boom or a Yo-Yo Device

To fully appreciate the issues involved in the design of a passive stabilization system for a spacecraft, we shall revisit the problem of gravity-gradient stability once again.

The Euler equations of motion of a rigid body, i.e. nonlinear equations of motion of the spacecraft, about a body-fixed frame, with the roll axis in the direction of flight, the pitch axis perpendicular to the orbital plane, and the yaw axis directed toward the Earth, and with the controls fixed, may be expressed as,

$$\mathbf{I}\dot{\boldsymbol{\omega}} + \boldsymbol{\omega} \times \mathbf{I}\boldsymbol{\omega} \equiv \mathbf{I}\dot{\boldsymbol{\omega}} + \boldsymbol{\omega}_\times \mathbf{I}\boldsymbol{\omega} = \mathbf{M} + \mathbf{M}_{gg}, \quad (3.57)$$

or as

$$\mathbf{I}\dot{\boldsymbol{\omega}} + \boldsymbol{\Omega}\mathbf{I}\boldsymbol{\omega} = \mathbf{M} + \mathbf{M}_{gg}, \quad (3.58)$$

where $\mathbf{I}$ is the symmetric 3×3 inertia matrix of the spacecraft, given by,

$$\mathbf{I} = \begin{bmatrix} I_{11} & I_{12} & I_{13} \\ I_{12} & I_{22} & I_{23} \\ I_{13} & I_{23} & I_{33} \end{bmatrix}. \quad (3.59)$$

$\boldsymbol{\omega} = [\omega_1 \ \omega_2 \ \omega_3]^T$ is the body axis referred absolute angular velocity vector of the spacecraft, $\mathbf{M}_{gg} = 3n^2 \mathbf{C} \mathbf{I} \mathbf{c}$ is the gravity-gradient torque acting on the spacecraft, $\mathbf{c} = [c_1 \ c_2 \ c_3]^T$ is a component of the unit vector directed from the Earth's center to the spacecraft's mass center with direction cosine components, c_i, $\mathbf{M}$ is the vector sum of the external disturbance and fixed control torque vectors, and n is the orbital speed about the pitch axis. The quantities $\boldsymbol{\Omega}$ and $\mathbf{C}$ are matrix representations of $\boldsymbol{\omega}$ and $\mathbf{c}$, respectively, defined by,

$$\boldsymbol{\omega}_\times \equiv \boldsymbol{\Omega} = \begin{bmatrix} 0 & -\omega_3 & \omega_2 \\ \omega_3 & 0 & -\omega_1 \\ -\omega_2 & \omega_1 & 0 \end{bmatrix}, \quad \mathbf{c}_\times \equiv \mathbf{C} = \begin{bmatrix} 0 & -c_3 & c_2 \\ c_3 & 0 & -c_1 \\ -c_2 & c_1 & 0 \end{bmatrix}. \quad (3.60)$$

The Euler angles are defined by employing a 2–3–1 body axis sequence. In terms of the three attitudes of the spacecraft, $[\theta_1 \ \theta_2 \ \theta_3]^T$,

$$\mathbf{c} = \begin{bmatrix} c_1 \\ c_2 \\ c_3 \end{bmatrix} = \begin{bmatrix} -\sin\theta_2 \cos\theta_3 \\ \sin\theta_3 \sin\theta_2 \cos\theta_1 + \sin\theta_1 \cos\theta_2 \\ -\sin\theta_1 \sin\theta_2 \sin\theta_3 + \cos\theta_1 \cos\theta_2 \end{bmatrix} \approx \begin{bmatrix} -\theta_2 \\ +\theta_1 \\ 1 \end{bmatrix}. \qquad (3.61)$$

The attitude kinematics for the chosen Euler angle sequence are defined by,

$$\begin{bmatrix} \dot\theta_1 \\ \dot\theta_2 \\ \dot\theta_3 \end{bmatrix} = \frac{1}{\cos\theta_3} \begin{bmatrix} \cos\theta_3 & -\cos\theta_1 \sin\theta_3 & \sin\theta_1 \sin\theta_3 \\ 0 & \cos\theta_1 & -\sin\theta_1 \\ 0 & \sin\theta_1 \cos\theta_3 & \cos\theta_1 \cos\theta_3 \end{bmatrix} \begin{bmatrix} \omega_1 \\ \omega_2 \\ \omega_3 \end{bmatrix} + \begin{bmatrix} 0 \\ n \\ 0 \end{bmatrix}. \qquad (3.62)$$

The gravity-gradient torque acting on the spacecraft is,

$$\mathbf{M}_{gg} = 3n^2 \mathbf{C} \mathbf{I} \mathbf{c} = 3n^2 \begin{bmatrix} 0 & -c_3 & c_2 \\ c_3 & 0 & -c_1 \\ -c_2 & c_1 & 0 \end{bmatrix} \begin{bmatrix} I_{11} & I_{12} & I_{13} \\ I_{12} & I_{22} & I_{23} \\ I_{13} & I_{23} & I_{33} \end{bmatrix} \begin{bmatrix} c_1 \\ c_2 \\ c_3 \end{bmatrix}. \qquad (3.63)$$

Since for small angles,

$$\mathbf{c} = \begin{bmatrix} c_1 \\ c_2 \\ c_3 \end{bmatrix} \approx \begin{bmatrix} -\theta_2 \\ +\theta_1 \\ 1 \end{bmatrix}, \quad \mathbf{C} = \begin{bmatrix} 0 & -c_3 & c_2 \\ c_3 & 0 & -c_1 \\ -c_2 & c_1 & 0 \end{bmatrix}, \qquad (3.64)$$

it follows that,

$$\mathbf{M}_{gg} = 3n^2 \mathbf{C} \mathbf{I} \mathbf{c} = 3n^2 \begin{bmatrix} 0 & -1 & \theta_1 \\ 1 & 0 & \theta_2 \\ -\theta_1 & -\theta_2 & 0 \end{bmatrix} \begin{bmatrix} I_{11} & I_{12} & I_{13} \\ I_{12} & I_{22} & I_{23} \\ I_{13} & I_{23} & I_{33} \end{bmatrix} \begin{bmatrix} -\theta_2 \\ \theta_1 \\ 1 \end{bmatrix}. \qquad (3.65)$$

Thus,

$$\begin{aligned}\mathbf{M}_{gg} &= 3n^2 \begin{bmatrix} 0 & 0 & \theta_1 \\ 0 & 0 & \theta_2 \\ -\theta_1 & -\theta_2 & 0 \end{bmatrix} \begin{bmatrix} I_{13} \\ I_{23} \\ I_{33} \end{bmatrix} + 3n^2 \begin{bmatrix} -I_{12} & -I_{22} & -I_{23} \\ I_{11} & I_{12} & I_{13} \\ 0 & 0 & 0 \end{bmatrix} \begin{bmatrix} -\theta_2 \\ \theta_1 \\ 0 \end{bmatrix} \\ &+ 3n^2 \begin{bmatrix} -I_{23} \\ +I_{13} \\ 0 \end{bmatrix} + 3n^2 \begin{bmatrix} 0 & 0 & \theta_1 \\ 0 & 0 & \theta_2 \\ -\theta_1 & -\theta_2 & 0 \end{bmatrix} \begin{bmatrix} I_{11} & I_{12} & I_{13} \\ I_{12} & I_{22} & I_{23} \\ I_{13} & I_{23} & I_{33} \end{bmatrix} \begin{bmatrix} -\theta_2 \\ \theta_1 \\ 0 \end{bmatrix}. \end{aligned} \qquad (3.66)$$

The last term is entirely made up of higher order terms and can be ignored. Thus,

$$\mathbf{M}_{gg} = 3n^2 \begin{bmatrix} I_{33} - I_{22} & I_{12} & 0 \\ I_{12} & I_{33} - I_{11} & 0 \\ -I_{13} & -I_{23} & 0 \end{bmatrix} \begin{bmatrix} \theta_1 \\ \theta_2 \\ \theta_3 \end{bmatrix} + 3n^2 \begin{bmatrix} -I_{23} \\ +I_{13} \\ 0 \end{bmatrix}. \qquad (3.67)$$

From the attitude kinematics equation given by Equation (3.62), we obtain for small angles,

$$\begin{bmatrix} \dot\theta_1 \\ \dot\theta_2 \\ \dot\theta_3 \end{bmatrix} \approx \begin{bmatrix} 1 & -\theta_3 & 0 \\ 0 & 1 & -\theta_1 \\ 0 & \theta_1 & 1 \end{bmatrix} \begin{bmatrix} \omega_1 \\ \omega_2 \\ \omega_3 \end{bmatrix} + \begin{bmatrix} 0 \\ n \\ 0 \end{bmatrix}. \qquad (3.68)$$

Hence, as an approximation,

$$\begin{bmatrix}\omega_1\\\omega_2\\\omega_3\end{bmatrix}\approx\begin{bmatrix}1&\theta_3&0\\0&1&\theta_1\\0&-\theta_1&1\end{bmatrix}\begin{bmatrix}\dot\theta_1\\\dot\theta_2-n\\\dot\theta_3\end{bmatrix}\approx\begin{bmatrix}\dot\theta_1+(\dot\theta_2-n)\theta_3\\(\dot\theta_2-n)+\dot\theta_3\theta_1\\\dot\theta_3-(\dot\theta_2-n)\theta_1\end{bmatrix}\approx\begin{bmatrix}\dot\theta_1-n\theta_3\\\dot\theta_2-n\\\dot\theta_3+n\theta_1\end{bmatrix},$$

(3.69)

$$\begin{bmatrix}\dot\omega_1\\\dot\omega_2\\\dot\omega_3\end{bmatrix}\approx\begin{bmatrix}\ddot\theta_1-n\dot\theta_3\\\ddot\theta_2\\\ddot\theta_3+n\dot\theta_1\end{bmatrix}=\begin{bmatrix}\ddot\theta_1\\\ddot\theta_2\\\ddot\theta_3\end{bmatrix}+n\begin{bmatrix}-\dot\theta_3\\0\\\dot\theta_1\end{bmatrix}. \tag{3.70}$$

Hence we obtain,

$$\mathbf{I}\begin{bmatrix}\dot\omega_1\\\dot\omega_2\\\dot\omega_3\end{bmatrix}\approx\begin{bmatrix}I_{11}&I_{12}&I_{13}\\I_{12}&I_{22}&I_{23}\\I_{13}&I_{23}&I_{33}\end{bmatrix}\begin{bmatrix}\ddot\theta_1\\\ddot\theta_2\\\ddot\theta_3\end{bmatrix}+n\begin{bmatrix}I_{11}&I_{12}&I_{13}\\I_{12}&I_{22}&I_{23}\\I_{13}&I_{23}&I_{33}\end{bmatrix}\begin{bmatrix}-\dot\theta_3\\0\\\dot\theta_1\end{bmatrix}. \tag{3.71}$$

Substituting for the angular velocities in the full equations of motion, and ignoring second-order terms in the expression, we have,

$$\begin{bmatrix}0&-\omega_3&\omega_2\\\omega_3&0&-\omega_1\\-\omega_2&\omega_1&0\end{bmatrix}\mathbf{I}\begin{bmatrix}\omega_1\\\omega_2\\\omega_3\end{bmatrix}=\left\{\begin{bmatrix}0&-\dot\theta_3-n\theta_1&\dot\theta_2\\\dot\theta_3+n\theta_1&0&-\dot\theta_1+n\theta_3\\-\dot\theta_2&\dot\theta_1-n\theta_3&0\end{bmatrix}+\begin{bmatrix}0&0&-n\\0&0&0\\n&0&0\end{bmatrix}\right\}$$
$$\times\begin{bmatrix}I_{11}&I_{12}&I_{13}\\I_{12}&I_{22}&I_{23}\\I_{13}&I_{23}&I_{33}\end{bmatrix}\left\{\begin{bmatrix}\dot\theta_1-n\theta_3\\\dot\theta_2\\\dot\theta_3+n\theta_1\end{bmatrix}+n\begin{bmatrix}0\\-1\\0\end{bmatrix}\right\}$$
$$\approx -n\begin{bmatrix}0&-\dot\theta_3-n\theta_1&\dot\theta_2\\\dot\theta_3+n\theta_1&0&-\dot\theta_1+n\theta_3\\-\dot\theta_2&\dot\theta_1-n\theta_3&0\end{bmatrix}\begin{bmatrix}I_{12}\\I_{22}\\I_{23}\end{bmatrix}$$
$$-n\begin{bmatrix}I_{13}&I_{23}&I_{33}\\0&0&0\\-I_{11}&-I_{12}&-I_{13}\end{bmatrix}\begin{bmatrix}\dot\theta_1-n\theta_3\\\dot\theta_2\\\dot\theta_3+n\theta_1\end{bmatrix}+n^2\begin{bmatrix}I_{23}\\0\\-I_{12}\end{bmatrix}.$$

(3.72)

Hence,

$$\begin{bmatrix}0&-\omega_3&\omega_2\\\omega_3&0&-\omega_1\\-\omega_2&\omega_1&0\end{bmatrix}\mathbf{I}\begin{bmatrix}\omega_1\\\omega_2\\\omega_3\end{bmatrix}\approx -n\begin{bmatrix}0&-\dot\theta_3-n\theta_1&\dot\theta_2\\\dot\theta_3+n\theta_1&0&-\dot\theta_1+n\theta_3\\-\dot\theta_2&\dot\theta_1-n\theta_3&0\end{bmatrix}\begin{bmatrix}I_{12}\\I_{22}\\I_{23}\end{bmatrix}$$
$$-n\begin{bmatrix}I_{13}&I_{23}&I_{33}\\0&0&0\\-I_{11}&-I_{12}&-I_{13}\end{bmatrix}\begin{bmatrix}\dot\theta_1-n\theta_3\\\dot\theta_2\\\dot\theta_3+n\theta_1\end{bmatrix}+n^2\begin{bmatrix}I_{23}\\0\\-I_{12}\end{bmatrix}$$
$$=-n\begin{bmatrix}I_{13}\dot\theta_1+2I_{23}\dot\theta_2+(I_{33}-I_{22})\dot\theta_3\\I_{12}\dot\theta_3-I_{23}\dot\theta_1\\(I_{22}-I_{11})\dot\theta_1-2I_{12}\dot\theta_2-I_{13}\dot\theta_3\end{bmatrix}-n^2\begin{bmatrix}(I_{33}-I_{22})\theta_1-I_{13}\theta_3\\I_{12}\theta_1+I_{23}\theta_3\\(I_{11}-I_{22})\theta_3-I_{13}\theta_1\end{bmatrix}+n^2\begin{bmatrix}I_{23}\\0\\-I_{12}\end{bmatrix}.$$

(3.73)

The full equations of motion reduce to,

$$\mathbf{I}\begin{bmatrix}\dot{\omega}_1\\\dot{\omega}_2\\\dot{\omega}_3\end{bmatrix}+\begin{bmatrix}0&-\omega_3&\omega_2\\\omega_3&0&-\omega_1\\-\omega_2&\omega_1&0\end{bmatrix}\mathbf{I}\begin{bmatrix}\omega_1\\\omega_2\\\omega_3\end{bmatrix}-3n^2\begin{bmatrix}0&-c_3&c_2\\c_3&0&-c_1\\-c_2&c_1&0\end{bmatrix}\mathbf{I}\begin{bmatrix}c_1\\c_2\\c_3\end{bmatrix}$$

$$=\begin{bmatrix}I_{11}&I_{12}&I_{13}\\I_{12}&I_{22}&I_{23}\\I_{13}&I_{23}&I_{33}\end{bmatrix}\begin{bmatrix}\ddot{\theta}_1\\\ddot{\theta}_2\\\ddot{\theta}_3\end{bmatrix}-n\begin{bmatrix}2I_{23}\dot{\theta}_2+(I_{11}+I_{33}-I_{22})\dot{\theta}_3\\2I_{12}\dot{\theta}_3-2I_{23}\dot{\theta}_1\\(I_{22}-I_{11}-I_{33})\dot{\theta}_1-2I_{12}\dot{\theta}_2\end{bmatrix} \quad (3.74)$$

$$-n^2\begin{bmatrix}(I_{33}-I_{22})\theta_1-I_{13}\theta_3\\I_{12}\theta_1+I_{23}\theta_3\\(I_{11}-I_{22})\theta_3-I_{13}\theta_1\end{bmatrix}+n^2\begin{bmatrix}I_{23}\\0\\-I_{12}\end{bmatrix}-\mathbf{M}_{gg}=\mathbf{M},$$

where,

$$\mathbf{M}_{gg}=-n^2\begin{bmatrix}3(I_2-I_3)&-3I_{12}&0\\-3I_{12}&3(I_1-I_3)&0\\3I_{13}&3I_{23}&0\end{bmatrix}\begin{bmatrix}\theta_1\\\theta_2\\\theta_3\end{bmatrix}+3n^2\begin{bmatrix}-I_{23}\\+I_{13}\\0\end{bmatrix}. \quad (3.75)$$

Hence, writing in matrix form, the linearized equations of motion are,

$$\mathbf{I}\begin{bmatrix}\ddot{\theta}_1\\\ddot{\theta}_2\\\ddot{\theta}_3\end{bmatrix}+n\begin{bmatrix}0&-2I_{23}&I_{22}-I_{33}-I_{11}\\2I_{23}&0&-2I_{12}\\I_{11}+I_{33}-I_{22}&2I_{12}&0\end{bmatrix}\begin{bmatrix}\dot{\theta}_1\\\dot{\theta}_2\\\dot{\theta}_3\end{bmatrix}$$
$$+n^2\begin{bmatrix}4(I_{22}-I_{33})&-3I_{12}&0\\-3I_{12}&3(I_{11}-I_{33})&0\\3I_{13}&3I_{23}&(I_{22}-I_{11})\end{bmatrix}\begin{bmatrix}\theta_1\\\theta_2\\\theta_3\end{bmatrix}=\mathbf{M}+n^2\begin{bmatrix}-4I_{23}\\+3I_{13}\\+I_{12}\end{bmatrix}. \quad (3.76)$$

Equation (3.76) can be rearranged as,

$$\mathbf{I}\begin{bmatrix}\ddot{\theta}_1\\\ddot{\theta}_2\\\ddot{\theta}_3\end{bmatrix}+n\begin{bmatrix}0&-2I_{23}&I_{22}-I_{33}-I_{11}\\2I_{23}&0&-2I_{12}\\I_{11}+I_{33}-I_{22}&2I_{12}&0\end{bmatrix}\begin{bmatrix}\dot{\theta}_1\\\dot{\theta}_2\\\dot{\theta}_3\end{bmatrix}$$
$$+n^2\begin{bmatrix}4(I_{22}-I_{33})&-3I_{12}&0\\-3I_{12}&3(I_{11}-I_{33})&0\\3I_{13}&3I_{23}&(I_{22}-I_{11})\end{bmatrix}\begin{bmatrix}\theta_1-\theta_{10}\\\theta_2-\theta_{20}\\\theta_3-\theta_{30}\end{bmatrix}=\mathbf{M}, \quad (3.77)$$

where

$$\begin{bmatrix}\theta_{10}\\\theta_{20}\\\theta_{30}\end{bmatrix}=\begin{bmatrix}4(I_{22}-I_{33})&-3I_{12}&0\\-3I_{12}&3(I_{11}-I_{33})&0\\3I_{13}&3I_{23}&(I_{22}-I_{11})\end{bmatrix}^{-1}\begin{bmatrix}-4I_{23}\\+3I_{13}\\+I_{12}\end{bmatrix}. \quad (3.78)$$

Thus, if one wishes to ensure that $[\theta_{10} \quad \theta_{20} \quad \theta_{30}] = [0 \quad 0 \quad 0]$, $I_{23} = I_{13} = I_{12} = 0$. When this is satisfied, the linear equations of motion for verifying gravity-gradient stability are,

$$\mathbf{I}\begin{bmatrix}\ddot{\theta}_1\\\ddot{\theta}_2\\\ddot{\theta}_3\end{bmatrix} + n\begin{bmatrix}0 & 0 & I_{22}-I_{33}-I_{11}\\0 & 0 & 0\\I_{11}+I_{33}-I_{22} & 0 & 0\end{bmatrix}\begin{bmatrix}\dot{\theta}_1\\\dot{\theta}_2\\\dot{\theta}_3\end{bmatrix}$$
$$+n^2\begin{bmatrix}4(I_{22}-I_{33}) & 0 & 0\\0 & 3(I_{11}-I_{33}) & 0\\0 & 0 & (I_{22}-I_{11})\end{bmatrix}\begin{bmatrix}\theta_1\\\theta_2\\\theta_3\end{bmatrix} = \mathbf{M}. \quad (3.79)$$

The pitch axis motion is independent of the motion in the other two axes, and stability can be checked by applying the DeBra-Delp conditions. To guarantee passive gravity-gradient stabilization, booms could be added to the spacecraft to change its principal inertia properties, without changing the products of inertia, so as to meet the DeBra-Delp conditions.

3.8.3 Reaction Wheel Stabilization

Reaction wheels are used to control spacecraft. They generally operate around zero spin and may come to a complete stop. Alternatively, they could be used as biased momentum systems, in which there is always a certain minimum angular momentum in a certain direction. A typical example of a reaction wheel developed by NASA is illustrated in Figure 3.6.

Consider, for example, a reaction wheel with a moment of inertia J_{wz} spinning at a speed r_w about the Z axis of the satellite. The equations of motion are,

Figure 3.6 Reaction/momentum wheel developed by NASA.
(courtesy: NASA)

$$I_1\dot{p}_B + ((I_3 + J_{wz})r_B + J_{wz}r_w - I_2 r_B)q_B = M_1,$$
$$I_2\dot{q}_B + (I_1 r_B - (I_3 + J_{wz})r_B - J_{wz}r_w)p_B = M_2, \quad (3.80)$$
$$(I_3 + J_{wz})\dot{r}_B + J_{wz}\dot{r}_w + (I_2 - I_1)p_B q_B = M_3.$$

Thus, if $I_3' = I_3 + J_{wz}$, the equations of motion are,

$$I_1\dot{p}_B + (I_3' r_B - I_2 r_B)q_B = M_1 - J_{wz}r_w q_B$$
$$I_2\dot{q}_B + (I_1 r_B - I_3' r_B)p_B = M_2 + J_{wz}r_w p_B \quad (3.81)$$
$$I_3'\dot{r}_B + (I_2 - I_1)p_B q_B = M_3 - J_{wz}\dot{r}_w.$$

Thus, if one interprets I_3 as the total moment of inertia about the Z axis of the satellite, the stability equations are,

$$I_1\dot{p}_B + (I_3 r_B - I_2 r_B)q_B = M_1 - J_{wz}r_w q_B,$$
$$I_2\dot{q}_B + (I_1 r_B - I_3 r_B)p_B = M_2 + J_{wz}r_w p_B, \quad (3.82)$$
$$I_3\dot{r}_B + (I_2 - I_1)p_B q_B = M_3 - J_{wz}\dot{r}_w.$$

Although the pitch axis and the roll axis are no longer decoupled, stabilization can be achieved by a suitable choice of the momentum of the reaction wheel, $J_{wz}r_w$. One may apply these equations to study the stability of a three-axis stabilized spacecraft, including the gravity-gradient moments. From Section 3.8.2, Equations (3.77), including the torques due to the reaction wheels arising due to their angular momentum components $h_j, j = 1, 2, 3$, one has,

$$\mathbf{I}\begin{bmatrix}\ddot{\theta}_1\\\ddot{\theta}_2\\\ddot{\theta}_3\end{bmatrix} + n\begin{bmatrix}0 & -2I_{23} & I_{22}-I_{33}-I_{11}\\2I_{23} & 0 & -2I_{12}\\I_{11}+I_{33}-I_{22} & 2I_{12} & 0\end{bmatrix}\begin{bmatrix}\dot{\theta}_1\\\dot{\theta}_2\\\dot{\theta}_3\end{bmatrix}$$
$$+ n^2\begin{bmatrix}4(I_{22}-I_{33}) & -3I_{12} & 0\\-3I_{12} & 3(I_{11}-I_{33}) & 0\\3I_{13} & 3I_{23} & (I_{22}-I_{11})\end{bmatrix}\begin{bmatrix}\theta_1-\theta_{10}\\\theta_2-\theta_{20}\\\theta_3-\theta_{30}\end{bmatrix}$$
$$= \mathbf{M} - \begin{bmatrix}\dot{h}_1\\\dot{h}_2\\\dot{h}_3\end{bmatrix} - \begin{bmatrix}0 & -\dot{\theta}_3-n\theta_1 & \dot{\theta}_2\\\dot{\theta}_3+n\theta_1 & 0 & -\dot{\theta}_1+n\theta_3\\-\dot{\theta}_2 & \dot{\theta}_1-n\theta_3 & 0\end{bmatrix}\begin{bmatrix}h_1\\h_2\\h_3\end{bmatrix} + \begin{bmatrix}0 & 0 & -n\\0 & 0 & 0\\n & 0 & 0\end{bmatrix}\begin{bmatrix}h_1\\h_2\\h_3\end{bmatrix}.$$
$$(3.83)$$

Ignoring the second-order terms on the right-hand side, the linearized equations of motion of a three-axis stabilized spacecraft, in terms of the angular momentum components of the wheels $h_j, j = 1, 2, 3$, are,

$$\mathbf{I}\begin{bmatrix}\ddot{\theta}_1\\\ddot{\theta}_2\\\ddot{\theta}_3\end{bmatrix} + n\begin{bmatrix}0 & -2I_{23} & I_{22}-I_{33}-I_{11}\\2I_{23} & 0 & -2I_{12}\\I_{11}+I_{33}-I_{22} & 2I_{12} & 0\end{bmatrix}\begin{bmatrix}\dot{\theta}_1\\\dot{\theta}_2\\\dot{\theta}_3\end{bmatrix}$$
$$+ n^2\begin{bmatrix}4(I_{22}-I_{33}) & -3I_{12} & 0\\-3I_{12} & 3(I_{11}-I_{33}) & 0\\3I_{13} & 3I_{23} & (I_{22}-I_{11})\end{bmatrix}\begin{bmatrix}\theta_1-\theta_{10}\\\theta_2-\theta_{20}\\\theta_3-\theta_{30}\end{bmatrix} = \mathbf{M} - \begin{bmatrix}\dot{h}_1-nh_3\\\dot{h}_2\\\dot{h}_3+nh_1\end{bmatrix}.$$
$$(3.84)$$

For the purposes of control law synthesis, they are usually expressed in terms of the applied torques to the wheels $u_j, j = 1, 2, 3$, in the form,

$$\mathbf{I}\begin{bmatrix}\ddot{\theta}_1\\\ddot{\theta}_2\\\ddot{\theta}_3\end{bmatrix} + n\begin{bmatrix}0 & -2I_{23} & I_{22}-I_{33}-I_{11}\\2I_{23} & 0 & -2I_{12}\\I_{11}+I_{33}-I_{22} & 2I_{12} & 0\end{bmatrix}\begin{bmatrix}\dot{\theta}_1\\\dot{\theta}_2\\\dot{\theta}_3\end{bmatrix}$$
$$+n^2\begin{bmatrix}4(I_{22}-I_{33}) & -3I_{12} & 0\\-3I_{12} & 3(I_{11}-I_{33}) & 0\\3I_{13} & 3I_{23} & (I_{22}-I_{11})\end{bmatrix}\begin{bmatrix}\theta_1-\theta_{10}\\\theta_2-\theta_{20}\\\theta_3-\theta_{30}\end{bmatrix}$$
$$= \mathbf{M} - \begin{bmatrix}-nh_3\\0\\nh_1\end{bmatrix} - \begin{bmatrix}u_1\\u_2\\u_3\end{bmatrix}, \quad \begin{bmatrix}\dot{h}_1\\\dot{h}_2\\\dot{h}_3\end{bmatrix} = \begin{bmatrix}u_1\\u_2\\u_3\end{bmatrix}. \tag{3.85}$$

3.8.4 Momentum Wheel and Dual-Spin Stabilization

Dual-spin stabilization is one approach to stabilization of an orbiting satellite. The idealized model defining a dual-spin spacecraft is shown in Figure 3.7. The system consists of two rigid bodies B and B' connected to each other via a torque motor and constrained so there can be relative motion between the two bodies about the spacecraft's spin axis, which is a principal axis of the two bodies passing through the centroid of both bodies. The angular displacement of the second body B' relative to the main body B is assumed to be θ and the relative angular velocity is $\dot{\theta}$.

Let mutually perpendicular axes x, y, z and x', y', z' with unit vectors $\mathbf{i}, \mathbf{j}, \mathbf{k}$ and $\mathbf{i}', \mathbf{j}', \mathbf{k}'$ be defined and assumed to be attached to the bodies, with the z and z' axes aligned with the common spin axis and the origins of the two frames co-located at the center of gravity of the combined system C. Let I_1, I_2, I_3 and I'_1, I'_2, I'_3 be the principal mass moments of inertia of the bodies B and B' about the axes x, y, z and x', y', z', respectively, which are assumed to be the principal axes of the two bodies.

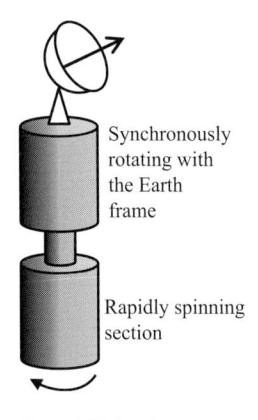

Synchronously rotating with the Earth frame

Rapidly spinning section

Figure 3.7 Dual spinner.

The body B is assumed to have body angular velocity components $\omega_1, \omega_2, \omega_3$ along the mutually perpendicular axes x, y, z. In addition, an inertial or space-fixed reference frame X, Y, Z is assumed to be co-located with its origin at the point C. The moments and products of inertia of the body B' are transformed to the frame x, y, z, which is fixed to the body B. The transformed components of the moment of inertia matrix of the body B',

$$\mathbf{I}' = \begin{bmatrix} I'_1 & -I'_{12} & -I'_{13} \\ -I'_{12} & I'_2 & -I'_{23} \\ -I'_{13} & -I'_{23} & I'_3 \end{bmatrix}, \tag{3.86}$$

are,

$$I_{1t} = I'_1 \cos^2\theta + I'_2 \sin^2\theta + I'_{12} \sin 2\theta,$$
$$I_{2t} = I'_1 \sin^2\theta + I'_2 \cos^2\theta - I'_{12} \sin 2\theta,$$
$$I_{12t} = I'_{12}(\cos^2\theta - \sin^2\theta) - (I'_1 - I'_2)\sin\theta\cos\theta,$$
$$I_{3t} = I'_3, \quad I_{13t} = I'_{13}\cos\theta - I'_{23}\sin\theta, \quad I_{23t} = I'_{23}\cos\theta + I'_{13}\sin\theta. \tag{3.87}$$

Given the total moment of momentum of the two bodies, $\mathbf{H}_T$, the time rate of change of the moment of momentum in the inertial frame, is related to the time rate of change of the moment of momentum in the body-fixed frame according to the relation,

$$^{(I)}\frac{d\mathbf{H}_T}{dt} = {^{(B)}}\frac{d\mathbf{H}_T}{dt} + \boldsymbol{\omega}_B \times \mathbf{H}_T, \tag{3.88}$$

Where,

$$\boldsymbol{\omega}_B = \omega_1 \mathbf{i} + \omega_2 \mathbf{j} + \omega_3 \mathbf{k}, \quad \mathbf{H}_T = \mathbf{H}_1 + \mathbf{H}_2,$$
$$\mathbf{H}_1 = I_1 \omega_1 \mathbf{i} + I_2 \omega_2 \mathbf{j} + I_3 \omega_3 \mathbf{k},$$
$$\mathbf{H}_2 = \mathbf{I}_t \cdot \boldsymbol{\omega}_B = \{(I'_1 \cos^2\theta + I'_2 \sin^2\theta)\omega_1 + (I'_1 - I'_2)\omega_2 \sin\theta\cos\theta\}\mathbf{i}$$
$$+ \{(I'_1 \sin^2\theta + I'_2 \cos^2\theta)\omega_2 + (I'_1 - I'_2)\omega_1 \sin\theta\cos\theta\}\mathbf{j} + I'_3(\omega_3 + \dot{\theta})\mathbf{k}. \tag{3.89}$$

The equation for $\mathbf{H}_2$ may also be expressed as,

$$\mathbf{H}_2 = \{(I'_1 - (I'_1 - I'_2)\sin^2\theta)\omega_1 + (I'_1 - I'_2)\omega_2 \sin\theta\cos\theta\}\mathbf{i}$$
$$+ \{(I'_2 + (I'_1 - I'_2)\sin^2\theta)\omega_2 + (I'_1 - I'_2)\omega_1 \sin\theta\cos\theta\}\mathbf{j} + I'_3(\omega_3 + \dot{\theta})\mathbf{k}, \tag{3.90}$$

and $\dot{\theta} = $ constant.

For each body, the equations of motion are,

$$^{(I)}\frac{d\mathbf{H}_1}{dt} = {^{(B)}}\frac{d\mathbf{H}_1}{dt} + \boldsymbol{\omega}_B \times \mathbf{H}_1 = \mathbf{T}_1,$$
$$^{(I)}\frac{d\mathbf{H}_2}{dt} = {^{(B)}}\frac{d\mathbf{H}_2}{dt} + \boldsymbol{\omega}_B \times \mathbf{H}_2 = -\mathbf{T}_1. \tag{3.91}$$

Thus,

$$I_1\dot{\omega}_1 + (I_3 - I_2)\omega_2\omega_3 = T_{11},$$
$$I_2\dot{\omega}_2 + (I_1 - I_3)\omega_3\omega_1 = T_{12}, \qquad (3.92)$$
$$I_3\dot{\omega}_3 + (I_2 - I_1)\omega_1\omega_2 = T_{13},$$

and

$$\begin{aligned}
&\left(I'_1 - (I'_1 - I'_2)\sin^2\theta\right)\dot{\omega}_1 + (I'_1 - I'_2)\dot{\omega}_2 \sin\theta\cos\theta \\
&\quad - (I'_1 - I'_2)\dot{\theta}\omega_1 \sin 2\theta + (I'_1 - I'_2)\omega_2\dot{\theta}\cos 2\theta \\
&+ I'_3\omega_2\dot{\theta} + \left(I'_3 - I'_2 - (I'_1 - I'_2)\sin^2\theta\right)\omega_2\omega_3 - (I'_1 - I'_2)\omega_1\omega_3 \sin\theta\cos\theta = -T_{11}, \\
&\left(I'_2 + (I'_1 - I'_2)\sin^2\theta\right)\dot{\omega}_2 + (I'_1 - I'_2)\dot{\omega}_1 \sin\theta\cos\theta \\
&\quad + (I'_1 - I'_2)\dot{\theta}\omega_2 \sin 2\theta + (I'_1 - I'_2)\omega_1\dot{\theta}\cos 2\theta \\
&- I'_3\omega_1\dot{\theta} + \left(I'_1 - I'_3 - (I'_1 - I'_2)\sin^2\theta\right)\omega_1\omega_3 + (I'_1 - I'_2)\omega_2\omega_3 \sin\theta\cos\theta = -T_{12}, \\
&I'_3(\dot{\omega}_3 + \ddot{\theta}) + (I'_1 - I'_2)(\omega_1^2 - \omega_2^2)\sin\theta\cos\theta - (I'_1 - I'_2)\omega_1\omega_2 \cos 2\theta = -T_{13}.
\end{aligned}$$
(3.93)

Another popular method of passive stabilization is to employ a spinning heavy flywheel to artificially increase the moment of momentum of the spacecraft in a particular direction without incurring any substantial weight penalty. This method of stabilization could be considered to be passive when no energy is delivered to the flywheel while the spacecraft is in orbit or active if the speed of the flywheel is actively controlled by a torque motor driven by an appropriate power source. The dynamics of such a system are considerably more complex due to the additional degree of freedom introduced by the flywheel. A typical illustration of a satellite requiring attitude control is shown in Figure 3.8. Some spin-stabilized satellites are dual spinners (Figure 3.7) or gyrostats. They have a rotating part and a stationary part (relative to the orbit frame) and rotate about the axis of minimum moment of inertia. The energy dissipated on the stationary part, which is stabilizing, must exceed the energy dissipation in the rotating part to ensure stability. The Earth-fixed portion of the spacecraft containing the platform-mounted antennae rotates in synchronism with the Earth's rotation, while the spinning portion provides the angular momentum necessary for gyro stabilization of the vehicle. After separation of the spacecraft from the launcher, both portions of the spacecraft are

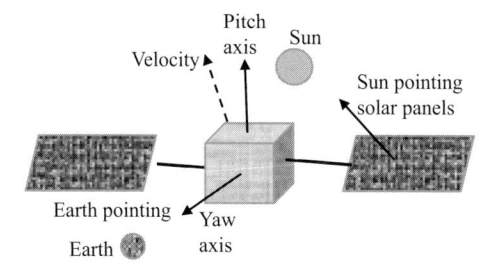

Figure 3.8 Typical example of a satellite requiring attitude control: The satellite points to the Earth, while the solar sail points to the Sun.

spun up as a unit using thrusters. The spacecraft is then injected into synchronous orbit. The platform is then "de-spun" until it is fixed in the Earth frame. In this type of spacecraft, any initial nutation angle between the spin axis and the angular momentum vector could grow, indicating instability, under certain circumstances. Conditions of stability of equilibria similar to the DeBra-Delp, Wilson, and Likins-Pringle [1] conditions, associated with Roberson, could be derived for representative dual-spin spacecraft. These conditions for stability must be met in the design of such a spacecraft. A typical example of this type of satellite is the British *Skynet1* military satellite that operated for a year and suffered from instability problems.

3.8.5 Momentum Wheel Approximation with MW along Axis 1

Let the wheel principal moments of inertia be,

$$\mathbf{J} = [J_1 \quad J_2 \quad J_3]. \tag{3.94}$$

From the symmetry of the wheel, it follows that,

$$J_3 = J_2. \tag{3.95}$$

Also,

$$\mathbf{I}_T = \mathbf{I} + \mathbf{J} = [I_1 \quad I_2 \quad I_3] + [J_1 \quad J_2 \quad J_3]. \tag{3.96}$$

Hence,

$$\begin{aligned} (I_{1T} - J_1)\dot{\omega}_1 + (I_{3T} - I_{2T})\omega_2\omega_3 + u &= 0, \\ I_{2T}\dot{\omega}_2 + (I_{1T} - I_{3T})\omega_3\omega_1 + \omega_3 h_w &= 0, \\ I_{3T}\dot{\omega}_3 + (I_{2T} - I_{1T})\omega_1\omega_2 - \omega_2 h_w &= 0. \end{aligned} \tag{3.97}$$

Along the wheel axis,

$$J_1\dot{\omega}_1 + \dot{h}_w - u = 0, \quad h_w = J_1\omega_{wh}. \tag{3.98}$$

Since $I_{1T} \gg J_1$,

$$\begin{aligned} I_{1T}\dot{\omega}_1 + (I_{3T} - I_{2T})\omega_2\omega_3 + u &= 0, \\ I_{2T}\dot{\omega}_2 + (I_{1T} - I_{3T})\omega_3\omega_1 + \omega_3 h_w &= 0, \\ I_{3T}\dot{\omega}_3 + (I_{2T} - I_{1T})\omega_1\omega_2 - \omega_2 h_w &= 0. \end{aligned} \tag{3.99}$$

Assuming $\dot{\omega}_{wh} \gg \dot{\omega}_1$,

$$\dot{h}_w - u = 0. \tag{3.100}$$

Thus,

$$\begin{aligned} I_{1T}\dot{\omega}_1 + (I_{3T} - I_{2T})\omega_2\omega_3 + \dot{h}_w &= 0, \\ I_{2T}\dot{\omega}_2 + (I_{1T} - I_{3T})\omega_3\omega_1 + \omega_3 h_w &= 0, \\ I_{3T}\dot{\omega}_3 + (I_{2T} - I_{1T})\omega_1\omega_2 - \omega_2 h_w &= 0. \end{aligned} \tag{3.101}$$

The equations of motion of a typical space vehicle with three light flywheels attached to it and spinning along the three principal axes are given by (see, for example, Hughes [4]),

$$I_1\dot{\omega}_1 + (I_3 - I_2)\omega_2\omega_3 = M_1 - (\omega_2 h_3^w - \omega_3 h_2^w) - M_1^W, \quad \dot{h}_1^W = M_1^W,$$
$$I_2\dot{\omega}_2 + (I_1 - I_3)\omega_3\omega_1 = M_2 - (\omega_3 h_1^w - \omega_1 h_3^w) - M_2^W, \quad \dot{h}_2^W = M_2^W, \quad (3.102)$$
$$I_3\dot{\omega}_3 + (I_2 - I_1)\omega_1\omega_2 = M_3 - (\omega_1 h_2^w - \omega_2 h_1^w) - M_3^W, \quad \dot{h}_3^W = M_3^W,$$

where h_i^W is the moment of momentum in the ith wheel, and $-M_i^W$ are the reaction torques acting on the satellite in a frame fixed in it. The reaction torque is equal in magnitude and opposite in direction to the total torque that appears in the wheel equations. The total wheel torque M_i^W, for each wheel, in principle is the sum of the torque applied to each wheel and the gyroscopic torque $(-\boldsymbol{\omega}_{wh_i} \times \mathbf{h}_i)$ acting on it. However, the latter term is zero, as each wheel is only rotating about its own as axis. Thus, the total wheel torque M_i^W, for each wheel, is the control torque applied to each wheel.

In the case of a passively stabilized spacecraft with one wheel spinning steadily with no external torques applied, the last terms in each of the three equations on the left and the three equations on the right are absent. h_i^W is the moment of momentum component in the direction of the ith principal axis. The stability analysis of the spacecraft is quite similar to the typical case considered in earlier sections.

3.8.6 Control Moment Gyroscopes

The operation of a control moment gyroscope (CMG) is very similar to a reaction wheel. Like a reaction wheel, it consists of a heavy flywheel driven by a brushless DC motor. In a reaction wheel, the torque generated is along the same axis as the momentum of the wheel. However, quite unlike a reaction wheel, the flywheel is mounted on one or more gimbals, which are each independently driven by DC motors. The control torque generated by a CMG is mutually perpendicular to the angular velocity vector of the gimbals and the angular momentum vector of the flywheel, and is obtained from the cross product of both. For this reason, the control torque vector generated on a spacecraft bears no simple relationship to the momentum vector, although this does offer greater flexibility in orienting the control torque vector. For this reason, single-gimbaled CMGs are preferred, as the control laws for these actuators are less complex to synthesize.

3.8.7 Example of Control System Based on Reaction Wheels

The *Cassini* spacecraft (Figure 3.9), which is over 5,600 kg in weight, is one of the largest robotic spacecraft built by NASA for the Cassini–Huygens mission to Saturn and is three-axis stabilized by three reaction wheels mounted on orthogonal axes, with a fourth one to spare.

The wheels are used for fine control, while coarse attitude control is achieved by hydrazine bipropellant thrusters using nitrogen tetroxide and monomethyl hydrazine. Power is provided by radioisotope thermoelectric generators (RTGs). The spacecraft propellant to mass ratio is about 55%. The control law synthesis is achieved by

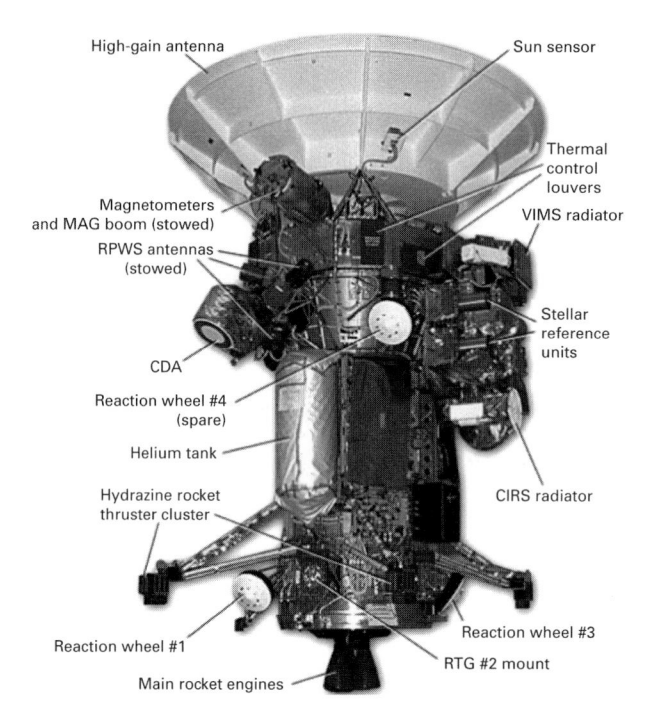

Figure 3.9 Example of a three-axis stabilized spacecraft: The *Cassini* spacecraft. (courtesy: NASA)

expressing Equations (3.102) in terms of the wheel moment of momentum $h_i = \omega_{wh_i} I_{wh_i}$ for each wheel and applied torques to the wheels u_j, $j = 1, 2, 3$, in the form,

$$\begin{bmatrix} \ddot{\theta}_1 \\ \ddot{\theta}_2 \\ \ddot{\theta}_3 \end{bmatrix} = -n\mathbf{I}^{-1} \begin{bmatrix} 0 & -2I_{23} & I_{22} - I_{33} - I_{11} \\ 2I_{23} & 0 & -2I_{12} \\ I_{11} + I_{33} - I_{22} & 2I_{12} & 0 \end{bmatrix} \begin{bmatrix} \dot{\theta}_1 \\ \dot{\theta}_2 \\ \dot{\theta}_3 \end{bmatrix}$$

$$-n^2 \mathbf{I}^{-1} \begin{bmatrix} 4(I_{22} - I_{33}) & -3I_{12} & 0 \\ -3I_{12} & 3(I_{11} - I_{33}) & 0 \\ 3I_{13} & 3I_{23} & (I_{22} - I_{11}) \end{bmatrix} \begin{bmatrix} \theta_1 - \theta_{10} \\ \theta_2 - \theta_{20} \\ \theta_3 - \theta_{30} \end{bmatrix}$$

$$-n\mathbf{I}^{-1} \begin{bmatrix} 0 & 0 & -I_{wh_3} \\ 0 & 0 & 0 \\ I_{wh_1} & 0 & 0 \end{bmatrix} \begin{bmatrix} \omega_{wh_1} \\ \omega_{wh_2} \\ \omega_{wh_3} \end{bmatrix} - \mathbf{I}^{-1} \begin{bmatrix} u_1 \\ u_2 \\ u_3 \end{bmatrix},$$

$$\begin{bmatrix} \dot{\omega}_{wh_1} \\ \dot{\omega}_{wh_2} \\ \dot{\omega}_{wh_3} \end{bmatrix} = \begin{bmatrix} I_{wh_1} & 0 & 0 \\ 0 & I_{wh_2} & 0 \\ 0 & 0 & I_{wh_3} \end{bmatrix}^{-1} \begin{bmatrix} u_1 \\ u_2 \\ u_3 \end{bmatrix}. \tag{3.103}$$

Introducing the state vector,

$$\mathbf{x} = \begin{bmatrix} \theta_1 - \theta_{10} & \theta_2 - \theta_{20} & \theta_3 - \theta_{30} & \dot{\theta}_1 & \dot{\theta}_2 & \dot{\theta}_3 & \omega_{wh_1} & \omega_{wh_2} & \omega_{wh_3} \end{bmatrix}^T, \tag{3.104}$$

and the auxiliary equations,

$$\begin{bmatrix} \dot{x}_1 & \dot{x}_2 & \dot{x}_3 \end{bmatrix} = \begin{bmatrix} x_4 & x_5 & x_6 \end{bmatrix}, \tag{3.105}$$

3.8 Momentum and Reaction Wheels

the equations are reduced to state space form given by,

$$\dot{\mathbf{x}} = \mathbf{A}\mathbf{x} + \mathbf{B}\mathbf{u}, \tag{3.106}$$

where **A** is a 9×9 matrix, **B** is a 3×9 matrix, and $\mathbf{u} = [u_1 \ u_2 \ u_3]^T$.

The design of the control laws to optimize a performance criterion is done by minimizing cost function,

$$J_{LQR} = \frac{1}{2}\int_0^\infty (\mathbf{x}^T\mathbf{Q}\mathbf{x} + \mathbf{u}^T\mathbf{R}\mathbf{u})dt = \frac{1}{2}\int_0^\infty \left(\|\mathbf{x}\|_\mathbf{Q} + \|\mathbf{u}\|_\mathbf{R}\right)dt. \tag{3.107}$$

The first term $(1/2)\int_0^\infty \|\mathbf{x}\|_\mathbf{Q}\, dt$ represents the energy of the state response and provides the energy of a combination of the states to assess the performance. The second term in the cost function $(1/2)\int_0^\infty \|\mathbf{u}\|_\mathbf{R}\, dt$ represents the energy contained in the control signal that is fed back into the plant by the controller. (For an introduction to quadratic optimal regulator systems [optimal control systems] design in the state-space domain, the reader is referred to Ogata [7]. For spacecraft applications, the reader is referred to Kaplan [8], Sidi [9], and Wei [10].) The linear quadratic regulator (LQR) uses a linear controller that minimizes the quadratic performance cost function of the states and control inputs. Thus, the optimal gain matrix **K** is chosen such that for a given continuous-time state-space model, the state-feedback control law $\mathbf{u} = -\mathbf{K}\mathbf{x}$ minimizes the quadratic cost function,

$$J_{LQR} = \frac{1}{2}\int_0^\infty (\mathbf{x}^T\mathbf{Q}\mathbf{x} + \mathbf{u}^T\mathbf{R}\mathbf{u})dt, \tag{3.108}$$

subject to the state vector satisfying the model's dynamics equations given by,

$$\dot{\mathbf{x}} = \mathbf{A}\mathbf{x} + \mathbf{B}\mathbf{u}. \tag{3.109}$$

A first choice for the matrices **Q** and **R** in the expression for the cost function is given by *Bryson's rule*: select **Q** and **R** as diagonal matrices with elements,

$$q_{ii} = 1/\text{maximum expected value of } x_i^2, \quad r_{ii} = 1/\text{maximum expected value of } u_i^2.$$

The solution to the optimal control problem may be conveniently expressed in terms of the constant *Riccati* matrix **P**, which is defined by,

$$d(\mathbf{x}^T\mathbf{P}\mathbf{x})/dt = -\mathbf{x}^T(\mathbf{Q} + \mathbf{K}^T\mathbf{R}\mathbf{K})\mathbf{x}. \tag{3.110}$$

Substituting the state feedback control input,

$$\mathbf{u} = -\mathbf{K}\mathbf{x}, \tag{3.111}$$

and using Equation (3.110),

$$J_{LQR} = \frac{1}{2}\int_0^\infty \mathbf{x}^T(\mathbf{Q} + \mathbf{K}^T\mathbf{R}\mathbf{K})\mathbf{x}\, dt = -\frac{1}{2}(\mathbf{x}^T\mathbf{P}\mathbf{x})\big|_0^\infty = \frac{1}{2}(\mathbf{x}^T(0)\mathbf{P}\mathbf{x}(0)). \tag{3.112}$$

It then follows that J_{LQR} has a minimum that may be obtained by expanding Equation (3.110) and solving the resulting algebraic matrix *Riccati* equation,

$$\mathbf{A}^T\mathbf{P} + \mathbf{P}\mathbf{A}^T - \mathbf{P}\mathbf{B}\mathbf{R}^{-1}\mathbf{B}^T\mathbf{P} + \mathbf{Q} = 0, \tag{3.113}$$

for **P**, by eigenvalue decomposition of the equations representing the optimal solution. The optimal gain is given by,

$$\mathbf{K} = \mathbf{R}^{-1}\mathbf{B}^T\mathbf{P}. \tag{3.114}$$

The actual calculation of the gain matrix **K** is accomplished by using the MATLAB function for applying the LQR algorithm. A crucial property of an LQR controller is that this closed-loop is asymptotically stable as long as the system is at least controllable. Furthermore, LQR controllers are inherently robust with respect to process uncertainty. Moreover, the gain margin of an LQR controller is infinite for gain increase and −6 dB for gain decrease, and the phase margin is at least 60 degrees. Thus, the LQR controllers have some very desirable robustness properties and can be tuned to meet other requirements, and are thus extremely well suited for spacecraft attitude control system design.

A block diagram of the closed-loop attitude control system implemented in practice is shown in Figure 3.10, where we have assumed that all the states are measured and available for feedback and that $\mathbf{C} = \mathbf{I}_{3\times 3}$ and $\mathbf{u} = \mathbf{u}_d$ is the control input to obtain the desired output state $\mathbf{y} = \mathbf{y}_d$. When only some of the states are measured, such as the attitude rates using rate gyroscopes and the wheel speeds using tachometers, all the states are constructed using appropriate full state observers or Kalman filters.

3.8.8 Quaternion Representation of Attitude

The quaternion parameters are used for computation of attitude relative to the orbiting frame to avoid the singularity problem that arises when Euler angles are used to represent the attitude. A quaternion representation is a four-part complex number, which can describe the sequential three-axes rotations as a single rotation about some direction or axis. The quaternion parameters are related to the rotation angle by the relation,

$$\vec{q} = [q_1 \ q_2 \ q_3]^T = \mathbf{n}\sin(\theta_r/2), \quad q_4 = \cos(\theta_r/2), \tag{3.115}$$

where the angle θ_{rr} is the total, single rotation angle about a direction defined by the unit vector **n**.

The attitude quaternion is then defined by the four parameters,

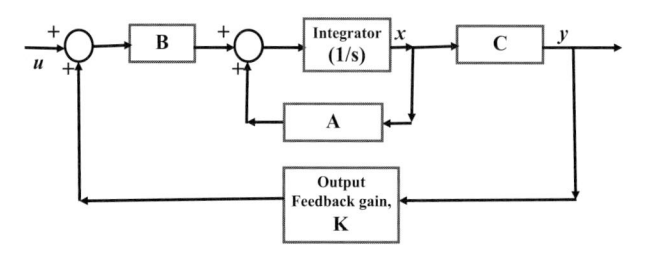

Figure 3.10 Full state feedback control implementation of the spacecraft attitude controller.

$$\mathbf{q} = [q_1 \quad q_2 \quad q_3 \quad q_4]^T. \tag{3.116}$$

Quaternions also satisfy the constraint,

$$q_4^2 + q_1^2 + q_2^2 + q_3^2 = \mathbf{q}^T\mathbf{q} = 1. \tag{3.117}$$

It can be shown that the four quaternion parameters are related to the three (3–2–1 sequence) Euler angles by,

$$\begin{bmatrix} q_1 \\ q_2 \\ q_3 \\ q_4 \end{bmatrix} = \begin{bmatrix} \cos\left(\frac{\psi}{2}\right)\cos\left(\frac{\theta}{2}\right)\sin\left(\frac{\phi}{2}\right) - \sin\left(\frac{\psi}{2}\right)\sin\left(\frac{\theta}{2}\right)\cos\left(\frac{\phi}{2}\right) \\ \cos\left(\frac{\psi}{2}\right)\sin\left(\frac{\theta}{2}\right)\cos\left(\frac{\phi}{2}\right) + \sin\left(\frac{\psi}{2}\right)\cos\left(\frac{\theta}{2}\right)\sin\left(\frac{\phi}{2}\right) \\ -\cos\left(\frac{\psi}{2}\right)\sin\left(\frac{\theta}{2}\right)\sin\left(\frac{\phi}{2}\right) + \sin\left(\frac{\psi}{2}\right)\cos\left(\frac{\theta}{2}\right)\cos\left(\frac{\phi}{2}\right) \\ \cos\left(\frac{\psi}{2}\right)\cos\left(\frac{\theta}{2}\right)\cos\left(\frac{\phi}{2}\right) + \sin\left(\frac{\psi}{2}\right)\sin\left(\frac{\theta}{2}\right)\sin\left(\frac{\phi}{2}\right) \end{bmatrix}. \tag{3.118}$$

The Euler angles may be uniquely extracted from a quaternion as,

$$\begin{aligned} \tan\phi &= 2\frac{q_1 q_4 + q_3 q_2}{q_4^2 - q_1^2 - q_2^2 + q_3^2}, \\ \sin\theta &= -2 \cdot (q_1 q_3 - q_2 q_4), \\ \tan\psi &= 2\frac{q_1 q_2 + q_3 q_4}{q_4^2 + q_1^2 - q_2^2 - q_3^2}. \end{aligned} \tag{3.119}$$

However, since the quaternion components satisfy the normalization equation given by Equation (3.117),

$$\begin{bmatrix} \phi \\ \theta \\ \psi \end{bmatrix} = \begin{bmatrix} \arctan\left(2\frac{q_1 q_4 + q_2 q_3}{1 - 2(q_1^2 + q_2^2)}\right) \\ \arcsin\left(-2(q_1 q_3 - q_2 q_4)\right) \\ \arctan\left(2\frac{q_1 q_2 + q_3 q_4}{1 - 2(q_2^2 + q_3^2)}\right) \end{bmatrix}. \tag{3.120}$$

In terms of the attitude quaternion, the transformation from space-fixed to body-fixed axes is,

$$\mathbf{R}_{IB} = \mathbf{I} + 2q_4 \mathbf{S}(\vec{q}) + 2\mathbf{S}^2(\vec{q}) = \mathbf{I} + \sin\theta_r \mathbf{S}(\mathbf{n}) + (1 - \cos\theta_r)\mathbf{S}^2(\mathbf{n}), \tag{3.121}$$

where $\mathbf{I}$ is the 3×3 unit matrix and $\mathbf{S}(\vec{q})$ is the cross-product operator and is defined by,

$$\mathbf{S}(\vec{q}) = -\mathbf{S}(\vec{q})^T = \begin{bmatrix} 0 & -q_3 & q_2 \\ q_3 & 0 & -q_1 \\ -q_2 & q_1 & 0 \end{bmatrix}. \tag{3.122a}$$

The cross-product operator is also denoted as,

$$\mathbf{S}(\vec{q}) = \vec{q} \times. \tag{3.122b}$$

Thus,

$$\mathbf{R}_{IB} = \mathbf{R}_{BI}^T = \mathbf{I} + \begin{bmatrix} -2(q_2^2 + q_3^2) & 2(q_1 q_2 + q_4 q_3) & 2(q_1 q_3 + q_4 q_2) \\ 2(q_1 q_2 + q_4 q_3) & -2(q_1^2 + q_3^2) & 2(q_2 q_3 - q_4 q_1) \\ 2(q_1 q_3 - q_4 q_2) & 2(q_2 q_3 + q_4 q_1) & -2(q_1^2 + q_2^2) \end{bmatrix}. \quad (3.123)$$

Note that this is the same as,

$$\mathbf{R}_{IB}(\mathbf{q}) = \begin{bmatrix} q_4^2 + q_1^2 - q_2^2 - q_3^2 & 2(q_1 q_2 - q_3 q_4) & 2(q_1 q_3 + q_2 q_4) \\ 2(q_1 q_2 + q_3 q_4) & q_4^2 - q_1^2 + q_2^2 - q_3^2 & 2(q_2 q_3 - q_1 q_4) \\ 2(q_1 q_3 - q_2 q_4) & 2(q_2 q_3 + q_1 q_4) & q_4^2 - q_1^2 - q_2^2 + q_3^2 \end{bmatrix}. \quad (3.124)$$

3.8.9 The Relations between the Quaternion Rates and Angular Velocities

The quaternion rates are related to the angular velocity components by,

$$\frac{d\mathbf{q}}{dt} = \frac{1}{2}\mathbf{A}_\omega(\omega)\mathbf{q}, \quad \omega = \begin{bmatrix} p \\ q \\ r \end{bmatrix}, \quad \omega \times \equiv \omega_\times = \begin{bmatrix} 0 & -r & q \\ r & 0 & -p \\ -q & p & 0 \end{bmatrix}, \quad \mathbf{q} = \begin{bmatrix} q_1 \\ q_2 \\ q_3 \\ q_4 \end{bmatrix},$$

$$\mathbf{A}_\omega(\omega) = \begin{bmatrix} -\omega_\times & \omega \\ -\omega^T & 0 \end{bmatrix} = \begin{bmatrix} 0 & r & -q & p \\ -r & 0 & p & q \\ q & -p & 0 & r \\ -p & -q & -r & 0 \end{bmatrix}. \quad (3.125)$$

Thus,

$$\frac{d}{dt} \begin{bmatrix} q_1 \\ q_2 \\ q_3 \\ q_4 \end{bmatrix} = \frac{1}{2}\mathbf{A}_\omega(\omega) \begin{bmatrix} q_1 \\ q_2 \\ q_3 \\ q_4 \end{bmatrix}, \quad (3.126)$$

which is also expressed,

$$\frac{d}{dt} \begin{bmatrix} q_1 \\ q_2 \\ q_3 \\ q_4 \end{bmatrix} = \frac{1}{2} \begin{bmatrix} q_4 & -q_3 & q_2 \\ q_3 & q_4 & -q_1 \\ -q_2 & q_1 & q_4 \\ -q_1 & -q_2 & -q_3 \end{bmatrix} \begin{bmatrix} p \\ q \\ r \end{bmatrix}. \quad (3.127)$$

This relation may be compactly expressed as,

$$\frac{d}{dt}[\vec{q}\ q_4] = \frac{1}{2}\begin{bmatrix} q_4 \mathbf{I} + \mathbf{S}(\vec{q}) \\ -\vec{q} \end{bmatrix} \begin{bmatrix} p \\ q \\ r \end{bmatrix} \equiv \mathbf{\Gamma}(\mathbf{q}) \begin{bmatrix} p \\ q \\ r \end{bmatrix}, \quad \mathbf{\Gamma}(\mathbf{q}) = \frac{1}{2}\begin{bmatrix} q_4 \mathbf{I} + \mathbf{S}(\vec{q}) \\ -\vec{q} \end{bmatrix}. \quad (3.128)$$

Linearization about $q_4 = q_{40} = 1$ and $\vec{q} = \vec{q}_0 = [0\ 0\ 0]^T$, which corresponds to $\theta_r = 0$, leads to,

3.8 Momentum and Reaction Wheels

$$\frac{d}{dt}[\vec{q}\; q_4] = \frac{1}{2}\begin{bmatrix} \mathbf{I} \\ 0 \end{bmatrix}\begin{bmatrix} p \\ q \\ r \end{bmatrix}. \qquad (3.129)$$

Hence, it follows that,

$$\frac{d}{dt}\vec{q} = \frac{d}{dt}\begin{bmatrix} q_1 \\ q_2 \\ q_3 \end{bmatrix} = \frac{1}{2}\mathbf{I}\begin{bmatrix} p \\ q \\ r \end{bmatrix}. \qquad (3.130)$$

One may compare these with the linearized relations between the Euler angle rates and the angular velocity components, which are,

$$\frac{d}{dt}\begin{bmatrix} \phi \\ \theta \\ \psi \end{bmatrix} = \mathbf{I}\begin{bmatrix} p \\ q \\ r \end{bmatrix}. \qquad (3.131)$$

Thus, we have an important useful relationship between the linearized (small angle) Euler angle components and quaternion components, given by,

$$\begin{bmatrix} q_1 \\ q_2 \\ q_3 \end{bmatrix} = \frac{1}{2}\begin{bmatrix} \phi \\ \theta \\ \psi \end{bmatrix}, \quad \begin{bmatrix} \phi \\ \theta \\ \psi \end{bmatrix} = 2\begin{bmatrix} q_1 \\ q_2 \\ q_3 \end{bmatrix}. \qquad (3.132)$$

These relationships permit the generalization of the *linear* control laws derived in the Euler angle domain with the small angle assumption to the case of linear control laws with direct quaternion (perturbation) feedback. When using these linear relations, the quaternion components must be renormalized at every time step using the relations,

$$\mathbf{q} = [q_1 \; q_2 \; q_3 \; q_4]^T / \sqrt{(\mathbf{q}^T\mathbf{q})}. \qquad (3.133)$$

The transformation matrix may also be linearized. The transformation matrix is,

$$\mathbf{R}_{IB} = \mathbf{I} + 2q_4\mathbf{S}(\bar{q}) + 2\mathbf{S}^2(\vec{q}). \qquad (3.134)$$

Since, $\mathbf{S}^T(\bar{q}) = -\mathbf{S}(\bar{q})$,

$$\mathbf{R}_{BI} = \mathbf{I} - 2q_4\mathbf{S}(\bar{q}) + 2\mathbf{S}^2(\bar{q}). \qquad (3.135)$$

Hence, upon linearization,

$$\mathbf{R}_{IB} = \mathbf{I} + 2q_4\mathbf{S}(\bar{q}) + 2\mathbf{S}^2(\bar{q}) \approx \mathbf{I} + 2\mathbf{S}(\bar{q}), \qquad (3.136)$$
$$\mathbf{R}_{BI} = \mathbf{I} - 2q_4\mathbf{S}(\bar{q}) + 2\mathbf{S}^2(\bar{q}) \approx \mathbf{I} - 2\mathbf{S}(\bar{q}). \qquad (3.137)$$

The transformation relating the body axes to the space-fixed inertial reference axes is expressed in terms of three angles, known as the Euler angles, by successively rotating the body x axis by the negative roll angle $-\phi$, the roll-rotated body y axis by $-\theta$, and the roll and pitch-rotated body z axis by $-\psi$. The transformation is defined by,

$$\begin{bmatrix} x_I \\ y_I \\ z_I \end{bmatrix} = \mathbf{T}_{IB} \times \begin{bmatrix} x_B \\ y_B \\ z_B \end{bmatrix}, \qquad (3.138)$$

where

$$\mathbf{T}_{IB} = \begin{bmatrix} \cos\psi & -\sin\psi & 0 \\ \sin\psi & \cos\psi & 0 \\ 0 & 0 & 1 \end{bmatrix} \begin{bmatrix} \cos\theta & 0 & \sin\theta \\ 0 & 1 & 0 \\ -\sin\theta & 0 & \cos\theta \end{bmatrix} \begin{bmatrix} 1 & 0 & 0 \\ 0 & \cos\phi & -\sin\phi \\ 0 & \sin\phi & \cos\phi \end{bmatrix}$$

$$= \begin{bmatrix} c\psi c\theta & c\psi s\theta s\phi - s\psi c\phi & c\psi s\theta c\phi + s\psi s\phi \\ s\psi c\theta & s\psi s\theta s\phi - c\psi c\phi & s\psi s\theta c\phi - c\psi s\phi \\ -s\theta & c\theta s\phi & c\theta c\phi \end{bmatrix},$$ (3.139)

and $c\theta = \cos\theta$, $c\phi = \cos\phi$, $c\psi = \cos\psi$, $s\theta = \sin\theta$, $s\phi = \sin\phi$, and $s\psi = \sin\psi$, and the subscript B refers to the body axes, while the subscript I refers to the space-fixed inertial axes. Each of the three component matrices in $\mathbf{T}_{IB}$ is an orthogonal matrix and so is their product. The transformations represent the yaw, pitch, and roll sequence of transformations from inertial to body frame and the reverse rotations from the body to inertial frames.

Comparing with,

$$\mathbf{R}_{IB}(\mathbf{q}) = \begin{bmatrix} q_4^2 + q_1^2 - q_2^2 - q_3^2 & 2(q_1 q_2 - q_3 q_4) & 2(q_1 q_3 + q_2 q_4) \\ 2(q_1 q_2 + q_3 q_4) & q_4^2 - q_1^2 + q_2^2 - q_3^2 & 2(q_2 q_3 - q_1 q_4) \\ 2(q_1 q_3 - q_2 q_4) & 2(q_2 q_3 + q_1 q_4) & q_4^2 - q_1^2 - q_2^2 + q_3^2 \end{bmatrix},$$ (3.140)

$$\sin\theta = -2(q_1 q_3 - q_2 q_4),$$
$$\tan\psi = 2(q_1 q_2 + q_3 q_4)/(q_4^2 + q_1^2 - q_2^2 - q_3^2),$$ (3.141)
$$\tan\phi = 2(q_2 q_3 + q_1 q_4)/(q_4^2 - q_1^2 - q_2^2 + q_3^2).$$

The inverse transformation is defined by,

$$\begin{bmatrix} x_B \\ y_B \\ z_B \end{bmatrix} = \mathbf{T}_{IB}^{-1} \times \begin{bmatrix} x_I \\ y_I \\ z_I \end{bmatrix} = \mathbf{T}_{BI} \times \begin{bmatrix} x_I \\ y_I \\ z_I \end{bmatrix},$$ (3.142)

where

$$\mathbf{T}_{BI} = \begin{bmatrix} 1 & 0 & 0 \\ 0 & \cos\phi & \sin\phi \\ 0 & -\sin\phi & \cos\phi \end{bmatrix} \begin{bmatrix} \cos\theta & 0 & -\sin\theta \\ 0 & 1 & 0 \\ \sin\theta & 0 & \cos\theta \end{bmatrix} \begin{bmatrix} \cos\psi & \sin\psi & 0 \\ -\sin\psi & \cos\psi & 0 \\ 0 & 0 & 1 \end{bmatrix} = \mathbf{T}_{IB}^T.$$

(3.143)

In terms of the quaternion components,

$$\mathbf{R}_{BI}(\mathbf{q}) = \begin{bmatrix} q_1^2 - q_2^2 - q_3^2 + q_4^2 & 2(q_1 q_2 + q_3 q_4) & 2(q_1 q_3 - q_2 q_4) \\ 2(q_1 q_2 - q_3 q_4) & -q_1^2 + q_2^2 - q_3^2 + q_4^2 & 2(q_2 q_3 + q_1 q_4) \\ 2(q_1 q_3 + q_2 q_4) & 2(q_2 q_3 - q_1 q_4) & -q_1^2 - q_2^2 + q_3^2 + q_4^2 \end{bmatrix}.$$ (3.144)

The inverse relations are not unique and are,

$$q_1 = \frac{1}{4q_4}(R_{23} - R_{32}) = \pm\frac{1}{2}\sqrt{1 + R_{11} - R_{22} - R_{33}},$$
$$q_2 = \frac{1}{4q_1}(R_{12} + R_{21}) = \frac{1}{4q_4}(R_{23} - R_{32}),$$
$$q_3 = \frac{1}{4q_4}(R_{12} - R_{21}) = \frac{1}{4q_1}(R_{31} + R_{13}),\qquad(3.145)$$
$$q_4 = \frac{1}{4q_1}(R_{23} - R_{32}) = \pm\frac{1}{2}\sqrt{1 + R_{11} + R_{22} + R_{33}},$$

where $\mathbf{R}_{BI} = \{R_{ij}\}$.

3.8.10 The Gravity-Gradient Stability Equations in Terms of the Quaternion

The gravity-gradient stability equations may be expressed in a compact form using the quaternion representation of attitude. Thus, the attitude dynamics and the gravity-gradient torques are, respectively, given by,

$$\mathbf{I}\dot{\boldsymbol{\omega}} + \boldsymbol{\omega} \times \mathbf{I}\boldsymbol{\omega} = \mathbf{M} + \mathbf{M}_{gg},\quad \mathbf{M}_{gg} \equiv 3n^2\mathbf{c}_\times(\mathbf{q})\mathbf{I}\mathbf{c}(\mathbf{q}).\qquad(3.146)$$

If we let,

$$\boldsymbol{\omega} \to \boldsymbol{\omega} + \delta\boldsymbol{\omega},\quad \mathbf{q} \to \mathbf{q} + \delta\mathbf{q},\qquad(3.147)$$

then it follows that the perturbations $\delta\boldsymbol{\omega}$ and $\delta\mathbf{q}$ satisfy,

$$\mathbf{I}\delta\dot{\boldsymbol{\omega}} + \boldsymbol{\omega}_\times\mathbf{I}\delta\boldsymbol{\omega} - (\mathbf{I}\boldsymbol{\omega})_\times\delta\boldsymbol{\omega} = \mathbf{I}\delta\dot{\boldsymbol{\omega}} + (\boldsymbol{\omega}_\times\mathbf{I} - (\mathbf{I}\boldsymbol{\omega})_\times)\delta\boldsymbol{\omega} = \delta\mathbf{M} + \delta\mathbf{M}_{gg},\qquad(3.148)$$

and

$$\delta\mathbf{M}_{gg} \equiv 3n^2\mathbf{c}_\times(\mathbf{q})\mathbf{I}\delta\mathbf{c}(\mathbf{q}) - 3n^2(\mathbf{I}\mathbf{c}(\mathbf{q}))_\times\delta\mathbf{c}(\mathbf{q}) = 3n^2\big(\mathbf{c}_\times(\mathbf{q})\mathbf{I} - (\mathbf{I}\mathbf{c}(\mathbf{q}))_\times\big)\delta\mathbf{c}(\mathbf{q}),\qquad(3.149)$$

where $\boldsymbol{\omega}$ and $\mathbf{q}$ are evaluated in an equilibrium state.

The direction vector of the center of gravity of the satellite pointing to the Earth is given by the last column of $\mathbf{T}_{BR}$, the transformation from the Earth orbiting frame to the body-fixed frame of the satellite, as,

$$\mathbf{c} = [c_1\ c_2\ c_3]^T.\qquad(3.150)$$

The corresponding cross-product operator $\mathbf{c}_\times$ is defined as,

$$\mathbf{c}_\times = \begin{bmatrix} 0 & -c_3 & c_2 \\ c_3 & 0 & -c_1 \\ -c_2 & c_1 & 0 \end{bmatrix}.\qquad(3.151)$$

Depending on the choice of the body axes of the satellite, the CG of each link pointing to the Earth is given either by the first, second, or last column of $\mathbf{T}_{BR}(\mathbf{q})$. Setting $\mathbf{T}_{BR}(\mathbf{q}) \equiv \mathbf{R}_{BI}(\mathbf{q})$ from the preceding section and from $\mathbf{T}_{BR}(\mathbf{q})$, we have,

$$\mathbf{c} = \begin{bmatrix} q_4^2 + q_1^2 - q_2^2 - q_3^2 \\ 2(q_1 q_2 + q_3 q_4) \\ 2(q_1 q_3 - q_2 q_4) \end{bmatrix} \quad \mathbf{c} = \begin{bmatrix} 2(q_1 q_2 - q_3 q_4) \\ q_4^2 - q_1^2 + q_2^2 - q_3^2 \\ 2(q_2 q_3 + q_1 q_4) \end{bmatrix} \quad \text{or} \quad \mathbf{c} = \begin{bmatrix} 2(q_1 q_3 - q_2 q_4) \\ 2(q_2 q_3 + q_1 q_4) \\ q_4^2 - q_1^2 - q_2^2 + q_3^2 \end{bmatrix}.$$

(3.152)

In what follows we assume for simplicity that,

$$\mathbf{c} = \begin{bmatrix} 2(q_1 q_3 - q_2 q_4) \\ 2(q_2 q_3 + q_1 q_4) \\ q_4^2 - q_1^2 - q_2^2 + q_3^2 \end{bmatrix}. \tag{3.153}$$

Hence, the gravity-gradient torques may be linearized and expressed as,

$$\delta \mathbf{M}_{gg} \equiv 3n^2 \mathbf{c}_\times(\mathbf{q}) \mathbf{I} \delta \mathbf{c}(\mathbf{q}) - 3n^2 (\mathbf{Ic}(\mathbf{q}))_\times \delta \mathbf{c}(\mathbf{q}) = 3n^2 \left(\mathbf{c}_\times(\mathbf{q}) \mathbf{I} - (\mathbf{Ic}(\mathbf{q}))_\times \right) \frac{d\mathbf{c}(\mathbf{q})}{d\mathbf{q}} \delta \mathbf{q},$$

(3.154)

with,

$$\mathbf{C} \equiv \frac{d\mathbf{c}(\mathbf{q})}{d\mathbf{q}} = 2 \begin{bmatrix} q_3 & -q_4 & q_1 & -q_2 \\ q_4 & q_3 & q_2 & q_1 \\ -q_1 & -q_2 & q_3 & q_4 \end{bmatrix}. \tag{3.155}$$

Thus, the vector equations for $\delta \mathbf{q}$ and $\delta \boldsymbol{\omega}$ are, respectively, given by,

$$\frac{d\delta \mathbf{q}}{dt} = \frac{1}{2} \mathbf{A}_\omega(\omega) \delta \mathbf{q} + \frac{1}{2} \Gamma(\mathbf{q}) \delta \boldsymbol{\omega}, \tag{3.156}$$

$$\delta \dot{\boldsymbol{\omega}} = 3n^2 \mathbf{I}^{-1} \left(\mathbf{c}_\times(\mathbf{q}) \mathbf{I} - (\mathbf{Ic}(\mathbf{q}))_\times \right) \frac{d\mathbf{c}(\mathbf{q})}{d\mathbf{q}} \delta \mathbf{q} - \mathbf{I}^{-1} \left(\boldsymbol{\omega}_\times \mathbf{I} - (\mathbf{I}\boldsymbol{\omega})_\times \right) \delta \boldsymbol{\omega} + \mathbf{I}^{-1} \delta \mathbf{M}.$$

(3.157)

In matrix form,

$$\begin{bmatrix} \delta \dot{\mathbf{q}} \\ \delta \dot{\boldsymbol{\omega}} \end{bmatrix} = \frac{1}{2} \begin{bmatrix} \mathbf{A}_\omega(\omega) & \Gamma(\mathbf{q}) \\ 6n^2 \mathbf{I}^{-1} \left(\mathbf{c}_\times(\mathbf{q}) \mathbf{I} - (\mathbf{Ic}(\mathbf{q}))_\times \right) \frac{d\mathbf{c}(\mathbf{q})}{d\mathbf{q}} & -2\mathbf{I}^{-1} \left(\boldsymbol{\omega}_\times \mathbf{I} - (\mathbf{I}\boldsymbol{\omega})_\times \right) \end{bmatrix} \begin{bmatrix} \delta \mathbf{q} \\ \delta \boldsymbol{\omega} \end{bmatrix} + \begin{bmatrix} \mathbf{0} \\ \mathbf{I}^{-1} \delta \mathbf{M} \end{bmatrix}.$$

(3.158)

These equations could be transformed to the discrete time domain and used to synthesize attitude control laws for regulating the orientation of a satellite. Their application to the attitude regulation of a satellite with an attached manipulator in an equilibrium state is discussed in the next chapter.

3.9   Definition of the General Control Problem with CMG Actuation

Consider a rigid spacecraft with the spacecraft's body denoted by B and controlled by a single-gimbaled variable speed CMG, as shown in Figure 3.11.

3.9 Definition of the General Control Problem with CMG Actuation

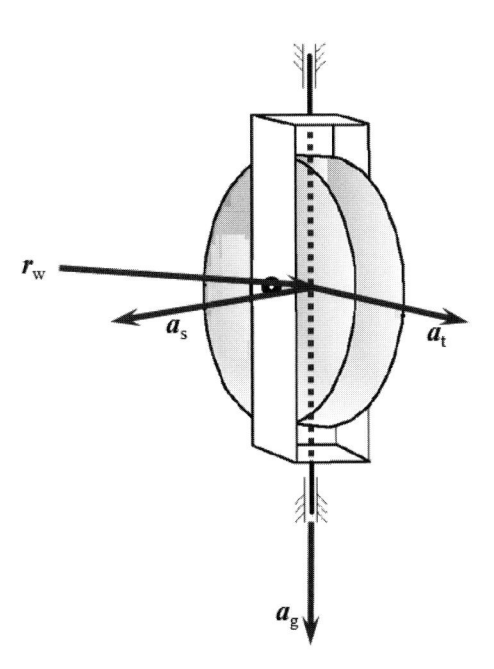

Figure 3.11 A single-gimbaled CMG showing the rotor spin axis, the gimbal rotation axis, and the third axis mutually perpendicular to the other two. Also shown is the position vector r_w of the origin of the CMG fixed three-axis reference frame.

The system is free to move space, and the position coordinates of a point in space are expressed in terms of a frame fixed in space and denoted in vectrix notion by $\mathbf{F_i}$ (it may be recalled that a vectrix is simply a matrix or vector with vectorial elements.). To develop the equations of motion of the spacecraft, the CMG, and its rotor we follow the methodology of Romano and Agrawal [11]. Consider also a reference frame fixed at the center of mass of the CMG gimbal with a rotor spin axis $\mathbf{a}_s$, the gimbal rotation axis $\mathbf{a}_g$, and the third axis mutually perpendicular to the other two and defined by $\mathbf{a}_t = \mathbf{a}_s \times \mathbf{a}_g$. In vectrix notation, the frame fixed to the gimbal is defined by,

$$\mathbf{F_g} = \begin{bmatrix} \mathbf{a}_s & \mathbf{a}_g & \mathbf{a}_t \end{bmatrix}^T. \tag{3.159}$$

To develop the nonlinear equations of motion of the spacecraft controlled by the CMG consider the moment of momentum of the spacecraft and the CMG. The total moment of momentum of the spacecraft and the CMG about their center of mass is given by,

$$\mathbf{h}_O = \mathbf{h}_b + \mathbf{h}_{rg} + m_{CMG} \vec{\mathbf{I}} \; (\mathbf{r}_w \cdot \mathbf{r}_w - \mathbf{r}_w \mathbf{r}_w) \cdot \boldsymbol{\omega}, \tag{3.160}$$

where $\vec{\mathbf{I}}$ is a unit dyadic, $\mathbf{r_w}$ is the vector position of the origin of the CMG fixed three-axis reference frame relative to the combined mass of spacecraft and the CMG about their center of mass O, $\mathbf{h}_b$ is the moment of momentum of the spacecraft's body about O, $\mathbf{h}_{rg} = \mathbf{h}_r + \mathbf{h}_g$ is the sum of the moment of momentum of the CMG rotor, the moment of momentum of the gimbal about the combined center of mass O_w, m_{CMG} is the total mass of the CMG, and $\boldsymbol{\omega}$ is the body angular velocity vector of the spacecraft's body relative to the inertial frame. Furthermore, the moment of momentum components are given by,

$$\mathbf{h}_b = \mathbf{I}_b \cdot \boldsymbol{\omega}, \quad \mathbf{h}_g = \mathbf{I}_g \cdot (\boldsymbol{\omega} + \boldsymbol{\omega}_{gb}); \quad \text{and} \quad \mathbf{h}_r = \mathbf{I}_r \cdot (\boldsymbol{\omega} + \boldsymbol{\omega}_{gb} + \boldsymbol{\omega}_{rg}), \quad (3.161)$$

where $\mathbf{I}_b$, $\mathbf{I}_g$, and $\mathbf{I}_r$ are the moments of inertia dyadics, respectively, of the spacecraft's body, the CMG gimbal, and the CMG rotor, $\boldsymbol{\omega}_{gb}$ and $\boldsymbol{\omega}_{rg}$ are the gimbal angular velocity vector relative to spacecraft's body and the CMG rotor's angular velocity vector relative to the gimbal, respectively. The center of masses of the gimbal and rotor are assumed to coincide with the center of mass of the combined CMG and, as a consequence, the center of mass of the overall spacecraft does not change during the motion of the CMG rotor with respect to the gimbal. Thus,

$$\mathbf{h}_O = \mathbf{h}_b + \mathbf{h}_r + \mathbf{h}_g + m_{CMG} \vec{\mathbf{I}} \ (\mathbf{r}_w \cdot \mathbf{r}_w - \mathbf{r}_w \mathbf{r}_w) \cdot \boldsymbol{\omega}. \quad (3.162)$$

Given the transformations from the gimbal frame to the spacecraft's body frame $\mathbf{T}_{bg}$ and the transformation from the CMG rotor frame to the gimbal frame $\mathbf{T}_{gr}$, one may express $\mathbf{h}_b$, $\mathbf{h}_g$, and $\mathbf{h}_r$ as,

$$\mathbf{h}_b = \mathbf{I}_b \boldsymbol{\omega}, \quad \mathbf{h}_g = \mathbf{T}_{bg} \mathbf{I}_g (\mathbf{T}_{gb} \boldsymbol{\omega} + \boldsymbol{\omega}_{gb}),$$
$$\mathbf{h}_r = \mathbf{T}_{bg} (\mathbf{T}_{gr} \mathbf{I}_r (\mathbf{T}_{rg} \mathbf{T}_{gb} \boldsymbol{\omega} + \mathbf{T}_{rg} \boldsymbol{\omega}_{gb} + \boldsymbol{\omega}_{rg})) = \mathbf{T}_{bg} (\mathbf{I}_{rg} (\mathbf{T}_{gb} \boldsymbol{\omega} + \boldsymbol{\omega}_{gb} + \mathbf{T}_{gr} \boldsymbol{\omega}_{rg})),$$
$$(3.163)$$

where we have assumed the CMG rotor to be rotating about an axisymmetric axis so,

$$\mathbf{I}_{rg} = \mathbf{T}_{gr} \mathbf{I}_r \mathbf{T}_{rg} = \mathbf{I}_r. \quad (3.164)$$

Further,

$$\boldsymbol{\omega}_{rg} = [1 \ 0 \ 0]^T \Omega \quad \text{and} \quad \mathbf{T}_{rg}[1 \ 0 \ 0]^T \Omega = [1 \ 0 \ 0]^T \Omega, \quad (3.165)$$

where Ω is spin angular velocity of the CMG rotor about the spin axis $\mathbf{a}_s$ with respect to the gimbal. Thus,

$$\mathbf{h}_r = \mathbf{T}_{bg} \mathbf{I}_r \mathbf{T}_{gb} (\boldsymbol{\omega} + \mathbf{T}_{bg} \boldsymbol{\omega}_{gb} + \mathbf{T}_{bg}[1 \ 0 \ 0]^T \Omega). \quad (3.166)$$

Furthermore, we may assume that,

$$\boldsymbol{\omega}_{gb} = [0 \ 1 \ 0]^T \dot{\delta}. \quad (3.167)$$

where $\dot{\delta}$ is the angular velocity of the gimbal about the gimbal axis $\mathbf{a}_g$. Thus,

$$\mathbf{h}_r = \mathbf{T}_{bg} \mathbf{I}_r \mathbf{T}_{gb} (\boldsymbol{\omega} + \mathbf{T}_{bg}[0 \ 1 \ 0]^T \dot{\delta} + \mathbf{T}_{bg}[1 \ 0 \ 0]^T \Omega),$$
$$\mathbf{h}_g = \mathbf{T}_{bg} \mathbf{I}_g (\mathbf{T}_{gb} \boldsymbol{\omega} + \boldsymbol{\omega}_{gb}) = \mathbf{T}_{bg} \mathbf{I}_g \mathbf{T}_{gb} (\boldsymbol{\omega} + \mathbf{T}_{bg}[0 \ 1 \ 0]^T \dot{\delta}), \quad (3.168)$$

and,

$$\mathbf{h}_O = \mathbf{I}_b \boldsymbol{\omega} + m_{CMG} \vec{\mathbf{I}} \ (\mathbf{r}_w \cdot \mathbf{r}_w - \mathbf{r}_w \mathbf{r}_w) \cdot \boldsymbol{\omega} + \mathbf{h}_r + \mathbf{h}_g. \quad (3.169)$$

If we let,

$$\mathbf{I}_B \boldsymbol{\omega} = \mathbf{I}_b \boldsymbol{\omega} + m_{CMG} \vec{\mathbf{I}} \ (\mathbf{r}_w \cdot \mathbf{r}_w - \mathbf{r}_w \mathbf{r}_w) \cdot \boldsymbol{\omega}, \quad (3.170)$$

and $\mathbf{I}_{rg} = \mathbf{I}_r + \mathbf{I}_g$, the total moment of momentum of the spacecraft and CMG in the spacecraft's body axes is,

$$\mathbf{h}_O = \left(\mathbf{I}_B + \mathbf{T}_{bg}\mathbf{I}_{rg}\mathbf{T}_{gb}\right)\boldsymbol{\omega} + \mathbf{T}_{bg}\mathbf{I}_{rg}[0 \ 1 \ 0]^T\dot{\delta} + \mathbf{T}_{bg}\mathbf{I}_r[1 \ 0 \ 0]^T\Omega, \quad (3.171)$$

while the total moment of momentum of the CMG in the spacecraft's body axes is,

$$\mathbf{h}_{rg} = \mathbf{T}_{bg}\mathbf{I}_{rg}\mathbf{T}_{gb}\boldsymbol{\omega} + \mathbf{T}_{bg}\mathbf{I}_{rg}[0 \ 1 \ 0]^T\dot{\delta} + \mathbf{T}_{bg}\mathbf{I}_r[1 \ 0 \ 0]^T\Omega. \quad (3.172)$$

The moment of momentum of the CMG rotor alone in the spacecraft's body axes is,

$$\mathbf{h}_r = \mathbf{T}_{bg}\mathbf{I}_r\mathbf{T}_{gb}\boldsymbol{\omega} + \mathbf{T}_{bg}\mathbf{I}_r[0 \ 1 \ 0]^T\dot{\delta} + \mathbf{T}_{bg}\mathbf{I}_r[1 \ 0 \ 0]^T\Omega. \quad (3.173)$$

The equations of motion of the combined spacecraft and the CMG, the equations of motion of the combined CMG gimbal and rotor, and the equations of motion of the CMG rotor alone, respectively, are,

$$\dot{\mathbf{h}}_O + \boldsymbol{\omega} \times \mathbf{h}_O = \boldsymbol{\tau}_O, \quad \dot{\mathbf{h}}_{rg} + \boldsymbol{\omega} \times \mathbf{h}_{rg} = \boldsymbol{\tau}_{rg}, \quad \text{and} \quad \dot{\mathbf{h}}_r + \boldsymbol{\omega} \times \mathbf{h}_r = \boldsymbol{\tau}_r. \quad (3.174)$$

In these equations, $\boldsymbol{\tau}_O$, $\boldsymbol{\tau}_{rg}$, and $\boldsymbol{\tau}_r$ are, respectively, the control torques acting on the spacecraft body, the CMG gimbal, and the CMG rotor alone.

Also, since,

$$\mathbf{T}_{bg} = \mathbf{T}_{gb}^T = [\mathbf{a}_s \ \mathbf{a}_g \ \mathbf{a}_t], \quad \dot{\mathbf{T}}_{bg} = \mathbf{T}_{bg}\boldsymbol{\omega}_{gb\times} \quad \text{and} \quad \dot{\mathbf{T}}_{gb} = -\boldsymbol{\omega}_{gb\times}\mathbf{T}_{gb}, \quad (3.175)$$

where,

$$\boldsymbol{\omega}_{gb\times} \equiv \begin{bmatrix} 0 & -\omega_{gb,3} & \omega_{gb,2} \\ \omega_{gb,3} & 0 & -\omega_{gb,1} \\ -\omega_{gb,2} & \omega_{gb,1} & 0 \end{bmatrix} = \dot{\delta}\begin{bmatrix} 0 & 0 & 1 \\ 0 & 0 & 0 \\ -1 & 0 & 0 \end{bmatrix}, \quad (3.176)$$

it follows that,

$$\dot{\mathbf{T}}_{bg} = \mathbf{T}_{bg}\boldsymbol{\omega}_{gb\times} = \dot{\delta}[-\mathbf{a}_t \ 0 \ \mathbf{a}_s] \quad \text{and} \quad \dot{\mathbf{T}}_{gb} = -\boldsymbol{\omega}_{gb\times}\mathbf{T}_{gb} = \dot{\delta}[-\mathbf{a}_t \ 0 \ \mathbf{a}_s]^T. \quad (3.177)$$

Thus if, $\bar{\mathbf{T}} \equiv [-\mathbf{a}_t \ 0 \ \mathbf{a}_s]$, then

$$\dot{\mathbf{T}}_{bg} = \dot{\delta}\bar{\mathbf{T}} \quad \text{and} \quad \dot{\mathbf{T}}_{gb} = \dot{\delta}\bar{\mathbf{T}}^T. \quad (3.178)$$

Thus, the equations of motion of the combined spacecraft and the CMG, the equations of motion of the combined CMG gimbal and rotor, and the equations of motion of the CMG rotor alone, respectively, may be expressed as,

$$\dot{\mathbf{h}}_O + \boldsymbol{\omega} \times \mathbf{h}_O \equiv \left(\mathbf{I}_B + \mathbf{T}_{bg}\mathbf{I}_{rg}\mathbf{T}_{gb}\right)\dot{\boldsymbol{\omega}} + \mathbf{T}_{bg}\mathbf{I}_{rg}\mathbf{e}_2\ddot{\delta} + \mathbf{T}_{bg}\mathbf{I}_r\mathbf{e}_1\dot{\Omega}$$
$$+ \left(\left(\bar{\mathbf{T}}\mathbf{I}_{rg}\mathbf{T}_{gb} + \mathbf{T}_{bg}\mathbf{I}_{rg}\bar{\mathbf{T}}^T\right)\boldsymbol{\omega} + \bar{\mathbf{T}}\mathbf{I}_{rg}\mathbf{e}_2\dot{\delta} + \bar{\mathbf{T}}\mathbf{I}_r\mathbf{e}_1\Omega\right)\dot{\delta}$$
$$+ \boldsymbol{\omega}_\times\left(\mathbf{I}_B + \mathbf{T}_{bg}\mathbf{I}_{rg}\mathbf{T}_{gb}\right)\boldsymbol{\omega} + \boldsymbol{\omega}_\times\mathbf{T}_{bg}\mathbf{I}_{rg}\mathbf{e}_2\dot{\delta} + \boldsymbol{\omega}_\times\mathbf{T}_{bg}\mathbf{I}_r\mathbf{e}_1\Omega = \boldsymbol{\tau}_O,$$
$$\dot{\mathbf{h}}_{rg} + \boldsymbol{\omega} \times \mathbf{h}_{rg} \equiv \mathbf{T}_{bg}\mathbf{I}_{rg}\mathbf{T}_{gb}\dot{\boldsymbol{\omega}} + \mathbf{T}_{bg}\mathbf{I}_{rg}\mathbf{e}_2\ddot{\delta} + \mathbf{T}_{bg}\mathbf{I}_r\mathbf{e}_1\dot{\Omega}$$

$$+\left(\left(\bar{\mathbf{T}}\mathbf{I}_{rg}\mathbf{T}_{gb}+\mathbf{T}_{bg}\mathbf{I}_{rg}\bar{\mathbf{T}}^T\right)\boldsymbol{\omega}+\bar{\mathbf{T}}\mathbf{I}_{rg}\mathbf{e}_2\dot{\delta}+\bar{\mathbf{T}}\mathbf{I}_r\mathbf{e}_1\Omega\right)\dot{\delta}$$
$$+\boldsymbol{\omega}_\times\left(\mathbf{T}_{bg}\mathbf{I}_{rg}\mathbf{T}_{gb}\right)\boldsymbol{\omega}+\boldsymbol{\omega}_\times\mathbf{T}_{bg}\mathbf{I}_{rg}\mathbf{e}_2\dot{\delta}+\boldsymbol{\omega}_\times\mathbf{T}_{bg}\mathbf{I}_r\mathbf{e}_1\Omega=\boldsymbol{\tau}_{rg},$$
$$\dot{\mathbf{h}}_r+\boldsymbol{\omega}\times\mathbf{h}_r\equiv\mathbf{T}_{bg}\mathbf{I}_r\mathbf{T}_{gb}\dot{\boldsymbol{\omega}}+\mathbf{T}_{bg}\mathbf{I}_r\mathbf{e}_2\ddot{\delta}+\mathbf{T}_{bg}\mathbf{I}_r\mathbf{e}_1\dot{\Omega}$$
$$+\left(\left(\bar{\mathbf{T}}\mathbf{I}_r\mathbf{T}_{gb}+\mathbf{T}_{bg}\mathbf{I}_r\bar{\mathbf{T}}^T\right)\boldsymbol{\omega}+\bar{\mathbf{T}}\mathbf{I}_r\mathbf{e}_2\dot{\delta}+\bar{\mathbf{T}}\mathbf{I}_r\mathbf{e}_1\Omega\right)\dot{\delta}$$
$$+\boldsymbol{\omega}_\times\left(\mathbf{T}_{bg}\mathbf{I}_r\mathbf{T}_{gb}\right)\boldsymbol{\omega}+\boldsymbol{\omega}_\times\mathbf{T}_{bg}\mathbf{I}_r\mathbf{e}_2\dot{\delta}+\boldsymbol{\omega}_\times\mathbf{T}_{bg}\mathbf{I}_r\mathbf{e}_1\Omega=\boldsymbol{\tau}_r, \quad (3.179)$$

with $\mathbf{e}_1=[1\ 0\ 0]^T$, $\mathbf{e}_2=[0\ 1\ 0]^T$.

The first of these equations may be expressed as,

$$\mathbf{I}_B\dot{\boldsymbol{\omega}}+\boldsymbol{\omega}_\times\mathbf{I}_B\boldsymbol{\omega}=\boldsymbol{\tau}_O-\boldsymbol{\tau}_{rg}. \quad (3.180)$$

Similarly, the second of these three equations is expressed as,

$$\mathbf{T}_{bg}\mathbf{I}_g\mathbf{T}_{gb}\dot{\boldsymbol{\omega}}+\mathbf{T}_{bg}\mathbf{I}_g\mathbf{e}_2\ddot{\delta}+\left(\left(\bar{\mathbf{T}}\mathbf{I}_g\mathbf{T}_{gb}+\mathbf{T}_{bg}\mathbf{I}_g\bar{\mathbf{T}}^T\right)\boldsymbol{\omega}+\bar{\mathbf{T}}\mathbf{I}_{rg}\mathbf{e}_2\dot{\delta}\right)\dot{\delta}$$
$$+\boldsymbol{\omega}_\times\left(\mathbf{T}_{bg}\mathbf{I}_g\mathbf{T}_{gb}\right)\boldsymbol{\omega}+\boldsymbol{\omega}_\times\mathbf{T}_{bg}\mathbf{I}_g\mathbf{e}_2\dot{\delta}=\boldsymbol{\tau}_{rg}-\boldsymbol{\tau}_r. \quad (3.181)$$

Following Romano and Agrawal [11], by projecting the equations of motion of the combined CMG gimbal and rotor and the equations of motion of the CMG rotor alone along the gimbal axis and the spin axis, respectively, one obtains,

$$\mathbf{a}_g^T\mathbf{T}_{bg}\mathbf{I}_g\mathbf{T}_{gb}\dot{\boldsymbol{\omega}}+\mathbf{a}_g^T\mathbf{T}_{bg}\mathbf{I}_g\mathbf{e}_2\ddot{\delta}+\mathbf{a}_g^T\left(\left(\bar{\mathbf{T}}\mathbf{I}_g\mathbf{T}_{gb}+\mathbf{T}_{bg}\mathbf{I}_g\bar{\mathbf{T}}^T\right)\boldsymbol{\omega}+\bar{\mathbf{T}}\mathbf{I}_{rg}\mathbf{e}_2\dot{\delta}\right)\dot{\delta}$$
$$+\mathbf{a}_g^T\boldsymbol{\omega}_\times\left(\mathbf{T}_{bg}\mathbf{I}_g\mathbf{T}_{gb}\right)\boldsymbol{\omega}+\mathbf{a}_g^T\boldsymbol{\omega}_\times\mathbf{T}_{bg}\mathbf{I}_g\mathbf{e}_2\dot{\delta}=\mathbf{a}_g^T\boldsymbol{\tau}_{rg}-\mathbf{a}_g^T\boldsymbol{\tau}_r,$$
$$\mathbf{a}_s^T\mathbf{T}_{bg}\mathbf{I}_r\mathbf{T}_{gb}\dot{\boldsymbol{\omega}}+\mathbf{a}_s^T\mathbf{T}_{bg}\mathbf{I}_r\mathbf{e}_2\ddot{\delta}+\mathbf{a}_s^T\mathbf{T}_{bg}\mathbf{I}_r\mathbf{e}_1\dot{\Omega} \quad (3.182)$$
$$+\mathbf{a}_s^T\left(\left(\bar{\mathbf{T}}\mathbf{I}_r\mathbf{T}_{gb}+\mathbf{T}_{bg}\mathbf{I}_r\bar{\mathbf{T}}^T\right)\boldsymbol{\omega}+\bar{\mathbf{T}}\mathbf{I}_r\mathbf{e}_2\dot{\delta}+\bar{\mathbf{T}}\mathbf{I}_r\mathbf{e}_1\Omega\right)\dot{\delta}$$
$$+\mathbf{a}_s^T\boldsymbol{\omega}_\times\left(\mathbf{T}_{bg}\mathbf{I}_r\mathbf{T}_{gb}\right)\boldsymbol{\omega}+\mathbf{a}_s^T\boldsymbol{\omega}_\times\mathbf{T}_{bg}\mathbf{I}_r\mathbf{e}_2\dot{\delta}+\mathbf{a}_s^T\boldsymbol{\omega}_\times\mathbf{T}_{bg}\mathbf{I}_r\mathbf{e}_1\Omega=\mathbf{a}_s^T\boldsymbol{\tau}_r.$$

Equations (3.180)–(3.182) constitute a set of five equations for defining the state of the spacecraft. When it is necessary to maintain the attitude of the satellite in an orbital frame, it is important to append to the above set the kinematic equations for the quaternion attitude. The kinematic equations for the quaternion attitude are expressed in terms of the relative angular velocity vector, $\boldsymbol{\omega}_{br}=\boldsymbol{\omega}-\boldsymbol{\omega}_R$, relative to the orbital angular velocity $\boldsymbol{\omega}_R$, which are given by Equations (3.127) with $\boldsymbol{\omega}$ set equal to $\boldsymbol{\omega}_{br}$, from Section 3.8.9. In the presence of several *CMGs*, the corresponding equations of motion can be derived in a manner similar to this set of five equations, with every CMG contributing to the reaction torques on the satellite's body and two additional equations on motion for the two additional degrees of freedom introduced by each CMG.

3.9.1 Nonlinear Attitude Control Laws

One approach to designing a stabilizing control law is based on Lyapunov's second method. One begins with a choice of a Lyapunov function V such as,

3.9 Definition of the General Control Problem with CMG Actuation

$$V = \frac{1}{2}(\boldsymbol{\omega} - \boldsymbol{\omega}_d)^T \mathbf{J}_1 (\boldsymbol{\omega} - \boldsymbol{\omega}_d) + \frac{1}{2}(\mathbf{q} - \mathbf{q}_d)^T \mathbf{J}_2 (\mathbf{q} - \mathbf{q}_d), \qquad (3.183)$$

where $\mathbf{J}_1$ and $\mathbf{J}_2$ are positive definite weighting matrices. Then,

$$\dot{V} = (\boldsymbol{\omega} - \boldsymbol{\omega}_d)^T \mathbf{J}_1 \dot{\boldsymbol{\omega}} + (\mathbf{q} - \mathbf{q}_d)^T \mathbf{J}_2 \dot{\mathbf{q}}. \qquad (3.184)$$

From the expression for $\dot{V}$, the equations of motion and the equations relating the angular velocities to the attitude quaternion, Equation (3.127), $\dot{\boldsymbol{\omega}}$ and $\dot{\mathbf{q}}$ are eliminated. The time derivative of the Lyapunov function V, $\dot{V}$, is then manipulated by introducing at least three CMGs and suitably choosing the control input torques to the CMG gimbal motor and the CMG rotor, $\boldsymbol{\tau}_{rg}$ and $\boldsymbol{\tau}_r$, respectively, for each CMG, so $\dot{V}$ is a negative semi-definite function of $\boldsymbol{\omega}$. The procedure then leads to control laws such that, $\boldsymbol{\omega} \to \boldsymbol{\omega}_d$ and $\mathbf{q} \to \mathbf{q}_d$ as $t \to \infty$. The proof of convergence of $\boldsymbol{\omega} \to \boldsymbol{\omega}_d$ and $\mathbf{q} \to \mathbf{q}_d$ as $t \to \infty$ is based on LaSalle's invariant set theorem, which is discussed in Chapter 7 and applied in chapter 12 of Vepa [12]. Romano and Agrawal [11] have adopted the Lyapunov approach to design a control law for a dual-body spacecraft.

3.9.2 Minimum Time Maneuvers

One of the major problems associated with spacecraft is that maneuvers must be completed in time and fast enough. Thus, it is often important to design optimal controls that achieve a particular maneuver in minimum time. Thus, a performance cost function such as,

$$J = \frac{1}{2}(\boldsymbol{\omega} - \boldsymbol{\omega}_d)^T \mathbf{J}_1 (\boldsymbol{\omega} - \boldsymbol{\omega}_d) + \frac{1}{2}(\mathbf{q} - \mathbf{q}_d)^T \mathbf{J}_2 (\mathbf{q} - \mathbf{q}_d) + \frac{1}{2}\mathbf{u}^T \mathbf{R} \mathbf{u} + \int_0^T dt, \qquad (3.185)$$

is minimized, while also minimizing the final time T, subject to the states satisfying the satellite's equations of motion. To solve this problem, it is often transformed to a fixed final time optimization problem by scaling the time variable. Thus,

$$x_{n+1} = x_{n+2}^2 t, \quad \dot{x}_{n+1} = x_{n+2}^2, \quad \dot{x}_{n+2} = 0. \qquad (3.186)$$

Scaling the time variable is a technique often adopted in integrating orbital dynamical equations. By introducing the scaled time variable x_{n+1}, the cost function J is defined with a fixed final time, and the scale factor is chosen to complete the maneuver in minimum time subject to practical constraints. Moreover, since the cost function J is a quadratic cost function, although the equations of motion are nonlinear, the problem can be solved iteratively using methods associated with constrained function minimization techniques.

3.9.3 Passive Damping Systems

Passively stabilized satellites that rely on intrinsic feedback to generate a restoring moment and maintain stability are not sufficiently damped, and consequently they can oscillate about one or more of the three body axes. A gravity-gradient stabilized satellite oscillates around the nadir vector. (The nadir vector at a given point is the local vertical direction pointing in the direction of the force of gravity at that location.) To aerodynamically stabilize the satellite, aerodynamic sails are used to develop stabilizing restoring torques. In the case of an aerodynamically stabilized satellite, the drag force is in a direction opposite to the local velocity vector, and consequently the satellite oscillates about the velocity vector when the damping is inadequate. Magnetically stabilized satellites experience a force normal to the magnetic field lines, and consequently they can oscillate about the field lines. Magnetic hysteresis damping can provide just enough damping to amp such oscillations about the field lines by dissipating energy in the onboard magnetic material. The damping coefficient of these dampers, which often function as passive feedback elements, are determined by the application of optimal control theory.

3.9.4 Spin Rate Damping

Passive spin rate damping of the satellite using dampers known as nutation dampers is often employed to reduce the satellite's spin rate or damp it out altogether. Many types of nutation dampers have been employed for spin-stabilized spacecraft that generally dissipate just the right amount of the kinetic energy of rotational oscillations of a satellite about a specific axis. The main objective of energy dissipation is to align the principal axis of the angular momentum vector with the axis of the satellite's largest moment of inertia. There are a variety of these dampers, including viscous ring dampers, ball in tube dampers, pendulum dampers, wheel dampers, and spring-mass dampers. Apart from magnetic hysteresis dampers, a rotational viscous damper, commonly known as a rotational dashpot, is often used to dampen spin rates or spin-related oscillations. Slosh dampers containing viscous fluids can also be effectively used to dampen spin rates.

3.10 Magnetic Actuators

The use of magnetic actuators is advantageous for spacecraft attitude control due to their lightweight nature, cheapness, and high reliability. This method of attitude control is employed by using three magnetic coils aligned with the spacecraft body frame. A current is passed through each coil, generating a net magnetic field that interacts with the Earth's magnetic field to produce torques in three mutually perpendicular directions. The Earth's magnetic field is generally modeled using an accepted standard such as the latest IGRF model. The directions of the torques are normal to both the field generated by the coils and the Earth's magnetic field, and are proportional to the currents in each of the coils. The direction of the magnetic dipole generated by the current in each coil is normal to the plane of the coil and, consequently, the torque is produced about an axis in the plane of the coil that is also orthogonal to the Earth's magnetic field. For this reason, no torque is generally produced in the local direction of the Earth's magnetic field vector. Thus, the major drawback of the method of control based on magneto-torquers is that the torques applied to the spacecraft to control the attitude are always constrained to lie in a plane orthogonal to the magnetic field vector. All spacecraft can usually be fitted with both magnetic torquers and mechanical actuators, such as reaction wheels. Generally, magnetic torquers are usually used only for de-tumbling the spacecraft after it has been released from the launch vehicle and for dumping the momentum of the reaction wheels. Reaction wheels are generally used for precision attitude control. However, magnetic torquers and reaction wheels are generally not both installed on the same satellite and are not designed to provide complementary control torques. When both actuation systems are present and can provide complementary control inputs, there can be substantial power savings as well as reduced reaction wheel torque requirements.

3.10.1 Active Control with Magnetic Actuators

The magneto-torquers exert a control torque in a local direction that depends on the local direction of the Earth's magnetic field. Thus, a geomagnetic frame must be introduced with the origin at the satellite's center of mass and one axis parallel to the geomagnetic field **B**. This geomagnetic frame orientation can be calculated using the orbit frame and the geomagnetic field's rotation matrix, which in turn must be calculated from the principal direction of the reference magnetic field. Moreover, magnetometers onboard a spacecraft are used to continuously provide three axis measurements of the Earth's true magnetic field, and the geomagnetic field's rotation matrix can then be estimated in real time. However, torques generated may not be accurate due to the unpredictable nature of the true magnetic field and the probabilistic nature of the estimated magnetic field. Thus, the control system must be sufficiently robust and insensitive to small changes in the direction of the Earth's magnetic field.

References

[1] Kane, T. R., Likins, P. W., and Levinson, D. A. (1983) *Spacecraft Dynamics*, New York: McGraw-Hill.
[2] Rimrott, F. P. J. (1989) *Introductory Attitude Dynamics*, New York: Springer-Verlag.
[3] Wertz, J. R. and Larson, W. J. (Eds.) (1999) *Space Mission Analysis and Design*, Dordrecht: Kluwer Academic.
[4] Hughes, P. C. (2004) *Spacecraft Attitude Dynamics*, Dover Paperback Edition, New York: Dover Publications Inc.
[5] Wertz, J. R. (Ed.) (1978) *Spacecraft Attitude Determination and Control*, Dordrecht: Kluwer Academic.
[6] DeBra, D. B. and Delp, R. H. (1961) Rigid body attitude stability and natural frequencies in a circular orbit. *Journal of the Astronautical Sciences*, 8: 14–17.
[7] Ogata, K. (2001) *Modern Control Engineering*, 4th edn, chapters 11 and 12, Upper Saddle River, NJ: Prentice Hall.
[8] Kaplan, M. H. (1986) *Modern Spacecraft Dynamics & Control*, New York: John Wiley.

[9] Sidi, M. J. (1997) *Spacecraft Dynamics and Control*, Cambridge: Cambridge University Press.
[10] Wie, B. (2015) *Space Vehicle Guidance Control and Astrodynamics*, 3rd edn, New York: AIAA Education Series.
[11] Romano, M. and Agrawal, B. N. (2004) Attitude dynamics/control of dual-body spacecraft with variable-speed control moment gyros. *Journal of Guidance, Control, and Dynamics*, 27(4): 513–525.
[12] Vepa, R. (2016) *Nonlinear Control of Robots and Unmanned Aerial Vehicles: An Integrated Approach*, Boca Raton, FL: CRC Press.

4 Manipulators on Space Platforms
Dynamics and Control

4.1 Review of Robot Kinematics

The representation of the position and orientation of a rigid manipulator link and the definition of its relationship to the translation and angular velocity relative to a point on the manipulator is a fundamental problem of robot kinematics. The pose (i.e. relative position and orientation) of a frame with respect to another frame can be represented by (i) the relative position vector of the origin of the first frame with respect to the origin of the latter and expressed with respect to the first and (ii) the relative orientation of the first frame with respect to the latter. Both these quantities are combined and expressed as a single transformation matrix known as the *homogeneous transformation* matrix. The construction of this transformation matrix is progressively done by invoking the *Denavit and Hartenberg (DH) decomposition* and defining four parameters as one sequentially transforms the pose from a frame located at a particular joint of a manipulator to another frame located at the next joint. A typical example is a free-flying space robot manipulator generally consisting of a base satellite rigid body and a multi-link manipulator mounted on it. The definition of the homogeneous transformations for such a manipulator may be accomplished by applying the same principles as a fixed-base manipulator. The DH methodology has been adequately documented in several texts on robotics (see, for example, Vepa [1]) and will not be repeated here. However, in the case of space robotic manipulators it is often essential to define the total momentum vector of the spacecraft, and for this reason it is extremely important to define the angular velocity and translational velocity vectors of the manipulator at a particular instant as it operates in three-dimensional space. The composite angular velocity and translational velocity vectors are usually expressed as a single vector known as a *screw vector*. Thus, the definition of the screw vector in terms of the joint velocities and its subsequent application in defining the momenta of the various manipulator components is of fundamental importance in space robotics.

4.1.1 The Total Moment of Momentum and Translational Momentum

Consider a free-flying space robot manipulator generally consisting of a base satellite rigid body and a multi-link manipulator mounted on it. Let $\vec{V}_{i,j}$ and $\vec{\Omega}_{i,j}$ represent the translational and angular velocities of the jth body of the ith manipulator with respect to a space-fixed (inertial) frame and let $\vec{v}_{i,j}$ and $\vec{\omega}_{i,j}$ represent the translational and angular velocities of the *j*th body of the *i*th manipulator with respect to a body-fixed frame, fixed in the spacecraft's main body. Then it follows that the angular and translational velocities are related to each other by,

$$\vec{\Omega}_{i,j} = \vec{\omega}_{i,j} + \vec{\Omega}_0, \quad \vec{V}_{i,j} = \vec{v}_{i,j} + \vec{V}_0 + \vec{\Omega}_0 \times \vec{r}_{i,j}. \tag{4.1}$$

$\vec{V}_0$ and $\vec{\Omega}_0$ represent the translational and angular velocities of the centroid of the base of the satellite. However, the velocity components of the manipulator links in the body coordinates could be related to the joint angular velocities by,

$$\begin{bmatrix} \vec{\omega}_{i,j}^T & \vec{v}_{i,j}^T \end{bmatrix}^T = \mathbf{J}_m \dot{\Theta}. \tag{4.2}$$

The total moment of momentum and translational momentum of the spacecraft are given by,

$$L = \begin{bmatrix} \mathbf{I}_0 & m_0 \vec{R}_0 \times \end{bmatrix} \begin{bmatrix} \vec{\Omega}_0 \\ \vec{V}_0 \end{bmatrix} + \sum_{i=1}^{M} \sum_{j=1}^{J} \begin{bmatrix} \mathbf{I}_{i,j} & m_{i,j} \vec{r}_{i,j} \times \end{bmatrix} \begin{bmatrix} \vec{\omega}_{i,j} \\ \vec{v}_{i,j} \end{bmatrix}, \tag{4.3}$$

$$P = \begin{bmatrix} \mathbf{0} & m_0 \end{bmatrix} \begin{bmatrix} \vec{\Omega}_0 \\ \vec{V}_0 \end{bmatrix} + \sum_{i=1}^{M} \sum_{j=1}^{J} \begin{bmatrix} \mathbf{0} & m_{i,j} \end{bmatrix} \begin{bmatrix} \vec{\omega}_{i,j} \\ \vec{v}_{i,j} \end{bmatrix}. \tag{4.4}$$

One may combine these relations and, adopting a vector notation, express them compactly for any particular body that is part of the manipulator as,

$$\begin{bmatrix} \vec{\omega}_{i,j}^T & \vec{v}_{i,j}^T \end{bmatrix}^T \equiv \begin{bmatrix} \boldsymbol{\omega} \\ \mathbf{v} \end{bmatrix} = \begin{bmatrix} \omega_1 \\ \omega_2 \\ \omega_3 \\ \dot{x} \\ \dot{y} \\ \dot{z} \end{bmatrix} \equiv \dot{\mathbf{x}} = \begin{bmatrix} \mathbf{s}_1 \ \mathbf{s}_2 \cdots \mathbf{s}_{M-1} \ \mathbf{s}_M \end{bmatrix} \begin{bmatrix} \dot{\theta}_1 \\ \dot{\theta}_2 \\ \cdots \\ \dot{\theta}_{M-1} \\ \dot{\theta}_M \end{bmatrix} \equiv \mathbf{J}_m \dot{\Theta}, \tag{4.5}$$

where the first three components $[\omega_1 \ \omega_2 \ \omega_3]$ are the components of the angular velocity vector, the last three $[\dot{x} \ \dot{y} \ \dot{z}]$ are the translational velocities in a Cartesian frame generally fixed at the base of the manipulator that is initially assumed to be fixed, and $\mathbf{J}_m$ is the Jacobian matrix of the manipulator. The total moment of momentum and translational momentum of the spacecraft are given by,

$$\mathbf{h} = \mathbf{I}_0 \begin{bmatrix} \boldsymbol{\omega}_0 \\ \mathbf{v}_0 \end{bmatrix} + \sum_{i=1}^{M} \begin{bmatrix} \mathbf{I}_{3\times 3} & \mathbf{r}_i \times \mathbf{I}_{3\times 3} \\ \mathbf{0}_{3\times 3} & \mathbf{I}_{3\times 3} \end{bmatrix} \mathbf{M}_i \mathbf{s}_i \theta_i, \tag{4.6}$$

where the generalized inertia matrix is defined by,

$$\mathbf{M}_i = \begin{bmatrix} \mathbf{I}_i & \mathbf{0}_{3\times 3} \\ \mathbf{0}_{3\times 3} & m_i \mathbf{I}_{3\times 3} \end{bmatrix}, \tag{4.7}$$

and m_i and $\mathbf{I}_i$ are the mass of a link and its moment of inertia matrix, while $\mathbf{I}_{3\times 3}$ is a 3×3 unit matrix and $\mathbf{0}_{3\times 3}$ is a 3×3 matrix of zeros. In many applications, the total momentum is conserved and therefore a constant. It is also an entity that can be controlled. The total momentum is a vector of six elements that essentially describes the behavior of the entire spacecraft, including the manipulator. Since it bears a linear relationship with joint angular velocities, one could estimate the desired joint angular velocities needed to realize the desired momentum vector. Thus, it is possible to balance the manipulator as well as execute other desired motions.

4.1.2 The Screw Vector and the Generalized Jacobian Matrix of the Manipulator

Considering the manipulator, the instantaneous screw motion of the end effector is related to the joint angular velocities and is defined as the vector,

$$\begin{bmatrix} \boldsymbol{\omega} \\ \mathbf{v} \end{bmatrix} \equiv \mathbf{J}_m \dot{\boldsymbol{\Theta}}. \tag{4.8}$$

The angular and translation velocity components refer to the end effector and all of the joint angular velocities of the manipulator contribute to them. It is assumed that the free-flying manipulator consists of $M + 1$ rigid bodies connected together by M revolute joints.

One may then define a total momentum vector about an inertially fixed reference point as,

$$\mathbf{h} = \begin{bmatrix} \boldsymbol{\alpha}^T & \mathbf{m}^T \end{bmatrix}^T, \tag{4.9}$$

where $\boldsymbol{\alpha}$ is the moment of momentum vector and $\mathbf{m}$ is the translational momentum vector, which are, respectively, given by,

$$\boldsymbol{\alpha} = \sum_{i=1}^{M} (\mathbf{I}_i \boldsymbol{\omega}_i + \mathbf{r}_i \times m_i \mathbf{v}_i), \quad \mathbf{m} = \sum_{i=1}^{M} m_i \mathbf{v}_i. \tag{4.10}$$

Hence, it follows that,

$$\mathbf{h} = \sum_{i=1}^{M} \begin{bmatrix} \mathbf{I}_{3\times 3} & \mathbf{r}_i \times \mathbf{I}_{3\times 3} \\ \mathbf{0}_{3\times 3} & \mathbf{I}_{3\times 3} \end{bmatrix} \mathbf{M}_i \mathbf{s}_i \theta_i. \tag{4.11}$$

If it is now assumed that the base of the manipulator is in fact in motion and is indeed the main body of a satellite, one may define an extended total momentum vector as,

$$\mathbf{h} = \begin{bmatrix} \mathbf{I}_{3\times 3} & \mathbf{r}_0 \times \mathbf{I}_{3\times 3} \\ \mathbf{0}_{3\times 3} & \mathbf{I}_{3\times 3} \end{bmatrix} \mathbf{M}_0 \mathbf{s}_0 \begin{bmatrix} \boldsymbol{\omega}_0 \\ \mathbf{v}_0 \end{bmatrix} + \sum_{i=1}^{M} \begin{bmatrix} \mathbf{I}_{3\times 3} & \mathbf{r}_i \times \mathbf{I}_{3\times 3} \\ \mathbf{0}_{3\times 3} & \mathbf{I}_{3\times 3} \end{bmatrix} \mathbf{M}_i \mathbf{s}_i \theta_i, \tag{4.12}$$

where the subscript "0" refers to the main body of the satellite. The matrix $\mathbf{s}_0$ defines the contribution of the satellite rigid body motion to the screw motion of the end effector, $\boldsymbol{\omega}_0$ is the rigid body angular velocity vector of the main body of the satellite, and $\mathbf{v}_0$ are the translational velocity components of the main body of the satellite. From the preceding section, the extended total momentum of the manipulator and of the satellite may also be expressed as,

$$\mathbf{h} = \mathbf{I}_0 \begin{bmatrix} \boldsymbol{\omega}_0 \\ \mathbf{v}_0 \end{bmatrix} + \sum_{i=1}^{M} \begin{bmatrix} \mathbf{I}_{3\times 3} & \mathbf{r}_i \times \mathbf{I}_{3\times 3} \\ \mathbf{0}_{3\times 3} & \mathbf{I}_{3\times 3} \end{bmatrix} \mathbf{M}_i \mathbf{s}_i \theta_i. \quad (4.13)$$

From Equations (4.1), the screw motion of the end effector may be expressed as,

$$\begin{bmatrix} \boldsymbol{\omega} \\ \mathbf{v} \end{bmatrix} = \mathbf{J}_0 \begin{bmatrix} \boldsymbol{\omega}_0 \\ \mathbf{v}_0 \end{bmatrix} + \mathbf{J}_m \dot{\boldsymbol{\Theta}}. \quad (4.14)$$

This relation defines how the manipulator motion alters the base motion of the satellite and vice versa, and represents the interaction between the robot manipulator and the satellite's base.

From the definition of the extended total momentum, since,

$$\begin{bmatrix} \boldsymbol{\omega}_0 \\ \mathbf{v}_0 \end{bmatrix} = \mathbf{I}_0^{-1} \mathbf{h} - \mathbf{I}_0^{-1} \sum_{i=1}^{M} \begin{bmatrix} \mathbf{I}_{3\times 3} & \mathbf{r}_i \times \mathbf{I}_{3\times 3} \\ \mathbf{0}_{3\times 3} & \mathbf{I}_{3\times 3} \end{bmatrix} \mathbf{M}_i \mathbf{s}_i \theta_i, \quad (4.15)$$

the screw motion of the end effector may be expressed as,

$$\begin{bmatrix} \boldsymbol{\omega} \\ \mathbf{v} \end{bmatrix} = \mathbf{J}_0 \mathbf{I}_0^{-1} \mathbf{h} + \mathbf{J}_m \dot{\boldsymbol{\Theta}} - \mathbf{J}_0 \mathbf{I}_0^{-1} \sum_{i=1}^{M} \begin{bmatrix} \mathbf{I}_{3\times 3} & \mathbf{r}_i \times \mathbf{I}_{3\times 3} \\ \mathbf{0}_{3\times 3} & \mathbf{I}_{3\times 3} \end{bmatrix} \mathbf{M}_i \mathbf{s}_i \theta_i = \mathbf{J}_0 \mathbf{I}_0^{-1} \mathbf{h} + \mathbf{J}_m^e \dot{\boldsymbol{\Theta}}, \quad (4.16)$$

where $\mathbf{J}_m^e$ is the generalized Jacobian matrix. The generalized Jacobian Matrix approach is particularly useful when, in addition to the end-effector dynamics and control, the motion of the spacecraft must also be computed or controlled for purposes of the attitude control of the space manipulator. Furthermore, if one wishes to minimize the influence of the manipulator motion on the main body of the satellite one could try and minimize both $\boldsymbol{\omega}_0$ and $\mathbf{v}_0$. Thus, the design of the controller would be purely based on a kinematic approach. Yet the complete design of a controller would necessarily require an understanding of the dynamics of the combined spacecraft and manipulator system.

Sometimes, when only the control of the end-effector motion is of interest, one could treat the end-effector as the "main body" and the satellite as just another manipulator link. The computation of the generalized Jacobian matrix in this case is considerably simplified.

4.2 Fundamentals of Robot Dynamics: The Lagrangian Approach

The Lagrangian approach is a very useful method for formulating the equations of motion of any space robot manipulator system. The method, which was originally introduced by Joseph L. Lagrange, provides a systematic method to derive the equations of motion of a space robot manipulator by considering the total kinetic and potential energies of the system. The Lagrange equation can be written in the form,

$$\frac{d}{dt}\left(\frac{\partial L}{\partial \dot{q}_i}\right) - \frac{\partial L}{\partial q_i} = Q_i + \Lambda_i^T(q_i)\lambda_i, \quad i = 1, 2, 3 \ldots N, \quad (4.17)$$

where, $L = T - V$ is the Lagrangian function, $T(\dot{q}_i, q_i)$ is the kinetic energy of the system, $V(q_i)$ is the potential energy of the system, q_i are the generalized coordinates, Q_i are the generalized forces, Λ_i is a vector of constraints, and λ_i is a vector of Lagrange multipliers associated with the constraints. The kinetic energy of the space robotic manipulator is the sum of the kinetic energies of the main satellite body that serves as a platform for the manipulator, the manipulator arms, and any other controller hardware such as reaction and momentum wheels. The kinetic energy expressions take the form,

$$T = (1/2) \sum_{j=0}^{M+1} \left(m_j v_{j,cm}^2 + \boldsymbol{\omega}_{j,cm}^T \mathbf{I}_{j,cm} \boldsymbol{\omega}_{j,cm} \right), \tag{4.18}$$

while the potential energy takes the form, $V = V(q_i)$. Generally, the translational velocities $v_{j,cm}$ and vector angular velocities $\boldsymbol{\omega}_{j,cm}$ are linear functions of the generalized velocities and can be expressed as,

$$v_{j,cm} = \sum_{i=1}^{N} v_{j,i}(q_k) \dot{q}_i = \mathbf{v}_j^T(\mathbf{q}) \dot{\mathbf{q}}, \tag{4.19}$$

$$\boldsymbol{\omega}_{j,cm} = \sum_{i=1}^{N} \boldsymbol{\omega}_{j,i}(q_k) \dot{q}_i = \vec{\boldsymbol{\omega}}_j T(\mathbf{q}) \dot{\mathbf{q}}. \tag{4.20}$$

The kinetic energy then takes the form,

$$T = (1/2) \dot{\mathbf{q}}^T \left\{ \sum_{j=0}^{M+1} \left(\mathbf{v}_j(\mathbf{q}) m_j \mathbf{v}_j^T(\mathbf{q}) + \vec{\boldsymbol{\omega}}_j(\mathbf{q}) \mathbf{I}_{j,cm} \vec{\boldsymbol{\omega}}_j^T(\mathbf{q}) \right) \right\} \dot{\mathbf{q}}. \tag{4.21}$$

Thus,

$$T = (1/2) \dot{\mathbf{q}}^T \mathbf{H}(\mathbf{q}) \dot{\mathbf{q}}, \tag{4.22}$$

where, $\mathbf{H}(\mathbf{q}) = \sum_{j=0}^{M+1} \left(\mathbf{v}_j(\mathbf{q}) m_j \mathbf{v}_j^T(\mathbf{q}) + \vec{\boldsymbol{\omega}}_j(\mathbf{q}) \mathbf{I}_{j,cm} \vec{\boldsymbol{\omega}}_j^T(\mathbf{q}) \right)$.

Following Kane and Levinson [2], $\mathbf{v}_j(\mathbf{q})$ and $\vec{\boldsymbol{\omega}}_j(\mathbf{q})$ are referred to as partial translational and partial angular velocity vectors. The complete Euler-Lagrange equations may be expressed as,

$$\mathbf{H}(\mathbf{q}) \ddot{\mathbf{q}} + \mathbf{C}(\mathbf{q}\dot{\mathbf{q}}) \dot{\mathbf{q}} + \mathbf{G}(\mathbf{q}) = \mathbf{Q} + \Lambda \lambda, \tag{4.23}$$

where, $\mathbf{H}(\mathbf{q})$ is a symmetric matrix and,

$$\mathbf{C}(\mathbf{q}\dot{\mathbf{q}}) = \sum_{j=1}^{N} \dot{q}_j (\partial \mathbf{H}(\mathbf{q})/\partial q_j) - \frac{1}{2} \dot{\mathbf{q}}^T (\partial \mathbf{H}(\mathbf{q})/\partial \mathbf{q}) = \frac{1}{2} \dot{\mathbf{q}}^T (\partial \mathbf{H}(\mathbf{q})/\partial \mathbf{q}), \tag{4.24}$$

$$\mathbf{G}(\mathbf{q}) = [\partial V(\mathbf{q})/\partial q_i]. \tag{4.25}$$

The constraints are generally eliminated, if they are present, by transforming into the null space of the constraint matrix Λ. The application of Lagrangian method to a number of robotic manipulators may be found in Vepa [3]. A typical example of a

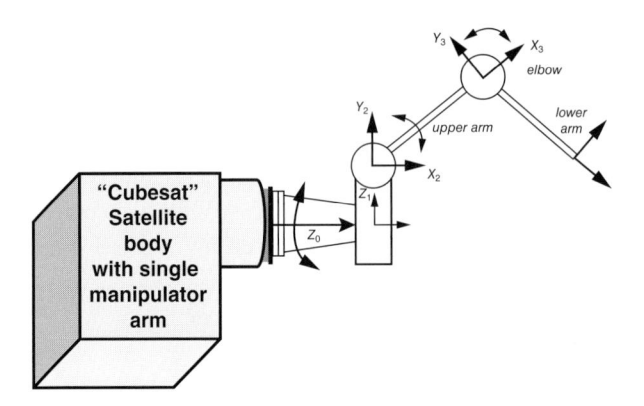

Figure 4.1 A typical example of a biomimetic model of a manipulator arm attached to satellite.

biomimetic model of a manipulator arm is shown in Figure 4.1. In this example, there are four joints and four links. The base frame is assumed to be $[X_0 \quad Y_0 \quad Z_0]$, where Y_0 and Z_0 are in the plane of the page and Y_0 is in the direction of the vertical axis. A position vector in this frame is denoted by, $[x_0 \quad y_0 \quad z_0]$.

The equations of motion may be conveniently expressed in terms of the partial velocities and partial angular velocities.

The potential energy is,

$$V = \sum_{i=1}^{4} m_i g y_0^i. \tag{4.26}$$

The translational kinetic energy is,

$$T_1 = \frac{1}{2} \sum_{i=1}^{4} m_i \left(\sum_{j=0}^{i-1} \dot{\theta}_j v_{ij} \right)^T \left(\sum_{j=0}^{i-1} \dot{\theta}_j v_{ij} \right). \tag{4.27}$$

The kinetic energy of rotation is,

$$T_2 = \frac{1}{2} \sum_{i=1}^{4} \sum_{k=1}^{3} I_{i,k} \left(\sum_{j=0}^{i-1} \dot{\theta}_j \omega_{ij,k} \right)^2. \tag{4.28}$$

The Lagrangian may be defined as,

$$L = T - V = T_1 + T_2 - V. \tag{4.29}$$

The Euler-Lagrange equations are,

$$\frac{d}{dt}\frac{\partial L}{\partial \dot{q}_i} - \frac{\partial L}{\partial q_i} = Q_i; \quad \frac{d}{dt}\frac{\partial T_1}{\partial \dot{\theta}_i} - \frac{\partial T_1}{\partial \theta_i} + \frac{d}{dt}\frac{\partial T_2}{\partial \dot{\theta}_i} - \frac{\partial T_2}{\partial \theta_i} + \frac{\partial V}{\partial \theta_i} = Q_i, \tag{4.30}$$

where, $q_i = \theta_i$; Q_i are the generalized forces other than those accounted for by the potential energy function and are equal to the torques applied by the joint servo motors τ_i,

4.2 Fundamentals of Robot Dynamics: The Lagrangian Approach

$$\frac{\partial V}{\partial \theta_j} = \sum_{i=1}^{4} m_i g \frac{\partial y_0^i}{\partial \theta_j}, \quad \text{for } j = 0, 1, \ldots \quad (4.31)$$

$$\frac{\partial T_1}{\partial \dot{\theta}_j} = \frac{1}{2} \sum_{i=1}^{4} m_i \frac{\partial}{\partial \dot{\theta}_j} \left(\sum_{k=0}^{i-1} \dot{\theta}_k v_{ik} \right)^T \left(\sum_{k=0}^{i-1} \dot{\theta}_k v_{ik} \right) = \sum_{i=1}^{4} m_i \left(\sum_{k=0}^{i-1} \dot{\theta}_k v_{ik} \right)^T v_{ij}, \quad (4.32)$$

$$\frac{d}{dt} \frac{\partial T_1}{\partial \dot{\theta}_j} = \frac{1}{2} \sum_{i=1}^{4} m_i \frac{d}{dt} \frac{\partial}{\partial \dot{\theta}_j} \left(\sum_{k=0}^{i-1} \dot{\theta}_k v_{ik} \right)^T \left(\sum_{k=0}^{i-1} \dot{\theta}_k v_{ik} \right)$$

$$= \sum_{i=1}^{4} m_i v_{ij}^T \left(\sum_{k=0}^{i-1} \ddot{\theta}_k v_{ik} \right)^T v_{ij} + \sum_{i=1}^{4} m_i \left(\sum_{k=0}^{i-1} \dot{\theta}_k \sum_{m=0}^{i-1} \frac{\partial v_{ik}}{\partial \theta_m} \dot{\theta}_m \right)^T v_{ij}$$

$$+ \sum_{i=1}^{4} m_i \left(\sum_{k=0}^{i-1} \dot{\theta}_k v_{ik} \right)^T \left(\sum_{m=0}^{i-1} \frac{\partial v_{ij}}{\partial \theta_m} \dot{\theta}_m \right), \quad (4.33)$$

$$\frac{\partial T_1}{\partial \theta_j} = \frac{1}{2} \sum_{i=1}^{4} m_i \frac{\partial}{\partial \theta_j} \left\{ \left(\sum_{k=0}^{i-1} \dot{\theta}_k v_{ik} \right)^T \left(\sum_{k=0}^{i-1} \dot{\theta}_k v_{ik} \right) \right\} = \sum_{i=1}^{4} m_i \left(\sum_{k=0}^{i-1} \dot{\theta}_k v_{ik} \right)^T \left(\sum_{k=0}^{i-1} \dot{\theta}_k \frac{\partial v_{ik}}{\partial \theta_j} \right),$$

$$(4.34)$$

$$\frac{d}{dt} \frac{\partial T_1}{\partial \dot{\theta}_j} - \frac{\partial T_1}{\partial \theta_j} = \frac{1}{2} \sum_{i=1}^{4} m_i \frac{d}{dt} \frac{\partial}{\partial \dot{\theta}_j} \left(\sum_{k=0}^{i-1} \dot{\theta}_k v_{ik} \right)^T \left(\sum_{k=0}^{i-1} \dot{\theta}_k v_{ik} \right)$$

$$= \sum_{i=1}^{4} m_i v_{ij}^T \left(\sum_{k=0}^{i-1} \ddot{\theta}_k v_{ik} \right)^T v_{ij} + \sum_{i=1}^{4} m_i \left(\sum_{k=0}^{i-1} \dot{\theta}_k \sum_{m=0}^{i-1} \frac{\partial v_{ik}}{\partial \theta_m} \dot{\theta}_m \right)^T v_{ij}$$

$$+ \sum_{i=1}^{4} m_i \left(\sum_{k=0}^{i-1} \dot{\theta}_k v_{ik} \right)^T \left(\sum_{m=0}^{i-1} \dot{\theta}_m \frac{\partial v_{ij}}{\partial \theta_m} - \dot{\theta}_m \frac{\partial v_{im}}{\partial \theta_j} \right). \quad (4.35)$$

Similarly,

$$\frac{d}{dt} \frac{\partial T_2}{\partial \dot{\theta}_i} - \frac{\partial T_2}{\partial \theta_i} = \frac{1}{2} \sum_{i=1}^{4} \sum_{n=1}^{3} I_{in} \frac{d}{dt} \frac{\partial}{\partial \dot{\theta}_j} \left(\sum_{k=0}^{i-1} \dot{\theta}_k \omega_{ik,n} \right)^2$$

$$= \sum_{i=1}^{4} \sum_{n=1}^{3} I_{in} \left(\sum_{k=0}^{i-1} \ddot{\theta}_k \omega_{ik,n} \right) \omega_{ij,n} + \sum_{i=1}^{4} \sum_{n=1}^{3} I_{in} \left(\sum_{k=0}^{i-1} \dot{\theta}_k \sum_{m=0}^{i-1} \frac{\partial \omega_{ik,n}}{\partial \theta_m} \dot{\theta}_m \right) \omega_{ij,n}$$

$$+ \sum_{i=1}^{4} \sum_{n=1}^{3} I_{in} \left(\sum_{k=0}^{i-1} \dot{\theta}_k \omega_{ik,n} \right) \left(\sum_{m=0}^{i-1} \dot{\theta}_m \frac{\partial \omega_{ij,n}}{\partial \theta_m} - \dot{\theta}_m \frac{\partial \omega_{im,n}}{\partial \theta_j} \right), \quad (4.36)$$

where $\omega_{ij,n}$ are components of the vector ω_{ij}, which is expressed as,

$$\omega_{ij} = \begin{bmatrix} \omega_{ij,1} & \omega_{ij,2} & \omega_{ij,3} \end{bmatrix}^T. \quad (4.37)$$

Hence all the terms in,

$$\frac{d}{dt}\frac{\partial T_1}{\partial \dot{\theta}_i} - \frac{\partial T_1}{\partial \theta_i} + \frac{d}{dt}\frac{\partial T_2}{\partial \dot{\theta}_i} - \frac{\partial T_2}{\partial \theta_i} + \frac{\partial V}{\partial \theta_i} = \tau_i, \qquad (4.38)$$

are obtained. The inertia terms alone are given by,

$$\sum \mathbf{H}_{jk}\ddot{\theta}_k = \sum_{i=1}^{4} m_i \left(\sum_{k=0}^{i-1} \ddot{\theta}_k v_{ik} \right)^T v_{ij} + \sum_{i=1}^{4} \sum_{n=1}^{3} I_{in} \left(\sum_{k=0}^{i-1} \ddot{\theta}_k \omega_{ik,n} \right) \omega_{ij,n}. \qquad (4.39)$$

Hence,

$$\mathbf{H}_{jk} = \sum_{i=1}^{4} m_i v_{ij}^T v_{ik} + \sum_{n=1}^{3} I_{in}\omega_{ij,n}\omega_{ik,n}, \qquad (4.40)$$

where I_{in}, $n = 1, 2, 3$ are the principal moments of inertia of link i. All other inertia components are assumed to be zero.

The full equations of motion are:

$$\sum_{i=1}^{4} m_i \left(\sum_{k=0}^{i-1} \ddot{\theta}_k v_{ik} \right)^T v_{ij} + \sum_{i=1}^{4} \sum_{n=1}^{3} I_{in} \left(\sum_{k=0}^{i-1} \ddot{\theta}_k \omega_{ik,n} \right) \omega_{ij,n} + \sum_{i=1}^{4} m_i g \frac{\partial y_0^i}{\partial \theta_j}$$

$$+ \sum_{i=1}^{4} m_i \left(\sum_{k=0}^{i-1} \dot{\theta}_k \sum_{m=0}^{i-1} \frac{\partial v_{ik}}{\partial \theta_m} \dot{\theta}_m \right)^T v_{ij} + \sum_{i=1}^{4} m_i \left(\sum_{k=0}^{i-1} \dot{\theta}_k v_{ik} \right)^T \sum_{m=0}^{i-1} \dot{\theta}_m \left(\frac{\partial v_{ij}}{\partial \theta_m} - \frac{\partial v_{ik}}{\partial \theta_j} \right)$$

$$+ \sum_{i=1}^{4} \sum_{n=1}^{3} I_{in} \left(\sum_{k=0}^{i-1} \dot{\theta}_k \sum_{m=0}^{i-1} \frac{\partial \omega_{ik,n}}{\partial \theta_m} \dot{\theta}_m \right) \omega_{ij,n}$$

$$+ \sum_{i=1}^{4} \sum_{n=1}^{3} I_{in} \left(\sum_{k=0}^{i-1} \dot{\theta}_k \omega_{ik,n} \right)^T \sum_{m=0}^{i-1} \dot{\theta}_m \left(\frac{\partial \omega_{ij,n}}{\partial \theta_m} - \frac{\partial \omega_{im,n}}{\partial \theta_j} \right) = \tau_j. \qquad (4.41)$$

We shall verify these equations by applying them to just the last two links. This is equivalent to a two-link manipulator, for which the equations are well known.

Consider the case of a manipulator with just the last two links of the previously described manipulator model. In this case, the frame $[X_2 \ Y_2 \ Z_2]$ can be assumed to be fixed. In this case,

$$\omega_2 = \dot{\theta}_2 k_2 = \dot{\theta}_2 \omega_{10}, \quad \omega_3 = (\dot{\theta}_2 + \dot{\theta}_3)k_4 = \dot{\theta}_3 \omega_{21} + \dot{\theta}_2 \omega_{20}. \qquad (4.42)$$

The CG velocity and position of link 4, the outer most link, referred to the fixed frame are,

$$\begin{bmatrix} \dot{x}_2 \\ \dot{y}_2 \\ \dot{z}_2 \end{bmatrix} = (\dot{\theta}_3 + \dot{\theta}_2) \begin{bmatrix} -L_{2cg} \sin(\theta_2 + \theta_3) \\ L_{2cg} \cos(\theta_2 + \theta_3) \\ 0 \end{bmatrix} + \dot{\theta}_2 \begin{bmatrix} -L_1 \sin \theta_2 \\ L_1 \cos \theta_2 \\ 0 \end{bmatrix}, \qquad (4.43)$$

4.2 Fundamentals of Robot Dynamics: The Lagrangian Approach

$$\begin{bmatrix} x_2 \\ y_2 \\ z_2 \end{bmatrix} = \begin{bmatrix} (L_1 + L_{2cg} \cos \theta_3) \cos \theta_2 - L_{2cg} \sin \theta_3 \sin \theta_2 \\ (L_1 + L_{2cg} \cos \theta_3) \sin \theta_2 + L_{2cg} \sin \theta_3 \cos \theta_2 \\ 0 \end{bmatrix}$$

$$= \begin{bmatrix} L_1 \cos \theta_2 + L_{2cg} \cos (\theta_2 + \theta_3) \\ L_1 \sin \theta_2 + L_{2cg} \sin (\theta_2 + \theta_3) \\ 0 \end{bmatrix}. \quad (4.44)$$

Let,

$$\begin{bmatrix} \dot{x}_2 \\ \dot{y}_2 \\ \dot{z}_2 \end{bmatrix} = \dot{\theta}_3 v_{21} + \dot{\theta}_2 v_{20} = \sum_{j=0}^{1} \dot{\theta}_j v_{2j}. \quad (4.45)$$

$$v_{21} = \begin{bmatrix} -L_{2cg} \sin \varphi_3 \\ L_{2cg} \cos \varphi_3 \\ 0 \end{bmatrix}, \quad v_{20} = \begin{bmatrix} -L_1 \sin \theta_2 \\ L_1 \cos \theta_2 \\ 0 \end{bmatrix} + \begin{bmatrix} -L_{2cg} \sin \varphi_3 \\ L_{2cg} \cos \varphi_3 \\ 0 \end{bmatrix}, \quad \varphi_3 = \theta_3 + \theta_2.$$

(4.46)

For link 4, let the position vector of the CG in the base frame ($[x_2 \ y_2 \ z_2]$) be expressed as,

$$\begin{bmatrix} x_2 \\ y_2 \\ z_2 \end{bmatrix} = \begin{bmatrix} x_2^4 \\ y_2^4 \\ z_2^4 \end{bmatrix}, \quad \text{where} \quad \begin{bmatrix} x_2^4 \\ y_2^4 \\ z_2^4 \end{bmatrix} = \begin{bmatrix} L_1 \cos \theta_2 + L_{2cg} \cos \varphi_3 \\ L_1 \sin \theta_2 + L_{2cg} \sin \varphi_3 \\ 0 \end{bmatrix}. \quad (4.47)$$

We may use this expressions for finding the CG velocity of link 3, which adjoins the outermost link, by setting $L_{2cg} \equiv 0$, $L_1 \equiv L_{1cg}$.

Similar substitutions are used for the position coordinates of the CG.

Thus for link 3,

$$\begin{bmatrix} \dot{x}_2 \\ \dot{y}_2 \\ \dot{z}_2 \end{bmatrix} = \dot{\theta}_2 v_{10} = \dot{\theta}_2 v_{10} = \sum_{j=0}^{0} \dot{\theta}_j v_{1j}, \quad v_{10} = \begin{bmatrix} -L_{1cg} \sin \theta_2 \\ L_{1cg} \cos \theta_2 \\ 0 \end{bmatrix}. \quad (4.48)$$

For link i, $i = 3$ assume that the corresponding position vectors of the CGs in base coordinates are given by,

$$\begin{bmatrix} x_2 \\ y_2 \\ z_2 \end{bmatrix} = \begin{bmatrix} x_2^3 \\ y_2^3 \\ z_2^3 \end{bmatrix}, \quad \text{where} \quad \begin{bmatrix} x_2^3 \\ y_2^3 \\ z_2^3 \end{bmatrix} = \begin{bmatrix} L_{1cg} \cos \theta_2 \\ L_{1cg} \sin \theta_2 \\ 0 \end{bmatrix}. \quad (4.49)$$

The potential energy is,

$$V = \sum_{i=1}^{2} m_i g y_2^{2+i}. \quad (4.50)$$

The full equations of motion with appropriate masses and inertias of the two links, and making the appropriate changes to the indices, are,

$$\sum_{i=1}^{2} m_{i+2} \left(\sum_{k=0}^{i-1} \ddot{\theta}_{k+2} v_{ik} \right)^T v_{ij-1} + \sum_{i=1}^{2} \sum_{n=3}^{3} I_{i+2,n} \left(\sum_{k=0}^{i-1} \ddot{\theta}_{k+2} \omega_{ik,n} \right) \omega_{ij-1,n}$$

$$+ \sum_{i=1}^{2} m_{i+2} \left(\sum_{k=0}^{i-1} \dot{\theta}_{k+2} \sum_{m=0}^{i-1} \frac{\partial v_{ik}}{\partial \theta_{m+2}} \dot{\theta}_{m+2} \right)^T v_{ij-1}$$

$$+ \sum_{i=1}^{2} m_{i+2} \left(\sum_{k=0}^{i-1} \dot{\theta}_{k+2} v_{ik} \right)^T \sum_{m=0}^{i-1} \dot{\theta}_{m+2} \left(\frac{\partial v_{ij-1}}{\partial \theta_{m+2}} - \frac{\partial v_{im}}{\partial \theta_{j+1}} \right) + \sum_{i=1}^{2} m_i g \frac{\partial y_2^{2+i}}{\partial \theta_{j+1}} = \tau_j. \quad (4.51)$$

For $j = 1$,

$$m_3 \ddot{\theta}_2 v_{10}^T v_{10} + m_4 (\ddot{\theta}_2 v_{20}^T + \ddot{\theta}_3 v_{21}^T) v_{20} + I_{33} \ddot{\theta}_2 + I_{43} (\ddot{\theta}_2 + \ddot{\theta}_3) + m_3 \dot{\theta}_2^2 \frac{\partial v_{10}^T}{\partial \theta_2} v_{10}$$

$$+ m_4 \left(\dot{\theta}_2^2 \frac{\partial v_{20}^T}{\partial \theta_2} + \dot{\theta}_2 \dot{\theta}_3 \left(\frac{\partial v_{20}^T}{\partial \theta_3} + \frac{\partial v_{21}^T}{\partial \theta_2} \right) + \dot{\theta}_3^2 \frac{\partial v_{21}^T}{\partial \theta_3} \right) v_{20}$$

$$+ m_4 (\dot{\theta}_2 v_{20}^T + \dot{\theta}_3 v_{21}^T) \dot{\theta}_3 \left(\frac{\partial v_{20}}{\partial \theta_3} - \frac{\partial v_{21}}{\partial \theta_2} \right) + \sum_{i=1}^{2} m_i g \frac{\partial y_2^{2+i}}{\partial \theta_2} = \tau_1, \quad (4.52)$$

$$v_{21} = \begin{bmatrix} -L_{2cg} \sin \varphi_3 \\ L_{2cg} \cos \varphi_3 \\ 0 \end{bmatrix}, \quad v_{20} = \begin{bmatrix} -L_1 \sin \theta_2 \\ L_1 \cos \theta_2 \\ 0 \end{bmatrix} + \begin{bmatrix} -L_{2cg} \sin \varphi_3 \\ L_{2cg} \cos \varphi_3 \\ 0 \end{bmatrix}, \quad v_{10} = \begin{bmatrix} -L_{1cg} \sin \theta_2 \\ L_{1cg} \cos \theta_2 \\ 0 \end{bmatrix},$$

$$(4.53)$$

$$\frac{\partial v_{10}^T}{\partial \theta_2} = [-L_{1cg} \cos \theta_2 \quad -L_{1cg} \sin \theta_2 \quad 0],$$

$$\frac{\partial v_{20}}{\partial \theta_2} = \begin{bmatrix} -L_1 \cos \theta_2 \\ -L_1 \sin \theta_2 \\ 0 \end{bmatrix} + \begin{bmatrix} -L_{2cg} \cos \varphi_3 \\ -L_{2cg} \sin \varphi_3 \\ 0 \end{bmatrix}, \quad \frac{\partial v_{20}}{\partial \theta_3} = \begin{bmatrix} -L_{2cg} \cos \varphi_3 \\ -L_{2cg} \sin \varphi_3 \\ 0 \end{bmatrix},$$

$$\frac{\partial v_{21}}{\partial \theta_2} = \begin{bmatrix} -L_{2cg} \cos \varphi_3 \\ -L_{2cg} \sin \varphi_3 \\ 0 \end{bmatrix}, \quad \frac{\partial v_{21}}{\partial \theta_3} = \begin{bmatrix} -L_{2cg} \cos \varphi_3 \\ -L_{2cg} \sin \varphi_3 \\ 0 \end{bmatrix} \quad (4.54)$$

$$v_{10}^T v_{10} = L_{1cg}^2, \quad v_{20}^T v_{20} = L_1^2 + L_{2cg}^2 + 2 L_1 L_{2cg} \cos \theta_3, \quad v_{21}^T v_{20} = L_{2cg}^2 + L_1 L_{2cg} \cos \theta_3,$$

$$(4.55)$$

$$\frac{\partial v_{10}^T}{\partial \theta_2} v_{10} = 0, \quad \frac{\partial v_{20}^T}{\partial \theta_2} v_{20} = 0, \quad \frac{\partial v_{20}^T}{\partial \theta_3} v_{20} = -L_1 L_{2cg} \sin \theta_3,$$

$$\frac{\partial v_{21}^T}{\partial \theta_2} v_{20} = -L_1 L_{2cg} \sin \theta_3, \quad \frac{\partial v_{21}^T}{\partial \theta_3} v_{20} = -L_1 L_{2cg} \sin \theta_3,$$

$$v_{21}^T \frac{\partial v_{20}}{\partial \theta_2} = L_1 L_{2cg} \sin \theta_3, \quad v_{20}^T \frac{\partial v_{20}}{\partial \theta_3} = -L_1 L_{2cg} \sin \theta_3, \quad (4.56)$$

$$\frac{\partial v_{20}}{\partial \theta_3} - \frac{\partial v_{21}}{\partial \theta_2} = 0, \quad \frac{\partial v_{20}^T}{\partial \theta_3} v_{20} + \frac{\partial v_{21}^T}{\partial \theta_2} v_{20} = -2 L_1 L_{2cg} \sin \theta_3. \quad (4.57)$$

4.2 Fundamentals of Robot Dynamics: The Lagrangian Approach

Thus Equation (4.52) reduces to,

$$m_3\ddot{\theta}_2 L_{1cg}^2 + m_4\left(\ddot{\theta}_2\left(L_1^2 + L_{2cg}^2 + 2L_1L_{2cg}\cos\theta_3\right) + \ddot{\theta}_3\left(L_{2cg}^2 + L_1L_{2cg}\cos\theta_3\right)\right)$$

$$+ I_{33}\ddot{\theta}_2 + I_{43}(\ddot{\theta}_2 + \ddot{\theta}_3) + m_4 L_1 L_{2cg}\sin\theta_3\left(-2\dot{\theta}_2\dot{\theta}_3 - \dot{\theta}_3^2\right) + \sum_{i=1}^{2} m_i g \frac{\partial y_2^{2+i}}{\partial \theta_2} = \tau_1. \tag{4.58}$$

For the second equation,

$$\sum_{i=1}^{2} m_{i+2}\left(\sum_{k=0}^{i-1}\ddot{\theta}_{k+2}v_{ik}\right)^T v_{ij-1} + \sum_{i=1}^{2}\sum_{n=3}^{3} I_{i+2,n}\left(\sum_{k=0}^{i-1}\ddot{\theta}_{k+2}\omega_{ik,n}\right)\omega_{ij-1,n}$$

$$+ \sum_{i=1}^{2} m_{i+2}\left(\sum_{k=0}^{i-1}\dot{\theta}_{k+2}\sum_{m=0}^{i-1}\frac{\partial v_{ik}}{\partial \theta_{m+2}}\dot{\theta}_{m+2}\right)^T v_{ij-1}$$

$$+ \sum_{i=1}^{2} m_{i+2}\left(\sum_{k=0}^{i-1}\dot{\theta}_{k+2}v_{ik}\right)^T \sum_{m=0}^{i-1}\dot{\theta}_{m+2}\left(\frac{\partial v_{ij-1}}{\partial \theta_{m+2}} - \frac{\partial v_{im}}{\partial \theta_{j+1}}\right) + \sum_{i=1}^{2} m_i g \frac{\partial y_2^{2+i}}{\partial \theta_{j+1}} = \tau_j. \tag{4.59}$$

For $j = 2$,

$$m_3\ddot{\theta}_2 v_{10}^T v_{11} + m_4\left(\ddot{\theta}_2 v_{20}^T + \ddot{\theta}_3 v_{21}^T\right)v_{21} + I_{43}(\ddot{\theta}_2 + \ddot{\theta}_3) + m_3\dot{\theta}_2^2 \frac{\partial v_{10}^T}{\partial \theta_2}v_{11}$$

$$+ m_4\left(\dot{\theta}_2^2 \frac{\partial v_{20}^T}{\partial \theta_2} + \dot{\theta}_2\dot{\theta}_3\left(\frac{\partial v_{20}^T}{\partial \theta_3} + \frac{\partial v_{21}^T}{\partial \theta_2}\right) + \dot{\theta}_3^2 \frac{\partial v_{21}^T}{\partial \theta_3}\right)v_{21}$$

$$+ m_4\left(\dot{\theta}_2 v_{20}^T + \dot{\theta}_3 v_{21}^T\right)\left\{\dot{\theta}_2\left(\frac{\partial v_{11}}{\partial \theta_2} - \frac{\partial v_{10}}{\partial \theta_3}\right) + \dot{\theta}_3\left(\frac{\partial v_{21}}{\partial \theta_3} - \frac{\partial v_{21}}{\partial \theta_3}\right)\right\}$$

$$+ \sum_{i=1}^{2} m_i g \frac{\partial y_2^{2+i}}{\partial \theta_3} = \tau_2, \tag{4.60}$$

which reduces to,

$$m_4\left(\ddot{\theta}_2\left(L_{2cg}^2 + L_1 L_{2cg}\cos\theta_3\right) + \ddot{\theta}_3 L_{2cg}^2\right) + I_{43}(\ddot{\theta}_2 + \ddot{\theta}_3) + m_4\left(\dot{\theta}_2^2 L_1 L_{2cg}\sin\theta_3\right)$$

$$+ \sum_{i=1}^{2} m_i g \frac{\partial y_2^{2+i}}{\partial \theta_3} = \tau_2. \tag{4.61}$$

The two Euler-Lagrange equations are,

$$\ddot{\theta}_2\left(m_3 L_{1cg}^2 + m_4 L_1^2 + I_{33} + m_4 L_1 L_{2cg}\cos\theta_3\right)$$

$$+ (\ddot{\theta}_2 + \ddot{\theta}_3)\left(m_4 L_{2cg}^2 + I_{43} + m_4 L_1 L_{2cg}\cos\theta_3\right)$$

$$- m_4 L_1 L_{2cg}\sin\theta_3\left(2\dot{\theta}_2\dot{\theta}_3 + \dot{\theta}_3^2\right) + \sum_{i=1}^{2} m_i g \frac{\partial y_2^{2+i}}{\partial \theta_2} = \tau_1, \tag{4.62}$$

$$\ddot{\theta}_2 \left(m_4 L_1 L_{2cg} \cos \theta_3\right) + \left(\ddot{\theta}_2 + \ddot{\theta}_3\right)\left(m_4 L_{2cg}^2 + I_{43}\right)$$

$$+ m_4 L_1 L_{2cg} \sin \theta_3 \left(\dot{\theta}_2^2\right) + \sum_{i=1}^{2} m_i g \frac{\partial y_2^{2+i}}{\partial \theta_3} = \tau_2. \tag{4.63}$$

If we let,

$$I_{33} = m_3 k_{1cg}^2, \quad I_{43} = m_4 k_{4cg}^2, \quad I_{11} = m_3 \left(L_{1cg}^2 + k_{1cg}^2\right) + m_4 L_1^2,$$

$$I_{21} = m_4 L_{2cg} L_1 = \Gamma_{22} L_1, \quad I_{22} = m_4 \left(L_{2cg}^2 + k_{2cg}^2\right),$$

$$\Gamma_{11} = \left(m_3 L_{1cg} + m_4 L_1\right), \quad \Gamma_{22} = m_4 L_{2cg}, \tag{4.64}$$

$$\begin{bmatrix} I_{11} + I_{21} \cos(\theta_3) & I_{22} + I_{21} \cos(\theta_3) \\ I_{21} \cos(\theta_3) & I_{22} \end{bmatrix} \begin{bmatrix} \ddot{\theta}_2 \\ \ddot{\theta}_2 + \ddot{\theta}_3 \end{bmatrix}$$

$$- I_{21} \sin(\theta_3) \begin{bmatrix} 2\dot{\theta}_2 \dot{\theta}_3 + \dot{\theta}_3^2 \\ -\dot{\theta}_2^2 \end{bmatrix} + g \begin{bmatrix} \Gamma_1 \\ \Gamma_2 \end{bmatrix} = \begin{bmatrix} \tau_1 \\ \tau_2 \end{bmatrix}. \tag{4.65}$$

The equations are the well-known equations of a two link manipulator (Vepa [1]).

4.3 Other Approaches to Robot Dynamics Formulation

While the equations of motion could be obtained by the use of the Lagrangian method to determine the Euler-Lagrange equations directly in terms of the body's translational and angular velocities, a simpler approach to obtain the equations of motion in terms of the body's translational and angular velocities is the method of quasi-coordinates developed by Hamel [4] and Boltzmann [5]. The term "quasi-coordinates" is used as one cannot directly integrate the angular velocities to get the generalized coordinates. Thus, transformations are applied before deriving Lagrange's equations to obtain the equations of motion. The method is explained in Vepa [3].

Kane's method [2] is particularly suited for deriving equations of motion of a system of interconnected rigid bodies moving under the influence of gravitational fields, such as spacecraft carrying robotic manipulators, as they are made up of several individual bodies connected by joints in a chain-like structure. The method is quite different from the Lagrangian method, although it shares a number of features with the Lagrangian formulation and with D'Alembert's principle. Kane's method is based on the principles of the conservation of linear and angular momentum. Kane's equations are very similar to D'Alembert's principle, except that they use an independent set of generalized speeds. D'Alembert's principle is based on selecting an independent set of generalized coordinates. A form of Kane's equations can also be used for systems with constraints by selecting an independent set of generalized speeds to ensure that the equations of motion are independent. One of the features of Kane's method is the elimination of contact equations, and this offers a great advantage over the Euler-Newton formulation.

Although Lagrange's formulation requires the computation of the partial derivatives of the Lagrangian, a function of the total kinetic and potential energy, with respect to each configuration coordinate, use of the concepts of partial translational velocities and partial angular velocities, introduced by Kane, greatly simplifies the implementation of the Lagrangian method for systems with many degrees of freedom. Kane's equations have been compared with the earlier results of Gibbs and Appell [6], Jourdain [7], and Maggi [8]. These have been compared with each other and with Lagrange's equations [9,10]. An example of the application of Kane's method to spacecraft has been presented by Stoneking [11].

4.4 Fundamentals of Manipulator Deployment and Control

Before the manipulator can be used, the satellite must be placed right behind the target and in the same orbit of a target object. Initially, a launch vehicle puts the satellite into an elliptical transfer orbit or into a circular low Earth orbit.

If the orbit is highly eccentric with apogee at the desired final orbit, then as the satellite passes through apogee, additional velocity is given to the satellite by an Apogee Kick Motor (AKM), and the satellite changes orbits. When the orbit after launch is circular, two further maneuvers are required, first a maneuver at the perigee to transfer from the circular orbit into an elliptic orbit, and then a second maneuver at the apogee to transfer from the elliptic orbit into the required final orbit. Thereafter, additional adjustments are required to correct the position in the orbit and to alter the orbit inclination. Placing the satellite in a circular orbit initially allows the transfer orbit to be implemented, so its apogee is right behind the target body. Then only inclination changes are essential to ensure the satellite is following the same orbit as the target. At this stage, the attitude of the satellite is altered to point the satellite toward the target or the desired direction. At this stage, the manipulator is ready for deployment.

The controller for the satellite and manipulator belongs to one of three classes. In the first instance, the base satellite continues to be controlled so both its orbital position and its orientation can be considered to be fixed in space. The manipulator is also controlled, and in this case the design of the manipulator controller is similar to the design of the controller in a ground-based configuration, on the Earth's surface. The design of controllers for ground-based manipulators is discussed in some detail in Vepa [3]. For such manipulators, Cao, Modi, de Silva, and Misra [12] proposed a controller that was based on a Linear Quadratic Regulator (LQR) to augment the stability of the manipulator while it was in fixed configuration involving no deployment, slew, or translation, and a feedback linearization-based nonlinear controller while the manipulator was in translation, performing a slewing maneuver, or being deployed. In the second class, the controller on board the satellite only regulates the attitude of the satellite. Additional control torques are applied to the manipulator. In the third class, the satellite is uncontrolled and is orbiting quite freely. All of the control torques are applied only to the manipulator. In this class, the base of the manipulator floats in space, and any motion of the manipulator's arms produces a motion of the base, which in turn affects the motion of its end-effector. A coupling between the attitude and the orbital dynamics will follow, with special requirements for the controller. In order to compensate the attitude variation of the end effector or the base of the grasper due to the satellite's motion, a control torque has to be applied to the satellite by the manipulator, with joint-control torques being applied to the manipulator. An optimal strategy and constraints may be defined in order to reduce the necessary torques for reorienting the arms of the manipulator and to regulate the orientation of the free-floating base. Thus an optimal control law to continuously guide and deploy the arms of the manipulator for grasping a target body may be designed, in principle.

Returning to the first class of controllers, the use of these controllers could be limited, due to the relatively high fuel requirements, which could in turn saturate the reaction-jet controller, as noted by Dubowsky, Vance, and Torres [13]. To minimize the fuel requirements for regulating the base satellite, Nenchev, Umetani, and Yoshida [14] and Quinn, Chen, and Lawrence [15] proposed to control the manipulator arms without disturbing the attitude of the spacecraft. For this type of manipulator and its controller, Torres and Dubowsky [16] defined an Enhanced Disturbance Map (EDM), which could synthesize paths for a given manipulator that result in low-attitude fuel consumption. Thus, with two points P and Q both defined within the reachable workspace of the manipulator, where P is the initial position of the end effector and Q is the final desired position of the end effector, a path from P to Q can be synthesized that does not disturb the attitude of the base satellite.

When only the base satellite's attitude is controlled and its position is not regulated, this concept is very useful for reducing fuel consumption while keeping the orientation fixed when necessary by employing electrically driven reaction wheels. Alexander and Cannon [17] showed that the manipulator end effector can be regulated by solving the inverse dynamics that include the base's motion. However, the controller synthesis is more complicated to implement than for the first class, as the disturbances in the satellite's base motion and in the motion of the target object being grasped must both be taken into consideration (Longman, Lindberg, and Zedd [18], Vafa and Dubowsky [19–21], Lindberg, Longman, and Zedd [22]).

In the case of the free-floating base satellite, no actuators are used to control the position and orientation of the satellite. Therefore, the satellite base is free to move in response to the manipulator motion. The manipulator controllers have the advantages that no fuel is required to regulate the spacecraft and the impact of a collision of the robot end-effector with an object about to be grasped, due to the base satellite's thrusters suddenly firing, is eliminated. However, path planning for this class of controllers is more involved, as shown by Lindberg, Longman, and Zedd [22], since the position of the robot end-effector is no longer just a function of the present robot joint angles, but is a function of the whole time history of the joint angles. The inverse kinetics problem (instead of inverse kinematics for ground-fixed robots) is very complicated and generally has an infinite number of solutions. The correct solution depends on the joint-space history, and Longman [23] showed how one of these solutions could be obtained. For the case of a freely floating satellite base, with a manipulator mounted on it, Vafa and Dubowsky [19–21,24] developed the virtual manipulator approach that permits modeling of a free-floating manipulator as a fixed-base manipulator. The virtual manipulator is a fixed-base manipulator with its base frame located at the satellite's center of mass. If the satellite's attitude and translation are not controlled and if it is not subjected to external forces, the center of mass of the satellite maintains its initial position (if its initial velocity is null). This allows use of the center of mass of the combined satellite and manipulator as the origin of the base frame of the virtual manipulator. The geometry of the virtual manipulator arms is directly related to the geometrical and inertial properties of the combined satellite and manipulator. The joint rotations of the virtual manipulator arms are exactly the same as the real joint rotations, in the case of revolute joints. The virtual manipulator approach can also be applied to the case of when only the attitude of the spacecraft is regulated, if the attitude controller uses only actuators that only exchange internal forces with the rest of the spacecraft like gyroscopes and reaction wheels that inherently meet these constraints. Torres and Dubowsky [16] and then Papadopoulos [25] and Yoshida and Umetani [26] discussed the possible occurrence of dynamic singularities for the free-floating case.

A free-floating manipulator is a non-holonomic mechanical system, so the evolution of the joint motion is directly dependent on the previous motion history of the manipulator. Non-holonomic mechanical systems in the context of space robotic manipulators are considered in detail in the next chapter. For such non-holonomic mechanical systems a closed trajectory in the joint space does not produce a closed trajectory in the task-Cartesian space. Thus, at some configurations, the manipulator arms could become uncontrollable for certain joint positions and velocities. In this case, the Jacobian matrix of the manipulator is singular, and the manipulator arms are forced to execute large joint movements in order to perform small displacements or velocities in the task space. The singularities in the Jacobian show up as "dynamic singularities," which are path-dependent, and their location in the workspace depends on the manipulator's dynamic configuration parameters (Papadopoulos [27–29]). However Papadopoulos and Dubowsky [30] have suggested that nearly any control algorithm devoted to regulating a fixed-base manipulator can also be implemented for free-floating space robots with a small additional set of restrictions.

In other studies, coordinated controllers were designed to control both the spacecraft and the end-effector, and allow the tracking of a desirable manipulator configuration and the planning of the manipulator arm motion with the use of thrusters (Papadopoulos and Dubowsky [31], Papadopoulos, Moosavian, and Ali [32], Moosavian, Ali, and Papadopoulos [33]). Masutani, Miyazaki, and Arimoto [34] propose a sensory feedback control scheme for space manipulators based on an artificial potential defined in task-oriented coordinates. In this scheme, the controller determines the input torques of each joint from the deviations of the position and orientation errors of the end effector tip, multiplied by positive control gains and the transposes of the generalized position and orientation Jacobians.

A very successful robotic space manipulator was the CANADARM (also known as SRMS – Shuttle Remote Manipulator System), which was commissioned onboard the

Figure 4.2 The CANADARM Manipulator. (credit: Stocktrek Images / Getty Images)

Figure 4.3 The CANADARM2 Manipulator. (credit: Stocktrek Images / Getty Images)

International Space Station (ISS) (Figure 4.2). It is a six-degree of freedom 15 m long arm, weighing nearly 400 kg. CANADARM2, shown in Figure 4.3, is a result of a joint cooperative effort between the Canadian Space Agency and NASA. The primary functions of this massive manipulator are deploying or retrieving satellites and space modules in orbit. These two tele-operated manipulator systems operate in low earth orbit and are limited to their own reach, thereby requiring very long links to create a large workspace.

4.5 Free-Flying Multi-Link Serial Manipulator in Three Dimensions

Modeling of a free-flying multi-link manipulator can be achieved by adopting the Lagrangian formulation. With the correct choice of reference frames, the dynamics can be reduced to a standard form. The most appropriate choice of the reference frames are not the traditional frames defined by the DH convention. A typical three-link serial manipulator is illustrated in Figure 4.4. The environment is assumed to be gravity-free. The origin of the first link is assumed to be attached to the satellite at a point fixed to the satellite. The components of the position vector of the satellite's CM $\mathbf{p}_s$ relative to this point are given by p_{sx}, p_{sy}, and p_{sz} in the satellite's body-fixed axes. The link lengths are not assumed to be constant.

The positions of the link centers of mass in planar Cartesian coordinates for the first, second, third, and Nth link with respect to the origin of the first link are, respectively, given by,

$$x_1 = l_{C1} \cos \theta_1, \quad y_1 = l_{C1} \sin \theta_1, \quad z_1 = 0 \tag{4.66}$$

$$x_2 = l_1 \cos \theta_1 + l_{C2} \cos \theta_2, \tag{4.67}$$

$$y_2 = l_1 \sin \theta_1 + l_{C2} \sin \theta_2, \quad z_2 = 0 \tag{4.68}$$

$$x_3 = l_1 \cos \theta_1 + l_2 \cos \theta_2 + l_{C3} \cos \theta_3, \tag{4.69}$$

$$y_3 = l_1 \sin \theta_1 + l_2 \sin \theta_2 + l_{C3} \sin \theta_3, \quad z_3 = 0 \tag{4.70}$$

$$x_i = \sum_{j=1}^{i-1} l_j \cos \theta_j + l_{Ci} \cos \theta_i, \quad y_i = \sum_{j=1}^{i-1} l_j \sin \theta_j + l_{Ci} \sin \theta_i, \quad z_i = 0. \tag{4.71}$$

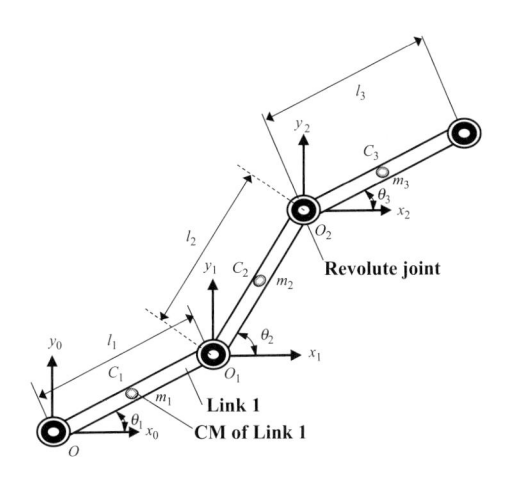

Figure 4.4 Typical three-link manipulator attached to a satellite, showing the definitions of the degrees of freedom.

The position vector of the system's CM is,

$$\mathbf{p} = \frac{m_{sat}\mathbf{p}_s + \sum_{i=1}^{N} m_i \mathbf{p}_i}{m_{sat} + \sum_{i=1}^{N} m_i} = \frac{m_{sat}\mathbf{p}_s + \sum_{i=1}^{N} m_i \mathbf{p}_i}{m_{total}}, \quad (4.72)$$

where $\mathbf{p}_i = [x_i \ y_i \ z_i]^T$, $m_{total} = m_{sat} + \sum_{i=1}^{N} m_i$, $\mathbf{p}_s$ is the position vector of the satellite's center of mass, m_{sat} is the mass of the satellite, and m_i is the mass of the ith link.

The velocity vector of the satellite and manipulator system CM is assumed to be $\mathbf{v}_{CM}$, and let $\boldsymbol{\omega} = [p \ q \ r]$ be the body's angular velocity vector of the satellite.

The velocities of the link centers of mass in planar Cartesian coordinates about the origin of the first link for the ith link are, respectively, given by,

$$v_{xi} = \dot{l}_{Ci} \cos \theta_i - \dot{\theta}_i l_{Ci} \sin \theta_i - \sum_{j=1}^{i-1} \left(\dot{\theta}_j l_j \sin \theta_j - \dot{l}_j \cos \theta_j \right), \quad (4.73)$$

$$v_{yi} = \dot{l}_{Ci} \sin \theta_i + \dot{\theta}_i l_{Ci} \cos \theta_i + \sum_{j=1}^{i-1} \left(\dot{\theta}_j l_j \cos \theta_j + \dot{l}_j \sin \theta_j \right), \quad v_{zi} = 0, \quad (4.74)$$

and $\mathbf{v}_i = \dot{\mathbf{p}}_i = [v_{xi} \ v_{yi} \ v_{zi}]$. Hence,

$$\mathbf{v}_{CM} = \mathbf{v} = \dot{\mathbf{p}} = \frac{m_{sat}\dot{\mathbf{p}}_s + \sum_{i=1}^{N} m_i \dot{\mathbf{p}}_i}{m_{sat} + \sum_{i=1}^{N} m_i} = \frac{m_{sat}\dot{\mathbf{p}}_s + \sum_{i=1}^{N} m_i \mathbf{v}_i}{m_{total}}. \quad (4.75)$$

The kinetic energy for the satellite's main body is,

$$T_{sys} = \frac{1}{2} m_{total} \mathbf{v}_{CM}^2 + \frac{1}{2} \boldsymbol{\omega}^T \mathbf{I}_{sat\_CM} \boldsymbol{\omega} + \frac{1}{2} \sum_{i=1}^{N} \left(\boldsymbol{\omega} + \mathbf{t}_i \mathbf{e}_z \dot{\theta}_i \right)^T \mathbf{I}_{CMi} \left(\boldsymbol{\omega} + \mathbf{t}_i \mathbf{e}_z \dot{\theta}_i \right), \quad (4.76)$$

where $\mathbf{e}_z = [0 \ 0 \ 1]^T$ is a unit vector directed along the joint axis, and $\mathbf{t}_i$ is the rotational transformation from the link coordinate frame at the link CM to the system coordinates frame at the system CM. It is the rotational part of the DH transformation. The kinetic energy of translation is,

$$T_{tran} = \frac{1}{2} m_{total} \mathbf{v}_{CM}^2 = \frac{1}{2 m_{total}} \left(m_{sat} \dot{\mathbf{p}}_s + \sum_{i=1}^{N} m_i \mathbf{v}_i \right)^2. \quad (4.77)$$

The manipulator's contribution to the rotational kinetic energy is,

$$T_{rot,m} = \frac{1}{2} \sum_{i=1}^{N} \left(\boldsymbol{\omega} + \mathbf{t}_i \mathbf{e}_z \dot{\theta}_i \right)^T \mathbf{I}_{CMi} \left(\boldsymbol{\omega} + \mathbf{t}_i \mathbf{e}_z \dot{\theta}_i \right). \quad (4.78)$$

The total moment of momentum of the satellite and manipulator system is,

$$\mathbf{H}_{sys} = \mathbf{I}_{sat\_CM}\boldsymbol{\omega} + \sum_{i=1}^{N}\mathbf{I}_{CMi}\left(\boldsymbol{\omega} + \mathbf{t}_i\mathbf{e}_z\dot{\theta}_i\right), \tag{4.79}$$

with, $\boldsymbol{\omega} = \begin{bmatrix} p & q & r \end{bmatrix}^T$,

$$\mathbf{I}_{sat\_CM} = \mathbf{I}_{sat} + m_{sat}\mathbf{I}_{3\times 3}(\mathbf{p}_s - \mathbf{p})^T \cdot (\mathbf{p}_s - \mathbf{p}) - m_{sat}(\mathbf{p}_s - \mathbf{p})(\mathbf{p}_s - \mathbf{p})^T. \tag{4.80}$$

$\mathbf{I}_{sat}$ is the satellite's mass moment of inertia matrix in body-fixed coordinates. The quantity,

$$\mathbf{I}_{CMi} = \mathbf{t}_i\mathbf{I}_{Ci}\mathbf{t}_i^T + m_i\mathbf{I}_{3\times 3}(\mathbf{p}_i - \mathbf{p})^T \cdot (\mathbf{p}_i - \mathbf{p}) - m_i(\mathbf{p}_i - \mathbf{p})(\mathbf{p}_i - \mathbf{p})^T, \tag{4.81}$$

is the total mass moment of inertia matrix of the ith link about a frame parallel to the body axes, located at the center of mass of the link. Given that,

$$\begin{bmatrix} x_o \\ y_o \\ z_o \end{bmatrix} = \begin{bmatrix} \cos\theta_i & -\sin\theta_i & 0 \\ \sin\theta_i & \cos\theta_i & 0 \\ 0 & 0 & 1 \end{bmatrix} \begin{bmatrix} x_{ib} \\ y_{ib} \\ z_{ib} \end{bmatrix}, \quad \mathbf{t}_i = \begin{bmatrix} \cos\theta_i & -\sin\theta_i & 0 \\ \sin\theta_i & \cos\theta_i & 0 \\ 0 & 0 & 1 \end{bmatrix},$$

$$\dot{\mathbf{t}}_i = -\dot{\theta}_i \begin{bmatrix} \sin\theta_i & \cos\theta_i & 0 \\ -\cos\theta_i & \sin\theta_i & 0 \\ 0 & 0 & 0 \end{bmatrix}. \tag{4.82}$$

4.6 Application of the Principles of Momentum Conservation to Satellite-Manipulator Dynamics

We may now derive the satellite and manipulator attitude dynamics equations using either the Lagrangian approach with the quasi-coordinates $\boldsymbol{\omega}$ or the principle of conservation of momentum. To obtain the satellite's attitude dynamics, the principle of conservation of momentum is applied to the combined satellite and manipulator arms. Hence, assuming that the lengths of the links are constant, from the principle of conservation of momentum,

$$\frac{d}{dt}\left(\mathbf{I}_{sat\_CM}\boldsymbol{\omega} + \sum_{i=1}^{N}\mathbf{I}_{CMi}\left(\boldsymbol{\omega} + \mathbf{t}_i\mathbf{e}_z\dot{\theta}_i\right)\right) + \boldsymbol{\omega}_\times\mathbf{I}_{sat\_CM}\boldsymbol{\omega}$$

$$+ \sum_{i=1}^{N}\left(\boldsymbol{\omega} + \mathbf{t}_i\mathbf{e}_z\dot{\theta}_i\right)_\times\mathbf{I}_{CMi}\left(\boldsymbol{\omega} + \mathbf{t}_i\mathbf{e}_z\dot{\theta}_i\right) = \boldsymbol{\tau}_{sat}. \tag{4.83}$$

4.7 Application of the Lagrangian Approach to Satellite-Manipulator Dynamics

The manipulator dynamics on its own is obtained by the Lagrangian approach, since θ_i may be treated as a generalized coordinate. Hence the Euler-Lagrange equations are,

$$\frac{d}{dt}\left\{\mathbf{e}_z^T\mathbf{t}_i^T\mathbf{I}_{CMi}(\boldsymbol{\omega}+\mathbf{t}_i\mathbf{e}_z\dot{\theta}_i) + \frac{\partial T_{tran}}{\partial \dot{\theta}_i}\right\} - \frac{\partial T_{tran}}{\partial \theta_i} - \frac{\partial T_{rot,m}}{\partial \theta_i} = \mathbf{e}_z^T\mathbf{t}_i^T\boldsymbol{\tau}_i, \quad i=1,\cdots,N, \tag{4.84}$$

where $\boldsymbol{\tau}_{sat}$ is the control and other external torques acting on the satellite-manipulator system, $\boldsymbol{\tau}_i$ is the *net* control and other external torques acting at each link relative to the system CM, and,

$$T_{tran} = \frac{1}{2}m_{total}\mathbf{v}_{CM}^2 = \frac{1}{2m_{total}}\left(m_{sat}\dot{\mathbf{p}}_s + \sum_{i=1}^{N}m_i\mathbf{v}_i\right)^2, \tag{4.85}$$

$$\frac{\partial T_{tran}}{\partial \dot{\theta}_j} = \frac{1}{m_{total}}\left(m_{sat}\dot{\mathbf{p}}_s + \sum_{i=1}^{N}m_i\mathbf{v}_i\right)\left(\sum_{i=1}^{N}m_i\frac{\partial \mathbf{v}_i}{\partial \dot{\theta}_j}\right). \tag{4.86}$$

T_{tran} may be ignored only for a symmetrically actuated manipulator (balanced actuation). Further, the time derivative of the satellite's mass moment of inertia matrix at any instant is,

$$\dot{\mathbf{I}}_{sat\_CM} = \dot{\mathbf{I}}_{sat} - 2m_{sat}\mathbf{I}_{3\times 3}\dot{\mathbf{p}}^T\cdot(\mathbf{p}_s-\mathbf{p}) + m_{sat}\dot{\mathbf{p}}(\mathbf{p}_s-\mathbf{p})^T + m_{sat}(\mathbf{p}_s-\mathbf{p})\dot{\mathbf{p}}^T. \tag{4.87}$$

If $\mathbf{I}_{sat}$ is constant in the body axes, $\dot{\mathbf{I}}_{sat} = 0$, and it follows that,

$$\dot{\mathbf{I}}_{sat\_CM} = m_{sat}\dot{\mathbf{p}}(\mathbf{p}_s-\mathbf{p})^T + m_{sat}(\mathbf{p}_s-\mathbf{p})\dot{\mathbf{p}}^T - 2m_{sat}\mathbf{I}_{3\times 3}\dot{\mathbf{p}}^T\cdot(\mathbf{p}_s-\mathbf{p}). \tag{4.88}$$

Since

$$\mathbf{I}_{CMi} = \mathbf{t}_i\mathbf{I}_{Ci}\mathbf{t}_i^T + m_i\mathbf{I}_{3\times 3}(\mathbf{p}_i-\mathbf{p})^T\cdot(\mathbf{p}_i-\mathbf{p}) - m_i(\mathbf{p}_i-\mathbf{p})(\mathbf{p}_i-\mathbf{p})^T. \tag{4.89}$$

The time derivative of the ith link's mass moment of inertia matrix about a frame parallel to the body axes and located at the center of mass of the link, at any instant is,

$$\dot{\mathbf{I}}_{CMi} = \dot{\mathbf{t}}_i\mathbf{I}_{Ci}\mathbf{t}_i^T + \mathbf{t}_i\mathbf{I}_{Ci}\dot{\mathbf{t}}_i^T + 2m_i\mathbf{I}_{3\times 3}(\dot{\mathbf{p}}_i-\dot{\mathbf{p}})^T\cdot(\mathbf{p}_i-\mathbf{p})$$
$$- m_i(\dot{\mathbf{p}}_i-\dot{\mathbf{p}})(\mathbf{p}_i-\mathbf{p})^T - m_i(\mathbf{p}_i-\mathbf{p})(\dot{\mathbf{p}}_i-\dot{\mathbf{p}})^T. \tag{4.90}$$

Hence,

$$\dot{\mathbf{I}}_{CMi} = \dot{\mathbf{t}}_i\mathbf{I}_{Ci}\mathbf{t}_i^T + \mathbf{t}_i\mathbf{I}_{Ci}\dot{\mathbf{t}}_i^T + 2m_i\mathbf{I}_{3\times 3}(\mathbf{v}_i-\mathbf{v})^T\cdot(\mathbf{p}_i-\mathbf{p})$$
$$- m_i(\mathbf{v}_i-\mathbf{v})(\mathbf{p}_i-\mathbf{p})^T - m_i(\mathbf{p}_i-\mathbf{p})(\mathbf{v}_i-\mathbf{v})^T. \tag{4.91}$$

Let us now assume that link 1 is the satellite itself and that the origin of the base link coincides with the center of mass of the satellite as shown in Figure 4.5. Thus, $\mathbf{p}_s = \mathbf{0}$, $\mathbf{I}_{sat} = \mathbf{0}$, $m_{sat} = 0$, $\dot{\theta}_1 = 0$. The equations reduce to,

$$\frac{d}{dt}\left(\mathbf{I}_{CM1}\boldsymbol{\omega} + \sum_{i=2}^{N}\mathbf{I}_{CMi}(\boldsymbol{\omega}+\mathbf{t}_i\mathbf{e}_z\dot{\theta}_i)\right) + \boldsymbol{\omega}\times\mathbf{I}_{CM1}\boldsymbol{\omega}$$
$$+ \sum_{i=2}^{N}(\boldsymbol{\omega}+\mathbf{t}_i\mathbf{e}_z\dot{\theta}_i)\times\mathbf{I}_{CMi}(\boldsymbol{\omega}+\mathbf{t}_i\mathbf{e}_z\dot{\theta}_i) = \boldsymbol{\tau}_{sat}, \tag{4.92}$$

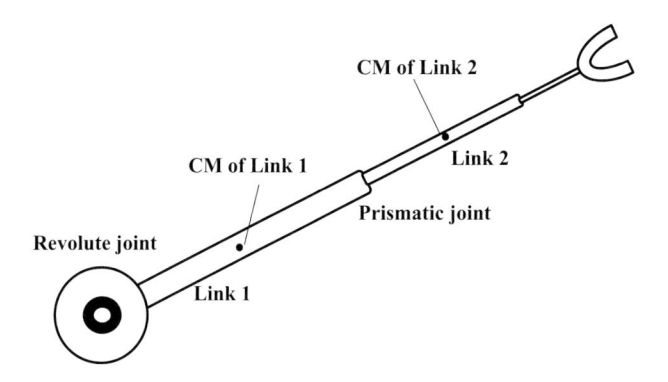

Figure 4.5 Satellite with a manipulator on board.

$$\frac{d}{dt}\left\{\mathbf{e}_z^T\mathbf{t}_i^T\mathbf{I}_{CMi}(\boldsymbol{\omega}+\mathbf{t}_i\mathbf{e}_z\dot{\theta}_i)+\frac{\partial T_{tran}}{\partial \dot{\theta}_i}\right\}-\frac{\partial T_{tran}}{\partial \theta_i}-\frac{\partial T_{rot,m}}{\partial \theta_i}=\mathbf{e}_z^T\mathbf{t}_i^T\boldsymbol{\tau}_i, \quad i=2,\cdots,N, \tag{4.93}$$

$$\mathbf{I}_{CMi}=\mathbf{t}_i\mathbf{I}_{Ci}\mathbf{t}_i^T+m_i\mathbf{I}_{3\times 3}(\mathbf{p}_i-\mathbf{p})^T\cdot(\mathbf{p}_i-\mathbf{p})-m_i(\mathbf{p}_i-\mathbf{p})(\mathbf{p}_i-\mathbf{p})^T, \tag{4.94}$$

$$\dot{\mathbf{I}}_{CMi}=\dot{\mathbf{t}}_i\mathbf{I}_{Ci}\mathbf{t}_i^T+\mathbf{t}_i\mathbf{I}_{Ci}\dot{\mathbf{t}}_i^T+2m_i\mathbf{I}_{3\times 3}(\mathbf{v}_i-\mathbf{v})^T\cdot(\mathbf{p}_i-\mathbf{p})$$
$$-m_i(\mathbf{v}_i-\mathbf{v})(\mathbf{p}_i-\mathbf{p})^T-m_i(\mathbf{p}_i-\mathbf{p})(\mathbf{v}_i-\mathbf{v})^T. \tag{4.95}$$

4.8 Gravity-Gradient Forces and Moments on an Orbiting Body

The gravity-gradient force and torque acting on an orbiting body can, respectively, be shown to be given by,

$$\vec{f}_g=-\mu m_s|\mathbf{r}|^{-2}\mathbf{c}-\frac{3}{2}\mu|\mathbf{r}|^{-4}(2\mathbf{I}_{sm}+Trace(\mathbf{I}_{sm})\mathbf{U}-5\mathbf{cc}\cdot\mathbf{I}_{sm})\cdot\mathbf{c}, \tag{4.96}$$

$$\vec{g}_g=3\mu|\mathbf{r}|^{-3}(\mathbf{c}\times\mathbf{I}_{sm}\cdot\mathbf{c}), \quad \mathbf{I}_{sm}=\mathbf{I}_s+\mathbf{I}_m=\mathbf{I}+\mathbf{I}_m, \tag{4.97}$$

where $\mu = GM_c$, $\mathbf{c} = \mathbf{r}/|\mathbf{r}|$, $m_r = m_s/M_c$, $\mathbf{r}$ is the vector position of the central body relative to the orbiting body's center of mass, and $\mathbf{U}$ is a 3×3 unit matrix.

It can be shown that $\mu|\mathbf{r}|^{-3} \equiv n^2$, where n is the orbital angular velocity. Since $n^2 = \mu|\mathbf{r}|^{-3}$,

$$|\mathbf{r}|^3 = \mu/n^2, \quad |\mathbf{r}| = (\mu/n^2)^{1/3}, \quad |\mathbf{r}|^{-1} = (n^2/\mu)^{1/3}. \tag{4.98}$$

The gravity-gradient force and torque about the link's joint axis, acting on each link are,

$$F_i = -m_i n^2 |\mathbf{r}|\mathbf{c}' - \frac{3}{2} n^2 |\mathbf{r}|^{-1} \left(2\mathbf{t}_{0i} \mathbf{I}_{Ci} \mathbf{t}_{0i}^T + Trace\left(\mathbf{t}_{0i} \mathbf{I}_{Ci} \mathbf{t}_{0i}^T\right) \mathbf{U} - 5\mathbf{c}\mathbf{c} \cdot \mathbf{t}_{0i} \mathbf{I}_{Ci} \mathbf{t}_{0i}^T\right) \cdot \mathbf{c}',$$

$$\mathbf{M}_i = 3n^2 \left(\mathbf{c}' \times \mathbf{t}_{0i} \mathbf{I}_{Ci} \mathbf{t}_{0i}^T \cdot \mathbf{c}'\right) + \left({}^R\mathbf{p}_{i0} - {}^R\mathbf{p}_{icm}\right) \times \mathbf{F}_i + \left({}^R\mathbf{p}_{i0} - {}^R\mathbf{p}_{ie}\right) \times \sum_{i=i+1}^{N} \mathbf{F}_i,$$

$$\mathbf{c}' = \left(\left(\mathbf{c}|\mathbf{r}| + {}^R\mathbf{p}_{i0} - {}^R\mathbf{p}_{icm}\right) / \left|\mathbf{c}|\mathbf{r}| + {}^R\mathbf{p}_{i0} - {}^R\mathbf{p}_{icm}\right|\right). \tag{4.99}$$

The gravitation forces of all the links ahead of the ith link in the chain also contribute to the moment. The force is approximated as,

$$F_i = -m_i n^2 |\mathbf{r}|\mathbf{c} - \frac{3}{2} n^2 |\mathbf{r}|^{-1} \left(2\mathbf{t}_{0i} \mathbf{I}_{Ci} \mathbf{t}_{0i}^T + Trace\left(\mathbf{t}_{0i} \mathbf{I}_{Ci} \mathbf{t}_{0i}^T\right) \mathbf{U} - 5\mathbf{c}\mathbf{c} \cdot \mathbf{t}_{0i} \mathbf{I}_{Ci} \mathbf{t}_{0i}^T\right) \cdot \mathbf{c}, \tag{4.100}$$

or as,

$$F_i = -m_i n (\mu\, n)^{\frac{1}{3}} \mathbf{c} - \frac{3}{2} n^2 \left(n^2/\mu\right)^{\frac{1}{3}} \left(2\mathbf{t}_{0i} \mathbf{I}_{Ci} \mathbf{t}_{0i}^T + Trace\left(\mathbf{t}_{0i} \mathbf{I}_{Ci} \mathbf{t}_{0i}^T\right) \mathbf{U} - 5\mathbf{c}\mathbf{c} \cdot \mathbf{t}_{0i} \mathbf{I}_{Ci} \mathbf{t}_{0i}^T\right) \cdot \mathbf{c}, \tag{4.101}$$

and moment is approximated as,

$$\mathbf{M}_i \cong 3n^2 \left(\mathbf{c} \times \mathbf{t}_{0i} \mathbf{I}_{Ci} \mathbf{t}_{0i}^T \cdot \mathbf{c}\right) + \left({}^R\mathbf{p}_{i0} - {}^R\mathbf{p}_{icm}\right) \times \mathbf{F}_i + \left({}^R\mathbf{p}_{i0} - {}^R\mathbf{p}_{ie}\right) \times \sum_{i=i+1}^{N} \mathbf{F}_i. \tag{4.102}$$

We have assumed that the forces are acting at the centers of mass of each link. $\mathbf{t}_{0i} = \mathbf{t}_i$ in this equation. ${}^R\mathbf{p}_{i0} - {}^R\mathbf{p}_{icm}$ is the vector position of the joint relative to the link's CM in the satellite's orbital frame. ${}^R\mathbf{p}_{ie}$ is the vector position of the next joint.

4.8.1 Gravity-Gradient Moment Acting on the Satellite Body and Manipulator Combined

We obtain expressions for the gravity-gradient moment assuming that z axis of the satellite body is nominally pointing to the Earth. The direction vector from the CG of each link pointing to the Earth is denoted as, $\mathbf{c} = \begin{bmatrix} c_1 & c_2 & c_3 \end{bmatrix}^T$.

Depending on the choice of the body axes of the satellite, the direction vector from the CG of each link pointing to the Earth is given either by the first, second or last column of $\mathbf{R}_{IB}$ as,

$$\mathbf{c} = \begin{bmatrix} q_4^2 + q_1^2 - q_2^2 - q_3^2 \\ 2(q_1 q_2 + q_3 q_4) \\ 2(q_1 q_3 - q_2 q_4) \end{bmatrix} \quad \mathbf{c} = \begin{bmatrix} 2(q_1 q_2 - q_3 q_4) \\ q_4^2 - q_1^2 + q_2^2 - q_3^2 \\ 2(q_2 q_3 + q_1 q_4) \end{bmatrix} \quad \text{or} \quad \mathbf{c} = \begin{bmatrix} 2(q_1 q_3 - q_2 q_4) \\ 2(q_2 q_3 + q_1 q_4) \\ q_4^2 - q_1^2 - q_2^2 + q_3^2 \end{bmatrix}. \tag{4.103}$$

The corresponding cross-product operator $\mathbf{c}_\times$ is defined as,

$$\mathbf{c}_\times = \begin{bmatrix} 0 & -c_3 & c_2 \\ c_3 & 0 & -c_1 \\ -c_2 & c_1 & 0 \end{bmatrix}. \tag{4.104}$$

Hence, the gravity-gradient moments acting on the satellite and manipulator body are,

$$\mathbf{M}_{sm} = 3n^2 \mathbf{c}_\times (\mathbf{I} + \mathbf{I}_m)\mathbf{c} \equiv 3n^2 \mathbf{c}_\times \mathbf{I}_c \mathbf{c} = \begin{bmatrix} L_{gg} & M_{gg} & N_{gg} \end{bmatrix}^T. \quad (4.105)$$

$$\frac{1}{3n^2}\begin{bmatrix} L_{gg} \\ M_{gg} \\ N_{gg} \end{bmatrix} = \begin{bmatrix} 0 & -c_3 & c_2 \\ c_3 & 0 & -c_1 \\ -c_2 & c_1 & 0 \end{bmatrix} \mathbf{I}_c \begin{bmatrix} c_1 \\ c_2 \\ c_3 \end{bmatrix} = \begin{bmatrix} 0 & -c_3 & c_2 \\ c_3 & 0 & -c_1 \\ -c_2 & c_1 & 0 \end{bmatrix} \begin{bmatrix} I_{c1}c_1 - I_{c12}c_2 \\ I_{c2}c_2 - I_{c12}c_1 \\ I_{c3}c_3 \end{bmatrix}. \quad (4.106)$$

$$\mathbf{M}_{sm} = \begin{bmatrix} L_{gg} \\ M_{gg} \\ N_{gg} \end{bmatrix} = 3n^2 \begin{bmatrix} (I_{c3} - I_{c2})c_3 c_2 - I_{c12} c_3 c_1 \\ (I_{c1} - I_{c3})c_1 c_3 - I_{c12} c_3 c_2 \\ (I_{c2} - I_{c1})c_2 c_1 + I_{c12}(c_2^2 - c_1^2) \end{bmatrix}. \quad (4.107)$$

4.9 Application to Satellite-Manipulator Dynamics

The equations of motion of the satellite and manipulator and the manipulator links reduce to,

$$\frac{d}{dt}\left(\mathbf{I}_{CM1}\boldsymbol{\omega} + \sum_{i=2}^{N} \mathbf{I}_{CMi}\left(\boldsymbol{\omega} + \mathbf{t}_i \mathbf{e}_z \dot{\theta}_i\right)\right) + \boldsymbol{\omega}_\times \mathbf{I}_{CM1}\boldsymbol{\omega} + \sum_{i=2}^{N}\left(\boldsymbol{\omega} + \mathbf{t}_i \mathbf{e}_z \dot{\theta}_i\right)_\times \mathbf{I}_{CMi}\left(\boldsymbol{\omega} + \mathbf{t}_i \mathbf{e}_z \dot{\theta}_i\right)$$
$$= \boldsymbol{\tau}_{sat} + \mathbf{M}_{sm} + \left({}^R\mathbf{p}_{s0} - {}^R\mathbf{p}_{se}\right) \times \sum_{i=1}^{N} \mathbf{F}_i, \quad (4.108)$$

$$\frac{d}{dt}\left\{\mathbf{e}_z^T \mathbf{t}_i^T \mathbf{I}_{CMi}\left(\boldsymbol{\omega} + \mathbf{t}_i \mathbf{e}_z \dot{\theta}_i\right) + \frac{\partial T_{tran}}{\partial \dot{\theta}_i}\right\} - \frac{\partial T_{tran}}{\partial \theta_i} - \frac{\partial T_{rot,m}}{\partial \theta_i}$$
$$= \mathbf{e}_z^T \mathbf{t}_i^T \boldsymbol{\tau}_i + \mathbf{e}_z^T \mathbf{t}_i^T\left(\mathbf{M}_i + \left({}^R\mathbf{p}_{i0} - {}^R\mathbf{p}_{icm}\right) \times \mathbf{F}_i + \left({}^R\mathbf{p}_{i0} - {}^R\mathbf{p}_{ie}\right) \times \sum_{i=i+1}^{N} \mathbf{F}_i\right), \quad i = 2,\ldots,N. \quad (4.109)$$

Further,

$$\frac{\partial T_{tran}}{\partial \dot{\theta}_j} = \frac{1}{m_{total}}\left(\sum_{i=1}^{N} m_i \mathbf{v}_i\right)\left(\sum_{i=1}^{N} m_i \frac{\partial \mathbf{v}_i}{\partial \dot{\theta}_j}\right) = \frac{1}{m_{total}}\left(\sum_{i=1}^{N} m_i \mathbf{v}_{i,j}\right)\left(\sum_{i=1}^{N} m_i \sum_k \mathbf{v}_{i,k}\dot{\theta}_k\right), \quad (4.110)$$

$$\frac{d}{dt}\frac{\partial T_{tran}}{\partial \dot{\theta}_j} = \sum_k \left(\ddot{\theta}_k \frac{\partial}{\partial \dot{\theta}_k}\frac{\partial T_{tran}}{\partial \dot{\theta}_j} + \dot{\theta}_k \frac{\partial}{\partial \theta_k}\frac{\partial T_{tran}}{\partial \dot{\theta}_j}\right)$$
$$= \sum_k \left(\ddot{\theta}_k \frac{\partial}{\partial \dot{\theta}_k}\frac{\partial T_{tran}}{\partial \dot{\theta}_j} + \dot{\theta}_k \frac{\partial}{\partial \dot{\theta}_j}\frac{\partial T_{tran}}{\partial \theta_k}\right), \quad (4.111)$$

with

$$\frac{\partial^2 T_{tran}}{\partial \dot\theta_k \partial \dot\theta_j} = \frac{1}{m_{total}} \left(\sum_{i=1}^{N} m_i \mathbf{v}_{i,j} \right) \left(\sum_{i=1}^{N} m_i \mathbf{v}_{i,k} \right).$$

These derivatives may be evaluated compactly using the concept of partial translational velocities, $\mathbf{v}_{i,k}$.

The last set of equations may also be expressed as,

$$\mathbf{e}_z^T \mathbf{t}_i^T \mathbf{I}_{CMi} \frac{d}{dt}(\boldsymbol{\omega} + \mathbf{t}_i \mathbf{e}_z \dot\theta_i) + \frac{d}{dt} \frac{\partial T_{tran}}{\partial \dot\theta_i} + \frac{d}{dt}(\mathbf{e}_z^T \mathbf{t}_i^T) \cdot \mathbf{I}_{CMi}(\boldsymbol{\omega} + \mathbf{t}_i \mathbf{e}_z \dot\theta_i) - \frac{\partial T_{tran}}{\partial \theta_i} - \frac{\partial T_{rot,m}}{\partial \theta_i}$$

$$= \mathbf{e}_z^T \mathbf{t}_i^T \boldsymbol{\tau}_i + \mathbf{e}_z^T \mathbf{t}_i^T \left(\mathbf{M}_i + \left(^R\mathbf{p}_{i0} - {}^R\mathbf{p}_{icm}\right) \times \mathbf{F}_i + \left(^R\mathbf{p}_{i0} - {}^R\mathbf{p}_{ie}\right) \times \sum_{i=i+1}^{N} \mathbf{F}_i \right).$$

(4.112)

But,

$$\frac{d}{dt} \mathbf{e}_z^T \mathbf{t}_i^T = \sum_k \frac{\partial (\mathbf{e}_z^T \mathbf{t}_i^T)}{\partial \theta_k} \dot\theta_k, \quad \frac{d}{dt}(\mathbf{t}_i \mathbf{e}_z) = \sum_k \frac{\partial (\mathbf{t}_i \mathbf{e}_z)}{\partial \theta_k} \dot\theta_k. \quad (4.113)$$

Hence,

$$\mathbf{e}_z^T \mathbf{t}_i^T \mathbf{I}_{CMi}(\dot{\boldsymbol{\omega}} + \mathbf{t}_i \mathbf{e}_z \ddot\theta_i) + \frac{d}{dt} \frac{\partial T_{tran}}{\partial \dot\theta_i} - \frac{\partial T_{tran}}{\partial \theta_i} + \sum_k \dot\theta_k \frac{\partial (\mathbf{e}_z^T \mathbf{t}_i^T)}{\partial \theta_k} \cdot \mathbf{I}_{CMi} \boldsymbol{\omega}$$

$$+ \mathbf{e}_z^T \mathbf{t}_i^T \mathbf{I}_{CMi} \sum_k \frac{\partial (\mathbf{t}_i \mathbf{e}_z)}{\partial \theta_k} \dot\theta_k \dot\theta_i + \sum_k \dot\theta_k \frac{\partial (\mathbf{e}_z^T \mathbf{t}_i^T)}{\partial \theta_k} \cdot \mathbf{I}_{CMi} \mathbf{t}_i \mathbf{e}_z \dot\theta_i - \frac{\partial T_{rot,m}}{\partial \theta_i}$$

$$= \mathbf{e}_z^T \mathbf{t}_i^T \boldsymbol{\tau}_i + \mathbf{e}_z^T \mathbf{t}_i^T \left(\mathbf{M}_i + \left(^R\mathbf{p}_{i0} - {}^R\mathbf{p}_{icm}\right) \times \mathbf{F}_i + \left(^R\mathbf{p}_{i0} - {}^R\mathbf{p}_{ie}\right) \times \sum_{i=i+1}^{N} \mathbf{F}_i \right). \quad (4.114)$$

With,

$$\frac{d}{dt} \frac{\partial T_{tran}}{\partial \dot\theta_i} - \frac{\partial T_{tran}}{\partial \theta_i} = \sum_k \left(\ddot\theta_k \frac{\partial^2 T_{tran}}{\partial \dot\theta_k \partial \dot\theta_i} + \dot\theta_j \dot\theta_k \left(\frac{\partial}{\partial \theta_k} \frac{\partial^2 T_{tran}}{\partial \dot\theta_j \partial \dot\theta_i} - \frac{1}{2} \frac{\partial}{\partial \theta_i} \frac{\partial^2 T_{tran}}{\partial \dot\theta_j \partial \dot\theta_k} \right) \right). \quad (4.115)$$

Finally, one would like to separate the rigid body and manipulator contributions to the rotational kinetic energy. Thus,

$$T_{rot,m} = \frac{1}{2} \sum_{i=1}^{N} (\boldsymbol{\omega} + \mathbf{t}_i \mathbf{e}_z \dot\theta_i)^T \mathbf{I}_{CMi}(\boldsymbol{\omega} + \mathbf{t}_i \mathbf{e}_z \dot\theta_i)$$

$$= \frac{1}{2} \sum_{i=1}^{N} \boldsymbol{\omega}^T \mathbf{I}_{CMi} \boldsymbol{\omega} + \frac{1}{2} \sum_{i=1}^{N} \boldsymbol{\omega}^T \mathbf{I}_{CMi}(\mathbf{t}_i \mathbf{e}_z) \dot\theta_i + \frac{1}{2} \sum_{i=1}^{N} \dot\theta_i^T (\mathbf{t}_i \mathbf{e}_z)^T \mathbf{I}_{CMi} \boldsymbol{\omega}$$

$$+ \frac{1}{2} \sum_{i=1}^{N} \dot\theta_i^T (\mathbf{t}_i \mathbf{e}_z)^T \mathbf{I}_{CMi}(\mathbf{t}_i \mathbf{e}_z) \dot\theta_i. \quad (4.116)$$

Consequently,

$$\frac{\partial T_{rot,m}}{\partial \theta_j} = \frac{1}{2}\sum_{i=1}^{N} \boldsymbol{\omega}^T \mathbf{I}_{CMi} \frac{\partial (\mathbf{t}_i \mathbf{e}_z)}{\partial \theta_j} \dot{\theta}_i + \frac{1}{2}\sum_{i=1}^{N} \dot{\theta}_i^T \frac{\partial (\mathbf{t}_i \mathbf{e}_z)^T}{\partial \theta_j} \mathbf{I}_{CMi} \boldsymbol{\omega} + \frac{\partial T_{rot,m0}}{\partial \theta_j}, \qquad (4.117)$$

or,

$$\frac{\partial T_{rot,m}}{\partial \theta_j} = \sum_{i=1}^{N} \dot{\theta}_i^T \frac{\partial (\mathbf{t}_i \mathbf{e}_z)^T}{\partial \theta_j} \mathbf{I}_{CMi} \boldsymbol{\omega} + \frac{\partial T_{rot,m0}}{\partial \theta_j}, \qquad (4.118)$$

with,

$$\frac{\partial T_{rot,m0}}{\partial \theta_j} = \frac{1}{2}\sum_{i=1}^{N} \dot{\theta}_i^T \frac{\partial (\mathbf{t}_i \mathbf{e}_z)^T}{\partial \theta_j} \mathbf{I}_{CMi} (\mathbf{t}_i \mathbf{e}_z) \dot{\theta}_i + \frac{1}{2}\sum_{i=1}^{N} \dot{\theta}_i^T (\mathbf{t}_i \mathbf{e}_z)^T \mathbf{I}_{CMi} \frac{\partial (\mathbf{t}_i \mathbf{e}_z)}{\partial \theta_j} \dot{\theta}_i.$$

It follows that,

$$\mathbf{e}_z^T \mathbf{t}_i^T \mathbf{I}_{CMi} \left(\dot{\boldsymbol{\omega}} + \mathbf{t}_i \mathbf{e}_z \ddot{\theta}_i \right) + \frac{d}{dt}\frac{\partial T_{tran}}{\partial \dot{\theta}_i} - \frac{\partial T_{tran}}{\partial \theta_i} + \sum_k \dot{\theta}_k \left(\frac{\partial \left(\mathbf{e}_z^T \mathbf{t}_i^T \right)}{\partial \theta_k} - \frac{\partial \left(\mathbf{e}_z^T \mathbf{t}_k^T \right)}{\partial \theta_i} \right) \cdot \mathbf{I}_{CMi} \boldsymbol{\omega}$$

$$+ \mathbf{e}_z^T \mathbf{t}_i^T \mathbf{I}_{CMi} \sum_k \left(\frac{\partial (\mathbf{t}_i \mathbf{e}_z)}{\partial \theta_k} - \frac{1}{2}\frac{\partial (\mathbf{t}_k \mathbf{e}_z)}{\partial \theta_i} \right) \dot{\theta}_k \dot{\theta}_i + \sum_k \dot{\theta}_k \left(\frac{\partial \left(\mathbf{e}_z^T \mathbf{t}_i^T \right)}{\partial \theta_k} - \frac{1}{2}\frac{\partial \left(\mathbf{e}_z^T \mathbf{t}_k^T \right)}{\partial \theta_i} \right) \cdot \mathbf{I}_{CMi} \mathbf{t}_i \mathbf{e}_z \dot{\theta}_i$$

$$= \mathbf{e}_z^T \mathbf{t}_i^T \boldsymbol{\tau}_i + \mathbf{e}_z^T \mathbf{t}_i^T \left(\mathbf{M}_i + \left({}^R\mathbf{p}_{i0} - {}^R\mathbf{p}_{icm} \right) \times \mathbf{F}_i + \left({}^R\mathbf{p}_{i0} - {}^R\mathbf{p}_{ie} \right) \times \sum_{i=i+1}^{N} \mathbf{F}_i \right),$$

(4.119)

with,

$$\frac{d}{dt}\frac{\partial T_{tran}}{\partial \dot{\theta}_i} - \frac{\partial T_{tran}}{\partial \theta_i} = \sum_k \left(\ddot{\theta}_k \frac{\partial^2 T_{tran}}{\partial \dot{\theta}_k \partial \dot{\theta}_i} + \dot{\theta}_j \dot{\theta}_k \left(\frac{\partial}{\partial \theta_k}\frac{\partial^2 T_{tran}}{\partial \dot{\theta}_j \partial \dot{\theta}_i} - \frac{1}{2}\frac{\partial}{\partial \theta_i}\frac{\partial^2 T_{tran}}{\partial \dot{\theta}_j \partial \dot{\theta}_k} \right) \right)$$

and,

$$\frac{\partial^2 T_{tran}}{\partial \dot{\theta}_k \partial \dot{\theta}_j} = \frac{1}{m_{total}} \left(\sum_{i=1}^{N} m_i \mathbf{v}_{i,j} \right) \left(\sum_{i=1}^{N} m_i \mathbf{v}_{i,k} \right).$$

4.10 Dynamic Stability of Satellite-Manipulator Dynamics with Gravity-Gradient Forces and Moment

The stability of the satellite and the onboard manipulator, when the latter is stationary with respect to the satellite and when gravity gradient forces and moments are present, is considered in this section. The methodology is based on the DeBra-Delp stability analysis discussed in the preceding chapter. To apply the DeBra-Delp stability theory, it is essential that the combined moment of inertia matrix of both the satellite and the manipulator is obtained about a body-fixed reference frame located at the center of mass of the satellite together with the manipulator.

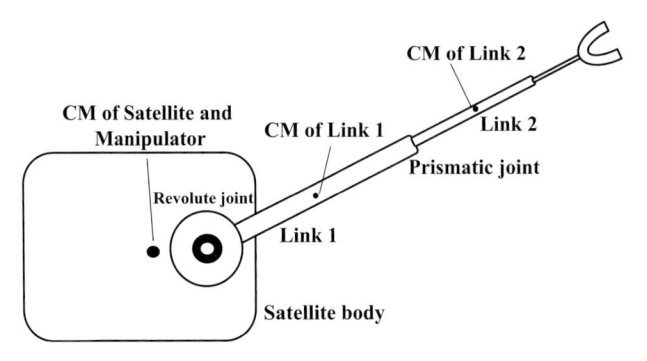

Figure 4.6 Typical deployable manipulator considered to analyze gravity-gradient stability.

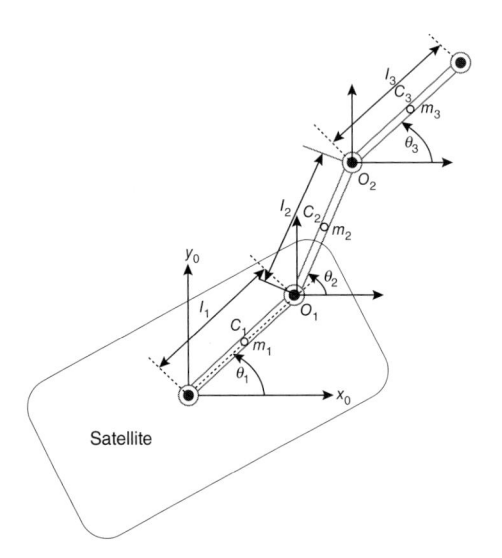

Figure 4.7 Manipulator arm attached to the satellite.

To this end, we consider a stowable two-degree of freedom manipulator arm with a single revolute joint and a single prismatic joint. The deployable manipulator is shown in Figure 4.6, while the arm itself is shown in Figure 4.7. The manipulator dynamics are assumed to be relatively faster than the spacecraft's slower dynamics. Thus, it is assumed that slow-fast separation of the dynamics could be applied. The manipulator is assumed to reach the equilibrium state rapidly. Consequently, the manipulator's degrees of freedom are assumed to be in a steady state. So only the steady equilibrium state of the manipulator is considered in the analysis, although it is assumed that the manipulator could be in one of several equilibrium configurations. The satellite and manipulator are required to be in an equilibrium state, and it is the stability of this equilibrium state that is being assessed.

To apply the DeBra-Delp stability conditions one must estimate the mass moments of inertia of the combined satellite and manipulator in a variety of equilibrium configurations at their joint center of mass. The moments of inertia are obtained by successive

4.10 Dynamic Stability of Satellite-Manipulator Dynamics

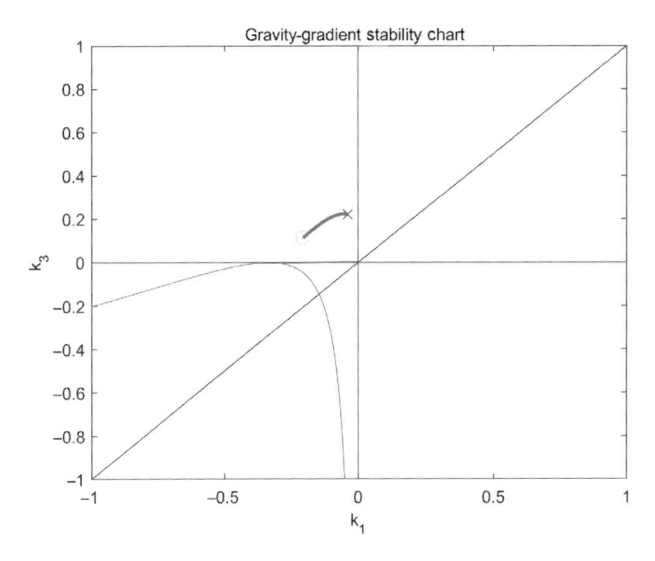

Figure 4.8 The locus of the Smelt parameters on the DeBra-Delp diagram with an extending outer arm.

application of a coordinate transformation for the rotation of axes followed by the application of parallel axes theorem. Once mass moments of inertia are estimated, the stability conditions can be applied. The manipulator considered has two degrees of freedom: the angle of rotation of the base link β and the extension of the outer link δ. The joint axis of the revolute joint is assumed to be parallel to the axis pointing toward the center of gravity of the Earth. Each pair of the variables β and δ defines a specific equilibrium configuration. For each equilibrium configuration, the mass moment of inertia matrix is obtained and the stability conditions applied.

In the example considered, the angle β is from 20 to 40 degrees, while the extension δ from a minimum of half the length of the outer link to twice the length (nominal length of 1 m). The locus of the Smelt parameters is plotted on the DeBra-Delp stability chart, shown in Figure 4.8, and is in the unstable region. The product of inertia is ignored, for now. The corresponding characteristic root locus plot, which is symmetric about both axes, is also shown in Figure 4.9. The example shows the instability along the locus corresponding to an unstable region on the stability diagram.

In the next example, the angle β is again varied from 20 to 40 degrees, while δ is retracted from 10% of the nominal length of the arm to 40% of the arm length. In this case, the corresponding DeBra-Delp diagram is shown in Figure 4.10 and it is seen that the satellite is only marginally stable.

Before considering the methods of stability augmentation, the regions in the β-δ space that map into one of the stable regions of the stability chart are obtained. Thus, a wide range of values are considered for β and δ, and using a Monte Carlo approach, a region traditionally known as the DeBra-Delp stability region is dealt with. This is shown in Figure 4.11, and the corresponding stability region in the β-δ space is shown in Figures 4.12 and 4.13.

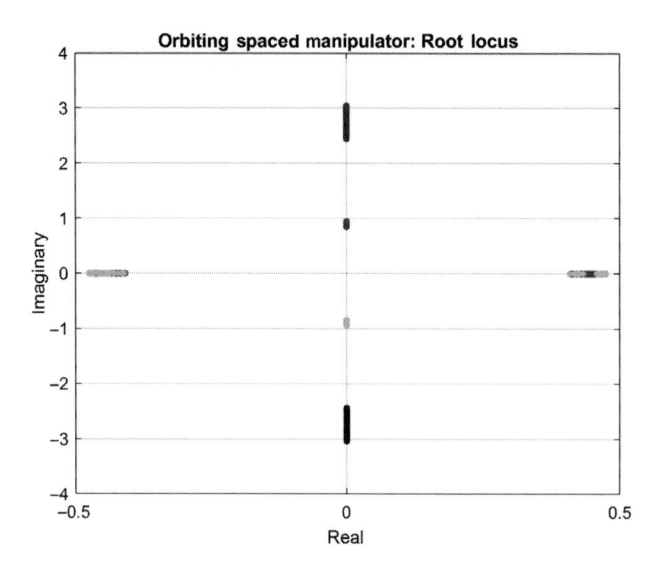

Figure 4.9 The locus of the roots of the characteristic polynomial on the complex "s" plane.

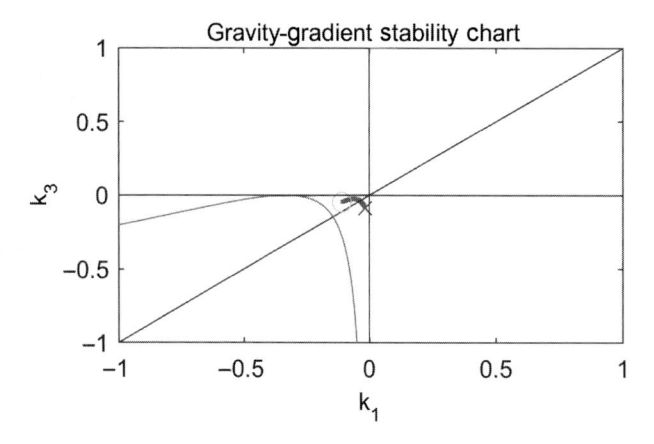

Figure 4.10 The locus of the Smelt parameters on the DeBra-Delp diagram with a retracting outer arm.

The DeBra-Delp approach assumes the moments of inertia are principle values. Direct analysis stability using the equations in Section 3.8 allows one to analyze cases with products of inertia included. The resulting stability regions are similar to those in Figures 4.12 and 4.13 but are slightly more restrictive.

The results of this study illustrate the importance of considering the gravity-gradient stability of a satellite with a robot manipulator arm, where the manipulator could be parked in any feasible configuration within its workspace. However, for satellites that orbit relatively far from the Earth, gravity-gradient stability becomes less and less important, and the satellite can be actively stabilized to maintain its orientation. Yet for satellites in low Earth orbits it is an important performance feature that must considered.

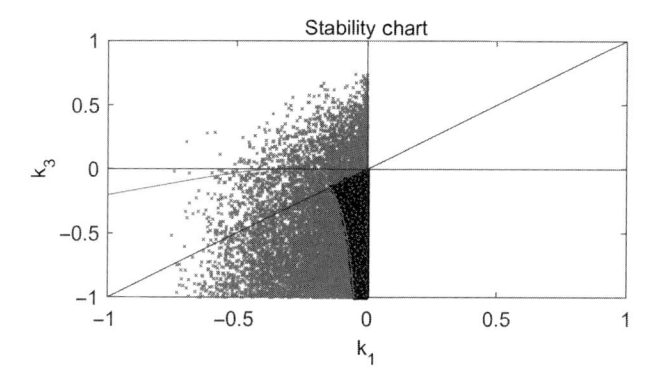

4.10 Dynamic Stability of Satellite-Manipulator Dynamics 195

Figure 4.11 Range of values of Smelt parameters considered on the stability chart. (The shaded area is the DeBra-Delp stability region).

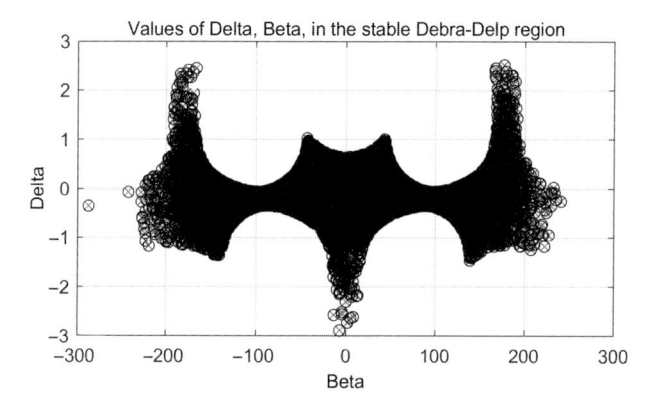

Figure 4.12 Stable values considered in Figure 4.11 in the β-δ space.

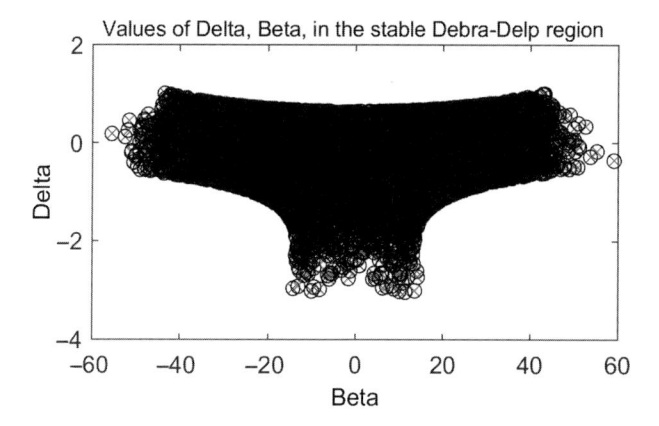

Figure 4.13 Stable values considered in Figure 4.11 over a smaller range of values in the β-δ space.

4.11 Three-Axis Control of a Satellite's Attitude with an Onboard Robot Manipulator

Typically, the attitude of a chaser spacecraft needs to be synchronized with that of the target body. This is achieved by considering a relative generic attitude control problem. From Section 3.8, in the absence of biases and noise, the attitude quaternion perturbation $\Delta\mathbf{q}$ satisfies a discrete update equation given by,

$$\Delta\mathbf{q}(k+1) \approx \mathbf{F}_1(\boldsymbol{\omega}_k, 0)\Delta\mathbf{q}(k) + \mathbf{G}_1(\Delta\mathbf{q}(k))\Delta\boldsymbol{\omega}(k). \tag{4.120}$$

The evaluation of $\mathbf{F}_1$ is done in accordance with the approach recommended by Markley [35]. Thus, given an attitude rate bias vector $\mathbf{b}$,

$$\mathbf{F}_1(\boldsymbol{\omega}, \mathbf{b}) = \begin{bmatrix} \alpha_k \mathbf{I} & \beta_k(\boldsymbol{\omega}+\mathbf{b}) \\ -\beta_k(\boldsymbol{\omega}+\mathbf{b})^{\mathrm{T}} & \alpha_k \end{bmatrix}, \quad \mathbf{G}_1(\Delta\mathbf{q}) = \frac{\Delta t}{2}\Gamma(\Delta\mathbf{q}),$$

$$\alpha_k = \cos\left(\frac{\Delta t}{2}\|\boldsymbol{\omega}+\mathbf{b}\|\right), \quad \beta_k = \sin\left(\frac{\Delta t}{2}\|\boldsymbol{\omega}+\mathbf{b}\|\right)\bigg/\|\boldsymbol{\omega}+\mathbf{b}\|. \tag{4.121}$$

The bias vector is particularly important in defining the estimation problem that is considered in a later chapter.

In the absence of disturbances, from the equation for the angular velocity vector in Section 3.8, the perturbation angular velocity update equations are,

$$\Delta\boldsymbol{\omega}(k+1) = \left(\mathbf{I}_{3\times 3} - \mathbf{I}^{-1}\mathbf{W}\Delta t\right)\Delta\boldsymbol{\omega}(k) + 3n^2\mathbf{I}^{-1}\mathbf{G}\mathbf{C}\Delta t\Delta\mathbf{q}(k) + \mathbf{I}^{-1}\Delta t\Delta\mathbf{M}_c, \tag{4.122}$$

where $\mathbf{W} = \boldsymbol{\omega}_\times\mathbf{I} - (\mathbf{I}\boldsymbol{\omega})_\times$, $\mathbf{G} = \mathbf{c}_\times(\mathbf{q})\mathbf{I} - (\mathbf{I}\mathbf{c}(\mathbf{q}))_\times$, $\mathbf{C} = d\mathbf{c}(\mathbf{q})/d\mathbf{q}$, and $\mathbf{I}_{3\times 3}$ is a 3×3 unit matrix. These equations may be used to construct the discrete time control law for attitude synchronization using the methodology of a discrete time linear quadratic regulator.

Typically, the use of the linearized optimal control law with nonlinear attitude dynamics results in small residual attitude motions. A typical case of this is illustrated in Figures 4.14 and 4.15. To reduce the residual response, a nonlinear control law is synthesized using a typical barrier Lyapunov function approach.

4.11.1 Rotation Rate Synchronization Control

Initially, the discrete time control law is used, and when the attitude error with respect to the desired attitude is within certain specified limits, the control law is smoothly switched to the attitude synchronization control law. The attitude synchronization control is designed based on the approach of He, Chen, and Yin [36]. Ignoring the disturbance torques, the equations describing the relative attitude and relative angular velocity dynamics of the chaser spacecraft are assumed to be in the state space form as,

$$\dot{\mathbf{x}}_1 = \mathbf{F}\mathbf{x}_1 + \mathbf{L}(\mathbf{x}_1)\mathbf{x}_2, \quad \mathbf{x}_1 \equiv \Delta\mathbf{q}, \quad \mathbf{F} = \frac{1}{2}\mathbf{A}_\omega(\boldsymbol{\omega}), \quad \mathbf{L}(\mathbf{x}_1) = \frac{1}{2}\Gamma(\Delta\mathbf{q}), \tag{4.123}$$

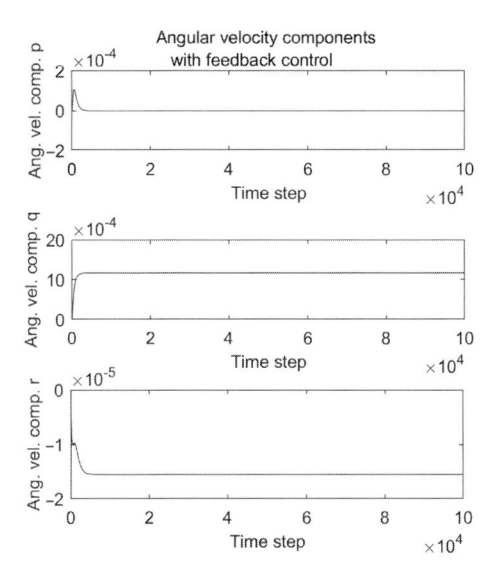

4.11 Three-Axis Control of a Satellite's Attitude with an Onboard Robot Manipulator

Figure 4.14 Typical angular velocity responses with an LQR-type discrete time linear control law.

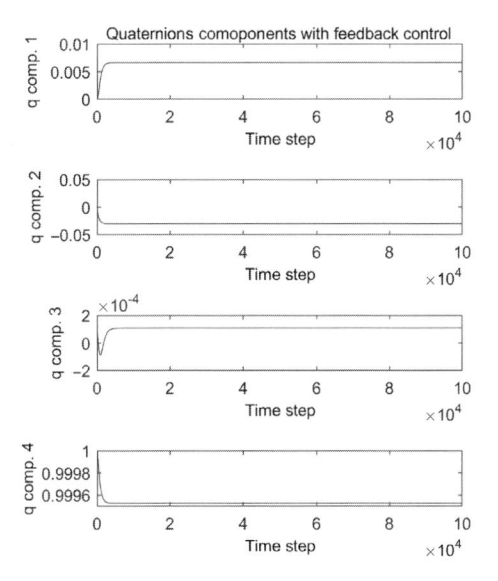

Figure 4.15 Typical quaternion responses with an LQR-type discrete time linear control law.

$$\dot{\mathbf{x}}_2 = \mathbf{I}^{-1}(\mathbf{x}_1)(\boldsymbol{\tau} - \boldsymbol{\tau}_{ct}), \quad \mathbf{x}_2 \equiv \Delta\boldsymbol{\omega}, \tag{4.124}$$

where $\boldsymbol{\tau}_{ct} = -3n^2\mathbf{I}^{-1}\big(\mathbf{c}_\times(\mathbf{q})\mathbf{I} - (\mathbf{Ic}(\mathbf{q}))_\times\big)\dfrac{d\mathbf{c}(\mathbf{q})}{d\mathbf{q}}x_1 + \mathbf{I}^{-1}\big(\boldsymbol{\omega}_\times\mathbf{I} - (\mathbf{I}\boldsymbol{\omega})_\times\big)x_2$.

In Equations (4.124), which are obtained from Equations (4.122), $\mathbf{I}$ is an inertia matrix and $\boldsymbol{\tau}$ is the applied control torque. We assume that $\mathbf{L}(\mathbf{x}_1)$ is invertible and when it is not square the inverse is interpreted as the Moore-Penrose generalized inverse. Recall that,

$$\mathbf{L}^{-1}(\mathbf{x}_1) = 2\mathbf{\Gamma}^{-1}(\Delta \mathbf{q}), \quad \mathbf{F} = \frac{1}{2}\mathbf{A}_\omega(\dot{\boldsymbol{\omega}}), \quad \dot{\mathbf{L}}(\mathbf{x}_1) = \frac{1}{2}\mathbf{\Gamma}(\Delta \dot{\mathbf{q}}), \qquad (4.125)$$

$$\mathbf{A}_\omega(\dot{\boldsymbol{\omega}}) = \begin{bmatrix} -\boldsymbol{\Omega}(\dot{\boldsymbol{\omega}}) & \dot{\boldsymbol{\omega}} \\ -\dot{\boldsymbol{\omega}}^T & 0 \end{bmatrix}, \quad \mathbf{\Gamma}(\Delta \dot{\mathbf{q}}) = \begin{bmatrix} \Delta \dot{\eta} \mathbf{I}_{3\times 3} + \mathbf{S}(\Delta \dot{\boldsymbol{\varepsilon}}) \\ -\Delta \dot{\boldsymbol{\varepsilon}}^T \end{bmatrix}. \qquad (4.126)$$

The desired values of $\mathbf{x}_1$ are specified to be given by $\mathbf{x}_d(t)$ but are yet to be specified, while the corresponding desired relative angular velocities are assumed to be $\boldsymbol{\alpha}$. In this approach, it is possible to specify either $\mathbf{x}_d(t)$ or $\boldsymbol{\alpha}$ but not both. Let the bounds on $\mathbf{x}_d(t)$ and its time derivative be defined as,

$$-\mathbf{X}_0 \leq \mathbf{x}_d(t) \leq \mathbf{X}_0, \quad -\mathbf{Y}_0 \leq \dot{\mathbf{x}}_d(t) \leq \mathbf{Y}_0. \qquad (4.127)$$

The bounds on $\mathbf{x}_1, \mathbf{x}_2$ are known to be, $|\mathbf{x}_1| \leq \mathbf{k}_{c1}, |\mathbf{x}_2| \leq \mathbf{k}_{c2}$. Furthermore, we let,

$$\mathbf{k}_a = \mathbf{k}_{c1} - |\mathbf{X}_0|, \quad \mathbf{k}_b = \mathbf{k}_{c2} - |\mathbf{Y}_0|. \qquad (4.128)$$

We define the error states $\mathbf{z}_1$ and $\mathbf{z}_2$ as,

$$\mathbf{z}_1 = \mathbf{x}_1 - \mathbf{x}_d; \quad \mathbf{z}_2 = \mathbf{x}_2 - \boldsymbol{\alpha}. \qquad (4.129)$$

Following Tee, Ge, and Tay [37], we define a barrier Lyapunov function (BLF) as follows: A BLF is a scalar function $V(\mathbf{x})$ defined with respect to the states of the system $\dot{\mathbf{x}} = \mathbf{f}(\mathbf{x})$ on an open region D containing the origin that is continuous, positive definite, has continuous first-order partial derivatives at every point of D, has the property $V(\vec{\mathbf{x}})\infty$ as $\mathbf{x}$ approaches the boundary of D, and satisfies $V(\mathbf{x}(t)) \leq b, \forall t \geq 0$ along the solution of $\dot{\mathbf{x}} = \mathbf{f}(\mathbf{x})$ for $\mathbf{x}(0) \in D$ and some positive constant b. Thus, we choose the asymmetric BLF as,

$$V_1 = \frac{1}{2}\sum_{i=1}^{n} \log\left(k_{ai}^2/\left(k_{ai}^2 - z_{1i}^2\right)\right). \qquad (4.130)$$

Differentiating with respect to time,

$$\dot{V}_1 = \sum_{i=1}^{n} \frac{z_{1i}\dot{z}_{1i}}{k_{ai}^2 - z_{1i}^2}. \qquad (4.131)$$

Using the relation $\dot{\mathbf{z}}_1 = \dot{\mathbf{x}}_1 - \dot{\mathbf{x}}_d$, we write $\dot{\mathbf{z}}_1$ as,

$$\dot{\mathbf{z}}_1 = \mathbf{F}\mathbf{z_1} + \mathbf{F}\mathbf{x}_d + \mathbf{L}(\mathbf{x}_1)\mathbf{z}_2 - \dot{\mathbf{x}}_d + \mathbf{L}(\mathbf{x}_1)\boldsymbol{\alpha}. \qquad (4.132)$$

We let,

$$\mathbf{L}(\mathbf{x}_1)\boldsymbol{\alpha} = \dot{\mathbf{x}}_d - \mathbf{A} - \mathbf{F}\mathbf{x}_d, \qquad (4.133)$$

with,

$$\mathbf{A} = \begin{bmatrix} (k_{a1}^2 - z_{11}^2)k_{11}z_{11} & (k_{a2}^2 - z_{12}^2)k_{12}z_{12} & \cdots & (k_{an}^2 - z_{1n}^2)k_{1n}z_{1n} \end{bmatrix}^T$$
$$= \Lambda_a(\mathbf{z}_1)\mathbf{K}_1\mathbf{z}_1, \qquad (4.134)$$

and k_{in} are positive constants yet to be chosen, and $\mathbf{K}_1$ and $\Lambda_a(\mathbf{z}_1)$ are diagonal matrices defined by,

4.11 Three-Axis Control of a Satellite's Attitude with an Onboard Robot Manipulator

$$\mathbf{K}_1 = \begin{bmatrix} k_{11} & 0 & 0 & 0 \\ 0 & k_{12} & 0 & 0 \\ 0 & 0 & \cdots & 0 \\ 0 & 0 & 0 & k_{1n} \end{bmatrix}, \quad \Lambda_a(\mathbf{z}_1) = \begin{bmatrix} k_{a1}^2 - z_{11}^2 & 0 & 0 & 0 \\ 0 & k_{a2}^2 - z_{12}^2 & 0 & 0 \\ 0 & 0 & \cdots & 0 \\ 0 & 0 & 0 & k_{an}^2 - z_{1n}^2 \end{bmatrix}. \quad (4.135)$$

It follows that,

$$\dot{\mathbf{z}}_1 = \mathbf{F}\mathbf{z}_1 + \mathbf{L}(\mathbf{x}_1)\mathbf{z}_2 - \mathbf{A}. \qquad (4.136)$$

The time derivative of V_1 can be expressed as,

$$\dot{V}_1 = \sum_{i=1}^{n} \frac{z_{1i}\left(\sum_k F_{ik}(\mathbf{x}_2)z_{1k} + L_{ik}(\mathbf{x}_1)z_{2k}\right)}{k_{ai}^2 - z_{1i}^2} - \sum_{i=1}^{n} \frac{z_{1i}(k_{ai}^2 - z_{1i}^2)k_{1i}z_{1i}}{k_{ai}^2 - z_{1i}^2}, \qquad (4.137)$$

which reduces to,

$$\dot{V}_1 = \sum_{i=1}^{n} \frac{z_{1i}\sum_k L_{ik}(\mathbf{x}_1)z_{2k}}{k_{ai}^2 - z_{1i}^2} - \mathbf{z}_1^T\left(\mathbf{K}_1 - \Lambda_a^{-1}(\mathbf{z}_1)\mathbf{F}\right)\mathbf{z}_1. \qquad (4.138)$$

Thus, $\mathbf{K}_1$ is chosen such that, $\mathbf{K}_1 - \Lambda_a^{-1}(\mathbf{z}_1)\mathbf{F} > 0$. It follows that $\mathbf{K}_1$ is set equal to,

$$\mathbf{K}_1 = \mathbf{K}_{10} + \Lambda_a^{-1}(\mathbf{z}_1)\mathbf{F}, \qquad (4.139)$$

and,

$$\mathbf{A} = (\Lambda_a(\mathbf{z}_1)\mathbf{K}_{10} + \mathbf{F})\mathbf{z}_1 \equiv \mathbf{K}_0\mathbf{z}_1,$$

$$\boldsymbol{\alpha} = \mathbf{L}^{-1}(\mathbf{x}_1)(\dot{\mathbf{x}}_d - \mathbf{F}\mathbf{x}_d - (\Lambda_a(\mathbf{z}_1)\mathbf{K}_{10} + \mathbf{F})\mathbf{z}_1), \qquad (4.140)$$

where $\mathbf{K}_{10}$ is a symmetric, positive definite (SPD) matrix. Thus, $\mathbf{K}_1$ is not a constant. Hence,

$$\dot{V}_1 = \sum_{i=1}^{n} \frac{z_{1i}\sum_k L_{ik}(\mathbf{x}_1)z_{2k}}{k_{ai}^2 - z_{1i}^2} - \mathbf{z}_1^T\mathbf{K}_{10}\mathbf{z}_1. \qquad (4.141)$$

Moreover, $\dot{\mathbf{z}}_1$ reduces to,

$$\dot{\mathbf{z}}_1 = \mathbf{F}\mathbf{z}_1 + \mathbf{L}(\mathbf{x}_1)\mathbf{z}_2 - \Lambda_a(\mathbf{z}_1)\mathbf{K}_{10}\mathbf{z}_1 - \mathbf{F}\mathbf{z}_1 = -\Lambda_a(\mathbf{z}_1)\mathbf{K}_{10}\mathbf{z}_1 + \mathbf{L}(\mathbf{x}_1)\mathbf{z}_2. \qquad (4.142)$$

Then we consider a second BLF as,

$$V_2 = V_1 + \frac{1}{2}\sum_{i=1}^{n}\log\left(\frac{k_{bi}^2}{k_{bi}^2 - z_{2i}^2}\right) + \frac{1}{2}\mathbf{z}_2^T\mathbf{I}\mathbf{z}_2. \qquad (4.143)$$

It follows that,

$$\dot{V}_2 = \dot{V}_1 + \sum_{i=1}^{n}\frac{z_{2i}\dot{z}_{2i}}{k_{bi}^2 - z_{2i}^2} + \mathbf{z}_2^T\mathbf{I}\dot{\mathbf{z}}_2. \qquad (4.144)$$

Consequently, we have,

$$\dot{V}_2 = -\mathbf{z}_1^T \mathbf{K}_{10} \mathbf{z}_1 + \sum_{i=1}^{n} \frac{z_{1i} \sum_k L_{ik}(\mathbf{x}_1) z_{2k}}{k_{ai}^2 - z_{1i}^2} + \sum_{i=1}^{n} \frac{z_{2i} \dot{z}_{2i}}{k_{bi}^2 - z_{2i}^2} + \mathbf{z}_2^T \mathbf{I} \dot{\mathbf{z}}_2. \quad (4.145)$$

Regrouping terms we obtain,

$$\dot{V}_2 = -\mathbf{z}_1^T \mathbf{K}_{10} \mathbf{z}_1 + \mathbf{z}_2^T \left(\mathbf{L}^T(\mathbf{x}_1) \Lambda_a^{-1}(\mathbf{z}_1) \mathbf{z}_1 + \left(\Lambda_b^{-1}(\mathbf{z}_2) + \mathbf{I} \right) \dot{\mathbf{z}}_2 \right), \quad (4.146)$$

where $\Lambda_b(\mathbf{z}_2)$ is a diagonal matrix given by,

$$\Lambda_b(\mathbf{z}_2) = \begin{bmatrix} k_{b1}^2 - z_{21}^2 & 0 & 0 & 0 \\ 0 & k_{b2}^2 - z_{22}^2 & 0 & 0 \\ 0 & 0 & \cdots & 0 \\ 0 & 0 & 0 & k_{bn}^2 - z_{2n}^2 \end{bmatrix}. \quad (4.147)$$

Let,

$$\mathbf{L}^T(\mathbf{x}_1) \Lambda_a^{-1}(\mathbf{z}_1) \mathbf{z}_1 + \left(\Lambda_b^{-1}(\mathbf{z}_2) + \mathbf{I} \right) \dot{\mathbf{z}}_2 = -\mathbf{K}_2 \mathbf{z}_2, \quad (4.148)$$

where $\mathbf{K}_2$ is a SPD matrix. Hence, $\mathbf{K}_2$ is any SPD matrix, chosen to satisfy,

$$\dot{\mathbf{z}}_2 = -\left(\Lambda_b^{-1}(\mathbf{z}_2) + \mathbf{I} \right)^{-1} \left(\mathbf{K}_2 \mathbf{z}_2 + \mathbf{L}^T(\mathbf{x}_1) \Lambda_a^{-1}(\mathbf{z}_1) \mathbf{z}_1 \right). \quad (4.149)$$

Hence, it follows that,

$$\dot{V}_2 = -\mathbf{z}_1^T \mathbf{K}_{10} \mathbf{z}_1 - \mathbf{z}_2^T \mathbf{K}_2 \mathbf{z}_2. \quad (4.150)$$

But, $\mathbf{z}_2 = \mathbf{x}_2 - \boldsymbol{\alpha}$; $\dot{\mathbf{z}}_2 = \dot{\mathbf{x}}_2 - \dot{\boldsymbol{\alpha}}$ and from the equations defining $\dot{\mathbf{x}}_2$,

$$\dot{\mathbf{z}}_2 = \mathbf{I}^{-1}(\boldsymbol{\tau} - \boldsymbol{\tau}_{ct}) - \dot{\boldsymbol{\alpha}}. \quad (4.151)$$

Hence, since we also have,

$$\mathbf{L}^T(\mathbf{x}_1) \Lambda_a(\mathbf{z}_1) \mathbf{z}_1 + (\Lambda_b(\mathbf{z}_2) + \mathbf{I}) \dot{\mathbf{z}}_2 = -\mathbf{K}_2 \mathbf{z}_2, \quad (4.152)$$

$$\dot{\mathbf{z}}_2 = -(\Lambda_b(\mathbf{z}_2) + \mathbf{I})^{-1} \left(\mathbf{K}_2 \mathbf{z}_2 + \mathbf{L}^T(\mathbf{x}_1) \Lambda_a(\mathbf{z}_1) \mathbf{z}_1 \right) = \mathbf{I}^{-1}(\boldsymbol{\tau} - \boldsymbol{\tau}_{ct}) - \dot{\boldsymbol{\alpha}}. \quad (4.153)$$

Solving for $\boldsymbol{\tau}$,

$$\boldsymbol{\tau} = -\mathbf{I} \left(\Lambda_b^{-1}(\mathbf{z}_2) + \mathbf{I} \right)^{-1} \left(\mathbf{K}_2 \mathbf{z}_2 + \mathbf{L}^T(\mathbf{x}_1) \Lambda_a^{-1}(\mathbf{z}_1) \mathbf{z}_1 \right) + \mathbf{I} \dot{\boldsymbol{\alpha}} + \boldsymbol{\tau}_{ct}. \quad (4.154)$$

Differentiating Equation (4.133),

$$\mathbf{L}(\mathbf{x}_1) \dot{\boldsymbol{\alpha}} = \ddot{\mathbf{x}}_d - \mathbf{F} \dot{\mathbf{x}}_d - \dot{\mathbf{F}} \mathbf{x}_d - \dot{\mathbf{L}}(\mathbf{x}_1) \boldsymbol{\alpha} - \dot{\mathbf{A}}. \quad (4.155)$$

If we define the diagonal matrix $\Lambda_{az}(\mathbf{z}_1)$ as,

$$\Lambda_{az}(\mathbf{z}_1) = \begin{bmatrix} k_{a1}^2 - 3z_{11}^2 & 0 & 0 & 0 \\ 0 & k_{a2}^2 - 3z_{12}^2 & 0 & 0 \\ 0 & 0 & \cdots & 0 \\ 0 & 0 & 0 & k_{an}^2 - 3z_{1n}^2 \end{bmatrix}, \quad (4.156)$$

4.11 Three-Axis Control of a Satellite's Attitude with an Onboard Robot Manipulator

using the earlier definitions of $\mathbf{A}$ and $\mathbf{K}_1$,

$$\dot{\mathbf{A}} = \dot{\mathbf{F}}\mathbf{z}_1 + (\Lambda_{az}(\mathbf{z}_1)\mathbf{K}_{10} + \mathbf{F})\dot{\mathbf{z}}_1. \tag{4.157}$$

Thus, the complete control law to achieve precise synchronization is,

$$\boldsymbol{\tau} = -\mathbf{I}\big(\Lambda_b^{-1}(\mathbf{z}_2) + \mathbf{I}\big)^{-1}\big(\mathbf{K}_2\mathbf{z}_2 + \mathbf{L}^T(\mathbf{x}_1)\Lambda_a^{-1}(\mathbf{z}_1)\mathbf{z}_1\big) + \mathbf{I}\dot{\boldsymbol{\alpha}} + \boldsymbol{\tau}_{ct}. \tag{4.158}$$

Moreover,

$$\dot{\mathbf{z}}_2 = \dot{\mathbf{x}}_2 - \dot{\boldsymbol{\alpha}} = \mathbf{I}^{-1}(\boldsymbol{\tau} - \boldsymbol{\tau}_{ct}) - \dot{\boldsymbol{\alpha}}, \quad \dot{\mathbf{z}}_1 = \dot{\mathbf{x}}_1 - \dot{\mathbf{x}}_d = \mathbf{F}\mathbf{z}_1 + \mathbf{L}(\mathbf{x}_1)\mathbf{z}_2 - \mathbf{K}_0\mathbf{z}_1, \tag{4.159}$$

with $\mathbf{K}_0 = \Lambda_a(\mathbf{z}_1)\mathbf{K}_{10} + \mathbf{F}$, which is obtained from Equation (4.139).

The complete nonlinear control torque resembles a nonlinear, proportional control law. The control law may also be expressed as,

$$\boldsymbol{\tau} = \mathbf{I}\dot{\boldsymbol{\alpha}} + \boldsymbol{\tau}_{ct} - \mathbf{K}_z\begin{bmatrix}\mathbf{z}_1^T & \mathbf{z}_2^T\end{bmatrix}^T. \tag{4.160}$$

Furthermore, $\mathbf{z}_1$ and $\mathbf{z}_2$ are obtained by integrating Equations (4.159). The vector $\mathbf{K}_z$ is a gain vector that is not a constant but is a function of $\mathbf{x}_1$ and $\mathbf{z}_1$. It is,

$$\mathbf{K}_z = \mathbf{I}\big(\Lambda_b^{-1}(\mathbf{z}_2) + \mathbf{I}\big)^{-1}\begin{bmatrix}\mathbf{L}^T(\mathbf{x}_1)\Lambda_a^{-1}(\mathbf{z}_1) & \mathbf{K}_2\end{bmatrix}. \tag{4.161}$$

One completely acceptable special choice of $\mathbf{K}_2$ is,

$$\mathbf{K}_2 = \mathbf{K}_{20} + \big(\Lambda_b^{-1}(\mathbf{z}_2) + \mathbf{I}\big)\mathbf{I}^{-1}\mathbf{K}_{21}. \tag{4.162}$$

In this case, $\mathbf{K}_z$ takes the form,

$$\mathbf{K}_z = \mathbf{I}\big(\Lambda_b^{-1}(\mathbf{z}_2) + \mathbf{I}\big)^{-1}\begin{bmatrix}\mathbf{L}^T(\mathbf{x}_1)\Lambda_a^{-1}(\mathbf{z}_1) & \mathbf{K}_{20}\end{bmatrix} + \begin{bmatrix}\mathbf{0} & \mathbf{K}_{21}\end{bmatrix} \equiv \mathbf{K}_{z0} + \begin{bmatrix}\mathbf{0} & \mathbf{K}_{21}\end{bmatrix}. \tag{4.163}$$

In our implementation, the gain vectors $\mathbf{K}_0$, $\mathbf{K}_z$, and the control torque vector are evaluated at the end of the preceding time step based on the estimated states. The method of estimation used is the Unscented Kalman Filter (UKF) proposed by Julier [38]. The UKF is similar in form to the filter implemented in Vepa [39], Vepa and Zhahir [40], and Vepa [41], and is briefly described in Section 7.6.1. The application of the UKF to the nonlinear state estimation problem associated with rotation rate synchronization is discussed in Section 8.5.

Experience with the above control law indicates that although stability is guaranteed, sometimes precise tracking is not achieved. Precise tracking of the desired response can then be achieved with the use of an additional integral control term in the control law, which is modified to,

$$\boldsymbol{\tau} = \mathbf{I}\dot{\boldsymbol{\alpha}} + \boldsymbol{\tau}_{ct} - \mathbf{K}_{z0}\begin{bmatrix}\mathbf{z}_1^T & \mathbf{z}_2^T\end{bmatrix}^T - \mathbf{K}_{21}\left(\mathbf{z}_2 + \frac{1}{T_i}\int_0^t \mathbf{z}_2\,dt\right), \tag{4.164}$$

where T_i is a small positive integral time constant. In this application, T_i was chosen in the range $0.01 \leq T_i \leq 0.3$ to ensure that the tracking errors were within acceptable limits. Furthermore, in our application $\boldsymbol{\alpha}(t) = \dot{\boldsymbol{\alpha}}(t) = 0$. If on the other hand, the perturbation quaternion $\mathbf{x}_d(t)$ is specified $\boldsymbol{\alpha}(t)$ and $\dot{\boldsymbol{\alpha}}(t)$ are computed. Typical examples of the angular velocity and quaternion component errors for attitude stabilization using the above control law (and without the integral term) are shown in Figures 4.16 and 4.17. The implementation of these attitude control laws is done by the use of magnetic actuators. It is important to recognize that the methodology based on the BLF is, in

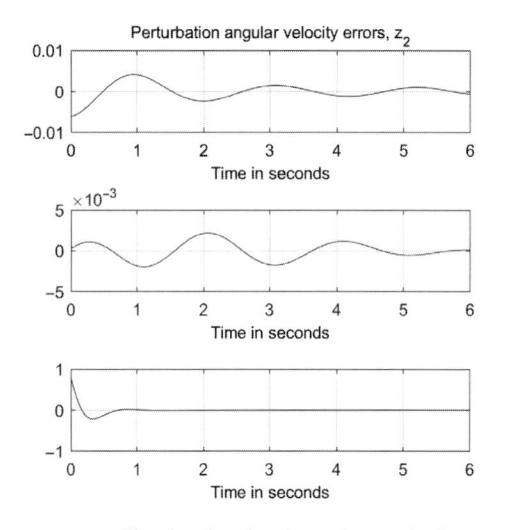

Figure 4.16 Simulated and estimated perturbation angular velocity component errors.

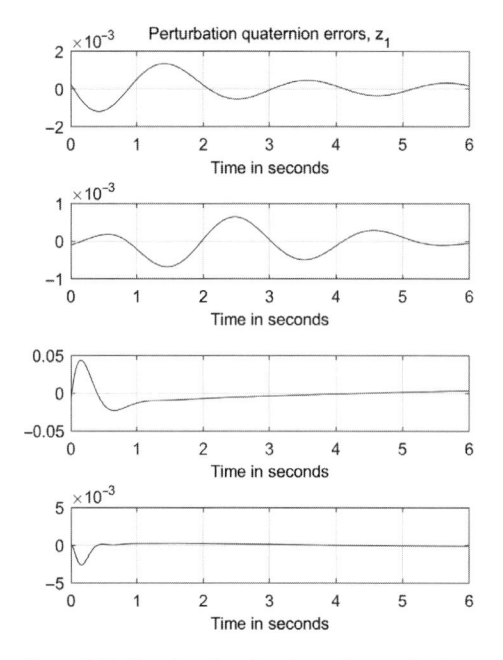

Figure 4.17 Simulated and estimated perturbation quaternion component errors.

principle, also suitable for the design of control systems for robot manipulators coupled with the attitude dynamics of the satellite, although the design process is quite lengthy in practice.

References

[1] Vepa, R. (2009) *Biomimetic Robotics: Mechanisms and Control*, New York: Cambridge University Press.
[2] Kane, T. R. and Levinson, D. A. (1985) *Dynamics: Theory and Applications*, McGraw-Hill Series in Mechanical Engineering, New York: McGraw-Hill Book Company.
[3] Vepa, R. (2016) *Nonlinear Control of Robots and Unmanned Aerial Vehicles: An Integrated Approach*, Boca Raton, FL: CRC Press.
[4] Hamel, A. (1904) Lagrange-Eulerschen Gleichungen der Mechanik. *Zeitschrift fur Mathematiks und Physiks (ZAMP)*, 50: 1–57.
[5] Boltzmann, R. (1902) *Sitzungsberichte*, Wien.
[6] Desloge, E. A. (1986) A comparison of Kane's equations of motion and the Gibbs-Appell equations of motion. *American Journal of Physics*, 54(5): 470–472.
[7] Roberson, R. E. and Scwertassek, R. (1988) *Dynamics of Multibody Systems*. Berlin/Heidelberg: Springer-Verlag.
[8] Borri, M., Bottasso, C., and Mantegazza, P. (1990) Equivalence of Kane's and Maggi's equations. *Meccanica*, 25(4): 272–274.
[9] Wang, L.S. and Pao, Y.H. (2003) Jourdain's variational equation and Appell's equation of motion for non-holonomic dynamical systems. *American Journal of Physics*, 71(1): 72–82.
[10] Townsend, M. A. (1992) Kane's equations, Lagrange's equations, and virtual work. *Journal of Guidance, Control, and Dynamics*, 15(1): 277–280.
[11] Stoneking, E. T. (2013) Implementation of Kane's method for a spacecraft composed of multiple rigid bodies, in *AIAA 2013–4649, AIAA Guidance, Navigation, and Control (GNC) Conference*, Boston, MA, 2013, *Guidance, Navigation, and Control and Co-located Conferences*, 1428–1440 https://doi.org/10.2514/6.2013-4649.
[12] Cao, Y., Modi, V. J., de Silva, C. W., and Misra, A. K. (2001) Dynamics and control of a novel space-based manipulator: Analyses and experiments, in *Proceedings of the Sixth International Symposium on Artificial Intelligence and Robotics & Automation in Space*: i-SAIRAS 2001, Paper No. AM068, Canadian Space Agency, June 18–22, St-Hubert, Quebec, Canada, 2001.
[13] Dubowsky, S., Vance, E. E., and Torres, M. A. (1989) The control of space manipulators subject to spacecraft attitude control saturation limits, in *Proceedings of the NASA Conference on Space Tekrobotics*, Jet Propulsion Lab., Pasadena, CA, Vol. IV, 109–418.
[14] Nenchev, D., Umetani, Y., and Yoshida, K. (1992) Analysis of a redundant free-flying spacecraft/manipulator system. *IEEE Transactions on Robotics and Automation*, 8(1): 1–6.
[15] Quinn, R. D., Chen, J. L., and Lawrence, C. (1994) Base reaction control for space-based robots operating in microgravity environment. *Journal of Guidance, Control, and Dynamics*, 17(2): 263–270.
[16] Torres, M. A. and Dubowsky, S. (1992) Minimizing spacecraft attitude disturbances in space manipulator systems. *Journal of Guidance, Control, and Dynamics*, 15(4): 1010–1017.
[17] Alexander, H. L. and Cannon, R. H. (1987) Experiments on the control of a satellite manipulator, in Proceedings of American Control Conference, Minneapolis, MN, USA, 1987.

[18] Longman, R. W., Lindberg, R. E., and Zedd, M. F. (1987) Satellite-mounted robot manipulators: New kinematics and reaction moment compensation. *The International Journal of Robotics Research*, Raleigh, NC, 6(3): 87–103.

[19] Vafa, Z. and Dubowsky, S. (1987) On the dynamics of manipulators in space using the virtual manipulator approach, in *Proceedings of the IEEE International Conference on Robotics and Automation*, Raleigh, NC, 579–585.

[20] Vafa, Z. and Dubowsky, S. (1990) The kinematics and dynamics of space manipulators: The virtual manipulator approach. *The International Journal of Robotics Research*, 9(4): 3–21.

[21] Vafa, Z. and Dubowsky, S. (1990) On the dynamics of space manipulator using the virtual manipulator with application to path planning. *The Journal of the Astronautical Sciences*, 38(4): 441–472.

[22] Lindberg, R. E., Longman, R. W., and Zedd, M. F. (1990) Kinematic and dynamic properties of an elbow manipulator mounted on a satellite. *The Journal of the Astronautical Sciences*, 38(4): 397–421.

[23] Longman, R. W. (1990) The kinetics and workspace of a satellite-mounted robot. *The Journal of the Astronautical Sciences*, 38(4): 423–440.

[24] Vafa, Z. and Dubowsky, S. (1987) A virtual manipulator model for space robotic systems, in *Proceedings of the Workshop on Space Telerobotics*, March 31–April 3, Pasadena, CA, 335–344.

[25] Papadopoulos, E. (1990) On the Dynamics and Control of Space Manipulators, PhD thesis, Massachusetts Institute of Technology, Cambridge, MA.

[26] Yoshida, K. and Umetani, Y. (1999) Resolved motion rate control of space manipulators with generalized Jacobian matrix. *IEEE Transactions on Robotics and Automation*, 5(3): 303–314.

[27] Papadopoulos, E. (1991) Path planning for space manipulators exhibiting nonholonomic behavior, in *Proceedings of IEEE International Conference on Intelligent Robots and Systems*: Raleigh, NC, 669–675.

[28] Papadopoulos, E. (1993) Nonholonomic behavior in free-floating space manipulators and its utilization. In *Nonhonomic Motion Planning*, Z. Li and J. F. Canny, eds., Boston: Kluwer Academic, pp. 423–445.

[29] Papadopoulos, E. and Dubowsky, S. (1993) Dynamic singularities in the control of free-floating space manipulators. *ASME Journal of Dynamic Systems, Measurement and Control*, 115(1): 44–52.

[30] Papadopoulos, E. and Dubowsky, S. (1991) On the nature of control algorithms for free-floating space manipulators. *IEEE Transactions on Robotics and Automation*, 7(6): 750–758.

[31] Papadopoulos, E. and Dubowsky, S. (1991) Coordinated manipulator/spacecraft motion control for space robotic systems, in *IEEE International Conference on Robotics and Automation*, Sacramento, CA, 1696–1701.

[32] Papadopoulos, E., Moosavian, S., and Ali, A. (1994) A comparison of motion control algorithms for space free-flyers, in *Proceedings of the Fifth International Conference on Adaptive Structures*, 1554–1561, Sendai, Japan.

[33] Moosavian, S., Ali, A., and Papadopoulos, E. (1997) Coordinated motion control of multiple manipulator space free-flyers, in *Proceedings of the Seventh AAS/AIAA Space Flight Mechanics Meeting*, Huntsville, AL.

[34] Masutani, Y., Miyazaki, F., and Arimoto, S. (1989) Modeling and sensory feedback for space manipulators, in *Proceedings of the NASA Conference on Space Telerobotics*, January 14–February 1, 1989, Pasadena, CA, 287–296.

[35] Markley, F. L. (1978) Matrix and vector algebra. In *Spacecraft Attitude Determination and Control*, J. R. Wertz, ed., Dordrecht: D. Reidel Publishing Co., pp. 754–755.

[36] He, W., Chen, Y., and Yin, Z. (2006) Adaptive neural network control of an uncertain robot with full-state constraints. *IEEE Transactions on Cybernetics*, 46(3): 620–629.

[37] Tee, K. P., Ge, S., and Tay, E. H. (2009) Barrier Lyapunov functions for the control of output-constrained nonlinear systems. *Automatica*, 45(4): 918–927.

[38] Julier, S. J. (2002) The scaled unscented transformation, in *Proceedings of the American Control Conference*, 6, 4555–4559.

[39] Vepa, R. (2010) Spacecraft large attitude estimation using a navigation sensor. *The Journal of Navigation*, 63(1): 89–104.

[40] Vepa, R. and Zhahir, A. (2011) High-precision kinematic satellite and doppler aided inertial navigation system. *The Journal of Navigation*, 64(1): 91–108.

[41] Vepa, R. (2017) Joint position localization of spacecraft and debris for autonomous navigation applications using angle measurements only. *Journal of Navigation*, 70(4): 748–760. doi: 10.1017/S0373463316000904.

5 Kinematics, Dynamics, and Control of Mobile Robot Manipulators

5.1 Kinematics of Wheeled Mobile Manipulators: Non-Holonomic Constraints

A mobile robot manipulator is a mobile platform hosting a robotic manipulator. The kinematics of such manipulators involve consideration of several issues that are quite unique to them. The primary feature is that wheeled mobile platforms are subject to non-holonomic constraints. These constraints have a significant impact on the coordination strategy as well as the dynamic interaction between the platform and the manipulator. In turn, this has a bearing on the control and path planning of mobile robot manipulators. A key feature of robot manipulators with non-holonomic constraints is that the number of independent position coordinates exceeds the number of independent velocities that may be used to describe the velocity kinematics and the dynamics of the manipulator. Normally, feedback control laws are designed to asymptotically stabilize the equilibrium position of a mechanical system. However, when non-holonomic constraints are present, Brockett [1] showed that a time-invariant full state feedback control law is inadequate if one wishes to asymptotically stabilize the equilibrium position of the mechanical system. To stabilize systems with non-holonomic constraints, a number of approaches, including time-varying control laws, piecewise smooth or discontinuous control laws, and combinations of these, have been proposed for the stabilization of non-holonomic control systems to equilibrium points, as outlined in a survey paper by Kolmanovsky and McClamroch [2], and in Godhavn and Egeland [3], Samson [4], Astolfi [5], Hespanha [6], and Aguiar and Pascoal [7]. The stability of such systems may be established using a Lyapunov approach, as outlined by Ye, Michel, and Hou [8].

Every mechanical system is subject to constraints. A standard technique that is adopted is to solve for some of the coordinates that arise in the constraint equations, in terms of the rest of the coordinates. Generally, if j constraints are present, one can solve for j of the coordinates in terms of the remaining coordinates, and eliminate them from the equations of motion. Some constraint equations, known as Pfaffian constraints, are constraints on the generalized velocities. However, they cannot be expressed as a differential of a function of the generalized position coordinates; consequently, the constraints are said to be non-integrable. Thus, non-holonomic constraints are expressed in terms of non-integrable functions of the generalized velocities.

A wheeled mobile platform is shown in Figure 5.1, and it is free to move in the horizontal plane. The mobile platform consists of one front steering wheel and two

5.1 Kinematics of Wheeled Mobile Manipulators: Non-Holonomic Constraints

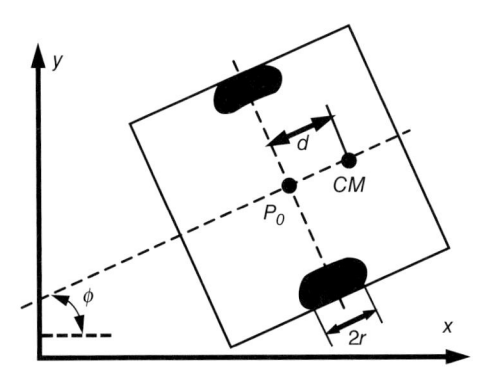

Figure 5.1 Typical mobile platform.

independent rear wheels driven by two motors. Since the platform is driven by the two driving wheels which are to roll without slipping, the velocity constraint is non-holonomic. A kinematic constraint equation can be written, relating the translation velocities in the Cartesian coordinates and the rotational angular velocity about an axis normal to the horizontal plane, for any point on the platform. Thus, for the center of mass of the platform, the constraint is expressed in terms of the virtual displacements as,

$$\delta y \cos \phi - \delta x \sin \phi = d \times \delta \phi. \tag{5.1}$$

In terms of velocities, one has,

$$\dot{y} \cos \phi - \dot{x} \sin \phi = d \times \dot{\phi}. \tag{5.2}$$

In matrix notation,

$$[-\sin \phi \quad \cos \phi \quad -d][\dot{x} \quad \dot{y} \quad \dot{\phi}]^T \equiv \mathbf{C}\dot{\mathbf{q}} = 0. \tag{5.3}$$

One method of eliminating the constraints is to determine the null space of the constraint matrix $\mathbf{C}$. The null space of the constraint matrix $\mathbf{C}$ can be written by inspection as,

$$\mathbf{N}_C = \begin{bmatrix} \cos \phi & -\sin \phi \\ \sin \phi & \cos \phi \\ 0 & 1/d \end{bmatrix}. \tag{5.4}$$

So it follows that,

$$\mathbf{C}\mathbf{N}_C = 0. \tag{5.5}$$

If we now introduce the transformation,

$$\dot{\mathbf{q}} = \mathbf{N}_C \mathbf{v}, \tag{5.6}$$

it follows that the constraints are satisfied. Moreover, it is possible to interpret the components of the vector $\mathbf{v}$ as the forward velocity and the tangential velocity of the mobile vehicle, respectively.

Yet if there is a need to establish the complete dynamics of the mobile platform, it is essential that we write the complete expression for the kinetic energy and generalized forces driving it. Thus, the kinetic energy is,

$$T = \frac{1}{2}m(\dot{x}^2 + \dot{y}^2) + \frac{1}{2}I\dot{\phi}^2. \tag{5.7}$$

The Euler-Lagrange equations are,

$$\begin{bmatrix} m & 0 & 0 \\ 0 & m & 0 \\ 0 & 0 & I \end{bmatrix} \begin{bmatrix} \ddot{x} \\ \ddot{y} \\ \ddot{\phi} \end{bmatrix} = \begin{bmatrix} Q_x \\ Q_y \\ Q_\phi \end{bmatrix} - \mathbf{C}^T \lambda, \tag{5.8}$$

which is expressed in matrix notation as,

$$\mathbf{m}\ddot{\mathbf{q}} = \mathbf{Q} - \mathbf{C}^T \lambda. \tag{5.9}$$

The effect of the constraints on the generalized forces acting on the platform is included by the term $\mathbf{C}^T \lambda$, where λ is an unknown parameter, known as a Lagrange multiplier. When the virtual work by the generalized forces is considered,

$$\delta W = \delta \mathbf{q}^T \left(\begin{bmatrix} Q_x \\ Q_y \\ Q_\phi \end{bmatrix} - \mathbf{C}^T \lambda \right) = \begin{bmatrix} Q_x & Q_y & Q_\phi \end{bmatrix} \delta \mathbf{q} = \mathbf{Q}\delta \mathbf{q}, \tag{5.10}$$

it is observed that the forces of constraint do not contribute to it. Differentiating the equation defining the coordinate transformation,

$$\ddot{\mathbf{q}} = \mathbf{N}_C \dot{\mathbf{v}} + \dot{\mathbf{N}}_C \mathbf{v}, \tag{5.11}$$

and projecting the equations of motion into the null space of the constraint matrix, the equations of motion can be expressed as,

$$\mathbf{N}_C^T \begin{bmatrix} m & 0 & 0 \\ 0 & m & 0 \\ 0 & 0 & I \end{bmatrix} (\mathbf{N}_C \dot{\mathbf{v}} + \dot{\mathbf{N}}_C \mathbf{v}) = \mathbf{N}_C^T \begin{bmatrix} Q_x \\ Q_y \\ Q_\phi \end{bmatrix} - \mathbf{N}_C^T \mathbf{C}^T \lambda = \mathbf{N}_C^T \begin{bmatrix} Q_x \\ Q_y \\ Q_\phi \end{bmatrix}. \tag{5.12}$$

In matrix notation, it is expressed as,

$$\mathbf{N}_C^T \mathbf{m} (\mathbf{N}_C \dot{\mathbf{v}} + \dot{\mathbf{N}}_C \mathbf{v}) = \mathbf{N}_C^T \mathbf{Q} - \mathbf{N}_C^T \mathbf{C}^T \lambda = \mathbf{N}_C^T \mathbf{Q}. \tag{5.13}$$

Thus, the unknown Lagrange multiplier has been successfully eliminated and the equations of motion may now be solved, provided the generalized forces driving and restraining the platform are known.

This simple example demonstrates the process one must adopt to eliminate the nonholonomic constraints. In the case of a mobile platform additionally hosting a robotic manipulator, the interaction between the degrees of freedom of the manipulator and constraints must be carefully considered.

5.2 Dynamics of Manipulators on a Moving Base

Consider a manipulator with at least two revolute joints on a moving platform. It is assumed for simplicity that it is a two-link serial manipulator with joint variables θ_1 and θ_2. There are two additional degrees of freedom. In matrix notation, the constraint equation is modified to,

$$[-\sin\phi \quad \cos\phi \quad -d \quad 0 \quad 0] [\dot{x} \quad \dot{y} \quad \dot{\phi} \quad \dot{\theta}_1 \quad \dot{\theta}_2]^T \equiv \mathbf{C}\dot{\mathbf{q}} \equiv \mathbf{C}[\mathbf{q}_1 \quad \mathbf{q}_2], \quad (5.14)$$

where the constraint matrix $\mathbf{C}$ has the form,

$$\mathbf{C} = [\mathbf{C}_{11} \quad \mathbf{0}], \quad \mathbf{q}_1 = [x \quad y \quad \phi]^T \quad \text{and} \quad \mathbf{q}_2 = [\theta_1 \quad \theta_2]^T. \quad (5.15)$$

The null space of the constraint matrix $\mathbf{C}$ can be written by inspection as,

$$\mathbf{N}_C = \begin{bmatrix} \mathbf{N}_{11} & \mathbf{0} \\ \mathbf{0} & \mathbf{I}_{2\times 2} \end{bmatrix}, \quad \text{where } \mathbf{N}_{11} = \begin{bmatrix} \cos\phi & -\sin\phi \\ \sin\phi & \cos\phi \\ 0 & 1/d \end{bmatrix}. \quad (5.16)$$

The transformation of coordinates needed to meet the constraints is,

$$\dot{\mathbf{q}}_1 = \mathbf{N}_{11}\mathbf{v}, \quad \dot{\mathbf{q}} = \mathbf{N}_C [\mathbf{v}^T \quad \mathbf{q}_2^T]^T \quad (5.17)$$

where $\mathbf{v}$ is the same pair of coordinates as in the preceding section.

The coupled Euler-Lagrange equations for the mobile platform and the manipulator are,

$$\mathbf{H}(\mathbf{q})\ddot{\mathbf{q}} + \mathbf{D}(\mathbf{q}\dot{\mathbf{q}})\dot{\mathbf{q}} + \mathbf{G}(\mathbf{q}) = [\mathbf{Q}^T \quad \boldsymbol{\tau}^T]^T + \mathbf{C}\lambda. \quad (5.18)$$

Eliminating $\ddot{\mathbf{q}}_1$ results in,

$$\mathbf{N}_C^T \mathbf{H}(\mathbf{q}) \mathbf{N}_C \begin{bmatrix} \dot{\mathbf{v}} \\ \ddot{\mathbf{q}}_2 \end{bmatrix} + \mathbf{N}_C^T (\mathbf{D}(\mathbf{q}\dot{\mathbf{q}})\mathbf{N}_C + \mathbf{H}(\mathbf{q})\dot{\mathbf{N}}_C) \begin{bmatrix} \mathbf{v} \\ \dot{\mathbf{q}}_2 \end{bmatrix} + \mathbf{N}_C^T \mathbf{G}(\mathbf{q}) = \begin{bmatrix} \mathbf{N}_{11}^T \mathbf{Q} \\ \boldsymbol{\tau} \end{bmatrix}. \quad (5.19)$$

Although one still has functions of the generalized coordinates $\mathbf{q}$ in the equations of motion, they no longer include the unknown Lagrange multiplier, which has been eliminated.

5.3 Dynamics of Wheeled Mobile Manipulators

The case of a mobile platform that is differentially driven by two motors, each powering one of the two rear wheels independently, is considered. The case of the rolling contact between the wheels and the surface of the ground is considered first. Both wheels are assumed to be of the same radius r_w and the wheel separation distance is assumed to be $2b$. Assuming pure rolling contact, the velocities of the left and right wheels are, respectively, given by,

$$v_{Lw} = r_w \dot{\theta}_L, \quad v_{Rw} = r_w \dot{\theta}_R. \tag{5.20}$$

These are the primary non-holonomic constraints.

The Cartesian velocity components at point P_0 are, respectively, given by,

$$\dot{x}_0 = \frac{1}{2}(v_{Lw} + v_{Rw})\cos\phi = \frac{r_w}{2}(\dot{\theta}_L + \dot{\theta}_R)\cos\phi, \tag{5.21}$$

$$\dot{y}_0 = \frac{1}{2}(v_{Lw} + v_{Rw})\sin\phi = \frac{r_w}{2}(\dot{\theta}_L + \dot{\theta}_R)\sin\phi. \tag{5.22}$$

The angular rotation rate of the mobile platform is,

$$\dot{\phi} = \frac{1}{b}(v_{Rw} - v_{Lw}) = \frac{r_w}{2b}(\dot{\theta}_R - \dot{\theta}_L). \tag{5.23}$$

The Cartesian velocity components at the center of mass of the platform are, respectively, given by,

$$\dot{x}_C = \dot{x}_0 - d\dot{\phi}\sin\phi = \frac{r_w}{2}(\dot{\theta}_L + \dot{\theta}_R)\cos\phi - d\frac{r_w}{2b}(\dot{\theta}_R - \dot{\theta}_L)\sin\phi, \tag{5.24}$$

$$\dot{y}_C = \dot{y}_0 + d\dot{\phi}\cos\phi = \frac{r_w}{2}(\dot{\theta}_L + \dot{\theta}_R)\sin\phi + d\frac{r_w}{2b}(\dot{\theta}_R - \dot{\theta}_L)\cos\phi. \tag{5.25}$$

It is assumed that one of the manipulator base joint axes intersects the horizontal plane of the mobile at the point B and that this point is located at distance L, ahead of the center of mass of the platform along the longitudinal axis of symmetry. The Cartesian velocity components at point B on the platform are, respectively, given by,

$$\dot{x}_B = \dot{x}_0 - (d+L)\dot{\phi}\sin\phi = \frac{r_w}{2}(\dot{\theta}_L + \dot{\theta}_R)\cos\phi - (d+L)\frac{r_w}{2b}(\dot{\theta}_R - \dot{\theta}_L)\sin\phi, \tag{5.26}$$

$$\dot{y}_B = \dot{y}_0 + (d+L)\dot{\phi}\cos\phi = \frac{r_w}{2}(\dot{\theta}_L + \dot{\theta}_R)\sin\phi + (d+L)\frac{r_w}{2b}(\dot{\theta}_R - \dot{\theta}_L)\cos\phi. \tag{5.27}$$

Moreover,

$$\dot{z}_B = 0. \tag{5.28}$$

These relations can expressed in matrix form as,

$$\mathbf{v_B} = \mathbf{J}_P \begin{bmatrix} \dot{\theta}_L & \dot{\theta}_R \end{bmatrix}^T. \tag{5.29}$$

Assume that the transformation relating the manipulator base joint axes to the inertial Cartesian frame at point B is given by $\mathbf{T}_{IB}$. Then a velocity vector in terms of the base frame located at the manipulator's base joint may be transformed to the inertial frame. It follows that the end effector inertial velocity components are given in terms of the joint angular velocities $\dot{\Theta}$ by,

$$\mathbf{v_E} = \mathbf{v_B} + \mathbf{T}_{IB}\mathbf{J}_{mV}\dot{\Theta}. \tag{5.30}$$

Thus we have,

$$\begin{bmatrix} \mathbf{v}_B \\ \mathbf{v}_E \end{bmatrix} = \begin{bmatrix} \mathbf{J}_P & \mathbf{0} \\ \mathbf{J}_P & \mathbf{T}_{IB}\mathbf{J}_{mV} \end{bmatrix} \begin{bmatrix} \dot{\theta}_L \\ \dot{\theta}_R \\ \dot{\Theta} \end{bmatrix}. \qquad (5.31)$$

In addition the relation,

$$\dot{\phi} = \frac{r_w}{2b}(\dot{\theta}_R - \dot{\theta}_L) = \frac{r_w}{2b}\begin{bmatrix} -1 & 1 & 0 \end{bmatrix}\begin{bmatrix} \dot{\theta}_L \\ \dot{\theta}_R \\ \dot{\Theta} \end{bmatrix}, \qquad (5.32)$$

is also required. In these equations, any rows with all zero entries as well as the corresponding velocities on the left-hand side are deleted. This relation may be compactly expressed as,

$$\dot{\mathbf{X}} = \mathbf{J}\begin{bmatrix} \dot{\theta}_L \\ \dot{\theta}_R \\ \dot{\Theta} \end{bmatrix} \quad \text{and} \quad \dot{\mathbf{q}} = \mathbf{N}_C\begin{bmatrix} \dot{\theta}_L \\ \dot{\theta}_R \\ \dot{\Theta} \end{bmatrix} \equiv \mathbf{N}_C\mathbf{v}, \qquad (5.33)$$

where $\dot{\mathbf{X}} = [x_B \ y_B \ x_E \ y_E \ z_E]^T$, $\mathbf{q} = [x_C \ y_C \ \phi \ \dot{\Theta}]^T$ and $\mathbf{v} = [\dot{\theta}_L \ \dot{\theta}_R \ \dot{\Theta}]^T$. Thus,

$$\dot{\mathbf{X}} = \mathbf{J}\begin{bmatrix} \dot{\theta}_L \\ \dot{\theta}_R \\ \dot{\Theta} \end{bmatrix} = \mathbf{J}\mathbf{v} \quad \text{and} \quad \mathbf{v} = \mathbf{J}^{-1}\dot{\mathbf{X}}, \qquad (5.34)$$

where we have tacitly assumed that the matrix $\mathbf{J}$ is square and invertible. This is related to the issue of manipulability.

The constraint matrix $\mathbf{C}$ satisfies,

$$\mathbf{C}\mathbf{N}_C = 0. \qquad (5.35)$$

The coupled Euler-Lagrange equations for the mobile platform and the manipulator are,

$$\mathbf{H}(\mathbf{q})\ddot{\mathbf{q}} + \mathbf{D}(\mathbf{q}\dot{\mathbf{q}})\dot{\mathbf{q}} + \mathbf{G}(\mathbf{q}) = \begin{bmatrix} \mathbf{Q}^T & \boldsymbol{\tau}^T \end{bmatrix}^T + \mathbf{C}\lambda. \qquad (5.36)$$

Eliminating $\ddot{\mathbf{q}}$, the following equation of motion is obtained,

$$\mathbf{N}_C^T\mathbf{H}(\mathbf{q})\mathbf{N}_C\dot{\mathbf{v}} + \mathbf{N}_C^T\left(\mathbf{D}(\mathbf{q}\dot{\mathbf{q}})\mathbf{N}_C + \mathbf{H}(\mathbf{q})\dot{\mathbf{N}}_C\right)\mathbf{v} + \mathbf{N}_C^T\mathbf{G}(\mathbf{q}) = \mathbf{N}_C^T\begin{bmatrix} \mathbf{Q} \\ \boldsymbol{\tau} \end{bmatrix}. \qquad (5.37)$$

5.3.1 Manipulability

Often the desired motion is specified in inertial coordinates and relative to the end effector's position coordinates. For this reason, we may eliminate the coordinates $\mathbf{v}$ and express the equations in terms of the inertial coordinates $\mathbf{X}$, where the relationship between **v** and **Ẋ** is given by, $\mathbf{v} = \mathbf{J}^{-1}\dot{\mathbf{X}}$. Thus, in order that the transformation of coordinates can be applied, it is essential that the Jacobian **J** is invertible. If **J** is not square, it is interpreted as the generalized inverse and given by,

$$\mathbf{J}^{+} = (\mathbf{J}^{T}\mathbf{J})^{-1}\mathbf{J}^{T}. \qquad (5.38)$$

In this case, the generalized inverse is feasible only if the matrix $\mathbf{J}^{T}\mathbf{J}$ is invertible. Thus, a measure of the manipulability, introduced by Yoshikawa [9] for mobile manipulators, is defined as the determinant of the matrix $\mathbf{J}^{T}\mathbf{J}$; i.e. $m = \det(\mathbf{J}^{T}\mathbf{J})$. Another measure of the manipulability is defined as the ratio of the smallest to the largest singular values of the matrix $\mathbf{J}^{T}\mathbf{J}$. Thus, $w_2 = \sigma_{\min}/\sigma_{\max}$ is another measure of manipulability; $\sigma_{\min}$ and $\sigma_{\max}$ are the minimum and maximum singular values of **J**. Bayle, Fourquet, and Renaud [10] proposed yet another measure of manipulability, which is given by,

$$w_5 = \sqrt{1 - (\sigma_{\min}/\sigma_{\max})^2}. \qquad (5.39)$$

Nait-Chabane, Hoppenot, and Colle [11] proposed yet another measure of manipulability, which is in fact also a function of the direction in the workspace in which the manipulator seeks to function.

5.3.2 Tip Over and Dynamic Stability Issues

To understand the basic issues related to tip over and dynamic stability, a typical example of a two-link manipulator on a moving platform is considered. The motion of the platform is restricted to one-dimension, so the non-holonomic constraints are implicitly satisfied. The platform carrying the manipulator is illustrated in Figure 5.2.

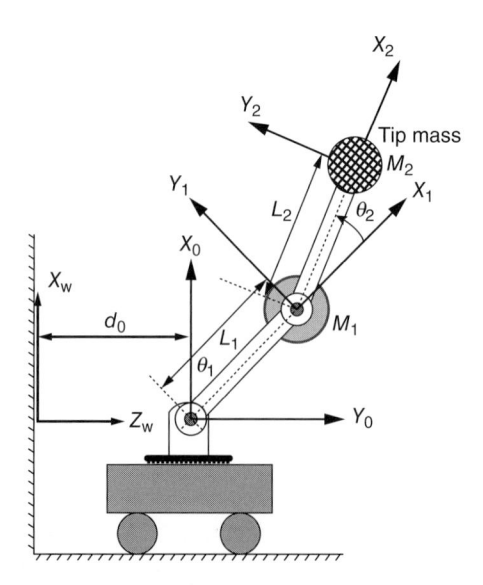

Figure 5.2 Typical mobile platform carrying a two-link manipulator.

5.3 Dynamics of Wheeled Mobile Manipulators

It is proposed to obtain the total kinetic and potential energies of the arm in terms of the total mass, moment of inertia, and mass-moment components,

$$M_{Total} = m_1 + m_2 + M_1 + M_2 + M_{cart}, \tag{5.40}$$

$$I_{11} = m_1 L_{1cm}^2 + M_1 L_1^2 + (m_2 + M_2)L_1^2 + m_1 k_1^2, \quad I_{22} = m_2 L_{2cm}^2 + M_2 L_2^2 + m_2 k_2^2,$$

$$\Gamma_1 = (m_1 L_{1cm} + m_2 L_1 + M_1 L_1 + M_2 L_1), \quad \Gamma_2 = (m_2 L_{2cm} + M_2 L_2). \tag{5.41}$$

In these expressions, M_2 is the tip mass and m_i, L_i, L_{icm}, and k_i are, respectively, the ith link mass, the ith link length, the ith link position of the center of mass with reference to the ith joint, and the ith link radius of gyration about an axis parallel to the joint axis at its center of mass.

The position coordinates of the centers of mass of link 1 and link 2, the position coordinates of the centers of mass of the lumped masses M_1 and M_2, and that of the cart are, respectively, given by,

$$x_{1,cm} = L_{1cm} \cos\theta_1, z_{1,cm} = d_0 + L_{1cm} \sin\theta_1, \tag{5.42}$$

$$x_{2,cm} = L_1 \cos\theta_1 + L_{2cm} \sin(90 - \theta_1 + \theta_2), \tag{5.43}$$

$$z_{2cm} = d_0 + L_1 \sin\theta_1 + L_{2cm} \cos(90 - \theta_1 + \theta_2), x_{3,cm} = L_1 \cos\theta_1, \tag{5.44}$$

$$z_{3,cm} = d_0 + L_1 \sin\theta_1, x_{4,cm} = L_1 \cos\theta_1 + L_2 \sin(90 - \theta_1 + \theta_2), \tag{5.45}$$

$$z_{4cm} = d_0 + L_1 \sin\theta_1 + L_2 \cos(90 - \theta_1 + \theta_2), x_{5,cm} = 0, \tag{5.46}$$

and,

$$z_{5,cm} = d_0. \tag{5.47}$$

The velocity components of the centers of mass of link 1 and link 2, the velocity components of the centers of mass of the lumped masses M_1 and M_2, and that of the cart are, respectively, given by,

$$\dot{x}_{1,cm} = -L_{1cm}\dot{\theta}_1 \sin\theta_1, \dot{z}_{1,cm} = \dot{d}_0 + L_{1cm}\dot{\theta}_1 \cos\theta_1, \tag{5.48}$$

$$\dot{x}_{2,cm} = -L_1\dot{\theta}_1 \sin\theta_1 + L_{2cm}(-\dot{\theta}_1 + \dot{\theta}_2) \cos(90 - \theta_1 + \theta_2), \tag{5.49}$$

$$\dot{z}_{2cm} = \dot{d}_0 + L_1\dot{\theta}_1 \cos\theta_1 - L_{2cm}(-\dot{\theta}_1 + \dot{\theta}_2) \sin(90 - \theta_1 + \theta_2), \tag{5.50}$$

$$\dot{x}_{3,cm} = -L_1\dot{\theta}_1 \sin\theta_1, \dot{z}_{3,cm} = \dot{d}_0 + L_1\dot{\theta}_1 \cos\theta_1, \tag{5.51}$$

$$\dot{x}_{4,cm} = -L_1\dot{\theta}_1 \sin\theta_1 + L_2(-\dot{\theta}_1 + \dot{\theta}_2) \cos(90 - \theta_1 + \theta_2), \tag{5.52}$$

$$\dot{z}_{4cm} = \dot{d}_0 + L_1\dot{\theta}_1 \cos\theta_1 - L_2(-\dot{\theta}_1 + \dot{\theta}_2) \sin(90 - \theta_1 + \theta_2), \tag{5.53}$$

$$\dot{x}_{5,cm} = 0, \tag{5.54}$$

and,

$$\dot{z}_{5,cm} = \dot{d}_0. \tag{5.55}$$

The total potential energy is given by,

$$V = (m_1 L_{1cm} + m_2 L_1 + M_1 L_1 + M_2 L_1) g \cos \theta_1 + (m_2 L_{2cm} + M_2 L_2) g \sin(90 - \theta_1 + \theta_2). \tag{5.56}$$

It is expressed as,

$$V \equiv \Gamma_1 g \cos \theta_1 + \Gamma_2 g \sin(90 - \theta_1 + \theta_2). \tag{5.57}$$

The cart does not contribute to the potential energy.

The total translational and rotational kinetic energy may be shown to be,

$$T = \frac{1}{2} M_{Total} \dot{d}_0^2 + \frac{1}{2} I_{11} \dot{\theta}_1^2 + \frac{1}{2} I_{22} (\dot{\theta}_1 - \dot{\theta}_2)^2 + \Gamma_1 \dot{d}_0 \dot{\theta}_1 \cos \theta_1$$
$$+ \Gamma_2 (\dot{\theta}_1 - \dot{\theta}_2) (\dot{d}_0 \cos(\theta_1 - \theta_2) + L_1 \dot{\theta}_1 \cos(\theta_2)). \tag{5.58}$$

The Euler-Lagrange equations may be shown to be,

$$I_{11} \ddot{\theta}_1 + I_{22} (\ddot{\theta}_1 - \ddot{\theta}_2) + \Gamma_2 (\ddot{d}_0 \cos(\theta_1 - \theta_2) + L_1 (2\ddot{\theta}_1 - \ddot{\theta}_2) \cos(\theta_2))$$
$$+ \Gamma_1 \ddot{d}_0 \cos \theta_1 - \Gamma_1 g \sin \theta_1 - \Gamma_2 L_1 (2\dot{\theta}_1 - \dot{\theta}_2) \dot{\theta}_2 \sin(\theta_2)$$
$$- \Gamma_2 g \cos(90 + \theta_2 - \theta_1) = \tau_1, \tag{5.59}$$

$$I_{22} (\ddot{\theta}_2 - \ddot{\theta}_1) - \Gamma_2 (\ddot{d}_0 \cos(\theta_1 - \theta_2) + L_1 \ddot{\theta}_1 \cos(\theta_2))$$
$$+ g \Gamma_2 \cos(90 + \theta_2 - \theta_1) + \Gamma_2 L_1 \dot{\theta}_1^2 \sin(\theta_2) = \tau_2, \tag{5.60}$$

$$M_{Total} \ddot{d}_0 + \Gamma_1 \ddot{\theta}_1 \cos \theta_1 + \Gamma_2 (\ddot{\theta}_1 - \ddot{\theta}_2) \cos(\theta_1 - \theta_2)$$
$$- \Gamma_1 \dot{\theta}_1^2 \sin \theta_1 - \Gamma_2 (\dot{\theta}_1 - \dot{\theta}_2)^2 \sin(\theta_1 - \theta_2) = f. \tag{5.61}$$

Examining the second equation for the outer link, the torques driving the link apart from the control torque are due to the d'Alembert forces generated by the translation acceleration of the cart, the tangential acceleration of the base link, the acceleration due to gravity, and the centrifugal force due to the rotation rate of the base link. However, none of these forces would contribute to the driving torque on the second link if the center of mass of the second link is such that,

$$\Gamma_2 = 0; \tag{5.62}$$

i.e. the joint is located at the center of mass of the second link and its tip mass. If this is indeed the case, the second equation reduces to,

$$I_{22} (\ddot{\theta}_2 - \ddot{\theta}_1) = \tau_2. \tag{5.63}$$

By a similar argument, if the base joint is also located at the center of mass of both links, i.e. if,

$$\Gamma_1 = (m_1 L_{1cm} + m_2 L_1 + M_1 L_1 + M_2 L_1) = 0, \qquad (5.64)$$

the other two equations reduce to,

$$I_{11}\ddot{\theta}_1 + I_{22}(\ddot{\theta}_1 - \ddot{\theta}_2) = \tau_1, \qquad (5.65)$$

and,

$$M_{Total}\ddot{d}_0 = f. \qquad (5.66)$$

In this case, the manipulator is said to be mass balanced and is referred to as a balanced manipulator. The simplicity of the equations of motion attests to the benefits of balancing the manipulator. There is no possibility of any static or tip over instability. Otherwise, due to the d'Alembert forces generated by the translation acceleration of the cart acting in the horizontal direction, by the tangential acceleration of the base link acting normal to it, the acceleration due to gravity acting vertically down, and the centrifugal force due to rotation rate of the base link acting radially to it, there is a real danger that the manipulator would tip over and fall to the ground. This is the problem of tip over instability.

A number of stability measures and criteria have been developed by various researchers to establish the stability of a mobile manipulator and predict the occurrence of tip over. These measures and criteria include the classical Zero-Moment Point, the Force-Angle stability measure, and the Moment-Height Stability measure. The Zero Moment Point, or ZMP, is a point on the ground where the sum of all the forces and moments acting on the robot platform can be replaced by a single force. It was originally defined for bipedal robots, so as to be able to assess their stability while they walk (see, for example, Vepa [12]), and it has been adapted to mobile manipulators by several researchers. A different approach to stability analysis was proposed by Papadopoulos and Rey [13], in which they referred to the Force-Angle stability measure (FA). The FA algorithm measures the angle of the total applied force on the center of mass of the entire mobile platform with reference to a support polygon, which is derived from the points of contact the mobile platform makes with the ground at any instant of time. The angle is considered to be positive when the force is acting within the support polygon and negative outside it. Tip over instability occurs when the angle is zero or negative. Moosavian and Alipour [14] proposed the Moment-Height Stability (MHS) measure, which also accounts for the manipulator's own inertia about each axis of the support polygon relative to the gravitation force acting at the manipulator's center of gravity.

5.4 Dynamic Control for Path Tracking by Wheeled Mobile Manipulators

Consider a typical two-link manipulator mounted on the mobile platform shown in Figure 5.2, via a capstan with its axis of rotation normal to the base of the mobile platform and pointing up. The inertial position coordinates of the centers of mass of the link 1 and link 2, the position coordinates of the centers of mass of the lumped masses M_1 and M_2, and that of the cart are, respectively, given by,

$$\bar{x}_{1,cm} = L_{1cm} \cos \theta_1, \tag{5.67}$$

$$\bar{z}_{1,cm} = d_{0z} + L \cos \phi + L_{1cm} \sin \theta_1 \cos (\phi + \phi_1), \tag{5.68}$$

$$\bar{y}_{1,cm} = d_{0y} + L \sin \phi + L_{1cm} \sin \theta_1 \sin (\phi + \phi_1), \tag{5.69}$$

$$\bar{x}_{2,cm} = L_1 \cos \theta_1 + L_{2cm} \sin (90 - \theta_1 + \theta_2), \tag{5.70}$$

$$\bar{z}_{2cm} = d_{0z} + L \cos \phi + (L_1 \sin \theta_1 + L_{2cm} \cos (90 - \theta_1 + \theta_2)) \cos (\phi + \phi_1), \tag{5.71}$$

$$\bar{y}_{2cm} = d_{0y} + L \sin \phi + (L_1 \sin \theta_1 + L_{2cm} \cos (90 - \theta_1 + \theta_2)) \sin (\phi + \phi_1), \tag{5.72}$$

$$\bar{x}_{3,cm} = L_1 \cos \theta_1, \tag{5.73}$$

$$\bar{z}_{3,cm} = d_{0z} + L \cos \phi + L_1 \sin \theta_1 \cos (\phi + \phi_1), \tag{5.74}$$

$$\bar{y}_{3,cm} = d_{0y} + L \sin \phi + L_1 \sin \theta_1 \sin (\phi + \phi_1), \tag{5.75}$$

$$\bar{x}_{4,cm} = L_1 \cos \theta_1 + L_2 \sin (90 - \theta_1 + \theta_2), \tag{5.76}$$

$$\bar{z}_{4cm} = d_{0z} + L \cos \phi + (L_1 \sin \theta_1 + L_2 \cos (90 - \theta_1 + \theta_2)) \cos (\phi + \phi_1), \tag{5.77}$$

$$\bar{y}_{4cm} = d_{0y} + L \sin \phi + (L_1 \sin \theta_1 + L_2 \cos (90 - \theta_1 + \theta_2)) \sin (\phi + \phi_1), \tag{5.78}$$

$$\bar{x}_{5,cm} = 0, \tag{5.79}$$

and

$$\bar{z}_{5,cm} = d_{0z} + L \cos \phi, \quad \bar{y}_{5,cm} = d_{0y} + L \sin \phi, \tag{5.80}$$

where d_{0z} and d_{0y} are position coordinates of the mobile platform in the horizontal plane, ϕ is the orientation angle of the platform to the horizontal, and ϕ_1 is the rotation angle of the plane of operation of the outer links of the manipulator with respect to the forward axis of the platform.

The velocity components of the centers of mass of the link 1 and link 2, the velocity components of the centers of mass of the lumped masses M_1 and M_2 and that of the cart are, respectively, given by,

$$\dot{\bar{x}}_{1,cm} = -\dot{\theta}_1 L_{1cm} \sin \theta_1, \tag{5.81}$$

$$\dot{\bar{z}}_{1,cm} = \dot{d}_{0z} - \dot{\phi} L \sin \phi + L_{1cm}\dot{\theta}_1 \cos \theta_1 \cos (\phi + \phi_1) - L_{1cm}(\dot{\phi} + \dot{\phi}_1) \sin \theta_1 \sin (\phi + \phi_1), \tag{5.82}$$

$$\dot{\bar{y}}_{1,cm} = \dot{d}_{0y} + \dot{\phi} L \cos \phi + L_{1cm}\dot{\theta}_1 \cos \theta_1 \sin (\phi + \phi_1) + L_{1cm}(\dot{\phi} + \dot{\phi}_1) \sin \theta_1 \cos (\phi + \phi_1), \tag{5.83}$$

$$\dot{\bar{x}}_{2,cm} = -\dot{\theta}_1 L_1 \sin \theta_1 - (\dot{\theta}_1 - \dot{\theta}_2) L_{2cm} \sin (\theta_1 - \theta_2), \tag{5.84}$$

$$\dot{\bar{z}}_{2cm} = \dot{d}_{0z} - \dot{\phi} L \sin \phi + (\dot{\theta}_1 L_1 \cos \theta_1 + (\dot{\theta}_1 - \dot{\theta}_2) L_{2cm} \cos (\theta_1 - \theta_2)) \cos (\phi + \phi_1) \\ - (\dot{\phi} + \dot{\phi}_1)(L_1 \sin \theta_1 + L_{2cm} \sin (\theta_1 - \theta_2)) \sin (\phi + \phi_1), \tag{5.85}$$

5.4 Dynamic Control for Path Tracking by Wheeled Mobile Manipulators

$$\dot{\bar{y}}_{2cm} = \dot{d}_{0y} + \dot{\phi}L\cos\phi + \left(\dot{\theta}_1 L_1 \cos\theta_1 + (\dot{\theta}_1 - \dot{\theta}_2)L_{2cm}\cos(\theta_1 - \theta_2)\right)\sin(\phi + \phi_1)$$
$$+ (\dot{\phi} + \dot{\phi}_1)(L_1\sin\theta_1 + L_{2cm}\sin(\theta_1 - \theta_2))\cos(\phi + \phi_1), \quad (5.86)$$

$$\dot{\bar{x}}_{3,cm} = -\dot{\theta}_1 L_1 \sin\theta_1, \quad (5.87)$$

$$\dot{\bar{z}}_{3,cm} = \dot{d}_{0z} - \dot{\phi}L\sin\phi + L_1\dot{\theta}_1\cos\theta\cos(\phi + \phi_1) - L_1(\dot{\phi} + \dot{\phi}_1)\sin\theta_1\sin(\phi + \phi_1), \quad (5.88)$$

$$\dot{\bar{y}}_{3,cm} = \dot{d}_{0y} + \dot{\phi}L\cos\phi + L_1\dot{\theta}_1\cos\theta\sin(\phi + \phi_1) + L_1(\dot{\phi} + \dot{\phi}_1)\sin\theta_1\cos(\phi + \phi_1), \quad (5.89)$$

$$\dot{\bar{x}}_{4,cm} = -\dot{\theta}_1 L_1 \sin\theta_1 - (\dot{\theta}_1 - \dot{\theta}_2)L_2\sin(\theta_1 - \theta_2), \quad (5.90)$$

$$\dot{\bar{z}}_{4cm} = \dot{d}_{0z} - \dot{\phi}L\sin\phi + \left(\dot{\theta}_1 L_1\cos\theta_1 + (\dot{\theta}_1 - \dot{\theta}_2)L_2\cos(\theta_1 - \theta_2)\right)\cos(\phi + \phi_1)$$
$$- (\dot{\phi} + \dot{\phi}_1)(L_1\sin\theta_1 + L_2\sin(\theta_1 - \theta_2))\sin(\phi + \phi_1), \quad (5.91)$$

$$\dot{\bar{y}}_{4cm} = \dot{d}_{0y} + \dot{\phi}L\cos\phi + \left(\dot{\theta}_1 L_1\cos\theta_1 + (\dot{\theta}_1 - \dot{\theta}_2)L_2\cos(\theta_1 - \theta_2)\right)\sin(\phi + \phi_1)$$
$$+ (\dot{\phi} + \dot{\phi}_1)(L_1\sin\theta_1 + L_2\sin(\theta_1 - \theta_2))\cos(\phi + \phi_1), \quad (5.92)$$

$$\dot{\bar{x}}_{5,cm} = 0 \text{ and } \dot{\bar{z}}_{5,cm} = \dot{d}_{0z} - \dot{\phi}L\sin\phi, \dot{\bar{y}}_{5,cm} = \dot{d}_{0y} + \dot{\phi}L\cos\phi. \quad (5.93)$$

The null space of the constraint is defined as,

$$\frac{d}{dt}\mathbf{q} \equiv \frac{d}{dt}\begin{bmatrix} d_{0y} \\ d_{0z} \\ \phi \\ \phi + \phi_1 \\ \theta_1 \\ \theta_1 - \theta_2 \end{bmatrix} = \mathbf{N}_C \begin{bmatrix} v_F \\ v_T \\ \dot{\phi} + \dot{\phi}_1 \\ \dot{\theta}_1 \\ \dot{\theta}_1 - \dot{\theta}_2 \end{bmatrix}, \mathbf{N}_C = \begin{bmatrix} \cos\phi & -\sin\phi & 0 & 0 & 0 \\ \sin\phi & \cos\phi & 0 & 0 & 0 \\ 0 & 1/d & 0 & 0 & 0 \\ 0 & 0 & 1 & 0 & 0 \\ 0 & 0 & 0 & 1 & 0 \\ 0 & 0 & 0 & 0 & 1 \end{bmatrix}. \quad (5.94)$$

Observe that $\mathbf{N}_C$ is of the form,

$$\mathbf{N}_C = \begin{bmatrix} \mathbf{N}_{C11} & \mathbf{0}_{3\times 3} \\ \mathbf{0}_{3\times 2} & \mathbf{I}_{3\times 3} \end{bmatrix}. \quad (5.95)$$

Thus the platform's horizontal plane velocity coordinates and the end effectors relative velocity components, relative to the platform, are given by,

$$\frac{d}{dt}\mathbf{x} \equiv \frac{d}{dt}\begin{bmatrix} \bar{y}_{5,cm} \\ \bar{z}_{5,cm} \\ \bar{x}_{4,cm} \\ \bar{y}_{4,cm} - \bar{y}_{5,cm} \\ \bar{z}_{4,cm} - \bar{z}_{5,cm} \end{bmatrix} = \mathbf{J}\frac{d}{dt}\mathbf{q} \equiv \mathbf{J}\frac{d}{dt}\begin{bmatrix} d_{0y} \\ d_{0z} \\ \phi \\ \phi + \phi_1 \\ \theta_1 \\ \theta_1 - \theta_2 \end{bmatrix} = \mathbf{J}\mathbf{N}_C\begin{bmatrix} v_F \\ v_T \\ \dot{\phi} + \dot{\phi}_1 \\ \dot{\theta}_1 \\ \dot{\theta}_1 - \dot{\theta}_2 \end{bmatrix},$$

$$\mathbf{J} = \begin{bmatrix} 1 & 0 & L\cos\phi & 0 & 0 & 0 \\ 0 & 1 & -L\sin\phi & 0 & 0 & 0 \\ 0 & 0 & 0 & 0 & -L_1\sin\theta_1 & -L_2\sin(\theta_1 - \theta_2) \\ 0 & 0 & 0 & y_{\phi_1} & L_1\cos\theta_1\sin(\phi + \phi_1) & L_2\cos(\theta_1 - \theta_2)\sin(\phi + \phi_1) \\ 0 & 0 & 0 & z_{\phi_1} & L_1\cos\theta_1\cos(\phi + \phi_1) & L_2\cos(\theta_1 - \theta_2)\cos(\phi + \phi_1) \end{bmatrix},$$

$$(5.96)$$

where,

$$y_{\phi_1} = (L_1 \sin\theta_1 + L_2 \sin(\theta_1 - \theta_2))\cos(\phi + \phi_1), \quad (5.97)$$

$$z_{\phi_1} = -(L_1 \sin\theta_1 + L_2 \sin(\theta_1 - \theta_2))\sin(\phi + \phi_1). \quad (5.98)$$

Moreover, **J** is of the form,

$$\mathbf{J} = \begin{bmatrix} \mathbf{J}_p & \mathbf{0}_{2\times 3} \\ \mathbf{0}_{3\times 3} & \mathbf{J}_m \end{bmatrix}. \quad (5.99)$$

It can now be confirmed that the Jacobian-based transformation $\mathbf{J}\mathbf{N}_C$ is non-singular and we can proceed to find the Euler-Lagrange equations. To this end let, $M_{Total} = m_1 + m_2 + M_1 + M_2 + M_{cart}$, $m_{11} = m_1 + M_1$, $m_{22} = m_2 + M_2$, $I_{11} = m_1 L_{1cm}^2 + M_1 L_1^2 + (m_2 + M_2)L_1^2 + m_1 k_1^2$, $I_{22} = m_2 L_{2cm}^2 + M_2 L_2^2 + m_2 k_2^2$, $\Gamma_{11} = m_1 L_{1cm} + M_1 L_1$, $\Gamma_2 = m_2 L_{2cm} + M_2 L_2$, $\Gamma_1 = \Gamma_{11} + m_{22} L_1$. The total translational and rotational kinetic energy may then be shown to be,

$$T = \frac{1}{2}M_{Total}\left(\dot{d}_{0y}^2 + \dot{d}_{0z}^2\right) + \frac{1}{2}\left(I_p + M_{cart}L^2\right)\dot\phi^2 + \frac{1}{2}I_m(\theta_1,\theta_{12})\dot\phi_t^2 + \frac{1}{2}I_{11}\dot\theta_1^2 + \frac{1}{2}I_{22}\dot\theta_{12}^2$$
$$+ M_{Total}L\dot\phi\left(\dot d_{0y}\cos\phi - \dot d_{0z}\sin\phi\right) + \Gamma_1 \dot\theta_1 \cos\theta_1\left(\dot d_{0y}\sin\phi_t + \dot d_{0z}\cos\phi_t\right)$$
$$+ \Gamma_2 \dot\theta_{12}\left\{\left(\dot d_{0y}\sin\phi_t + \dot d_{0z}\cos\phi_t\right)\cos\theta_{12} + L_1\dot\theta_1\cos\theta_2\right\}$$
$$+ \dot\phi L\left(\dot\theta_1 \Gamma_1 \cos\theta_1 + \dot\theta_{12}\Gamma_2\cos\theta_{12}\right)\sin\phi_1 + L\dot\phi\dot\phi_t(\Gamma_1\sin\theta_1 + \Gamma_2\sin\theta_{12})\cos\phi_1$$
$$+ \dot\phi_t(\Gamma_1\sin\theta_1 + \Gamma_2\sin\theta_{12})\left(\dot d_{0y}\cos\phi_t - \dot d_{0z}\sin\phi_t\right). \quad (5.100)$$

In the above expression, $\theta_{12} = \theta_1 - \theta_2$, $\phi_t = \phi + \phi_1$, I_p is the mass moment of inertia of the platform about an axis normal to the platform and about its center of mass, and I_m is the moment of inertia of the manipulator about an axis normal to the platform at the capstan, which is given by,

$$I_m(\theta_1,\theta_{12}) = \frac{1}{2}m_1\left(k_{1y}^2 + L_{1cm}^2\right)\sin^2\theta_1$$
$$+ \frac{1}{2}m_2\left(k_{2y}^2 \sin^2\theta_{12} + (L_1\sin\theta_1 + L_{2cm}\sin\theta_{12})^2\right) \quad (5.101)$$
$$+ \frac{1}{2}M_1 L_1^2 \sin^2\theta_1 + \frac{1}{2}M_2(L_1\sin\theta_1 + L_2\sin\theta_{12})^2.$$

In the previous equation, k_{iy} are the radii of gyration of the ith link at its center of mass, with respect to an axis that is mutually perpendicular to both the joint axis and the direction of the vector pointing to the next joint.

The total potential energy is given by,

$$V = (m_1 L_{1cm} + m_2 L_1 + M_1 L_1 + M_2 L_1)g\cos\theta_1 + (m_2 L_{2cm} + M_2 L_2)g\cos\theta_{12}. \quad (5.102)$$

It is expressed as,

$$V \equiv \Gamma_1 g \cos\theta_1 + \Gamma_2 g \cos\theta_{12}. \quad (5.103)$$

5.4 Dynamic Control for Path Tracking by Wheeled Mobile Manipulators

The platform does not contribute to the potential energy. It follows that the nonzero derivatives of the potential energy function are,

$$\frac{\partial}{\partial \theta_1} V \equiv -\Gamma_1 g \sin \theta_1,$$
$$\frac{\partial}{\partial \theta_{12}} V \equiv -\Gamma_2 g \sin \theta_{12}.$$
(5.104)

Let,

$$I_{pL} = I_p + M_{cart}L^2, \mathbf{H} = [\mathbf{H}_1 \quad \mathbf{H}_2], c() = \cos(), s() = \sin(),$$
(5.105)

where

$$\mathbf{H}_1 = \begin{bmatrix} M_{Total} & 0 & M_{Total}Lc\phi \\ 0 & M_{Total} & -M_{Total}Ls\phi \\ M_{Total}Lc\phi & -M_{Total}Ls\phi & I_{pL} \\ (\Gamma_1 s\theta_1 + \Gamma_2 s\theta_{12})c\phi_t & -(\Gamma_1 s\theta_1 + \Gamma_2 s\theta_{12})s\phi_t & L(\Gamma_1 s\theta_1 + \Gamma_2 s\theta_{12})c\phi_1 \\ \Gamma_1 c\theta_1 s\phi_t & \Gamma_1 c\theta_1 c\phi_t & \Gamma_1 c\theta_1 s\phi_1 \\ \Gamma_2 c\theta_{12} s\phi_t & \Gamma_2 c\theta_{12} c\phi_t & \Gamma_2 c\theta_{12} s\phi_1 \end{bmatrix}$$

$$\mathbf{H}_2 = \begin{bmatrix} (\Gamma_1 s\theta_1 + \Gamma_2 s\theta_{12})c\phi_t & \Gamma_1 c\theta_1 s\phi_t & \Gamma_2 c\theta_{12} s\phi_t \\ -(\Gamma_1 s\theta_1 + \Gamma_2 s\theta_{12})s\phi_t & \Gamma_1 c\theta_1 c\phi_t & \Gamma_2 c\theta_{12} c\phi_t \\ L(\Gamma_1 s\theta_1 + \Gamma_2 s\theta_{12})c\phi_1 & \Gamma_1 c\theta_1 s\phi_1 & \Gamma_2 c\theta_{12} s\phi_1 \\ I_m(\theta_1, \theta_{12}) & 0 & 0 \\ 0 & I_{11} & \Gamma_2 L_1 c\theta_2 \\ 0 & \Gamma_2 L_1 c\theta_2 & I_{22} \end{bmatrix}.$$

(5.106)

Furthermore,

$$\mathbf{H} \equiv [\mathbf{H}_1 \quad \mathbf{H}_2] \equiv \begin{bmatrix} \mathbf{H}_{11} & \mathbf{H}_{12} \\ \mathbf{H}_{21} & \mathbf{H}_{22} \end{bmatrix},$$
(5.107)

where $\mathbf{H}_{11}$ and $\mathbf{H}_{22}$ are square matrices.

Also, let the generalized coordinates, the gradient of the potential function, and the generalized forces be defined, respectively, as,

$$\mathbf{q} \equiv \begin{bmatrix} d_{0y} \\ d_{0z} \\ \phi \\ \phi_t \\ \theta_1 \\ \theta_{12} \end{bmatrix}, \frac{\partial}{\partial \mathbf{q}} V = -g \begin{bmatrix} 0 \\ 0 \\ 0 \\ 0 \\ \Gamma_1 s\theta_1 \\ \Gamma_2 s\theta_{12} \end{bmatrix} \equiv -g\mathbf{G}, \mathbf{Q} \equiv \begin{bmatrix} F_y \\ F_z \\ N \\ N_t \\ \tau_1 \\ \tau_{12} \end{bmatrix}.$$
(5.108)

The kinetic energy and its partial derivatives may be expressed as,

$$T = \frac{1}{2}\dot{\mathbf{q}}^T \mathbf{H} \dot{\mathbf{q}}, \frac{\partial}{\partial \dot{\mathbf{q}}} T = \mathbf{H} \dot{\mathbf{q}}, \frac{\partial}{\partial \mathbf{q}} T = \frac{1}{2} \dot{\mathbf{q}}^T \frac{\partial \mathbf{H}}{\partial \mathbf{q}} \dot{\mathbf{q}}, \frac{d}{dt} \frac{\partial}{\partial \dot{\mathbf{q}}} T = \mathbf{H} \ddot{\mathbf{q}} + \dot{\mathbf{q}}^T \frac{\partial \mathbf{H}}{\partial \mathbf{q}} \dot{\mathbf{q}}.$$
(5.109)

Thus the Euler-Lagrange equations are,

$$\frac{d}{dt}\frac{\partial}{\partial \dot{\mathbf{q}}}T - \frac{\partial}{\partial \mathbf{q}}T + \frac{\partial}{\partial \mathbf{q}}V = \mathbf{H}\ddot{\mathbf{q}} + \dot{\mathbf{q}}T\frac{\partial \mathbf{H}}{\partial \mathbf{q}}\dot{\mathbf{q}} - \frac{1}{2}\dot{\mathbf{q}}T\frac{\partial \mathbf{H}}{\partial \mathbf{q}}\dot{\mathbf{q}} + g\mathbf{G} = \mathbf{Q} - \lambda \mathbf{C}^{\mathbf{T}}, \quad (5.110)$$

where $\mathbf{C}$ is the non-holonomic constraint matrix.

The Euler-Lagrange equations simplify to,

$$\frac{d}{dt}\frac{\partial}{\partial \dot{\mathbf{q}}}T - \frac{\partial}{\partial \mathbf{q}}T + \frac{\partial}{\partial \mathbf{q}}V = \mathbf{H}\ddot{\mathbf{q}} + \frac{1}{2}\dot{\mathbf{q}}^T\frac{\partial \mathbf{H}}{\partial \mathbf{q}}\dot{\mathbf{q}} + g\mathbf{G} = \mathbf{Q} - \lambda \mathbf{C}^{\mathbf{T}}. \quad (5.111)$$

Hence,

$$\mathbf{H}\ddot{\mathbf{q}} + \frac{1}{2}\dot{\mathbf{q}}^T\frac{\partial \mathbf{H}}{\partial \mathbf{q}}\dot{\mathbf{q}} + g\mathbf{G} = \mathbf{Q} - \lambda \mathbf{C}^{\mathbf{T}}. \quad (5.112)$$

Making the coordinate transformation,

$$\dot{\mathbf{q}} = \mathbf{N}_C\begin{bmatrix} v_F & v_T & \dot{\phi} + \dot{\phi}_1 & \dot{\theta}_1 & \dot{\theta}_1 - \dot{\theta}_2 \end{bmatrix}^T \equiv \mathbf{N}_C \mathbf{v}, \quad (5.113)$$

and multiplying by $\mathbf{N}_C^T$ gives,

$$\mathbf{N}_C^T \mathbf{H} \mathbf{N}_C \dot{\mathbf{v}} + \mathbf{N}_C^T \mathbf{H} \dot{\mathbf{N}}_C \mathbf{v} + \frac{1}{2}\mathbf{N}_C^T \mathbf{v}^T \mathbf{N}_C^T \frac{\partial \mathbf{H}}{\partial \mathbf{q}} \mathbf{N}_C \mathbf{v} + \mathbf{N}_C^T g \mathbf{G} = \mathbf{N}_C^T \mathbf{Q}. \quad (5.114)$$

Making the similarity transformation, $\dot{\mathbf{x}} = \mathbf{J}\mathbf{N}_C \mathbf{v}$, with,

$$\mathbf{J}\mathbf{N}_C = \begin{bmatrix} \mathbf{J}_p & \mathbf{0}_{2\times 3} \\ \mathbf{0}_{3\times 3} & \mathbf{J}_m \end{bmatrix}\begin{bmatrix} \mathbf{N}_{C11} & \mathbf{0}_{3\times 3} \\ \mathbf{0}_{3\times 2} & \mathbf{I}_{3\times 3} \end{bmatrix} = \begin{bmatrix} \mathbf{J}_p \mathbf{N}_{C11} & \mathbf{0}_{2\times 3} \\ \mathbf{0}_{3\times 2} & \mathbf{J}_m \end{bmatrix}, \quad (5.115)$$

so,

$$\mathbf{v} = (\mathbf{J}\mathbf{N}_C)^{-1}\dot{\mathbf{x}}, \mathbf{v}^T = \dot{\mathbf{x}}^T\left(\mathbf{N}_C^T \mathbf{J}^T\right)^{-1}. \quad (5.116)$$

It follows that,

$$\mathbf{H}_x\ddot{\mathbf{x}} + \mathbf{B}\mathbf{V}\dot{\mathbf{x}} + \mathbf{B}g\mathbf{G} = \mathbf{B}\mathbf{Q} \equiv \boldsymbol{\tau}_x, \quad (5.117)$$

where,

$$\mathbf{H}_x = \mathbf{J}\mathbf{N}_C\left(\mathbf{N}_C^T \mathbf{H} \mathbf{N}_C\right)(\mathbf{J}\mathbf{N}_C)^{-1}, \mathbf{B} = \mathbf{J}\mathbf{N}_C \mathbf{N}_C^T,$$
$$\mathbf{V} = \mathbf{H}\mathbf{N}_C\left(\mathbf{J}\dot{\mathbf{N}}_C\right)^{-1} + \mathbf{H}\dot{\mathbf{N}}_C(\mathbf{J}\mathbf{N}_C)^{-1} + \frac{1}{2}\dot{\mathbf{x}}^T\left(\mathbf{N}_C^T \mathbf{J}^T\right)^{-1}\mathbf{N}_C^T \frac{\partial \mathbf{H}}{\partial \mathbf{q}}\mathbf{N}_C(\mathbf{J}\mathbf{N}_C)^{-1}.$$

Also,

$$\mathbf{N}_C^T \mathbf{H} \mathbf{N}_C = \begin{bmatrix} \mathbf{N}_{C11}^T & \mathbf{0}_{2\times 3} \\ \mathbf{0}_{3\times 3} & \mathbf{I}_{3\times 3} \end{bmatrix}\begin{bmatrix} \mathbf{H}_{11} & \mathbf{H}_{12} \\ \mathbf{H}_{21} & \mathbf{H}_{22} \end{bmatrix}\begin{bmatrix} \mathbf{N}_{C11} & \mathbf{0}_{3\times 3} \\ \mathbf{0}_{3\times 2} & \mathbf{I}_{3\times 3} \end{bmatrix}, \quad (5.118)$$

so,

$$\mathbf{N}_C^T \mathbf{H} \mathbf{N}_C = \begin{bmatrix} \mathbf{N}_{C11}^T \mathbf{H}_{11} \mathbf{N}_{C11} & \mathbf{N}_{C11}^T \mathbf{H}_{12} \\ \mathbf{H}_{21}\mathbf{N}_{C11} & \mathbf{H}_{22} \end{bmatrix} \text{ and } \mathbf{B} = \begin{bmatrix} \mathbf{J}_p \mathbf{N}_{C11} \mathbf{N}_{C11}^T & \mathbf{0}_{2\times 3} \\ \mathbf{0}_{3\times 2} & \mathbf{J}_m \end{bmatrix}. \quad (5.119)$$

Consequently, one could express $\mathbf{H}_x$ and $\mathbf{B}\mathbf{V}$ as,

5.4 Dynamic Control for Path Tracking by Wheeled Mobile Manipulators

$$\mathbf{H}_x \equiv \begin{bmatrix} \mathbf{H}_{x11} & \mathbf{H}_{x12} \\ \mathbf{H}_{x21} & \mathbf{H}_{x22} \end{bmatrix}, \quad \mathbf{BV} \equiv \begin{bmatrix} \mathbf{V}_{b11} & \mathbf{V}_{b12} \\ \mathbf{V}_{b21} & \mathbf{V}_{b22} \end{bmatrix}. \tag{5.120}$$

The structure of these equations suggests that we could split the state vector $\mathbf{x}$ into two components, $\mathbf{x}_1$ and $\mathbf{x}_2$, such that,

$$\mathbf{x} = \begin{bmatrix} \mathbf{x}_1 \\ \mathbf{x}_2 \end{bmatrix}, \quad \mathbf{x}_1 = \begin{bmatrix} \bar{y}_{5,cm} \\ \bar{z}_{5,cm} \end{bmatrix}, \quad \mathbf{x}_2 = \begin{bmatrix} \bar{x}_{4,cm} \\ \bar{y}_{4,cm} - \bar{y}_{5,cm} \\ \bar{z}_{4,cm} - \bar{z}_{5,cm} \end{bmatrix}. \tag{5.121}$$

As a result, the equations of motion in the $\mathbf{x}$ domain can be expressed in terms of two equations for $\mathbf{x}_1$ and $\mathbf{x}_2$. It is now feasible to design a control law to implement a desired motion plan.

Having established the equations of motion of a mobile manipulator in an inertial frame with no further constraints, the controller synthesis for tracking a desired motion plan is now considered. The coordinated control of a wheeled mobile manipulator has been considered in some detail by Yamamoto [15]. The control is assumed to be generated from the demanded acceleration, the demanded rate vector, the demanded position vector, and three feedbacks, (i) the gravity-gradient potential function V, (ii) a dissipation function similar to the Rayleigh dissipation function, and (iii) an artificial potential field. To a certain extent, such a methodology employs potentials that are similar in principle to the potentials encountered in the derivation of the Euler-Lagrange equations in the preceding section and would preserve the conservative nature of the robot dynamics, thus ensuring that the closed loop system is inherently stable.

The control torques are assumed to be,

$$\boldsymbol{\tau}_x = \mathbf{H}_x \ddot{\mathbf{x}}_d + \mathbf{BV}\dot{\mathbf{x}}_d + \mathbf{B}g\mathbf{G} - \frac{dD}{d\dot{\mathbf{x}}} - \frac{d\Phi}{d\mathbf{x}}, \tag{5.122}$$

where $\dot{\mathbf{x}}_d$ is the demanded rate vector, D is a dissipation function, and Φ is an artificial potential function. With an appropriate choice of D and Φ, the control law may be interpreted as a proportional-derivative control law.

The equations of motion reduce to,

$$\mathbf{H}_x(\ddot{\mathbf{x}} - \ddot{\mathbf{x}}_d) + \mathbf{BV}(\dot{\mathbf{x}} - \dot{\mathbf{x}}_d) + \frac{dD}{d\dot{\mathbf{x}}} + \frac{d\Phi}{d\mathbf{x}} = \mathbf{0}. \tag{5.123}$$

If D and Φ are, respectively, assumed to be,

$$D = \frac{1}{2}(\dot{\mathbf{x}} - \dot{\mathbf{x}}_d)^T \mathbf{K}_d (\dot{\mathbf{x}} - \dot{\mathbf{x}}_d) \quad \text{and} \quad \Phi = \frac{1}{2}(\mathbf{x} - \mathbf{x}_d)^T \mathbf{K}_p (\mathbf{x} - \mathbf{x}_d), \tag{5.124}$$

with $\mathbf{K}_p$ being positive definite, it follows that,

$$\mathbf{H}_x(\ddot{\mathbf{x}} - \ddot{\mathbf{x}}_d) + (\mathbf{K}_d + \mathbf{BV})(\dot{\mathbf{x}} - \dot{\mathbf{x}}_d) + \mathbf{K}_p(\mathbf{x} - \mathbf{x}_d) = \mathbf{0}. \tag{5.125}$$

If we define the error $\mathbf{e}$ as, $\mathbf{e} = \mathbf{x} - \mathbf{x}_d$, the error vector between the current position vector and the demanded position vector, the error satisfies the equation,

$$\mathbf{H}_x \ddot{\mathbf{e}} + (\mathbf{K}_d + \mathbf{BV})\dot{\mathbf{e}} + \mathbf{K}_p \mathbf{e} = \mathbf{0}. \tag{5.126}$$

To ensure stability, consider the Lyapunov function and its time derivative,

$$V = \frac{1}{2}(\dot{\mathbf{e}}^T \mathbf{H}_x \dot{\mathbf{e}} + \mathbf{e}\mathbf{K}_p\mathbf{e}), \tag{5.127}$$

$$\dot{V} = \dot{\mathbf{e}}^T(\mathbf{H}_x\ddot{\mathbf{e}} + \mathbf{K}_p\mathbf{e}) = -\dot{\mathbf{e}}^T(\mathbf{K}_d + \mathbf{BV})\dot{\mathbf{e}}. \tag{5.128}$$

Thus, to guarantee stability, $\mathbf{K}_d + \mathbf{BV}$ must be adequately positive definite.

5.5 Decoupled Control of the Mobile Platform and Manipulator

In general, a mobile manipulator cannot follow or track a desired trajectory unless the mobile platform is in motion, so the desired position of the end effector is within the local workspace of the manipulator. Moreover, the definition of the useful workspace of the manipulator is based on the manipulability of the manipulator within the workspace. Thus, it is assumed that there are no singular configurations of the manipulator within the local workspace. Moreover, the position of the end effector in three dimensions requires three coordinates. Thus, a manipulator with three or more links is generally required. However, when there are more than three links, there is a certain level of redundancy, which must be effectively managed. One approach to doing this is to decouple the manipulator and the mobile platform dynamics. Such a decoupled control approach is feasible, as both the transformations, $\mathbf{N}_C$ and $\mathbf{J}\mathbf{N}_C$, are decoupled. Thus, one may split the state vector $\mathbf{x}$ into two components, $\mathbf{x}_1$ and $\mathbf{x}_2$, and design independent controllers for positioning the mobile platform and the manipulator end-effector within the workspace. The mobile is positioned so the actual position of the end effector is closest to its desired position (for example, vertically below or just behind in the case of the manipulator and platform discussed in the preceding section!)

To illustrate the decoupling approach, consider the equations of motion given by Equation (5.117) in Section 5.4, which are,

$$\mathbf{H}_x\ddot{\mathbf{x}} + \mathbf{BV}\dot{\mathbf{x}} + \mathbf{B}g\mathbf{G} = \mathbf{BQ} \equiv \tau_x. \tag{5.129}$$

They may be expressed as,

$$\begin{bmatrix} \mathbf{H}_{x11} & \mathbf{H}_{x12} \\ \mathbf{H}_{x21} & \mathbf{H}_{x22} \end{bmatrix}\begin{bmatrix} \ddot{\mathbf{x}}_1 \\ \ddot{\mathbf{x}}_2 \end{bmatrix} + \begin{bmatrix} \mathbf{V}_{b11} & \mathbf{V}_{b12} \\ \mathbf{V}_{b21} & \mathbf{V}_{b22} \end{bmatrix}\begin{bmatrix} \dot{\mathbf{x}}_1 \\ \dot{\mathbf{x}}_2 \end{bmatrix} + g\begin{bmatrix} \mathbf{0} \\ \mathbf{J}_m\mathbf{G}_1 \end{bmatrix} = \mathbf{B}\begin{bmatrix} \mathbf{Q}_1 \\ \mathbf{Q}_2 \end{bmatrix},$$

$$\mathbf{B} = \begin{bmatrix} \mathbf{J}_p\mathbf{N}_{C11}\mathbf{N}_{C11}^T & \mathbf{0}_{2\times 3} \\ \mathbf{0}_{3\times 2} & \mathbf{J}_m \end{bmatrix}, \quad \mathbf{G}_1 = \begin{bmatrix} 0 \\ \Gamma_1 s\theta_1 \\ \Gamma_2 s\theta_{12} \end{bmatrix} \text{ and } \begin{bmatrix} \mathbf{Q}_1 \\ \mathbf{Q}_2 \end{bmatrix} = \mathbf{Q}. \tag{5.130}$$

The first step toward decoupling is to define auxiliary control inputs $\mathbf{U}_p$ and $\mathbf{U}_m$ such that,

$$\mathbf{U}_p = \mathbf{BQ}_1 - \mathbf{V}_{b12}\dot{\mathbf{x}}_2 - \mathbf{H}_{x12}\ddot{\mathbf{x}}_2, \quad \mathbf{U}_m = \mathbf{BQ}_2 - \mathbf{V}_{b21}\dot{\mathbf{x}}_1 - \mathbf{H}_{x21}\ddot{\mathbf{x}}_1. \tag{5.131}$$

Thus, the decoupled equations of motion are,

$$\mathbf{H}_{x11}\ddot{\mathbf{x}}_1 + \mathbf{V}_{b11}\dot{\mathbf{x}}_1 = \mathbf{U}_p, \tag{5.132}$$

$$\mathbf{H}_{x22}\ddot{\mathbf{x}}_2 + \mathbf{V}_{b22}\dot{\mathbf{x}}_2 + g\mathbf{J}_m\mathbf{G}_1 = \mathbf{U}_m. \tag{5.133}$$

5.6 Motion Planning for Mobile Manipulators

The control problem discussed in Section 5.5 has a few pitfalls, particularly when there are significant disturbances, and consequently the tracking error will not tend to zero asymptotically, although it may still be small. In general, one can identify three types of control problems: (i) maintain a particular equilibrium configuration with adequate stability margins; (ii) tracking a reference cart's position and velocity where the reference position and velocity vectors satisfy the same constraints as the actual cart's position and velocity vectors; (iii) a general path-following problem, where the end effector follows a general trajectory within its global workspace. It is possible in principle to solve all of the control problems in a unified framework. Thus, it is possible to adopt one of several alternative approaches that are used to control a typical fixed-base manipulator (see, for example, Vepa [16]).

Given that the objective of motion planning is to define a feasible path that is smooth and continuous, to reach a destination point without colliding with any obstacles, the generation of a motion plan for robots and mobile manipulators in unstructured, unknown, dynamic environments can be a complex task involving a variety of sensors and a number of algorithms for localization, navigation, and obstacle avoidance. Sensory information from a variety of sensors, including stereo cameras and laser, infrared, or ultrasonic ranging sensors such as a Red-Green-Blue depth (RGBD) sensors, which use a combination of a color camera, an infrared sensor for depth estimation, and a microphone array for sound measurements, must be used to collect basic data prior to implementing an algorithm for localization, navigation, and obstacle avoidance. Other complex sensor systems include laser scanners, Charge Coupled Device (CCD) cameras, Global Positioning System (GPS) sensors, and inertial measuring units. Typically, a robotic mobile manipulator must not only avoid dynamic obstacles autonomously at all times but must also deal with the emergence of new obstacles such as the sudden entry of an object, which requires the mobile planner to dynamically and reactively re-plan the motion paths. Obstacle avoidance may be implemented by first constructing the configuration space. Given a map of the obstacles on a two-dimensional space, one can define the actual workspace by identifying the most traversable domains. Once the workspace is identified, the configuration space is obtained from the workspace by a process known as image morphology dilation. The configuration space is obtained by eroding the edges around the free workspace. This step involves the creation of a morphology dilation of the robot footprint. Dilation is a basic operator in creating a new morphology of a binary image. The effect of applying the operator on a binary image is to enlarge the boundaries by specific amounts. Thus, areas of boundary pixels are grown in size while holes within those regions are made smaller.

To plan a path, two metrics are constructed over a grid in the configuration space; one representing the distance from the nearest obstacle, while the second is the weighted distance to the goal position. An artificial potential function Φ is constructed by subtracting the second metric from the first. One always moves forward such that the gradient of the artificial potential function Φ is always maximum, thus ensuring that one is always avoiding an obstacle while also moving toward the intended goal position.

The ability to re-plan the motion requires that localization, navigation, and obstacle avoidance must be implemented continuously and in real time. Thus, there is a need not only to estimate the current position of the mobile platform and its workspace but also the intended future motion of the platform and manipulator arms. This requires establishing a dynamic model or map of the environment, which in turn is used to define a dynamic reference model for the future motion. The model of the environment is established by several rounds of filtering and de-noising, followed by terrain mapping and decomposition, and obstacle detection to build a three-dimensional occupancy grid as well as a semantic representation of the known objects. The task of terrain mapping, which is performed extensively, is to establish whether the nature of the terrain is urban, natural with vegetation, desert, or canyons, to distinguish between a traversable terrain and a non-traversable terrain and to determine a feasible local plane path for the platform to track, from the characteristics of the traversable terrain most suited for the platform to follow. Thus, the search for a feasible motion plan is limited by the robot's sensory capabilities and its own kinematic constraints. When several feasible paths are available, the choice is made based on an optimal criterion.

An imaging sensor produces a two-dimensional projection of a three-dimensional space. This is known as the image space. The occupancy grid is initially generated in the image space and then projected on to the ground plane to generate a Cartesian map relative to the ground plane. A sparsely populated occupancy grid is a desirable feature of the domain in which the mobile platform can move forward. An algorithm such as the A* algorithm (see, for example, Vepa [12]) or a related algorithm is then used to find an optimal path through the occupancy grid between the platform's current position and the goal. Thus, we have succinctly described the entire process of motion planning, although it must be said that several algorithms must be implemented in real time to successfully implement a feasible motion-planning scheme. Such a qualitative description of the process is generally also applicable to a manipulator platform operating in space.

5.7 Non-Holonomic Space Manipulators

The maneuvers and dynamic interactions between a space robotic manipulator on a spacecraft while it captures an orbiting object is a function of several factors, including the proximity of the two and their behavior before and after docking. Thus, it involves four phases, the approach phase, freely floating, docking, and post-docking. While both the approach phase when the spacecraft is in flight and the freely floating phase involve dynamic coupling between the satellite and the manipulator, they both require motion planning so as to be able to complete the docking or capture of the orbiting object. Of the translational and angular momentum conservation equations, only the former can be integrated, while the latter are non-integrable or non-holonomic constraints.

5.7 Non-Holonomic Space Manipulators

Furthermore, while the total translational momentum of a spacecraft in orbit or the orbital moment of momentum is generally conserved, it is possible that in the absence of any applied torques, the total angular momentum is also conserved.

In Section 4.9 the equations of motion of a satellite were derived based on the principles of the conservation of moment of momentum or angular momentum. The moment of momentum of the satellite and the manipulator it is carrying was evaluated in a body-fixed coordinate system. In general, the total moment of momentum of a multi-body system about its center of mass in body-fixed coordinates may be expressed as,

$$\mathbf{H}_B = \sum_i \mathbf{H}_{Bi} = \sum_i \mathbf{I}_{Bi}\boldsymbol{\omega}_{Bi}. \tag{5.134}$$

The expression may be transformed to an inertial or space-fixed frame with its origin at the center of mass of the multibody system and consequently the moment of momentum in the space-fixed frame is,

$$\mathbf{H}_I = \mathbf{T}_{IB}\mathbf{H}_B = \mathbf{T}_{IB}\sum_i \mathbf{H}_{Bi} = \mathbf{T}_{IB}\sum_i \mathbf{I}_{Bi}\boldsymbol{\omega}_{Bi}. \tag{5.135}$$

As a result of the conservation of the moment of momentum in the space-fixed frame,

$$\mathbf{H}_I = \mathbf{T}_{IB}\mathbf{H}_B = \mathbf{T}_{IB}\sum_i \mathbf{H}_{Bi} = \mathbf{T}_{IB}\sum_i \mathbf{I}_{Bi}\boldsymbol{\omega}_{Bi} = \mathbf{T}_{IB_0}\sum_i \mathbf{I}_{Bi0}\boldsymbol{\omega}_{Bi0}, \tag{5.136}$$

where $\mathbf{T}_{IB}$ the transformation from the body to the inertial frame, $\mathbf{I}_{iB}$ the moment of inertia matrix of the ith body, and $\boldsymbol{\omega}_{Bi}$ the angular velocity vector in the body-fixed frame of the ith body are evaluated at time $t = 0$, on the right-hand side of Equation (5.136). Thus, the total moment of momentum of the multibody system about its center of mass in body-fixed coordinates is given by,

$$\mathbf{H}_B = \mathbf{T}_{BI}\mathbf{T}_{IB_0}\sum_i \mathbf{I}_{Bi0}\boldsymbol{\omega}_{Bi0}. \tag{5.137}$$

It follows that the left-hand side of Equation (5.137) is proportional to a linear combination of the column vectors of the inverse of $\mathbf{T}_{IB}$ or $\mathbf{T}_{BI}$, the transformation from the inertial frame to the body frame. However, when the multibody system has *zero initial angular momentum* at time $t = 0$, so $\boldsymbol{\omega}_{Bi0} = 0$ for all i, the total moment of momentum of the multibody system about its center of mass in body-fixed coordinates $\mathbf{H}_B$ is zero. This principle was first used for space robotic manipulators by Papadopoulos and Dubowsky [17] for the case of zero initial angular momentum and by Nanos and Papadopoulos [18]. Hence, from Section 4.9 it follows that the equations of motion of the main body of the satellite are given by,

$$\frac{d}{dt}\left(\mathbf{I}_{CM1}\boldsymbol{\omega} + \sum_{i=2}^N \mathbf{I}_{CMi}\left(\boldsymbol{\omega} + \mathbf{t}_i\mathbf{e}_z\dot{\theta}_i\right)\right) + \boldsymbol{\omega}\times\mathbf{I}_{CM1}\boldsymbol{\omega} + \sum_{i=2}^N \left(\boldsymbol{\omega} + \mathbf{t}_i\mathbf{e}_z\dot{\theta}_i\right)_\times\mathbf{I}_{CMi}\left(\boldsymbol{\omega} + \mathbf{t}_i\mathbf{e}_z\dot{\theta}_i\right)$$
$$= \boldsymbol{\tau}_{sat} + \mathbf{M}_{sm} + \left(^R\mathbf{p}_{s0} - ^R\mathbf{p}_{se}\right) \times \sum_{i=1}^N \mathbf{F}_i. \tag{5.138}$$

Assuming the total external moment acting on the main satellite body is zero, so,

$$\boldsymbol{\tau}_{sat} + \mathbf{M}_{sm} + \left(^R\mathbf{p}_{s0} - {}^R\mathbf{p}_{se}\right) \times \sum_{i=1}^{N} \mathbf{F}_i = 0, \qquad (5.139)$$

it can be seen that the total moment of momentum of the multibody system about its center of mass in body-fixed coordinates $\mathbf{H}_B$ is zero, and consequently,

$$\mathbf{I}_{CM1}\boldsymbol{\omega} + \sum_{i=2}^{N} \mathbf{I}_{CMi}\left(\boldsymbol{\omega} + \mathbf{t}_i\mathbf{e}_z\dot{\theta}_i\right) = 0. \qquad (5.140)$$

Hence,

$$\left(\sum_{i=1}^{N} \mathbf{I}_{CMi}\right)\boldsymbol{\omega} + \sum_{i=2}^{N} \mathbf{I}_{CMi}\mathbf{t}_i\mathbf{e}_z\dot{\theta}_i = 0, \qquad (5.141)$$

which can be solved for $\boldsymbol{\omega}$. Thus $\boldsymbol{\omega}$ is given by,

$$\boldsymbol{\omega} = -\left(\sum_{i=1}^{N} \mathbf{I}_{CMi}\right)^{-1} \sum_{i=2}^{N} \mathbf{I}_{CMi}\mathbf{t}_i\mathbf{e}_z\dot{\theta}_i. \qquad (5.142)$$

Note that in the case when the initial angular momentum is nonzero,

$$\mathbf{H}_B = \mathbf{T}_{BI}\mathbf{T}_{IB_0}\sum_i \mathbf{I}_{Bi0}\boldsymbol{\omega}_{Bi0} = \mathbf{T}_{BI}(q_j)\mathbf{H}_{I_0}. \qquad (5.143)$$

In this equation, $\mathbf{H}_{I_0}$ is a vector constant. It follows that $\mathbf{H}_B$ is not zero and the right-hand side is a function of the attitude quaternion components of the main body of the satellite, $q_j, j = 1, 2, 3, 4$. In this case, the equation for $\boldsymbol{\omega}$ includes additional terms that are functions of the attitude quaternion components of the main body of the satellite. Thus,

$$\boldsymbol{\omega} = -\left(\sum_{i=1}^{N} \mathbf{I}_{CMi}\right)^{-1} \left\{\sum_{i=2}^{N} \mathbf{I}_{CMi}\mathbf{t}_i\mathbf{e}_z\dot{\theta}_i - \mathbf{T}_{BI}(q_j)\mathbf{H}_{I_0}\right\}. \qquad (5.144)$$

From Section 4.9 it follows that the equations of motion of the manipulator arms are given by,

$$\mathbf{e}_z^T\mathbf{t}_i^T\mathbf{I}_{CMi}(\dot{\boldsymbol{\omega}} + \mathbf{t}_i\mathbf{e}_z\ddot{\theta}_i) + \frac{d}{dt}\frac{\partial T_{tran}}{\partial \dot{\theta}_i} - \frac{\partial T_{tran}}{\partial \theta_i} + \sum_k \dot{\theta}_k\left(\frac{\partial(\mathbf{e}_z^T\mathbf{t}_i^T)}{\partial \theta_k} - \frac{\partial(\mathbf{e}_z^T\mathbf{t}_k^T)}{\partial \theta_i}\right) \cdot \mathbf{I}_{CMi}\boldsymbol{\omega}$$
$$+ \mathbf{e}_z^T\mathbf{t}_i^T\mathbf{I}_{CMi}\sum_k \left(\frac{\partial(\mathbf{t}_i\mathbf{e}_z)}{\partial \theta_k} - \frac{1}{2}\frac{\partial(\mathbf{t}_k\mathbf{e}_z)}{\partial \theta_i}\right)\dot{\theta}_k\dot{\theta}_i$$
$$+ \sum_k \dot{\theta}_k\left(\frac{\partial(\mathbf{e}_z^T\mathbf{t}_i^T)}{\partial \theta_k} - \frac{1}{2}\frac{\partial(\mathbf{e}_z^T\mathbf{t}_k^T)}{\partial \theta_i}\right) \cdot \mathbf{I}_{CMi}\mathbf{t}_i\mathbf{e}_z\dot{\theta}_i$$
$$= \mathbf{e}_z^T\mathbf{t}_i^T\boldsymbol{\tau}_i + \mathbf{e}_z^T\mathbf{t}_i^T\left(\mathbf{M}_i + \left(^R\mathbf{p}_{i0} - {}^R\mathbf{p}_{icm}\right) \times \mathbf{F}_i + \left(^R\mathbf{p}_{i0} - {}^R\mathbf{p}_{ie}\right) \times \sum_{i=i+1}^{N} \mathbf{F}_i\right),$$
$$(5.145)$$

with

$$\frac{d}{dt}\frac{\partial T_{tran}}{\partial \dot{\theta}_i} - \frac{\partial T_{tran}}{\partial \theta_i} = \sum_k \left(\ddot{\theta}_k \frac{\partial^2 T_{tran}}{\partial \dot{\theta}_k \partial \dot{\theta}_i} + \dot{\theta}_j \dot{\theta}_k \left(\frac{\partial}{\partial \theta_k} \frac{\partial^2 T_{tran}}{\partial \dot{\theta}_j \partial \dot{\theta}_i} - \frac{1}{2} \frac{\partial}{\partial \theta_i} \frac{\partial^2 T_{tran}}{\partial \dot{\theta}_j \partial \dot{\theta}_k} \right) \right),$$

and

$$\frac{\partial^2 T_{tran}}{\partial \dot{\theta}_k \partial \dot{\theta}_j} = \frac{1}{m_{total}} \left(\sum_{i=1}^N m_i \mathbf{v}_{i,j} \right) \left(\sum_{i=1}^N m_i \mathbf{v}_{i,k} \right).$$

From these equations one may eliminate the satellite angular velocity vector ω and obtain the equations of motion of the manipulator in terms of its joint angles only. In the case of nonzero initial angular momentum, there are additional terms in the manipulator equations, which are functions of the attitude quaternion components of the main body of the satellite as well as the joint angles and the joint angle rates. These terms that appear in the manipulator equations may be cancelled by applying initial decoupling control torques to the joints of the manipulator, provided the attitude quaternion components are available for feedback; i.e. they can either be measured or estimated. Thus, the manipulator's dynamics may be transformed such that they are completely decoupled from the main satellite body's attitude dynamics, as long as the condition of the conservation of the *initial angular momentum* at time $t = 0$ is satisfied. Moreover, all of the methods of controller synthesis that can be applied to fixed-base manipulators may now be applied to the manipulator on board the satellite.

References

[1] Brockett, R. W. (1983) Asymptotic stability and feedback stabilization. In *Differential Geometric Control Theory*, R. W. Brockett, R. S. Millmann, H. J. Sussmann, eds., Boston: Birkhauser, 181–191.
[2] Kolmanovsky, I. and McClamroch, N. H. (1995) Developments in nonholonomic control problems. *IEEE Control Systems Magazine*, 15: 20–36.
[3] Godhavn, J. M. and Egeland, O. (1997) A Lyapunov approach to exponential stabilization of nonholonomic systems in power form. *IEEE Transactions on Automatic Control*, 42(7): 1028–1032.
[4] Samson, C. (1995) Control of chained systems: Application to path following and time-varying point-stabilization of mobile robots. *IEEE Transactions On Automatic Control*, 40(1): 64–77.
[5] Astolfi, A. (1999) Exponential stabilization of a wheeled mobile robot via discontinuous control. *Journal of Dynamic Systems Measurement and Control*, 121: 121–126.
[6] Hespanha, J. P. (1996) Stabilization of nonholonomic integrators via logic-based switching, in *Proceedings of the 13th World Congress of IFAC.*, E., San Francisco, CA. 467–472.
[7] Aguiar, A. P. and Pascoal, A. (2000) Stabilization of the Extended nonholonomic double integrator via logic-based hybrid control, *Proceedings of the SYROCO'00 – 6th IFAC Symposium on Robot Control*, Vienna, Austria.

[8] Ye, H., Michel, A. N., and Hou, L. (1998) Stability theory for hybrid dynamical systems. *IEEE Transactions on Automatic Control*, 43(4): 461–474.

[9] Yoshikawa, T. (1985) Manipulability of robotic mechanisms. *The International Journal of Robotics Research*, 4: 3–9.

[10] Bayle, B., Fourquet, J. Y., and Renaud, M. (2001) Manipulability analysis for mobile manipulators, in *ICRA'2001*, Seoul, South Korea, 1251–1256.

[11] Nait-Chabane, K., Hoppenot, P., and Colle, E. (2007) Directional manipulability for motion coordination of an assistive mobile arm, *ICINCO, May 2007*, Angers, France, hal00341319.

[12] Vepa, R. (2009) *Biomimetic Robotics: Mechanisms and Control*, New York: Cambridge University Press.

[13] Papadopoulos, E. and Rey, D. (2000) The force-angle measure of tip-over stability margin for mobile manipulators. *Vehicle System Dynamics*, 33: 29–48.

[14] Moosavian, S. A. and Alipour, K. (2006) Moment-height tip-over measure for stability analysis of mobile robotic systems, in *Proceedings of the IEEE/RSJ International Conference on Intelligent Robots and Systems,* October 2006, 5546–5551.

[15] Yamamoto, Y. (1994) Coordinated control of a mobile manipulator, University of Pennsylvania Department of Computer and Information Science *Technical Report No. MS-CIS-94–12, March 1994*.

[16] Vepa, R. (2016) *Nonlinear Control of Robots and Unmanned Aerial Vehicles: An Integrated Approach*, Boca Raton, FL: CRC Press.

[17] Papadopoulos, E. and Dubowsky, S. (1991) Coordinated manipulator/spacecraft motion control for space robotic systems, in *IEEE International Conference on Robotics and Automation* (Sacramento, CA), 1696–1701.

[18] Nanos, K. and Papadopoulos, E. G. (2017) On the dynamics and control of free-floating space manipulator systems in the presence of angular momentum. *Frontiers in Robotics and AI*, 4: 26. doi: 10.3389/frobt.2017.00026.

6 Planetary Rovers and Mobile Robotics

6.1 Planetary Rovers: Architecture

After successfully landing on the Moon and on Mars, NASA also successfully deployed a planetary rover to explore the Martian landscape. In 1997 Mars Pathfinder was launched and landed on Mars, demonstrating a new way to land on the planet's surface, using large protective airbags and bouncing safely to complete the landing. Pathfinder was equipped with a host of sensors, including an imaging camera and meteorology sensors. It not only operated on the surface of the planet for 85 days but also delivered the first planetary rover, Sojourner, to the surface of the planet. The rover carried cameras and a spectrometer to measure properties of the Martian surface and soil. The Martian exploration rover Opportunity (see Figure 6.1) was sent in 2003 to explore a smooth region near the Martian equator known as the Meridiani Planum [1]. The Mars exploration rover Spirit was sent in search of water on the Martian surface six months later and to explore the depths of what was considered to have been a lake, now a 140 km-wide crater. The Mars Science Laboratory rover Curiosity (Figure 6.2) was landed [2] on the edge of a crater known as the Gale Crater in 2012. Three years later it travelled a distance of 12 km on the surface of the crater. The primary task of an autonomous planetary rover is to safely move forward and explore the surrounding region, without direct contact with its mission operators. The rover is expected to sense, identify, and perceive information about its unstructured and unknown surroundings while operating in and adapting to an unexplored environment. In particular, it is required to identify and map traversable paths in the region around it as well as maintain a log of such paths for future reuse.

The Mars Exploration Rovers (MER) were essentially solar powered [1] with an onboard battery for storing and regulating the power supply, and an optional Radioisotope Thermoelectric Generator (RTG) power source. Power conditioning and regulation was performed by a redundant set of power avionics modules. A comprehensive sensor and instrument package included panoramic cameras, navigation cameras, and thermal emission spectrometers mounted on a mast. Cameras for hazard detection were mounted in the front and rear of the vehicle. A platform mounted on the vehicle carried a single robotic arm with an imager and Mössbauer and alpha particle X-ray spectrometers as well as a rock abrasion tool mounted on the arm. The rovers were fitted with telecommunications systems for communications directly with the Earth and with the mother satellite orbiting the planet, as well as high and low gain antennae and a UHF antenna. Communications were affected depending on the rover's operating position and orbit

Figure 6.1 NASA's Mars Exploration Rover. (courtesy: NASA)

Figure 6.2 NASA's Mars Science Laboratory. (courtesy: NASA)

schedules. An onboard computer that could be switched on when needed and off when not in use was employed for all data manipulations, system state monitoring, mission planning, and obstacle avoidance.

6.1.1 Vehicle Dynamics and Control

The design of a planetary rover involves not only the design of the mission and its analysis, but also the general architectural design of the rover, sensing and navigation, obstacle avoidance, and mission and path planning. These issues can be dealt with only if the complete motion kinematics, dynamic modeling, including the models of slip and traction mechanics, and the control of the vehicle are fully understood in some detail. However, very few published papers have dealt with motion dynamics of rovers in detail, mainly because the subject is in its infancy. In an effort to highlight these issues, we shall devote an entire section of this chapter to vehicle dynamic modeling and control. We will focus on the successful architectures of planetary rovers that have emerged over the last decade.

6.1.2 Mission Planning

Typically, to navigate through natural terrain an autonomous planetary rover must be endowed with the ability to sense and perceive its environment, plan and traverse a feasible course through that environment, and react appropriately to any unpredictable situations as and when they appear. The goal is to navigate autonomously from the current position or pose to an autonomously specified position or pose located beyond the sensing horizon of the rover. To achieve the goal, it is assumed that a global map of the terrain is available from satellite imagery, previous missions, or from data collected during descent. Based on the experience gained [3] and reported by successful planetary rover missions to Mars [4] and other tests on proposed planetary rovers [5], the goal is achieved in several distinct steps, which are broadly classified into two phases. The first phase is known as mission planning, which is shown in block diagram form in Figure 6.3, and involves seeking to derive a set of routes for the rover exploring the planetary surface. The second phase is path planning, path following, and control, and is shown in Figure 6.4.

Mission planning is intended to plan not only for obstacle avoidance but also for time, power, communication, and other resource limitations or constraints. A key step in mission planning is autonomous path selection from a terrain model acquired from sensor data. A key issue that must be considered is whether or not terrain is traversable. Whether terrain is traversable is affected by the rover's mobile performance capabilities and the actual terrain gradient. The main sensor suite used for terrain assessment by planetary rovers that have been successfully deployed on planets like Mars have been a set of stereo vision cameras. The terrain models obtained through stereo imagery are used for both autonomous terrain assessment and visual odometry. In the case of autonomous terrain assessment, a cluster of three-dimensional waypoints are used to

Figure 6.3 Mission planning in block diagram form.

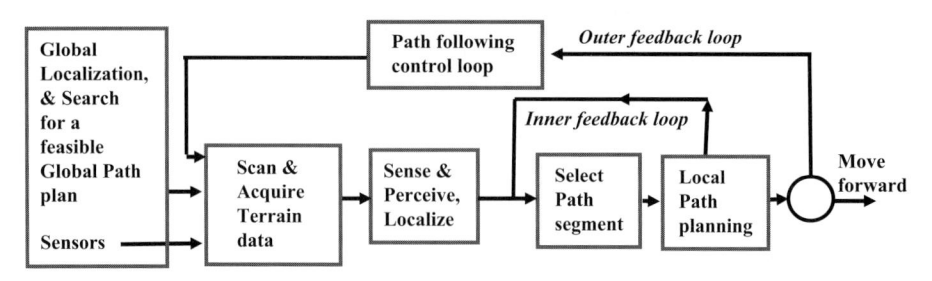

Figure 6.4 Path planning, path following, and feedback control loop.

evaluate the traversability of the terrain immediately in front of the rover, defined as a regular grid of triangular or rectangular patches. In the case of visual odometry, the model is used to match, estimate, identify, and track features of the terrain to mitigate the effect of slip. Once a set of feasible paths can be established from a starting location and with a particular orientation or pose to a goal pose, it becomes important to select one feasible, optimal path for moving the rover along. This is the first step in path planning. The main problem in traversing natural terrain autonomously is that the information available to the vehicle about its environment is necessarily noisy and incomplete. Therefore, the path planning algorithm must be as robust as possible in the face of incomplete and inaccurate data. Furthermore, the vehicle must react appropriately and in real time to unexpected situations that will inevitably arise. Thus, the objective of robot path planning is to search for a trajectory that achieves a goal state while avoiding contact with obstacles and to determine a feasible or optimal route between start waypoint and goal waypoint. The search is carried out by such algorithms as the $A*$, a heuristic search algorithm, $D*$ and its variants, and *Dijkstra's breadth-first shortest path algorithm* [6]. Other algorithms that have been proposed are based on *simulated annealing* and *genetic algorithms*. What may be considered feasible or optimal varies with the specific application region and planet. Once a feasible path that is deemed optimal is selected, the trajectory is planned through the representation of the local scan that the rover must follow to navigate to the next way-point. At this stage, path planning algorithms follow three basic approaches – cell decomposition into cells of rectangular shape, road-mapping methods through the feasible grid, and potential field techniques for path following. The latter technique was discussed in the preceding chapter. At the current state, the localization and pose estimation from the Inertial Measurement Unit (IMU) and the odometer, combined with the trajectory length in the order of tens of meters, allows safe navigation of the open loop before re-localizing prior to the next successive scan.

6.1.3 Propulsion and Locomotion

Rovers are expected to offer exceptional versatility in being able to negotiate any kind of terrain [7]. Thus, they are expected to be able to smoothly switch between walking and rolling, as and when required. A planetary rover must not only be able to traverse irregular and treacherous terrain in a timely and efficient manner but also to travel to and return from areas deemed too dangerous for human exploration. At the same time, the rover must also be able to adapt to the nature of the current terrain that is encountered. Although several rover architectures have been proposed by the Jet Propulsion Laboratory (JPL) and others, only a few of these have been successfully demonstrated in interplanetary missions to Mars.

Not only have Lithium-ion batteries [8] been successfully deployed on NASA's Mars Rovers, Spirit and Opportunity but they have also been combined with triple-junction solar cells to power the rovers and primarily assist in keeping the rover environment warm enough.

6.1.4 Planetary Navigation

NASA's JPL [9] has developed a navigation system for planetary rovers called Terrain Adaptive Navigation (TANav), which was designed to enable greater access to and more robust operations within terrains of widely varying slippage. The TANav system uses onboard stereo cameras to remotely classify surrounding terrain, predict the slippage of that terrain, and use this information in the planning of a path to the goal. This navigation system employs several new and integrated techniques for map generation, terrain classification, slip prediction, path planning, high-fidelity traversability analysis, and slip-compensated path following. Nonlinear model predictive techniques have since been successfully applied to this problem [10].

6.2 Dynamic Modeling of Planetary Rovers

6.2.1 Non-Holonomic Constraints

We consider a half-body model of a general front-wheel steer, rear-wheel drive vehicle as shown in Figure 6.5. The general non-holonomic constraint in terms of the planar wheel velocity components (x, y) and the yaw angle of the vehicle ψ is of the form,

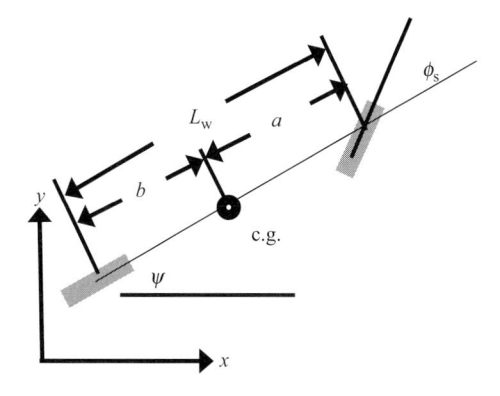

$$\dot{x} \sin \psi - \dot{y} \cos \psi = 0. \tag{6.1}$$

Figure 6.5 General coordinates of an autonomous rover.

In terms of the position coordinates of the center of gravity of the platform (x_c, y_c) and the vehicle yaw angle ψ the position coordinates of the mid-points of the rear and front axle are respectively given by,

$$(x_1, y_1) = (x_c, y_c) - b(\cos\psi, \sin\psi), \tag{6.2}$$

$$(x_2, y_2) = (x_c, y_c) + a(\cos\psi, \sin\psi). \tag{6.3}$$

The corresponding velocity components are,

$$(\dot{x}_1, \dot{y}_1) = (\dot{x}_c, \dot{y}_c) + b\dot\psi(\sin\psi, -\cos\psi), \tag{6.4}$$

$$(\dot{x}_2, \dot{y}_2) = (\dot{x}_c, \dot{y}_c) - a\dot\psi(\sin\psi, -\cos\psi). \tag{6.5}$$

The non-holonomic constraints for each wheel are,

$$\dot{x}_1 \sin\psi - \dot{y}_1 \cos\psi = 0, \quad \dot{x}_2 \sin(\psi + \phi_s) - \dot{y}_2 \cos(\psi + \phi_s) = 0. \tag{6.6}$$

In this equation, ϕ_s is the steering angle. Thus, in terms of the velocity components of the center of gravity of the platform,

$$\dot{x}_c \sin\psi - \dot{y}_c \cos\psi + b\dot\psi = 0, \tag{6.7}$$

$$\dot{x}_c \sin(\psi + \phi_s) - \dot{y}_c \cos(\psi + \phi_s) + a\dot\psi \cos\psi = 0. \tag{6.8}$$

The components of the velocity vector in the body axes are,

$$\dot{x}_c = u\cos\psi - v\sin\psi, \quad \dot{y}_c = u\sin\psi + v\cos\psi, \quad u = \dot{x}_c \cos\psi + \dot{y}_c \sin\psi,$$
$$v = -\dot{x}_c \sin\psi + \dot{y}_c \cos\psi. \tag{6.9}$$

The velocity components are,

$$(\dot{x}_1, \dot{y}_1) = (u\cos\psi - (v - b\dot\psi)\sin\psi, u\sin\psi + (v - b\dot\psi)\cos\psi), \tag{6.10}$$

$$(\dot{x}_2, \dot{y}_2) = (u\cos\psi - (v + a\dot\psi)\sin\psi, u\sin\psi + (v + a\dot\psi)\cos\psi). \tag{6.11}$$

Thus, in terms of the velocity components in the body axes of the center of gravity of the platform,

$$v - b\dot\psi = 0, \quad l\dot\psi - u\tan\phi_s = 0, \quad l = a + b. \tag{6.12}$$

Hence,

$$v = (b/l)u \tan\phi_s. \tag{6.13}$$

Thus, both v and $\dot\psi$ may be eliminated in terms of ϕ_s and u.

The velocity components reduce to,

$$(\dot{x}_1, \dot{y}_1) = u(\cos\psi, \sin\psi), \tag{6.14}$$

$$(\dot{x}_2, \dot{y}_2) = u(\cos\psi - \tan\phi_s \sin\psi, \sin\psi + \tan\phi_s \cos\psi). \tag{6.15}$$

In the (u, v) frame shown in Figure 6.6, the corresponding velocity components for the rear and front axles are $(u, 0)$ and $(u, u\tan\phi_s)$.

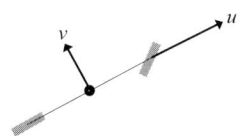

Figure 6.6 Rover body coordinates.

The rear and front wheel velocities are therefore,

$$u_r = u, \quad u_f = u \cos\phi_s + u \tan\phi_s \sin\phi_s = u_r / \cos\phi_s. \tag{6.16}$$

This provides yet another constraint for relating the wheel velocities of the front wheels to the rear wheels.

6.2.2 Vehicle Generalized Forces

One may now establish the generalized forces and moments acting on the vehicle. It is assumed that there is no friction force between the wheels and the vehicle, and that the rear wheels are locked to be in the same orientation as the vehicle. The forces acting on the rear wheel are the driving force F_D, the tangential friction force between the wheel and the ground F_{Rtf}, and the normal side force on the tire F_{Rnt}. The forces acting on the front wheel are the tangential friction force F_{Ftf} and the normal side force on the tire F_{Fnt}. The issue of wheel slippage will be dealt with later, when the issue of generating the driving force and its relationship to the wheel dynamics is considered. Resolving the rear wheel forces in the (u, v) frame,

$$F_{Ru} = F_D - F_{Rtf}, \quad F_{Rv} = F_{Rnt}. \tag{6.17}$$

Similarly, for the front wheel we have,

$$F_{Fu} = -F_{Ftf} \cos\phi - F_{Fnt} \sin\phi, \quad F_{Fv} = F_{Fnt} \cos\phi - F_{Ftf} \sin\phi. \tag{6.18}$$

The total driving and ground forces on the vehicle are,

$$F_u = F_D - F_{Rtf} - F_{Ftf} \cos\phi - F_{Fnt} \sin\phi, \tag{6.19}$$

$$F_v = F_{Rnt} + F_{Fnt} \cos\phi - F_{Ftf} \sin\phi. \tag{6.20}$$

The yawing moment acting on the vehicle is,

$$M_\psi = \left(F_{Fnt} \cos\phi - F_{Ftf} \sin\phi\right)a - bF_{Fnt}. \tag{6.21}$$

6.2.3 Modeling the Suspension System and Limbs

Consider a two-link manipulator on a mobile wheel base as shown in Figure 6.7, which is assumed to be a model of a typical limb. The rover or vehicle platform is assumed to be supported on four or more such limbs. The wheel base could be a simple rocker arm type or a swing arm type. The wheel base dynamics are generally decoupled from the rest of the vehicle and will be considered later.

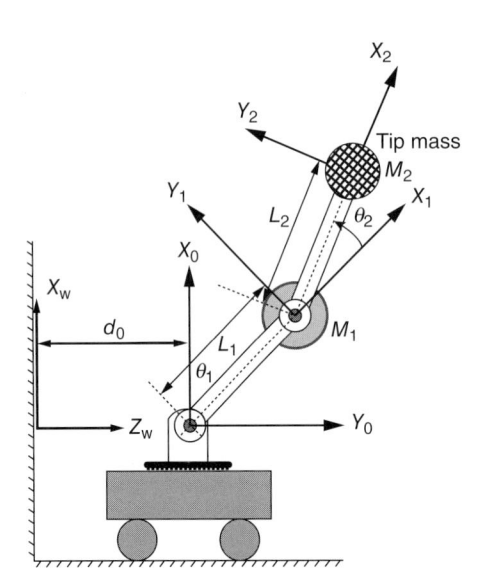

Figure 6.7 Two-link manipulator on a wheel base or cart.

The objective here is to obtain the total kinetic and potential energies of the arm in terms of the moment of inertia and mass-moment components. The local coordinate frames are already indicated. To adopt a step-by-step approach,

Step 1: Observe the coordinate frames;
Step 2: Obtain the position coordinates of the centers of mass of each mass.

There are five masses to consider:
The bottom link or link 1, the top link or link 2, the bottom lumped mass or lumped mass 1, the top lumped mass or lumped mass 2, and the wheel-base.

Step 3: Obtain the velocities of the centers of mass of each mass.

We may then obtain the velocities and the heights. Although the wheel base's forward path is assumed to be horizontal, the terrain may well be at a gradient to the surface normal to the direction of the local gravity vector. The terrain gradient angle is assumed to be θ_g;

Step 4: Obtain the energies and the Lagrangian.

The translational kinetic energy of the five masses is,

$$T_1 = \frac{1}{2}m_1\left(\dot{x}_{1cm}^2 + \dot{z}_{1cm}^2\right) + \frac{1}{2}m_2\left(\dot{x}_{2cm}^2 + \dot{z}_{2cm}^2\right) \\ + \frac{1}{2}M_1\left(\dot{x}_{3cm}^2 + \dot{z}_{3cm}^2\right) + \frac{1}{2}M_2\left(\dot{x}_{4cm}^2 + \dot{z}_{4cm}^2\right) + \frac{1}{2}M_{w\_base}\left(\dot{x}_{5cm}^2 + \dot{z}_{5cm}^2\right). \quad (6.22)$$

The rotational kinetic energy of the two rigid bodies, the manipulator links, is,

$$T_2 = \frac{1}{2}m_1 k_1^2 \dot{\theta}_1^2 + \frac{1}{2}m_2 k_2^2\left(-\dot{\theta}_1 + \dot{\theta}_2\right)^2. \quad (6.23)$$

After simplifying,

$$T_1 = \frac{1}{2}(m_1 + m_2 + M_1 + M_2 + M_{w\_base})\dot{d}_0^2$$
$$+ \frac{1}{2}\left(m_1 L_{1cm}^2 + M_1 L_1^2 + (m_2 + M_2)L_1^2\right)\dot{\theta}_1^2$$
$$+ \frac{1}{2}\left(m_2 L_{2cm}^2 + M_2 L_2^2\right)\left(\dot{\theta}_1 - \dot{\theta}_2\right)^2$$
$$+ \Gamma_1 \dot{d}_0 \dot{\theta}_1 \cos\theta_1 + \Gamma_2 (\dot{\theta}_1 - \dot{\theta}_2)\left(\dot{d}_0 \cos(\theta_1 - \theta_2) + L_1 \dot{\theta}_1 \cos(\theta_2)\right). \quad (6.24)$$

If we let,

$$M_{Leg} = m_1 + m_2 + M_1 + M_2 + M_{w\_base}, \quad (6.25)$$

$$I_{11} = m_1 L_{1cm}^2 + M_1 L_1^2 + (m_2 + M_2)L_1^2 + m_1 k_1^2, \quad (6.26)$$

$$I_{22} = m_2 L_{2cm}^2 + M_2 L_2^2 + m_2 k_2^2, \quad (6.27)$$

the total kinetic energy of the two-link manipulator model of a limb is,

$$T_l = T_1 + T_2 = \frac{1}{2} M_{Leg} \dot{d}_0^2 + \frac{1}{2} I_{11} \dot{\theta}_1^2 + \frac{1}{2} I_{22} (\dot{\theta}_1 - \dot{\theta}_2)^2$$
$$+ \Gamma_1 \dot{d}_0 \dot{\theta}_1 \cos\theta_1 + \Gamma_2 (\dot{\theta}_1 - \dot{\theta}_2)\left(\dot{d}_0 \cos(\theta_1 - \theta_2) + L_1 \dot{\theta}_1 \cos(\theta_2)\right). \quad (6.28)$$

The potential energies of the masses are evaluated from the heights of the masses from a reference plane normal to the gravity vector. Thus,

$$V_l = m_1 g\left[L_{1cm} \cos\theta_{1g}\right] + m_2 g\left[L_1 \cos\theta_{1g} + L_{2cm} \sin(90 - \theta_{1g} + \theta_2)\right]$$
$$+ M_1 g L_1 \cos\theta_{1g} + M_2 g\left[L_1 \cos\theta_{1g} + L_2 \sin(90 - \theta_{1g} + \theta_2)\right] \quad (6.29)$$
$$+ M_{w\_base} g d_0 \sin\theta_g.$$

It follows that,

$$V_l = (m_1 L_{1cm} + m_2 L_1 + M_1 L_1 + M_2 L_1) g \cos\theta_{1g}$$
$$+ (m_2 L_{2cm} + M_2 L_2) g \sin(90 - \theta_{1g} + \theta_2) + M_{w\_base} g \sin\theta_g \quad (6.30)$$
$$\equiv \Gamma_1 g \cos\theta_{1g} + \Gamma_2 g \sin(90 - \theta_{1g} + \theta_2) + M_{w\_base} g d_0 \sin\theta_g,$$

where $\theta_{1g} = \theta_1 + \theta_g$, $\Gamma_1 = (m_1 L_{1cm} + m_2 L_1 + M_1 L_1 + M_2 L_1)$, $\Gamma_2 = (m_2 L_{2cm} + M_2 L_2)$. The wheel-base does contribute to the potential energy due to the terrain gradient angle θ_g. Hence, the Lagrangian may be defined as, $L = T_l - V_l$ and it follows that,

$$L = \frac{1}{2} M_{Leg} \dot{d}_0^2 + \frac{1}{2} I_{11} \dot{\theta}_1^2 + \frac{1}{2} I_{22} (\dot{\theta}_1 - \dot{\theta}_2)^2$$
$$+ \Gamma_1 \dot{d}_0 \dot{\theta}_1 \cos\theta_1 + \Gamma_2 (\dot{\theta}_1 - \dot{\theta}_2)\left(\dot{d}_0 \cos(\theta_1 - \theta_2) + L_1 \dot{\theta}_1 \cos(\theta_2)\right) \quad (6.31)$$
$$- \Gamma_1 g \cos\theta_{1g} - \Gamma_2 g \sin(90 - \theta_{1g} + \theta_2) - M_{w\_base} g \sin\theta_g.$$

The space rover platform is assumed to be supported by four such "legs." The legs are attached to the platform by a universal or a ball and socket joint. Each leg is modeled as above the manipulator, and each leg height, for each of the four legs, $j = 1, 2...4$,

$$h_{L,j} = h_{4,cm,j} = L_{1j} \cos \theta_{1gj} + L_{2j} \sin \left(90 - \theta_{1gj} + \theta_{2j}\right), \tag{6.32}$$

where the addition subscript j is used to identify one of the four legs. The legs and their components are numbered in a clockwise manner, starting with the rear left leg and ending with the rear right leg. The mean height of the platform is,

$$h_p = \frac{1}{2L_w} \left(ah_{L,1,left} + bh_{L,2,left} + bh_{L,3,right} + ah_{L,4,right}\right), \tag{6.33}$$

where L_w is the longitudinal distance between the two left wheels, which is assumed for simplicity to also be the longitudinal distance between the two right wheels.

Based on the small angle assumption, the platform roll angle is,

$$\phi = \frac{1}{2b_w} \left(h_{L,1,left} + h_{L,2,left} - h_{L,3,right} - h_{L,4,right}\right), \tag{6.34}$$

where b_w is the distance between the two rear wheels, which is assumed for simplicity to also be the distance between the two front wheels. Similarly, based on the small angle assumption, the nose-up pitch angle is,

$$\theta = \frac{1}{2L_w} \left(h_{L,2,left} - h_{L,1,left} - h_{L,4,right} + h_{L,3,right}\right). \tag{6.35}$$

Now, as our interest is to develop the complete nonlinear dynamics of the rover, using the small angle assumption is inadequate. Thus, the orientation of the platform must be defined by an Euler angle sequence. Based on the Euler angle sequence, one can obtain the rotational transformation relating a frame fixed in space to the frame fixed in the platform.

The rotation from one frame to another can be visualized as a sequence of three simple rotations about base vectors. Each rotation is through an angle (Euler angle) about a specified axis. To describe the attitude of the platform we shall adopt a 3–1–2 Euler angle sequence. In the case of the 3–1–2 Euler angle sequence, the first rotation is about the third axis, through angle ψ, followed by the second rotation about the first axis through angle ϕ, and then followed by a final rotation again about the third axis by angle θ. Thus, considering an initial space-fixed reference frame $\eta = \frac{\partial u}{\partial z} - \frac{\partial w}{\partial x}$, and assuming that the latter frame is then rotated about Oz axis by an angle ψ, the transformation relating the transformed axes to the initial axes is,

$$\begin{bmatrix} x' \\ y' \\ z' \end{bmatrix} = \begin{bmatrix} \cos \psi & \sin \psi & 0 \\ -\sin \psi & \cos \psi & 0 \\ 0 & 0 & 1 \end{bmatrix} \begin{bmatrix} x \\ y \\ z \end{bmatrix}. \tag{6.36}$$

Following the rotation about the Oz axis, the Ox'-Oy'-Oz' is rotated about the Ox' axis by an angle ϕ. The transformed reference frame is denoted by the frame $Ox'\prime$-$Oy'\prime$-$Oz'\prime$. The expression for the three-axes transformation matrix representing the second rotation is,

$$\begin{bmatrix} x'' \\ y'' \\ z'' \end{bmatrix} = \begin{bmatrix} 1 & 0 & 0 \\ 0 & \cos \phi & \sin \phi \\ 0 & -\sin \phi & \cos \phi \end{bmatrix} \begin{bmatrix} x' \\ y' \\ z' \end{bmatrix}. \tag{6.37}$$

Following the successive rotations about the Oz and Ox' axes, the frame $Ox'\prime\text{-}Oy'\prime\text{-}Oz'\prime$ is rotated about the $Oy'\prime$ axis by an angle θ. The transformed reference frame is attached to a body and defines the body orientation. It is denoted by the frame $Ox_b\text{-}Oy_b\text{-}Oz_b$. The expression for the three-axes transformation matrix representing the third rotation is,

$$\begin{bmatrix} x_b \\ y_b \\ z_b \end{bmatrix} = \begin{bmatrix} \cos\theta & 0 & -\sin\theta \\ 0 & 1 & 0 \\ \sin\theta & 0 & \cos\theta \end{bmatrix} \begin{bmatrix} x'' \\ y'' \\ z'' \end{bmatrix}. \tag{6.38}$$

Hence,

$$\mathbf{T}_{BI} = \begin{bmatrix} \cos\theta & 0 & -\sin\theta \\ 0 & 1 & 0 \\ \sin\theta & 0 & \cos\theta \end{bmatrix} \begin{bmatrix} 1 & 0 & 0 \\ 0 & \cos\phi & \sin\phi \\ 0 & -\sin\phi & \cos\phi \end{bmatrix} \begin{bmatrix} \cos\psi & \sin\psi & 0 \\ -\sin\psi & \cos\psi & 0 \\ 0 & 0 & 1 \end{bmatrix} = \mathbf{T}_{IB}^T, \tag{6.39}$$

and

$$\mathbf{T}_{BI} \begin{bmatrix} 0 \\ 0 \\ 1 \end{bmatrix} = \begin{bmatrix} -\sin\theta\cos\phi \\ \sin\phi \\ \cos\theta\cos\phi \end{bmatrix} = ([0\ 0\ 1]\mathbf{T}_{IB})^T. \tag{6.40}$$

The angular velocity vector of the body in the body-fixed axes is given by,

$$\omega = \dot\theta [0\ 1\ 0] \begin{bmatrix} i_b \\ j_b \\ k_b \end{bmatrix} + \dot\phi [1\ 0\ 0] \begin{bmatrix} i'' \\ j'' \\ k'' \end{bmatrix} + \dot\psi [0\ 0\ 1] \begin{bmatrix} i' \\ j' \\ k' \end{bmatrix}, \tag{6.41}$$

or,

$$\omega = \dot\theta j_b + \dot\phi i'' + \dot\psi k', \tag{6.42}$$

where $[i_b\ j_b\ k_b]^T$ are unit vectors in the three-axes directions of the body-fixed frame, while $[i'\ j'\ k']^T$ are the corresponding unit vectors in the $[x'\ y'\ z']^T$ frame, while the unit vectors $[i''\ j''\ k''']^T$ correspond to the $[x''\ y''\ z'']^T$ frame. Now,

$$\begin{bmatrix} i'' \\ j'' \\ k'' \end{bmatrix} = \begin{bmatrix} c\theta & 0 & -s\theta \\ 0 & 1 & 0 \\ s\theta & 0 & c\theta \end{bmatrix} \begin{bmatrix} i_b \\ j_b \\ k_b \end{bmatrix}, \tag{6.43}$$

and

$$\begin{bmatrix} i' \\ j' \\ k' \end{bmatrix} = \begin{bmatrix} 1 & 0 & 0 \\ 0 & c\phi & s\phi \\ 0 & -s\phi & c\phi \end{bmatrix} \begin{bmatrix} c\theta & 0 & -s\theta \\ 0 & 1 & 0 \\ s\theta & 0 & c\theta \end{bmatrix} \begin{bmatrix} i_b \\ j_b \\ k_b \end{bmatrix}. \tag{6.44}$$

$$\omega = \dot\theta [0\ 1\ 0] \begin{bmatrix} i_b \\ j_b \\ k_b \end{bmatrix} + \dot\phi [1\ 0\ 0] \begin{bmatrix} c\theta & 0 & -s\theta \\ 0 & 1 & 0 \\ s\theta & 0 & c\theta \end{bmatrix} \begin{bmatrix} i_b \\ j_b \\ k_b \end{bmatrix}$$
$$+ \dot\psi [0\ 0\ 1] \begin{bmatrix} 1 & 0 & 0 \\ 0 & c\phi & s\phi \\ 0 & -s\phi & c\phi \end{bmatrix} \begin{bmatrix} c\theta & 0 & -s\theta \\ 0 & 1 & 0 \\ s\theta & 0 & c\theta \end{bmatrix} \begin{bmatrix} i_b \\ j_b \\ k_b \end{bmatrix}. \tag{6.45}$$

Hence,

$$\omega = \dot{\theta} j_b + \dot{\phi}(i_b \cos\theta - k_b \sin\theta) + \dot{\psi}\cos\phi(i_b\sin\theta + k_b\cos\theta) - \dot{\psi}j_b\sin\phi, \quad (6.46)$$

and it follows that,

$$\omega = \begin{bmatrix} p \\ q \\ r \end{bmatrix} = \begin{bmatrix} \dot{\phi}\cos\theta + \dot{\psi}\cos\phi\sin\theta \\ \dot{\theta} - \dot{\psi}\sin\phi \\ \dot{\psi}\cos\phi\cos\theta - \dot{\phi}\sin\theta \end{bmatrix} = \begin{bmatrix} \cos\phi\sin\theta & \cos\theta & 0 \\ -\sin\phi & 0 & 1 \\ \cos\phi\cos\theta & -\sin\theta & 0 \end{bmatrix} \begin{bmatrix} \dot{\psi} \\ \dot{\phi} \\ \dot{\theta} \end{bmatrix}. \quad (6.47)$$

Position coordinates $\mathbf{x}_{Lp,j}$ for $j = 1,2...4$ of the four leg attachment points on the platform in a frame fixed to the platform body are,

$$[-b \quad b_w/2 \quad 0], \quad [a \quad b_w/2 \quad 0], \quad [a \quad -b_w/2 \quad 0], \quad [-b \quad -b_w/2 \quad 0]. \quad (6.48)$$

Then it follows that in a space-fixed frame, and for $j = 1,2...4$, the leg attachment points are,

$$h_{ap,j} = h_p + [0 \quad 0 \quad 1]\mathbf{T}_{IB}\mathbf{x}_{Lp,j}. \quad (6.49)$$

For the 3–1–2 Euler angle sequence,

$$\mathbf{T}_{BI}\begin{bmatrix} 0 \\ 0 \\ 1 \end{bmatrix} = \begin{bmatrix} -\sin\theta\cos\phi \\ \sin\phi \\ \cos\theta\cos\phi \end{bmatrix} = ([0 \quad 0 \quad 1]\mathbf{T}_{IB})^T. \quad (6.50)$$

Thus,

$$h_{ap,j} = h_p + [-\sin\theta\cos\phi \quad \sin\phi \quad \cos\theta\cos\phi]\mathbf{x}_{Lp,j}. \quad (6.51)$$

Hence it follows that,

$$\sin\phi = \frac{1}{2b_w}\left(h_{ap,1,left} + h_{ap,2,left} - h_{ap,3,right} - h_{ap,4,right}\right), \quad (6.52)$$

and with $L_w = a + b$, that,

$$\sin\theta = \frac{1}{2L_w\cos\phi}\left(h_{ap,2,left} - h_{ap,1,left} - h_{ap,4,right} + h_{ap,3,right}\right). \quad (6.53)$$

Finally, the yaw angle is a function of the steering angle; assuming front or rear wheel steering as covered in an earlier subsection, it can be expressed in terms of other variables in the energy expressions.

6.2.4 Platform Kinetic and Potential Energies

Assuming that the velocities of the platform are defined in a frame fixed to the body of the platform, the translational kinetic energy of the platform is:

$$T_{uvw} = (1/2)m(u^2 + v^2 + w^2), \quad w = \dot{h}_p, \quad v = (b/L_w)u\tan\phi_s. \quad (6.54)$$

Hence,

$$T_{uvw} = (1/2)mu^2\left(1 + (b/L_w)^2 \tan^2\phi_s\right) + (1/2)m\dot{h}_p^2. \qquad (6.55)$$

Assuming a suitable moment of inertia matrix for the platform, the rotational kinetic energy of the platform is,

$$T_{pqr} = \frac{1}{2}\begin{bmatrix}p\\q\\r\end{bmatrix}^T\begin{bmatrix}I_{xx} & -I_{xy} & 0\\-I_{xy} & I_{yy} & 0\\0 & 0 & I_{zz}\end{bmatrix}\begin{bmatrix}p\\q\\r\end{bmatrix}. \qquad (6.56)$$

In this equation,

$$\omega = \begin{bmatrix}p\\q\\r\end{bmatrix} = \begin{bmatrix}\dot\phi\cos\theta + \dot\psi\cos\phi\sin\theta\\ \dot\theta - \dot\psi\sin\phi\\ \dot\psi\cos\phi\cos\theta - \dot\phi\sin\theta\end{bmatrix} = \begin{bmatrix}\cos\phi\sin\theta & \cos\theta & 0\\ -\sin\phi & 0 & 1\\ \cos\phi\cos\theta & -\sin\theta & 0\end{bmatrix}\begin{bmatrix}\dot\psi\\\dot\phi\\\dot\theta\end{bmatrix} \equiv \mathbf{L}(\phi,\theta)\begin{bmatrix}\dot\psi\\\dot\phi\\\dot\theta\end{bmatrix}. \qquad (6.57)$$

Hence,

$$T_{pqr} = \frac{1}{2}\begin{bmatrix}\dot\psi\\\dot\phi\\\dot\theta\end{bmatrix}^T \mathbf{L}^T(\phi,\theta)\begin{bmatrix}I_{xx} & -I_{xy} & 0\\-I_{xy} & I_{yy} & 0\\0 & 0 & I_{zz}\end{bmatrix}\mathbf{L}(\phi,\theta)\begin{bmatrix}\dot\psi\\\dot\phi\\\dot\theta\end{bmatrix}, \qquad (6.58)$$

with $\dot\psi = (u/L_w)\tan\phi_s$. The potential energy of the platform is:

$$V_p = mgh_p. \qquad (6.59)$$

Note that,

$$h_p = \frac{1}{2L_w}\left(ah_{ap,1} + bh_{ap,2} + bh_{ap,3} + ah_{ap,4}\right), \qquad (6.60)$$

$$\sin\phi = \frac{1}{2b_w}\left(h_{ap,1} + h_{ap,2} - h_{ap,3} - h_{ap,4}\right), \qquad (6.61)$$

and that,

$$\sin\theta = \frac{1}{2L_w\cos\phi}\left(h_{ap,2} - h_{ap,1} - h_{ap,4} + h_{ap,3}\right). \qquad (6.62)$$

We shall assume further that the platform is directly supported by the limbs and that there are no spring-supported suspensions between the limbs and the platform. These could easily be introduced if and when necessary, although one needs to introduce additional degrees of freedom and add the potential energy stored in the suspension system to the total potential energy. The additional potential energy due to the suspension system takes the form,

$$V_{sus} = \frac{1}{2}\sum_{j=1}^{N_b} K_j\left(h_{ap,j} - h_{L,j}\right)^2. \qquad (6.63)$$

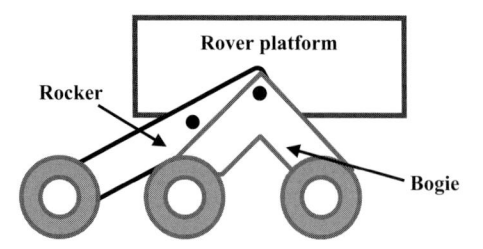

Figure 6.8 The rocker-bogie suspension modeled as a two-link manipulator.

Thus, in the absence of spring-supported suspensions between the limbs and the platform,

$$h_{ap,j} = h_{L,j} \equiv L_{1j} \cos\theta_{1gj} + L_{2j} \sin(90 - \theta_{1gj} + \theta_{2j}). \tag{6.64}$$

These constraint equations are used to eliminate the platform's orientation angles ϕ and θ, as well as the plunging displacement h_p. Thus, the platform introduces just one other new degree of freedom, u, the velocity of the rear axle of the rover.

The rocker-bogie suspension system (Figure 6.8), which was specifically designed for space exploration vehicles, has been extensively used on Mars exploration rovers. The "rocker" arm is one of the larger links on each side of the vehicle, and constitutes a suspension system that supports a "bogie" on either one or the other end of the arm. The two rockers on either side of the vehicle are connected to each other and the vehicle chassis through differential-like gears. The "bogie" refers to the conjoining links that have a drive wheel attached at each end. The bogie is attached to the rocker via a revolute joint. The rocker-bogie system is, therefore, kinematically equivalent to a two-link manipulator with three or more wheels, some of which are independently driven. A rocker-bogie suspension that is not spring-supported facilitates the rover to climb over obstacles.

6.2.5 Assembling the Vehicle's Kinetic and Potential Energies

It is now possible to assemble the complete Lagrangian for the rover or vehicle, assuming that there are N_b limbs supporting the platform. It will be assumed that N_b is four, although two are adequate if each is a rocker-bogie system. Thus, the complete Lagrangian is,

$$\begin{aligned} L = &\frac{1}{2}\sum_{j=1}^{N_b} M_{Leg,j} \dot{d}_{0,j}^2 + \sum_{j=1}^{N_b} \dot{d}_{0j}\left(\Gamma_{1j}\dot{\theta}_{1j}\cos\theta_{1j} + \Gamma_{2j}(\dot{\theta}_{1j} - \dot{\theta}_{2j})\cos(\theta_{1j} - \theta_{2j})\right) \\ &+ \frac{1}{2}\sum_{j=1}^{N_b}\left(I_{11j}\dot{\theta}_{1j}^2 + I_{22j}(\dot{\theta}_{1j} - \dot{\theta}_{2j})^2\right) + \frac{1}{2}\sum_{j=1}^{N_b}\Gamma_{2j}(\dot{\theta}_{1j} - \dot{\theta}_{2j})L_{1j}\dot{\theta}_{1j}\cos(\theta_{2j}) \\ &- g\sum_{j=1}^{N_b}\Gamma_{1j}\cos\theta_{1gj} - \Gamma_{2j}\sin(90 - \theta_{1gj} + \theta_{2j}) - g\sum_{j=1}^{N_b} M_{w\_base,j}\sin\theta_{gj} \\ &+ T_{uvw} + T_{pqr} - V_p. \end{aligned} \tag{6.65}$$

However,

$$\dot{d}_{0j} = u, \quad \text{for } j = 1, 4 \quad \text{and} \quad \dot{d}_{0j} = u/\cos\phi_s, \quad \text{for } j = 2, 3. \quad (6.66)$$

Thus, one has the complete Lagrangian of the planetary rover. Yet it is important to bear in mind that a number of simplifying assumptions were made.

6.2.6 Deriving the Dynamic Equations of Motion

One may now derive the Euler-Lagrange equations. Concepts such as partial translational velocities and partial angular velocities are generally extremely useful in mechanizing the process. Although the complete derivation of the Euler-Lagrange equations may be tedious, it can be done, and the resulting equations cast in the standard form for a robot manipulator as,

$$\mathbf{H}(\mathbf{q})\ddot{\mathbf{q}} + \mathbf{C}(\mathbf{q}\dot{\mathbf{q}})\dot{\mathbf{q}} + \mathbf{G}(\mathbf{q}) = \mathbf{Q}. \quad (6.67)$$

6.2.7 Considerations of Slip and Traction

With the exception of planetary environmental and gravity forces, all of the forces that affect the motion of a planetary rover are produced by the interaction of the wheels with the terrain and the soil. Wheel forces that provide the principal external influences are highly nonlinear and cause the largest variations in the net forces and moments restraining the motion of the vehicle. It is naturally important to use a realistic nonlinear wheel force model, especially for designing the appropriate drive motors for generating the torque on wheels and moving the vehicle forward. Given the radius r_w, $u = v_x$ the longitudinal velocity, and the angular speed ω of the wheel, the slip ratio s is defined as,

$$s = (r_w\omega/u) - 1, \quad \text{when } r_w\omega/u < 1, u \neq 0, \text{ for braking}, \quad (6.68)$$

$$s = 1 - (u/r_w\omega), \quad \text{when } u/r_w\omega < 1, \omega \neq 0, \text{ for driving}. \quad (6.69)$$

The forward velocity, which is driven by a wheel, and the wheel rotation dynamics driven by a motor are governed by equations of the form,

$$m\dot{u} = F_x, \quad J\dot{\omega} = \tau - r_w F_x. \quad (6.70)$$

Thus, the slip dynamics could be expressed, for a braking vehicle, as,

$$\dot{s} = -\frac{1}{u}\left(\frac{1+s}{m} + \frac{r_w^2}{J}\right)F_x + \frac{1}{u}\frac{r_w \tau}{J}. \quad (6.71)$$

The original model of the forces and radial stresses acting on a rigid wheel in a deformable terrain, from Bekker [12], are shown in Figure 6.9. In the figure, W is the net vertical load on the wheel due to the rover's mass and payload, while F_{dp} is the net horizontal restraining force, sometimes referred to as the *drawbar pull* acting on the wheel. T is the net driving torque applied by the drive motor about the wheel rotation axis. τ is the shear stress acting on the wheel surface in contact with the terrain and against the direction of rotation of the wheel, which is a function of the angle θ measured anti-clockwise from the vertical axis pointing down. The maximum contact

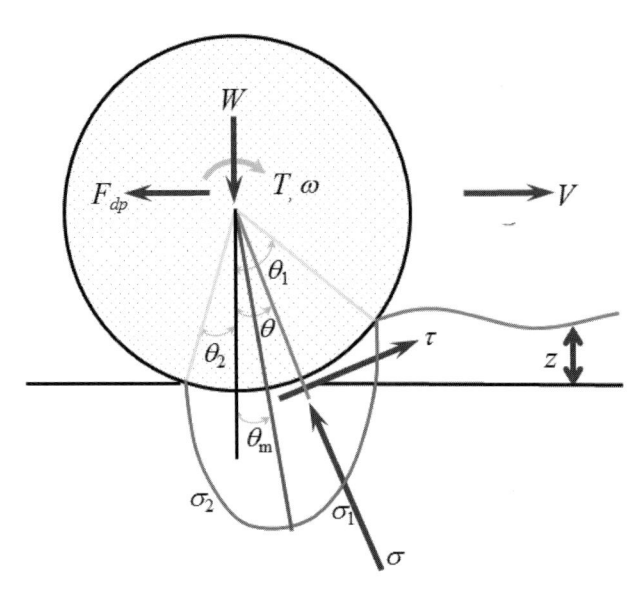

Figure 6.9 Forces and continuous radial stress distribution acting on a rigid wheel in a deformable terrain. (Ani et al. [11], Bekker [12])

angle is assumed to be θ_1, while the minimum contact angle is assumed to be $-\theta_2$. The normal stress directed toward the center of rotation of the wheel is assumed to be given by σ, which is assumed to be a maximum when $\theta = \theta_m$. The basic assumption is that wheel slip will occur when soil "fails" in shear, leading to the immobilization of the vehicle. When the shear stress exerted on the terrain by the wheel of the robot exceeds the maximum shear stress of the soil, the soil is deemed to have "failed." The maximum shear stress τ_m is given by the Mohr-Coulomb soil failure criterion,

$$\tau_m = (c + \sigma_m \tan \phi). \tag{6.72}$$

Following the work of Bekker [12], Janosi and Hanamoto [13] proposed a model for the variation of the soil shear stress τ as a function of deformation j, which may be related as follows,

$$\tau = \tau_m (1 - \exp(-j/k)), \tag{6.73}$$

given that k depends on soil properties and that τ_m is obtained from the Mohr-Coulomb soil failure criterion.

Quantitatively, the stress changes at various depths are often calculated on the assumption that the solid behaves as a homogeneous, isotropic elastic solid. To obtain an expression for the normal stress, σ, consider the stress distribution in a semi-infinite, homogeneous, isotropic, elastic medium when subjected to a vertical point load applied on the surface, which was first derived by Boussinesq in 1885. Boussinesq obtained an expression for the vertical stress σ_z at a point in the elastic medium, given by,

$$\sigma_z = \frac{3W}{2A} \left(\frac{z}{R}\right)^3 = \frac{3W}{2A} \cos^3 \theta. \tag{6.74}$$

6.2 Dynamic Modeling of Planetary Rovers

For a circular contact area A with radius r_0 and uniform pressure p_0, the vertical stress at depth z below the center of the contact area can be determined given that the load acting upon the contact area is represented by a number of discrete point loads. Thus, applying the Boussinesq formula,

$$d\sigma_z = \frac{3}{2\pi} \frac{p_0 r dr d\theta}{z^2 \left(1 + (r/z)^2\right)^{5/2}}. \tag{6.75}$$

Integrating this expression leads to an expression for the vertical stress, which is,

$$\sigma_z = 3p_0 \int_0^{r_0/z} \frac{u du}{(1+u^2)^{5/2}} = p_0 \left(1 - \frac{z^3}{(z^2 + r_0^2)^{3/2}}\right). \tag{6.76}$$

The horizontal stresses acting on an element of the soil at a depth z below the surface are much more difficult to estimate than the vertical stresses but can be assumed to be proportional to the vertical stresses at a particular depth. The distribution of normal stress could be expressed as a function of the angular location on the wheel rim denoted as θ, the radius of the wheel rim r_w, and the depth z. The normal stress distribution $\sigma(\theta)$ can be modeled based on the relationship between the pressure that acts on the wheel from the soil and the wheel's sinkage. The pressure–sinkage relationship developed by Terzaghi [14] is defined as follows,

$$p = kz^n, \tag{6.77}$$

where p denotes the pressure that acts on an object penetrated in soils, and z denotes the penetration depth. The parameters k and n are called the pressure–sinkage modulus and sinkage exponent, respectively. The parameters k and n are constant parameters that represent the bearing capability of the soil. Terzaghi's pressure–sinkage relationship was improved later by Reece (Wong and Reece [15, 16]) to take into account the effect of the dimension of the penetrated object as follows,

$$p = k_\sigma (r_w/b_w)^n, \quad k_\sigma = ck_c + \rho g b_w k_\phi, \tag{6.78}$$

where b_w represents the smaller of the two dimensions of the contact patch for rectangular objects, the radius for circular objects or the wheel-width for a wheel; c denotes the soil cohesion; k_c and k_ϕ denote the pressure–sinkage moduli; ρ denotes the soil bulk density; and g denotes gravitational acceleration. The sinkage $z(\theta)$ at the angle θ of the wheel can be geometrically calculated as,

$$z(\theta) = r_w (\cos\theta - \cos\theta_0). \tag{6.79}$$

The following normal stress distribution model can then be derived,

$$\sigma(\theta) = k_\sigma (r_w/b_w)^n (\cos\theta^* - \cos\theta_0)^n, \tag{6.80}$$

$$\theta^* = \theta \quad \text{for } \theta_m \leq \theta \leq \theta_0, \tag{6.81}$$

$$\theta^* = \theta_0 - (\theta_0 - \theta_m)\frac{\theta - \theta_r}{\theta_m - \theta_r} \quad \text{for } \theta_r \leq \theta \leq \theta_m, \tag{6.82}$$

where $\theta = \theta_m$ is the angular location of the maximum normal stress and when $\theta = \theta_r$, $\sigma(\theta) = 0$. Thus, following Shibly, Iagnemma, and Dubowsky [17], one could express the stress distribution as two functions, one prior to the angular location of the maximum normal stress and the other following the location of the maximum normal stress, which are, respectively, given by,

$$\sigma_1(\theta) = (k_1 + k_2 b_w)(r_w/b_w)^n (\cos\theta - \cos\theta_1)^n, \tag{6.83}$$

$$\sigma_2(\theta) = (k_1 + k_2 b_w)(r_w/b_w)^n \left(\cos\left(\theta_1 \frac{1-\theta}{\theta_m} + \theta\right) - \cos\theta_1\right)^n. \tag{6.84}$$

When $\theta = \theta_m$, the two expressions are equal to each other and are also a maximum.

For a rigid wheel in a deformable terrain, the shear stress distribution and the normal stress distribution over the wheel surface are important in evaluating the tractive effort and the driving torque necessary to move the vehicle forward. While the shear stress is a function of the shear displacement, the normal stress is maximum at a particular point on the rim of the wheel. An important variable in the development of shear displacement along the wheel–soil interface is the slip velocity at any point on the rim of the wheel. The slip velocities in the tangential and lateral directions, v_{jt} and v_{jl}, of a point on the rim relative to the terrain are the tangential and lateral components of the absolute velocity at the same point. The magnitudes of the slip velocities v_{jt} and v_{jl}, and the point at which the normal stress is a maximum of a point on the rim defined by angle θ can, respectively, be expressed by (Wong and Reece [15, 16]),

$$v_{jt}(\theta) = r_w\omega - v_x\cos\theta = r_w\omega\{1 - (1-s)\cos\theta\}, \tag{6.85}$$

$$v_{jl}(\theta) = -v_y = -r_w\omega(1-s)\tan\beta, \quad \tan\beta = v_y/v_x, \tag{6.86}$$

$$\theta_m = (c_1 + c_2 s)\theta_1, \tag{6.87}$$

where s is the slip, r_w is the radius of the wheel rim, ω the angular speed of the wheel, v_x is the longitudinal velocity, v_y is the lateral velocity, and c_1 and c_2 are constants. The slip velocities vary with angle θ and slip for a rigid wheel, so the tangential and lateral shear displacements j_t and j_l, and the net shear displacement j along the wheel–soil interface are given as,

$$j_t = \int_0^{\theta_0} v_{jt}\frac{d\theta}{\omega} = r_w\{(\theta_0 - \theta) - (1-s)(\sin\theta_0 - \sin\theta)\}, \tag{6.88}$$

$$j_l = \int_0^{\theta_0} v_{jl}\frac{d\theta}{\omega} = -r_w(1-s)(\theta_0 - \theta)\tan\beta, \tag{6.89}$$

$$j(\theta) = \sqrt{j_t^2(\theta) + j_l^2(\theta)}. \tag{6.90}$$

where θ_0 is the terrain contact entry angle that defines the angle where a point on the rim comes into contact with the terrain. The shear stress distribution along the contact sector of a rigid wheel may be determined on the basis of the relationship between the shear stress and shear displacement. From the Janosi and Hanamoto [13] model relating the shear stress distribution to the maximum shear stress, the total, tangential, and lateral shear stress distribution could be expressed as a function of the angular location on the wheel rim. Hence, one obtains the relation,

$$\tau = \tau_m(1 - \exp(-j/k)) = (c + \sigma \tan \phi)(1 - \exp(-j/k)), \quad (6.91)$$

$$\tau(\theta) = (c + \sigma(\theta) \tan \phi)(1 - \exp(-j/k)), \quad (6.92)$$

$$\tau_t(\theta) = \frac{v_{jt}(\theta)}{\sqrt{v_{jt}^2(\theta) + v_{jl}^2(\theta)}} \tau(\theta), \quad \tau_l(\theta) = \frac{v_{jl}(\theta)}{\sqrt{v_{jt}^2(\theta) + v_{jl}^2(\theta)}} \tau(\theta). \quad (6.93)$$

The total tractive effort F can be determined by integrating the horizontal component of the tangential stress over the entire contact sector. Similarly, the vertical load W, the drawbar pull F_{dp}, and the torque T may be obtained, integrating the relevant components of the normal and/or the tangential stress over the contact sector. To simplify the integrations, Shibly, Ingnemma, and Dubowsky [17] used Lagrange interpolation to replace the distributions of the shear and normal stresses. These expressions are,

$$\sigma_1(\theta) = \frac{\theta_1 - \theta}{\theta_1 - \theta_m} \sigma_m, \quad \sigma_2(\theta) = \frac{\theta}{\theta_m} \sigma_m, \quad \tau_1(\theta) = \frac{\theta_1 - \theta}{\theta_1 - \theta_m} \tau_m, \quad \tau_2(\theta) = \frac{\theta}{\theta_m} \tau_m. \quad (6.94)$$

With these approximations, expressions for the total tractive effort F, the vertical load W, the drawbar pull F_{dp}, and the torque T may be, respectively, given by,

$$F = \int_0^{\theta_0} \tau(\theta) \cos \theta d\theta, \quad W = r_w b_w \left(\int_0^{\theta_0} \sigma(\theta) \cos \theta d\theta + \int_0^{\theta_0} \tau(\theta) \sin \theta d\theta \right), \quad (6.95)$$

$$F_{dp} = r_w b_w \left(\int_0^{\theta_0} \tau(\theta) \cos \theta d\theta - \int_0^{\theta_0} \sigma(\theta) \sin \theta d\theta \right), \quad T_w = r_w^2 b_w \int_0^{\theta_0} \tau(\theta) d\theta. \quad (6.96)$$

The evaluation and optimization of these forces within the context of planetary rovers is discussed by Ani et al. [11]. A slightly different approach is adopted by Yoshida and Hamano [18] and by Yoshida [19].

In Yoshida and Hamano [18], the driving torque of the wheel is modeled, which is driven by a PWM-controlled DC motor, with known characteristics. The tire force components are composed of the normal force and the tangential force. Given the sinkage and the slip ratio of each wheel, the normal stress and force are obtained first. The tangential force is then obtained by multiplying the load-traction factor to the normal force. The motor is controlled so that the slip is always close to the low, desired, value of the slip. The proportional integral controller is implemented using an estimation of the slip. A block diagram of slip-based traction control is shown in Figure 6.10.

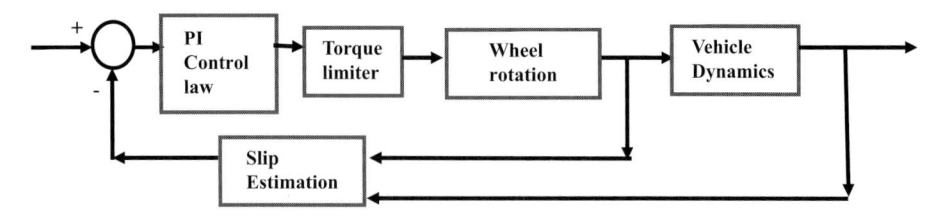

Figure 6.10 Block diagram of slip-based traction feedback control.

6.3 Control of Planetary Rovers

6.3.1 Path Following Control: Kinematic Modeling

The fundamental problem of controlling an autonomous mobile vehicle involves, besides point stabilization, trajectory tracking, where both the position and velocity vector of the vehicle track a desired position and a desired velocity vector, and path following, where the commanded path is followed to within a desired error tolerance. While a mobile vehicle is in motion, it must not only avoid all known obstacles but also be able to alter its forward velocity as required, and alter its angular velocity as well as follow a desired path. Although there are several methods of controlling the inputs to the vehicle to achieve the objectives of vehicle motion, the main problem can be reduced to three phases: (i) controlling the vehicle orientation, (ii) choosing the most appropriate forward velocity and (iii) generating a reference path for purposes of path following. Given a spatial path to be followed, the kinematics of a mobile robot can be described by,

$$\begin{bmatrix} \dot{x} & \dot{y} & \dot{\theta} \end{bmatrix}^T = \begin{bmatrix} u\cos\theta & u\sin\theta & \omega \end{bmatrix}^T, \tag{6.97}$$

where the state vector describing the vehicle's two-dimensional position and orientation in the world frame is defined by $\mathbf{x} = \begin{bmatrix} x & y & \theta \end{bmatrix}^T$, and u and ω are the vehicle's linear and angular velocities, respectively. An objective is to find a control input angular velocity ω such that the vehicle converges to the path while the specified forward velocity is also tracked by it. The path error with respect to the path frame [20] is defined by,

$$\begin{bmatrix} x_e \\ y_e \\ \theta_e \end{bmatrix} = \begin{bmatrix} \cos\theta & \sin\theta & 0 \\ -\sin\theta & \cos\theta & 0 \\ 0 & 0 & 1 \end{bmatrix} \begin{bmatrix} x - x_d \\ y - y_d \\ \theta - \theta_d \end{bmatrix}, \tag{6.98}$$

where the state vector of the reference path, $\mathbf{x}_d = \begin{bmatrix} x_d & y_d & \theta_d \end{bmatrix}^T$, based on a variety of different considerations from the robot's current state and other environmental conditions.

Mobile robots operating in off-road conditions suffer from problems associated with slip caused by the fact that rolling of a wheel is not perfect. Thereby, the ability of the mobile robot to follow a path is considerably influenced by the condition of the terrain. Thus, one often designs motion controllers for mobile robots that compensate for the

6.3 Control of Planetary Rovers

slip effects [21]. When wheel slip is not considered, the linear velocities of a point on the rim of the right and left wheels, at a time t, are, respectively, given by,

$$u_r(t) = r_w \omega_r(t), \quad u_l(t) = r_w \omega_l(t) \tag{6.99}$$

where $\omega_r(t)$ and $\omega_l(t)$ are the angular velocities of the right and left wheel axles, respectively, at a time t, and r_w is the radius of the wheel rim. In the presence of slip, both these velocities are reduced depending on the slip of the right and left wheels, respectively. Thus, given the slip of the right and left wheels is $s_r(t)$ and $s_l(t)$, respectively, the linear velocities of a point on the rim of the right and left wheels, at any time t, are, respectively, given by,

$$u_{rs}(t) = r_w \omega_r(t)(1 - s_r(t)), \quad u_{ls}(t) = r_w \omega_l(t)(1 - s_l(t)). \tag{6.100}$$

The actual values of the slip of the right and left wheels, $s_r(t)$ and $s_l(t)$, respectively, are assumed to be estimated in real time from continuous measurements of the forward velocity and estimates of the wheel's angular velocities obtained from measurements of both the wheel's angular positions and the wheel's angular velocities. Thus,

$$\begin{bmatrix} \dot{x} \\ \dot{y} \\ \dot{\theta} \end{bmatrix} = \begin{bmatrix} (u_{rs} + u_{ls}) \cos \theta(t)/2 \\ (u_{rs} + u_{ls}) \sin \theta(t)/2 \\ (u_{rs} - u_{ls})/b_w \end{bmatrix}$$
$$= \begin{bmatrix} (u_r + u_l) \cos \theta(t)/2 \\ (u_r + u_l) \sin \theta(t)/2 \\ (u_r - u_l)/b_w \end{bmatrix} - \begin{bmatrix} (u_r s_r + u_l s_l) \cos \theta(t)/2 \\ (u_r s_r + u_l s_l) \sin \theta(t)/2 \\ (u_r s_r - u_l s_l)/b_w \end{bmatrix}, \tag{6.101}$$

where b_w is the distance between the right and left wheel centers. The path error vector is defined by the vector, $\mathbf{x}_e = \begin{bmatrix} x_e & y_e & \theta_e \end{bmatrix}^T$. Thus, it follows that,

$$\dot{\theta} = \omega = (u_r - u_l)/b_w - (u_r s_r - u_l s_l)/b_w. \tag{6.102}$$

Differentiating the path error vector,

$$\begin{bmatrix} \dot{x}_e \\ \dot{y}_e \\ \dot{\theta}_e \end{bmatrix} = \omega \begin{bmatrix} y - y_d \\ -(x - x_d) \\ 0 \end{bmatrix} + \begin{bmatrix} \cos \theta & \sin \theta & 0 \\ -\sin \theta & \cos \theta & 0 \\ 0 & 0 & 1 \end{bmatrix} \begin{bmatrix} \dot{x} - \dot{x}_d \\ \dot{y} - \dot{y}_d \\ \omega - \dot{\theta}_d \end{bmatrix}, \tag{6.103}$$

with,

$$\begin{bmatrix} \dot{x}_d \\ \dot{y}_d \\ \dot{\theta}_d \end{bmatrix} = \begin{bmatrix} (u_{r,d} + u_{l,d}) \cos \theta_d(t)/2 \\ (u_{r,d} + u_{l,d}) \sin \theta_d(t)/2 \\ (u_{r,d} - u_{l,d})/b_w \end{bmatrix}. \tag{6.104}$$

Observe that,

$$\begin{bmatrix} \cos \theta & \sin \theta & 0 \\ -\sin \theta & \cos \theta & 0 \\ 0 & 0 & 1 \end{bmatrix} \begin{bmatrix} \dot{x} \\ \dot{y} \\ \dot{\theta} \end{bmatrix} = \begin{bmatrix} (u_{rs} + u_{ls})/2 \\ 0 \\ (u_{rs} - u_{ls})/b_w \end{bmatrix} = \begin{bmatrix} (u_r + u_l)/2 \\ 0 \\ (u_r - u_l)/b_w \end{bmatrix} - \begin{bmatrix} (u_r s_r + u_l s_l)/2 \\ 0 \\ (u_r s_r - u_l s_l)/b_w \end{bmatrix},$$
$$\tag{6.105}$$

and that,

$$\begin{bmatrix} \cos\theta & \sin\theta & 0 \\ -\sin\theta & \cos\theta & 0 \\ 0 & 0 & 1 \end{bmatrix} \begin{bmatrix} \dot{x}_d \\ \dot{y}_d \\ \dot{\theta}_d \end{bmatrix} = \begin{bmatrix} (u_{r,d}+u_{l,d})\cos\theta_e(t)/2 \\ (u_{r,d}+u_{l,d})\sin\theta_e(t)/2 \\ (u_{r,d}-u_{l,d})/b_w \end{bmatrix}. \tag{6.106}$$

Hence,

$$\begin{bmatrix} \dot{x}_e \\ \dot{y}_e \\ \dot{\theta}_e \end{bmatrix} = \omega \begin{bmatrix} 0 & 1 & 0 \\ -1 & 0 & 0 \\ 0 & 0 & 0 \end{bmatrix} \begin{bmatrix} x_e \\ y_e \\ \theta_e \end{bmatrix} + \begin{bmatrix} (u_r+u_l)/2 \\ 0 \\ (u_r-u_l)/b_w \end{bmatrix} - \begin{bmatrix} (u_{r,d}+u_{l,d})\cos\theta_e(t)/2 \\ (u_{r,d}+u_{l,d})\sin\theta_e(t)/2 \\ (u_{r,d}-u_{l,d})/b_w \end{bmatrix} - \begin{bmatrix} (u_r s_r + u_l s_l)/2 \\ 0 \\ (u_r s_r - u_l s_l)/b_w \end{bmatrix}. \tag{6.107}$$

Define the kinematic control inputs,

$$\begin{bmatrix} u_1 \\ u_2 \end{bmatrix} = \begin{bmatrix} (u_r+u_l)/2 \\ (u_r-u_l)/b_w \end{bmatrix} - \begin{bmatrix} (u_{r,d}+u_{l,d})/2 \\ (u_{r,d}-u_{l,d})/b_w \end{bmatrix} - \begin{bmatrix} (u_r s_r + u_l s_l)/2 \\ (u_r s_r - u_l s_l)/b_w \end{bmatrix}. \tag{6.108}$$

Thus, the nonlinear state space equations for the kinematic position and orientation errors may be expressed as,

$$\begin{bmatrix} \dot{x}_e \\ \dot{y}_e \\ \dot{\theta}_e \end{bmatrix} = \omega \begin{bmatrix} 0 & 1 & 0 \\ -1 & 0 & 0 \\ 0 & 0 & 0 \end{bmatrix} \begin{bmatrix} x_e \\ y_e \\ \theta_e \end{bmatrix} - \begin{bmatrix} -(u_{r,d}+u_{l,d})(1-\cos\theta_e(t))/2 \\ (u_{r,d}+u_{l,d})\sin\theta_e(t)/2 \\ 0 \end{bmatrix} + \begin{bmatrix} u_1 \\ 0 \\ u_2 \end{bmatrix}. \tag{6.109}$$

Assume that $\theta_e(t)$ is small, so the small angle approximations,

$$\cos\theta_e(t) \approx 1 \quad \text{and} \quad \sin\theta_e(t) \approx \theta_e(t), \tag{6.110}$$

apply. Thus, the linearized error equations are,

$$\begin{bmatrix} \dot{x}_e \\ \dot{y}_e \\ \dot{\theta}_e \end{bmatrix} = \omega_r \begin{bmatrix} 0 & 1 & 0 \\ -1 & 0 & 0 \\ 0 & 0 & 0 \end{bmatrix} \begin{bmatrix} x_e \\ y_e \\ \theta_e \end{bmatrix} + \frac{(u_{r,d}+u_{l,d})}{2} \begin{bmatrix} 0 & 0 & 0 \\ 0 & 0 & 1 \\ 0 & 0 & 0 \end{bmatrix} \begin{bmatrix} x_e \\ y_e \\ \theta_e \end{bmatrix} + \begin{bmatrix} u_1 \\ 0 \\ u_2 \end{bmatrix}, \tag{6.111}$$

with,

$$\omega_r = (u_{r,d}-u_{l,d})/b_w - (u_{r,d}s_r - u_{l,d}s_l)/b_w. \tag{6.112}$$

Thus, the state space equations for the kinematic position and orientation errors may be cast in the matrix form in terms of two control inputs and two scalar parameters as,

$$\begin{bmatrix} \dot{x}_e \\ \dot{y}_e \\ \dot{\theta}_e \end{bmatrix} = \mathbf{A}_e \begin{bmatrix} x_e \\ y_e \\ \theta_e \end{bmatrix} + \begin{bmatrix} 1 & 0 \\ 0 & 0 \\ 0 & 1 \end{bmatrix} \begin{bmatrix} u_1 \\ u_2 \end{bmatrix}, \tag{6.113}$$

where,

$$\mathbf{A}_e = \omega_r \begin{bmatrix} 0 & 1 & 0 \\ -1 & 0 & 0 \\ 0 & 0 & 0 \end{bmatrix} + \frac{(u_{r,d} + u_{l,d})}{2} \begin{bmatrix} 0 & 0 & 0 \\ 0 & 0 & 1 \\ 0 & 0 & 0 \end{bmatrix}. \tag{6.114}$$

6.3.2 Estimating Slip

Before considering the synthesis of the controller, the issue of estimating the slip at each wheel will be discussed. It is known that in a set body-fixed axis,

$$\begin{bmatrix} \dot{x}_b \\ \dot{y}_b \\ \dot{\theta} \end{bmatrix} = \begin{bmatrix} u \\ v \\ \omega \end{bmatrix} = \begin{bmatrix} (u_r + u_l)/2 \\ 0 \\ (u_r - u_l)/b_w \end{bmatrix} - \begin{bmatrix} (u_r s_r + u_l s_l)/2 \\ 0 \\ (u_r s_r - u_l s_l)/b_w \end{bmatrix}. \tag{6.115}$$

Hence,

$$s_r = 1 - \frac{u}{r_w \omega_r(t)} - \frac{b_w \omega}{2 r_w \omega_r(t)}, \tag{6.116}$$

and

$$s_l = 1 - \frac{u}{r_w \omega_l(t)} + \frac{b_w \omega}{2 r_w \omega_l(t)}. \tag{6.117}$$

Thus the slip of the right and left wheels s_r and s_l may be estimated using measurements u, which could be done by either *visual odometry* [22] or by using ultrasonic sensors, measurements of ω using a high precision rate gyroscope, and using real-time estimates of $\omega_r(t)$ and $\omega_l(t)$, which are obtained from measurements of both the wheel's angular positions and the wheel's angular velocities at both the right and left wheel axles and a first-order Kalman filter dedicated to each wheel.

One approach to visual odometry is the *template matching* method [23]. Gonzalez et al. [23] also introduce the idea of *visual compassing* to measure θ by template matching. It avoids the problem of finding and tracking features by simply estimating changes in the appearance of sections of the camera image. This is done by taking a template or a patch from an image in real time and matching it to the same patch at the previous instant. No system identification or tracking of features is needed and there is also no need to measure image velocities at several locations. The appearance-based method has been successfully applied by employing either single or multiple cameras.

There are also a number of tools available for feature extraction and classification using Artificial Neural Networks, Support Vector Machines, and discrete wavelet transforms. These tools are particularly suitable for preprocessing, de-noising, and re-scaling images before algorithms such as template matching are applied.

6.3.3 Slip-Compensated Path Following Control Law Synthesis

One approach to synthesizing the kinematic controller for slip-compensated path following control is to adopt the model predictive controller (MPC) synthesis approach.

To briefly describe the synthesis of linear optimal control based on the MPC approach, consider a linear discrete time system in the form,

$$\mathbf{x}(k+1) = \mathbf{A}(k)\mathbf{x}(k) + \mathbf{B}(k)\mathbf{u}(k), \quad \mathbf{y}(k) = \mathbf{C}(k)\mathbf{x}(k). \tag{6.118}$$

Our aim is to find an optimal control input sequence defined over a control prediction window $\mathbf{u}(j), j = 0, 1, 2 \ldots N - 1$ or the vector $\mathbf{U} = \begin{bmatrix} \mathbf{u}^T(0) & \mathbf{u}^T(1) & \cdots & \mathbf{u}^T(N-1) \end{bmatrix}^T$ so as to minimize the performance index, which now includes a terminal weighting matrix $\mathbf{Q}_N$,

$$J(\mathbf{x}(0), \mathbf{U}) = \sum_{k=0}^{N-1} \left\{ \mathbf{y}^T(k) q \mathbf{Q} \mathbf{y}(k) + \mathbf{u}^T(k) r \mathbf{R} \mathbf{u}(k) \right\} + \mathbf{y}^T(N) q \mathbf{Q}_N \mathbf{y}(N), \tag{6.119}$$

where q and r are scalar scaling parameters that are used to re-scale the relative contributions of the states and the control inputs to the cost function. When the magnitude of the control input is relatively large when compared with the magnitude of the state, r is assumed to be small relative to q and vice-versa.

Defining the vector, $\mathbf{X} = \begin{bmatrix} \mathbf{x}^T(1) & \mathbf{x}^T(2) & \cdots & \mathbf{x}^T(N-1) & \mathbf{x}^T(N) \end{bmatrix}$, we may write,

$$J(\mathbf{x}(0), \mathbf{U}) = q\mathbf{x}^T(0)\mathbf{C}^T(0)\mathbf{Q}\mathbf{C}(0)\mathbf{x}(0) + q\mathbf{X}^T \bar{\mathbf{Q}} \mathbf{X} + r\mathbf{U}^T \bar{\mathbf{R}} \mathbf{U}, \tag{6.120}$$

where $\bar{\mathbf{Q}}$ is a block diagonal matrix with matrix $\mathbf{C}^T(k)\mathbf{Q}\mathbf{C}(k), k = 1, 2 \ldots N - 1$ along the diagonal except the last element, which is $\mathbf{C}^T(N)\mathbf{Q}_N \mathbf{C}(N)$, and $\bar{\mathbf{R}}$ is a block diagonal matrix with matrix $\mathbf{R}$ along the diagonal. Using the state space model Equation (6.118) recursively, we may construct a prediction model in the form,

$$\mathbf{X} = \mathbf{S}\mathbf{U} + \mathbf{T}\mathbf{x}(0) \tag{6.121}$$

where $\mathbf{S} = \begin{bmatrix} \mathbf{B} & 0 & 0 & 0 \\ \mathbf{A}(1)\mathbf{B} & \mathbf{B} & 0 & 0 \\ \cdots & \cdots & \cdots & \cdots \\ \left(\prod_{k=N-1}^{1} \mathbf{A}(k) \right) \mathbf{B} & \left(\prod_{k=N-2}^{1} \mathbf{A}(k) \right) \mathbf{B} & \cdots & \mathbf{B} \end{bmatrix}, \mathbf{T} = \begin{bmatrix} \mathbf{A}(0) \\ \mathbf{A}(1)\mathbf{A}(0) \\ \cdots \\ \prod_{k=N-1}^{0} \mathbf{A}(k) \end{bmatrix}.$

$$\tag{6.122}$$

We have assumed for simplicity that $\mathbf{B}$ is constant but not $\mathbf{A}$. This formula could be generalized for the case when $\mathbf{B}$ is not constant. Thus, the cost function may be expressed as,

$$J(\mathbf{x}(0), \mathbf{U}) = \frac{1}{2}\mathbf{x}(0)^T \mathbf{G} \mathbf{x}(0) + \frac{1}{2}\mathbf{U}^T \mathbf{H} \mathbf{U} + \mathbf{x}^T(0)\mathbf{F}\mathbf{U}, \tag{6.123}$$

with, $\mathbf{G} = 2q(\mathbf{T}^T \bar{\mathbf{Q}} \mathbf{T} + \mathbf{C}^T(0)\mathbf{Q}\mathbf{C}(0))$, $\mathbf{H} = 2r\bar{\mathbf{R}} + 2q\mathbf{S}^T \bar{\mathbf{Q}} \mathbf{S}$ and $\mathbf{F} = q\mathbf{T}^T \bar{\mathbf{Q}} \mathbf{S}$.

The optimum control sequence is obtained by setting the gradient of $J(\mathbf{x}(0), \mathbf{U})$ to zero. Minimizing the cost function results in,

$$dJ(\mathbf{x}(0), \mathbf{U})/d\mathbf{U} = \mathbf{U}^T \mathbf{H} + \mathbf{x}^T(0)\mathbf{F} = 0$$
$$\Rightarrow \mathbf{HU} + \mathbf{F}^T \mathbf{x}(0) = 0 \Rightarrow \mathbf{U} = -\mathbf{H}^{-1} \mathbf{F}^T \mathbf{x}(0). \quad (6.124)$$

The state $\mathbf{x}(0)$, at the start of the prediction window, is assumed to represent the state at the next time instant, in real time. The control law based on the receding horizon is,

$$\mathbf{u}(k) = -[1 \quad 0 \quad \cdots \quad 0]\mathbf{H}^{-1}\mathbf{F}^T \mathbf{x}(k). \quad (6.125)$$

The control sequence is recursively calculated over successive control prediction windows. The product $\mathbf{H}^{-1}\mathbf{F}^T$ may be expressed as,

$$\mathbf{H}^{-1}\mathbf{F}^T = \frac{q}{2}\left(r\bar{\mathbf{R}} + q\mathbf{S}^T\bar{\mathbf{Q}}\mathbf{S}\right)^{-1}\mathbf{S}^T\bar{\mathbf{Q}}\mathbf{T} = \frac{q^2}{2}\left(\frac{r}{q}\bar{\mathbf{R}} + q\mathbf{S}^T\bar{\mathbf{Q}}\mathbf{S}\right)^{-1}\mathbf{S}^T\bar{\mathbf{Q}}\mathbf{T}. \quad (6.126)$$

If we let $q = 1$, Equation (6.126) reduces to,

$$\mathbf{H}^{-1}\mathbf{F}^T = \frac{1}{2}\left(r\bar{\mathbf{R}} + \mathbf{S}^T\bar{\mathbf{Q}}\mathbf{S}\right)^{-1}\mathbf{S}^T\bar{\mathbf{Q}}\mathbf{T}, \quad (6.127)$$

where r is treated as a free parameter to be chosen.

Once a control law $\mathbf{u} = \mathbf{u}_{mpc}$ is synthesized,

$$\mathbf{u}_{mpc} = \begin{bmatrix} u_{mpc,1} \\ u_{mpc,2} \end{bmatrix} = \begin{bmatrix} (u_r + u_l)/2 \\ (u_r - u_l)/b_w \end{bmatrix} - \begin{bmatrix} (u_{r,d} + u_{l,d})/2 \\ (u_{r,d} - u_{l,d})/b_w \end{bmatrix} - \begin{bmatrix} (u_r s_r + u_l s_l)/2 \\ (u_r s_r - u_l s_l)/b_w \end{bmatrix}, \quad (6.128)$$

one could solve for the desired $u_{r,d}$ and $u_{l,d}$, which are given by,

$$\begin{bmatrix} u_{r,d} \\ u_{l,d} \end{bmatrix} = \begin{bmatrix} u_r \\ u_l \end{bmatrix} - \begin{bmatrix} u_r s_r \\ u_l s_l \end{bmatrix} - \begin{bmatrix} 2u_{mpc,1} + b_w u_{mpc,2} \\ 2u_{mpc,1} - b_w u_{mpc,2} \end{bmatrix}. \quad (6.129)$$

A slightly different approach was adopted by Helmick et al. [24]. They adopt an algorithm called the "carrot heading algorithm" to calculate the desired heading, ϕ_{carrot}, of the vehicle given the desired path and the current rover position and orientation. The algorithm was employed by them for its robustness to path error. The desired path consists of a set of linear segments between waypoints. To allow for paths of arbitrary complexity, the waypoints can be spaced at any distance apart. The desired heading is obtained by computing the intersection of a circle with the desired heading direction. The radius of the circle is chosen from considerations of the path error and changes in the heading. Their slip compensation control algorithm is implemented as two separate control loops. The first control loop uses feedbacks of the heading error and heading slip rate, while the second loop uses feedbacks of the lateral component of slip rate.

Kanjanawanishkul, Hofmeister, and Zell [20] adopt a MPC approach to choose the desired heading rate and use a rule-based approach for forward velocity selection. Gonzalez et al. [21] present an approach for the trajectory tracking of mobile rovers under slip conditions to obtain the adaptive control law using the Lyapunov and Linear Matrix Inequality (LMI)-based synthesis to guarantee stability under input and state constraints.

6.3.4 The Focused D^* Algorithm

A primary problem in many applications in robotics, such as route planning for a mobile rover, is finding the lowest-cost path through a graph. A number of algorithms exist for producing optimal traverses given fixed arc costs such as the A^* algorithm (see, for example, Vepa [25]). If arc costs change during the traverse, then the remainder of the path must be re-planned. Re-planning is feasible for a mobile rover equipped with appropriate sensors and with imperfect information about its environment. When the rover has additional information from its sensors, it can revise its plan to reduce the total cost of the traverse. If the prior information is incomplete or even totally inadequate, the robot may discover useful additional information in newly acquired sensor data. During re-planning, the robot must either move in a completely new direction with no information about the feasibility of traverse or wait for the new optimal path to be computed. To avoid moving in a completely new direction with no information about the feasibility of traverse or wait, rapid re-planning is essential. The D^* algorithm (Dynamic A^*) plans optimal traverses in real time by incrementally altering the paths to the robot's state as new information is discovered. Stentz [26] describes an extension to D^* that focusses the altered paths to significantly reduce the total time required for re-planning the optimal path. The extension by Stentz [26] provides a development of the D^* algorithm that can be considered as a full generalization of A^* for dynamic environments, when arc costs can change during the traverse of the optimal path.

References

[1] Anon (2018) Mars Exploration Rover, Wikipedia, https://en.wikipedia.org/wiki/Mars_Exploration_Rover, accessed June 2018.

[2] Anon (2012) Ames' Involvement in Mars Science Laboratory/Curiosity, NASA AMES Research Center, 2012, www.nasa.gov/centers/ames/research/ames-msl-contributions.html, accessed June 2018.

[3] Rekleitis, I., Bedwani, J-L., Dupuis, E., and Allard, P. (2008) Path planning for planetary exploration, in Canadian Conference on Computer & Robot Vision, Windsor, ON, 61–68.

[4] Gat, E., Slack, M. G., Miller, D. P., Fiby, R. J. (1990) Path planning and execution monitoring for a planetary rover, *Tagungsband: IEEE International Conference on Robotics and Automation*, Seiten 20–25, Cincinnati, OH.

[5] Peng, W. U. and Hehua, J. U. (2013) Mission-integrated path planning for planetary rover exploration. *Journal of Software*, 8(10): 2620–2627.

[6] Johnson, A., Hoffman, J., Newman, D., Mazarico, E., and Zuber, M. (2010) An integrated traverse planner and analysis tool for planetary exploration, *AIAA SPACE 2010 Conference & Exposition*, Anaheim, CA, 2010, American Institute of Aeronautics and Astronautics, 1–28.

[7] Schenker, P. S., Huntsberger, T. L., Pirjanian, P., Baumgartner, E., Aghazarian, H., Trebi-Ollennu, A., Leger, P. C., Cheng, Y., Backes, P. G., Tunstel, E. W., Dubowsky, S.,

Iagnemma, K., and McKee, G. T. (2018) Robotic automation for space: planetary surface exploration, terrain-adaptive mobility, and multi-robot cooperative tasks, robots.mit.edu/people/Karl/Spie4572.pdf (Accessed March 31, 2018).
[8] Ratnakumar, B. V., Smart, M. C., Whitcanack, L. D., Ewell, R. C., and Surampudi, S. (2014) Li-ion rechargeable batteries on Mars exploration rovers. *AIAA Paper*, www.researchgate.net/publication/238794735. (Accessed March 31, 2018).
[9] Helmick, D., Angelova, A., and Matthies, L. (2009) Terrain adaptive navigation for planetary rovers. *Journal of Field Robotics*, 26(4): Special Issue: Special Issue on Space Robotics, Part II, 391–410.
[10] Ostafew, C. J., Schoellig, A. P., Barfoot, T. D., and Collier, J. (2016) Learning-based nonlinear model predictive control to improve vision-based mobile robot path tracking. *Journal of Field Robotics*, 33(1): 133–152.
[11] Ani, O. A., Xu, H., Shen, Y-P., Liu, S-G., and Xue, K. (2013) Modeling and multi-objective optimization of traction performance for autonomous wheeled mobile robot in rough terrain. *Journal of Zhejiang University-SCIENCE C (Computers & Electronics)*, 14(1): 11–29.
[12] Bekker, M. G. (1956) *Theory of Land Locomotion*, Ann Arbor, MI: University of Michigan Press.
[13] Janosi, Z. and Hanamoto, B. (1961) Analytical determination of drawbar pull as a function of slip for tracked vehicles in deformable soils, *Proceedings of the 1st International Conference on Terrain-Vehicle Systems*, 707–726.
[14] Terzaghi, K. T. (1943) *Theoretical Soil Mechanics*, New York: John Wiley & Sons.
[15] Wong, J. Y. and Reece, A. R. (1967) Prediction of rigid wheel performance based on the analysis of soil-wheel stresses part I. Performance of driven rigid wheels. *Journal of Terramechanics*, 4(1): 81–98. doi:10.1016/0022-4898(67)90105-X.
[16] Wong, J. Y. and Reece, A. R. (1967) Prediction of rigid wheel performance based on the analysis of soil-wheel stresses part II. Performance of towed rigid wheels. *Journal of Terramechanics*, 4(2): 7–25. doi:10.1016/0022-4898(67)90047-X.
[17] Shibly, H., Iagnemma, K., and Dubowsky, S. (2005) An equivalent soil mechanics formulation for rigid wheels in deformable terrain, with application to planetary exploration rovers. *Journal of Terramechanics*, 42(1): 1–13. doi:10.1016/j.jterra. 2004.05.002.
[18] Yoshida, K. and Hamano, H. (2002) Motion dynamics and control of a planetary rover with slip-based traction model. *Proceedings of SPIE*, 4715: 275–286. doi:10.1117/12.474459.
[19] Yoshida, K. (2003) Slip, traction control, and navigation of a lunar rover, *7th International Symposium on Artificial Intelligence, Robotics and Automation in Space*. Nara, Japan.
[20] Kanjanawanishkul, K., Hofmeister, M., and Zell, A. (2010) Path following with an optimal forward velocity for a mobile robot in *7th IFAC Symposium on Intelligent Autonomous Vehicles*, 19–24.
[21] Gonzalez, R., Fiacchini, M., Alamo, T., Guzman, J. L., and Rodriguez, F. (2010) Adaptive control for a mobile robot under slip conditions using an LMI-based approach. *European Journal of Control* 2: 144–155, doi:10.3166/EJC.16.144–155.
[22] Angelova, A., Matthies, L., Helmick, D., and Perona, P. (2007) Learning and prediction of slip from visual information. *Journal of Field Robotics* 24(3): 205–231.
[23] Gonzalez, R., Rodriguez, F., Guzman, J. L., Pradalier C., and Siegwart, R. (2012) Combined visual odometry and visual compass for off-road mobile robots localization. *Robotica*, 30(6): 1–14. doi: 10.1017/S026357471100110X.

[24] Helmick, D. M., Roumeliotis, S. I., Cheng, Y., Clouse, D. S., Bajracharya, M., and Matthies, L. H. (2006) Slip-compensated path following for planetary exploration rovers. *Advanced Robotics*, 20(11): 1257–1280.

[25] Vepa, R. (2009) Path planning, obstacle avoidance and navigation. Chapter 9 in *Biomimetic Robotics: Mechanisms and Control*, New York: Cambridge University Press, 178–179.

[26] Stentz, A. (1995) The focused D* algorithm for real-time re-planning, in IJCAI'95, *Proceedings of the 14th International Joint Conference on Artificial Intelligence* – Volume 2, Montreal, Quebec, Canada, August 20–25, 1652–1659.

7 Navigation and Localization

7.1 Introduction to Navigation

The subject of navigation has evolved over the ages. A basic problem that arises when an individual moves from his current position in a particular environment to a new position is the question of the path or route to be taken, as one moves from one's current position to an intended destination. One naturally seeks a traversable path that is devoid of obstacles, and is also optimal in some way, such as the shortest distance, minimum time of travel, or least cluttered pathway, or is based on some other holistic attribute of the entire planned route or pathway. The primary aim of navigation is to guide a vehicle from where it is currently located to its intended destination. Navigation thus includes the process of not only determining or planning a suitable course or trajectory to a goal location but also of maintaining a course as close as possible to the planned trajectory. Of course, this implies that the navigator or the system responsible for navigating the vehicle either already knows its current location or is able to determine it from measurements of landmarks, local or environmental, or other features, or speed, direction of motion, and/or acceleration. Furthermore, it is also inherent that the navigator is capable of planning the route, and this again implies that a perfect map of the region covering the domain from where the vehicle is currently located to its intended destination should be available. When no map is available, it is assumed that the navigator has the tools to build a map of the region as it traverses from its current location to the intended destination.

7.1.1 Basic Navigation Activities

One could summarize all the activities associated with navigation in terms of a set of leading questions, and answers to these questions:

- The primary question a navigation system seeks to answer is: How can I get to my destination from where I am?
    - Implicit route planning, route traversability analysis, and avoiding obstacles are essential features;
    - One assumes a perfect map is available;
    - Sensing and motion actuation are key steps in getting to the destination.

- But, where am I in relation to my local environment?
  - Localization will determine my current local position and orientation;
  - One assumes a perfect map is available; imperfections in sensing are accepted.
- But, I do not have a map. What have I observed during my travels?
  - One builds one's own map. (What is a map?)
  - One assumes localization is feasible, in principle.
- Can I build a map and localize at the same time?
  - Yes, you can! Using simultaneous localization and mapping (SLAM);
  - One assumes no prior knowledge of the world around oneself;
  - Any such knowledge is a bonus.

From these questions and answers it is clear that navigation implies a number of interrelated activities such as route planning; traversability analysis; obstacle avoidance; sensing for localization; path planning; actuation for moving forward along a desired path; localization to determine the current position and orientation from observed noisy or imperfect data; sensing and observation for map building; and SLAM. In the next section the nature of these activities will be briefly defined before they are discussed in some detail.

7.2 Localization, Mapping, and Navigation

Localization can be broadly classified into two categories: (i) dead reckoning, where one "reckons" or estimates one's own position and orientation from measurements of velocities and/or accelerations, generally with respect to some inertial frame of reference (a frame of reference fixed in some sense in space.) (ii) using landmarks. In this approach, one identifies certain landmarks, which must be detectable and identifiable with the available sensors, apart from the fact that they could be active or passive, or natural or artificial, and be able to measure some feature of it based on its geometry and its activity, such as the ranges of certain geometric feature points, bearings, and distances. In the case of active landmarks, which could move with one or more degrees of freedom, it would help to know their precise position and orientation at the time of localization.

Maps come in a wide variety of forms and shapes. Mapping in 3D is a three-dimensional perspective view of a map illustrated in two-dimensions. With the enhanced plotting tools that are now available, geo-spatial visualization is possible in different three-dimensional axes and formats. In simple terms, an "Open Street Map" is an exhaustive database of every known street, city, road, building, and landmark on the planet, and not just a map that can be displayed. A grid map is a two-dimensional map that consists of a network of superimposed, equi-spaced, horizontal, and vertical lines, which are used for locating a sparse collection of points, within each square in the grid. Each square is identified by assigning an ordered set of letters or numbers to the domains between each successive pair of parallel horizontal and vertical lines. Thus, a grid reference locates a unique square region on the map. The precision of the location

7.2 Localization, Mapping, and Navigation

can vary, depending on how close the equi-spaced, horizontal, and vertical lines are. A grid reference in this system, such as "G3," locates a particular square with coordinates (G, 3), where the first character could refer to either the vertical or the horizontal domains, while the second would refer to the other. A terrain map is a topographic map characterizing in detail and providing a quantitative representation of the terrain, usually using contour lines to represent all points at the same height or altitude with reference to a known surface of uniform height. There is indeed a number of different ways to represent the space around us, and space maps is a generic term used to denote one or more different representations of the spatial environment around a vehicle. It is usually a three-dimensional representation of a limited region of the three-dimensional space around a vehicle or user, including a wide range of obstacles to varying degrees of precision.

Navigation is referred to as *experience-based navigation* when one can localize oneself on a map and, based on previous experience, select a route that has been traversed before. This is also the basis for *route-based navigation*, where a documented route on a map already exists. *Vision-based navigation* uses cameras of different types to localize by using landmarks, identified by using visual sensory information obtained from the cameras. Given a situation when a vehicle is quite close to its goal position, it may be necessary for it to precisely position and orient itself relative to a desired position, orientated in terms of a reference frame fixed in the destination. It is in such circumstances or in unknown planetary environments that vision-based navigation becomes extremely relevant. *Navigation in cluttered environments* involves the use of obstacle avoidance algorithms being used to plan and re-plan a path in a cluttered environment. *Reactive navigation* requires reactive fast re-planning of the entire planned route, or parts of it, in response to current sensory inputs, which may signal the appearance of new obstacles or constraints. Typically, a reactive navigation algorithm for wheeled mobile robots in an unknown environment populated by obstacles uses the geometric footprint of the robot to determine the set of all possible collision-free steering angles. The steering angle that falls in the widest gap and is closest to the target is most often selected. It is also important to account for the non-holonomic constraints of differentially steered robots by determining curved trajectories with a set of feasible radii of curvature. Contrasting a reactive navigation system is the deliberate approach to navigation, which is based on the more traditional approach. It does involve re-planning as new sensory information becomes available but not necessarily fast. A hybrid navigation system involves the use of both reactive and deliberate approaches, depending on the context.

7.2.1 Introduction to Localization

A method of localization is based on the principles of trilateration. In trilateration, one measures the time of flight from two points (in a plane) to and back from an unknown location. The problem is that the measurements are corrupted by noise. There are two nonlinear equations to solve for the Cartesian position coordinates defined by the x and y coordinates of the user:

$$(x_A - x)^2 + (y_A - y)^2 = c^2 t_A^2, \quad (x_B - x)^2 + (y_B - y)^2 = c^2 t_B^2, \tag{7.1}$$

where (x_A, y_A) and (x_B, y_B) are the position coordinates of two landmarks, A and B, (x, y) are the position coordinates of the user that are yet to be determined, and c is the velocity of electromagnetic radiation used to measure the times of flight, t_A and t_B, from each of the two landmarks, respectively, to the user position. These equations are a pair of nonlinear equations for the position coordinates of the user, (x, y), which could be solved exactly in principle by a transformation of the frame of reference [1]. Yet an iterative approximate method is adopted in practice, as such a method allows for the recursive reduction of the error in the position coordinates of the user due to the presence of measurement errors in the times of flight, errors in the knowledge of the precise location of the landmarks, and biases in the clocks used to measure the times of flight. Moreover, the solution is not unique, as illustrated in Figure 7.1. The figure indicates that there are two points of intersection of the two circles that represent the loci of all points at a fixed distance from each of the landmarks.

To resolve the ambiguity, one could use the measurement from a third landmark, as shown in Figure 7.2.

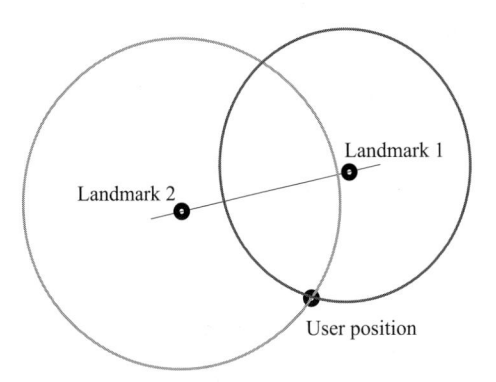

Figure 7.1 Trilateration in two dimensions.

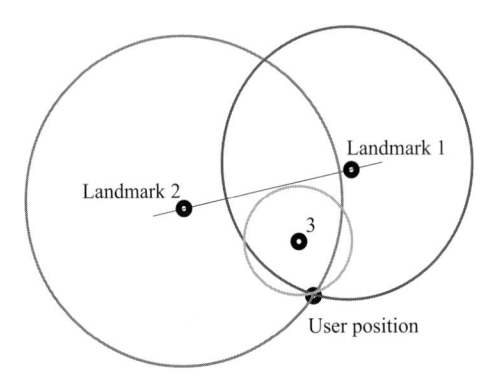

Figure 7.2 Ambiguity resolution by using a third landmark.

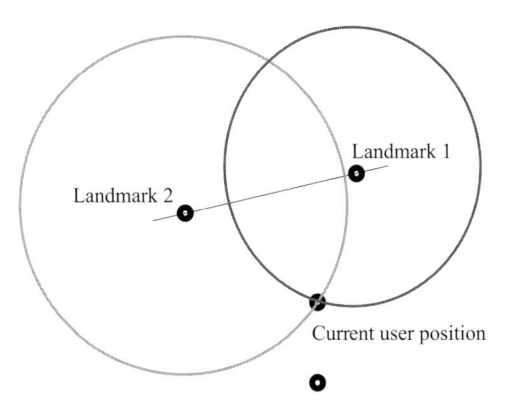

7.2 Localization, Mapping, and Navigation

Figure 7.3 Ambiguity resolution by monitoring the solution closest to the position in the previous instant.

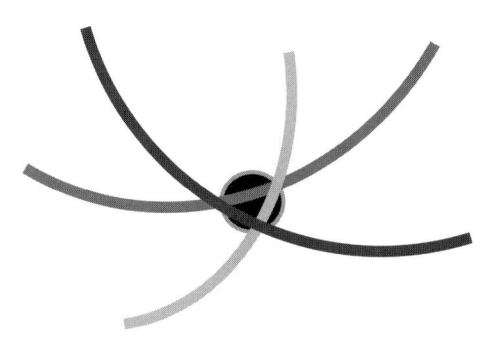

Figure 7.4 Close-up view of the user position in Figure 7.2.

Another method of ambiguity resolution is based on continuously monitoring the user position, so the solution closest to the position in the previous instant is chosen by evaluating the distance of all of the solutions to the position in the previous instant. This is illustrated in Figure 7.3.

Examining the user position closely in Figure 7.2, as shown in Figure 7.4, it may be noted that the three circles do not pass through the same point. Thus, there is a certain amount of uncertainty associated with the estimated position of the user.

One must therefore adopt a probabilistic approach to minimize the uncertainty associated with the estimated position of the user. Moreover, it may be also noted that, while the user position has been determined, albeit approximately, the orientation of the user has not. Two methods of determining the orientation are (i) monitoring the position over time to determine the heading angle of the vehicle or (ii) measuring the position of two or more points fixed in the body of the vehicle.

An approach that is commonly adopted for navigation of ships and aircraft involves the use of a master and a remotely located slave station that simultaneously transmit a signal. The difference between the arrival times of the signals from the two stations are accurately measured. Thus, the difference in distance of the user from the two stations is known. The locus of all points, which are always located such that the difference in distance from two reference stations is fixed, is a hyperbola. Consequently, the user may locate himself on a particular hyperbola on a chart consisting of a family of hyperbolae. Thus, a hyperbolic line of position along which the user is located is known on the chart. By considering the same master station and an alternative remotely located slave station, the user may locate himself on another hyperbola on the same chart. The intersection of the two hyperbolae enables the user to determine his position. A computer determines the position by finding the point of intersection of the two hyperbolae. The user is located at the latitude and longitude of the point of intersection. Here again, there is generally an ambiguity in the position of the user, which is resolved by using a third pair of master and remotely located slave stations. Thus, the user position is located to be within a region bounded by three hyperbolae, and the uncertainty is resolved by adopting probabilistic methods.

A similar approach may be adopted when the relative bearing of the user from two landmarks can be measured. The locus of a point that subtends the same angle with reference to two fixed points is a circle. Thus, by measuring the relative bearing of a user from two landmarks, the user is able to locate himself on a circle that passes through the locations of the two landmarks and the user's location. By using a third landmark, and measuring the relative bearings with respect to the other two, the user locates himself to be in the vicinity of a region bounded by three circles. Again, user position is located to be within a region bounded by three curves, and the uncertainty is resolved by adopting probabilistic methods. The uncertainty problem due to the uncertainty in the measurements is illustrated in Figure 7.5. The uncertainty problem is particularly difficult to resolve when the two landmarks and the user are almost along the same direction.

To deal with the problems of varying levels of uncertainty due to the geometry of the location and motion of the user relative to the landmarks, the notion of *dilution of precision* (DOP) is introduced. In particular, the loss of precision due to the geometry of

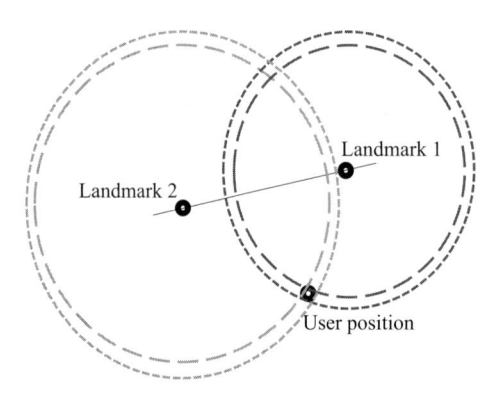

Figure 7.5 Uncertainty in locating the user due to the uncertainties in measurements.

7.2 Localization, Mapping, and Navigation

the location and motion of the user relative to the landmarks is referred to as the *geometric dilution of precision* (GDOP). In any range or bearing measurement system, the relative locations of the landmarks and the user influence the precision to which the user's position can be determined. In Figure 7.6 the uncertainty in the user's position is indicated by the shaded areas. In the first case, when the transmitters are far apart and not nearly co-linear with the user, the user's position uncertainty is small. Consequently, the GDOP can be said to be low. In the second case, the two landmarks are relatively close to each other and also almost co-linear. While the measurement uncertainty is the same, the user's position uncertainty is much larger. In this case, the GDOP can be said to be high. In three-dimensional range or bearing measurement systems one can distinguish between the *horizontal dilution of precision* (HDOP) and the *vertical dilution of precision* (VDOP). Moreover, one can also distinguish between the *positional dilution of precision* (PDOP) and the *temporal* or *time dilution of precision* (TDOP), as the range calculations are actually done in terms of the time of flight measurements.

The existence of clearly identifiable landmarks may not always be a reasonable assumption; in an unknown environment, a rover may have to select its own landmarks according to a set of metrics that will facilitate its localization. Alternatively, the rover may need to plant its own beacons as it moves forward, so the beacons may be used to localize the position of the rover as it moves forward. One method of establishing a landmark is to record a distinctive feature of it, such as its altitude, and one method of doing this is triangulation. The principle of triangulation is illustrated in Figure 7.7.

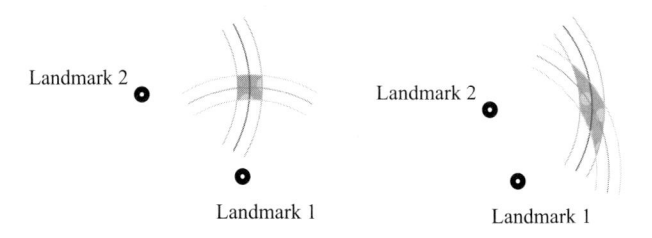

Figure 7.6 Illustration of the geometric dilution of precision.

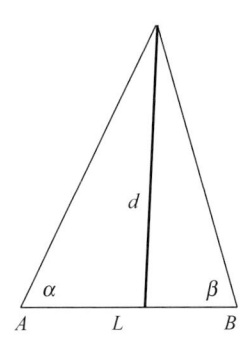

Figure 7.7 Principle of triangulation.

Triangulation is the process of determining the location of a point by measuring angles to it from known points at either end of a fixed baseline AB.

The baseline distance L is related to the height d. Thus,

$$L = \frac{d}{\tan \alpha} + \frac{d}{\tan \beta}, \quad d = \frac{L \tan \alpha \tan \beta}{\tan \alpha + \tan \beta}. \tag{7.2}$$

Once the height of the landmark is determined, its actual position relative to the points A and B could be determined by ranging. Thus, a landmark is established that may be used for localization of the rover. Triangulation was the method used by Sir George Everest to measure the height of the highest mountain in the Himalayan mountain range in the late 1800s.

7.3 Random Processes

A random, or stochastic, process is a sequence of events in which the outcome is probabilistic. The essential feature of a random process that distinguishes it from a deterministic process is that the state at any particular instant of time is not adequate to predict the future evolution of the process in time. One can imagine a large collection of physical processes, or an *ensemble*, each of which is indistinguishable from the others, in that it is statistically similar to the others and yet is different from all of the others due to slight variations in their responses. If all of the responses were available, one could determine the statistics of the ensemble. In particular, if $x_i(t)$ is a sample function of the ensemble of N members, one may define the following statistical properties by averaging over the ensemble,

$$\text{Mean:} \quad \bar{x}(t) = \sum_{i=1}^{N} x_i(t), \tag{7.3}$$

$$\text{Mean of the square:} \quad \overline{x^2}(t) = \sum_{i=1}^{N} x_i^2(t), \tag{7.4}$$

$$\text{the Variance,} \quad v(t) = \overline{x^2}(t) - (\bar{x}(t))^2, \tag{7.5}$$

$$\text{and the Correlation function,} \quad r(t, \tau) = \sum_{i=1}^{N} x_i(t) x_i(\tau). \tag{7.6}$$

In establishing these statistical properties, one has tacitly assumed that the sample functions are all equally probable. In reality, the sample functions may be characterized by a probability density function $p(x, t)$. The probability density function $p(x, t)$ defines the probability $p(x, t)dx$ of finding $x(t)$ within the limits $x + dx > x(t) > x$. When the probability density function is available, one may construct the ensemble averages defined by the use of the *expectation* operator, $E\{\cdot\}$, as,

Mean:
$$\mu(t) = E\{x(t)\} = \int_{-\infty}^{\infty} x(t) p(x(t), t) dx, \qquad (7.7)$$

Mean of the square:
$$E\{x^2(t)\} = \int_{-\infty}^{\infty} x^2(t) p(x(t), t) dx, \qquad (7.8)$$

and the variance,

$$v(t) = \sigma^2(t) = E\left\{(x(t) - \mu)^2\right\} = \int_{-\infty}^{\infty} (x(t) - \mu)^2 p(x(t), t) dx = E\{x^2(t)\} - \mu^2(t). \qquad (7.9)$$

To define the correlation function, one would need the joint probability density function $p(x_1(t), x_2(\tau); t, \tau)$. The correlation function follows and is defined as,

$$\rho(t, \tau) = E\{x(t)x(\tau)\} = \int_{-\infty}^{\infty} x_1(t) x_2(\tau) p(x_1(t), x_2(\tau); t, \tau) dx_1 dx_2. \qquad (7.10)$$

A generalization of the correlation function is the correlation matrix defined by,

$$\mathbf{R}_x(t, \tau) = E\{\mathbf{x}(t)\mathbf{x}^T(\tau)\}. \qquad (7.11)$$

The matrix $\mathbf{R}_x(t, t)$ is often called the covariance matrix of the vector process, $\mathbf{x}(t)$.

When the probability density function $p(x, t)$ is invariant following a translation of the time axis, such that $p(x, t) = p(x, t + \tau)$ for all τ, the process is said to be *stationary* (in the strict sense). When only the mean and the mean square are invariant with time, the process is said to *stationary* in a wide sense. Furthermore, when the time-averaged statistics are the same as the ensemble-averaged statistics, the process is said to be *ergodic*.

An important theoretical abstraction that is extremely useful in modeling other random processes is the concept of white noise. The correlation function of a white noise process is,

$$\rho(t, \tau) = \rho(t - \tau) = E\{x(t)x(\tau)\} = W\delta(t - \tau), \qquad (7.12)$$

where $\delta(t - \tau)$ is the Dirac delta function with the property,

$$\int_{-\infty}^{\infty} \delta(t - \tau) d\tau = 1. \qquad (7.13)$$

This implies that the correlation is a maximum, as $t \to \tau$ when $\rho(t - \tau) = \rho(0) \to \infty$ and zero everywhere else. Moreover, when W is constant, the White noise is said to be stationary and nonstationary when $W = W(t)$ is also a function of time.

Let us now suppose that white noise is the input to a linear system characterized by a known impulse response function. For a linear system with an input $\mathbf{u}(t)$ and an output $\mathbf{y}(t)$, the output can be obtained by applying the superposition integral and is given by,

$$\mathbf{y}(t) = \int_0^t \mathbf{H}(t,\tau)\mathbf{u}(\tau)d\tau, \tag{7.14}$$

where $\mathbf{H}(t,\tau)$ is the impulse response matrix of the system. The correlation matrix of the output process,

$$\mathbf{R}_y(t,\tau) = E\{\mathbf{y}(t)\mathbf{y}^T(\tau)\} = E\left\{\int_0^t \mathbf{H}(t,\tau)\mathbf{u}(\tau)d\tau \times \int_0^\tau \mathbf{u}^T(\lambda)\mathbf{H}^T(\tau,\lambda)d\lambda\right\}. \tag{7.15}$$

The product of integrals may be expressed as a double integral, and it follows that,

$$\mathbf{R}_y(t,\tau) = E\{\mathbf{y}(t)\mathbf{y}^T(\tau)\} = E\left\{\int_0^t\int_0^\tau \mathbf{H}(t,\tau)\mathbf{u}(\tau)\mathbf{u}^T(\lambda)\mathbf{H}^T(\tau,\lambda)d\lambda d\tau\right\}. \tag{7.16}$$

The expectation operator and the double integrals are interchangeable, as they are both linear operators and consequently we have,

$$\mathbf{R}_y(t,\tau) = E\{\mathbf{y}(t)\mathbf{y}^T(\tau)\} = \int_0^t\int_0^\tau \mathbf{H}(t,\tau)E\{\mathbf{u}(\tau)\mathbf{u}^T(\lambda)\}\mathbf{H}^T(\tau,\lambda)d\lambda d\tau. \tag{7.17}$$

When the input is white noise,

$$E\{\mathbf{u}(\tau)\mathbf{u}^T(\lambda)\} = \mathbf{Q}\delta(\tau-\lambda), \tag{7.18}$$

where the matrix $\mathbf{Q}$ is assumed to be constant. The property of the Dirac delta function is that, when $a < \tau < b$,

$$\int_a^b \delta(t-\tau)d\tau = 1. \tag{7.19}$$

The correlation matrix is then expressed as a single integral given by,

$$\mathbf{R}_y(t,\tau) = E\{\mathbf{y}(t)\mathbf{y}^T(\tau)\} = \int_0^t \mathbf{H}(t,\lambda)\mathbf{Q}\mathbf{H}^T(\tau,\lambda)d\lambda. \tag{7.20}$$

When the process is time-invariant, $\mathbf{H}(t,\tau) = \mathbf{H}(t-\tau)$, and by making the appropriate change in the variables, it follows that,

$$\mathbf{R}_y(t,t+\tau) = \int_0^t \mathbf{H}(t-\lambda)\mathbf{Q}\mathbf{H}^T(t+\tau-\lambda)d\lambda = \int_0^t \mathbf{H}(\lambda)\mathbf{Q}\mathbf{H}^T(\tau+\lambda)d\lambda. \tag{7.21}$$

In the limit, as $t \to \infty$, for a stationary process and for a stable system, the correlation function reaches a steady state and is,

$$\bar{\mathbf{R}}_y(\tau) = \int_0^\infty \mathbf{H}(\lambda)\mathbf{Q}\mathbf{H}^T(\tau+\lambda)d\lambda. \tag{7.22}$$

One may also consider the mean of the output of the linear system. Thus,

$$E\{\mathbf{y}(t)\} = E\left\{\int_0^t \mathbf{H}(t,\tau)\mathbf{u}(\tau)d\tau\right\} = \int_0^t \mathbf{H}(t,\tau)E\{\mathbf{u}(\tau)\}d\tau. \tag{7.23}$$

Thus, if the input is a zero-mean process, so is the output. These concepts are now applied to a system that can be represented by the state-space domain. Consider a system that is represented in the state-space domain by,

$$\dot{\mathbf{x}}(t) = \mathbf{A}\mathbf{x}(t) + \mathbf{B}\mathbf{w}, \quad \mathbf{y}(t) = \mathbf{C}\mathbf{x}(t). \tag{7.24}$$

In terms of the *transition matrix*, $\boldsymbol{\Phi}(t,\tau)$, the solution of the above state-space equations is,

$$\mathbf{x}(t) = \boldsymbol{\Phi}(t,t_0)\mathbf{x}(t_0) + \int_{t_0}^t \boldsymbol{\Phi}(t,\lambda)\mathbf{B}(\lambda)\mathbf{w}(\lambda)d\lambda. \tag{7.25}$$

It is assumed that the input $\mathbf{w}$ is a white noise process, and consequently it is assumed that,

$$E\{\mathbf{w}(\tau)\mathbf{w}^T(\lambda)\} = \mathbf{Q}\delta(\tau-\lambda), \tag{7.26}$$

where $\mathbf{Q}$ is assumed to be constant. It is also assumed that the covariance matrix of the initial conditions is known and given by,

$$E\{\mathbf{x}_0(t_0)\mathbf{x}_0^T(t_0)\} = \mathbf{P}(t_0). \tag{7.27}$$

It is also assumed that the vector process $\mathbf{x}_0(t_0)$ is independent of the input process $\mathbf{w}$. Thus, we may show that the correlation matrix may be expressed as,

$$\mathbf{R}_x(t,\tau) = \boldsymbol{\Phi}(t,t_0)\mathbf{P}(t_0)\boldsymbol{\Phi}^T(\tau,t_0) + \int_{t_0}^t \boldsymbol{\Phi}(t,\lambda)\mathbf{B}(\lambda)\mathbf{Q}\mathbf{B}^T(\lambda)\boldsymbol{\Phi}^T(\tau,\lambda)d\lambda. \tag{7.28}$$

For $\tau \geq t$, the above expression may be written as,

$$\mathbf{R}_x(t,\tau) = \left[\boldsymbol{\Phi}(t,t_0)\mathbf{P}(t_0)\boldsymbol{\Phi}^T(t,t_0) + \int_{t_0}^t \boldsymbol{\Phi}(t,\lambda)\mathbf{B}(\lambda)\mathbf{Q}\mathbf{B}^T(\lambda)\boldsymbol{\Phi}^T(t,\lambda)d\lambda\right]\boldsymbol{\Phi}^T(\tau,t). \tag{7.29}$$

However, the expression in the brackets is,

$$\mathbf{R}_x(t,t) = \boldsymbol{\Phi}(t,t_0)\mathbf{P}(t_0)\boldsymbol{\Phi}^T(t,t_0) + \int_{t_0}^t \boldsymbol{\Phi}(t,\lambda)\mathbf{B}(\lambda)\mathbf{Q}\mathbf{B}^T(\lambda)\boldsymbol{\Phi}^T(t,\lambda)d\lambda. \tag{7.30}$$

It is the covariance of the state vector process at time t and may be denoted as,

$$\mathbf{P}(t) = \mathbf{R}_x(t,t) = \mathbf{\Phi}(t,t_0)\mathbf{P}(t_0)\mathbf{\Phi}^T(t,t_0) + \int_{t_0}^{t} \mathbf{\Phi}(t,\lambda)\mathbf{B}(\lambda)\mathbf{Q}\mathbf{B}^T(\lambda)\mathbf{\Phi}^T(t,\lambda)d\lambda. \quad (7.31)$$

Furthermore, for $\tau \geq t$,

$$\mathbf{R}_x(t,\tau) = \mathbf{P}(t)\mathbf{\Phi}^T(\tau,t). \quad (7.32)$$

To find $\mathbf{R}_x(t,\tau)$ for $\tau \leq t$, recall that,

$$\mathbf{R}_x(t,\tau) = \mathbf{R}_x^T(\tau,t). \quad (7.33)$$

Also recall that the transition matrix satisfies the equation,

$$\frac{d}{dt}\mathbf{\Phi}(t,\tau) = \mathbf{A}\mathbf{\Phi}(t,\tau), \quad \mathbf{\Phi}(\tau,\tau) = \mathbf{I}. \quad (7.34)$$

Thus, differentiating the expression for $\mathbf{P}(t)$ and using Leibniz's rule for differentiation under the integral sign, $\mathbf{P}(t)$ satisfies the differential equation,

$$\dot{\mathbf{P}}(t) = \mathbf{A}\mathbf{P} + \mathbf{P}\mathbf{A}^T + \mathbf{F}\mathbf{Q}\mathbf{F}^T. \quad (7.35)$$

The initial conditions are given by,

$$\mathbf{P}(t)|_{t=t_0} = \mathbf{P}(t_0). \quad (7.36)$$

The correlation matrix of the output is given by,

$$\mathbf{R}_y(t,\tau) = E\{\mathbf{y}(t)\mathbf{y}^T(\tau)\} = E\{\mathbf{C}\mathbf{x}(t)\mathbf{x}^T(\tau)\mathbf{C}^T\} = \mathbf{C}\mathbf{R}_x(t,\tau)\mathbf{C}^T. \quad (7.37)$$

Thus,

$$\mathbf{P}_y(t) = \mathbf{C}\mathbf{P}(t)\mathbf{C}^T. \quad (7.38)$$

Consider a discrete-time system that is represented in the state-space domain by,

$$\mathbf{x}(k+1) = \mathbf{F}\mathbf{x}(k) + \mathbf{G}\mathbf{w}(k), \quad \mathbf{y}(k) = \mathbf{H}\mathbf{x}(k). \quad (7.39)$$

It is assumed that $\mathbf{w}$ is a discrete time White noise process and consequently it is assumed that,

$$E\{\mathbf{w}(k)\mathbf{w}^T(j)\} = \mathbf{Q}, \text{ when } j = k, \quad (7.40)$$

$$E\{\mathbf{w}(k)\mathbf{w}^T(j)\} = \mathbf{0}, \text{ when } j \neq k, \quad (7.41)$$

where $\mathbf{Q}$ is assumed to be constant. From the preceding analysis,

$$\mathbf{P}(k+1) = \mathbf{F}\mathbf{P}(k)\mathbf{F}^T + \mathbf{G}\mathbf{Q}\mathbf{G}^T, \quad (7.42)$$

$$\mathbf{P}_y(k) = \mathbf{H}\mathbf{P}(k)\mathbf{H}^T. \quad (7.43)$$

The matrix $\mathbf{F}$ of the discrete time system is indeed the transition matrix. In fact, the discrete time system driven by the white noise process has the *Markov* property.

A random process is said to have the *Markov* property if the conditional probability distribution of future states of the process (conditional on both past and present states) depends only upon the present state, and not on the sequence of events that preceded it. A process with this property is called a *Markov process*. Thus, a *Markov process* is a finite state stochastic process where the outcome at any stage depends only on the outcome of the previous stage with probabilities that are constant over time.

7.3.1 Basics of Probability

So far, the role of the probability density function, which has been assumed to be a generic function, $p(x,t)$ has not been considered. The role of the joint probability density function in multi-dimensional random processes, which was mentioned in passing in the preceding section, has not been discussed. There is a need to formally extend the concept of joint densities (or multivariate distributions) to the case of n variables,

$$\mathbf{x} = \begin{bmatrix} x_1 & x_2 & \cdots & x_{n-1} & x_n \end{bmatrix}^T. \tag{7.44}$$

Let us assume that the vector $\mathbf{x}$ is composed of two vectors,

$$\mathbf{x}_1 = \begin{bmatrix} x_1 & x_2 & \cdots & x_{r-1} & x_r \end{bmatrix}^T, \tag{7.45}$$

and,

$$\mathbf{x}_2 = \begin{bmatrix} x_{r+1} & x_{r+2} & \cdots & x_{n-1} & x_n \end{bmatrix}^T. \tag{7.46}$$

Thus,

$$\mathbf{x} = \begin{bmatrix} \mathbf{x}_1^T & \mathbf{x}_2^T \end{bmatrix}^T. \tag{7.47}$$

The probability distribution of the vector $\mathbf{x}_2$ is then expressed as,

$$p(\mathbf{x}_2) = \int_{\mathbf{x}_1 \in \mathcal{R}^r} p(\mathbf{x}_1, \mathbf{x}_2) d\mathbf{x}_1, \tag{7.48}$$

where $\mathcal{R}$ is the set of all real values from $-\infty$ to ∞. The joint probability density function $p(\mathbf{x}_1, \mathbf{x}_2)$ is defined as,

$$p(\mathbf{x}_1, \mathbf{x}_2) = p(\mathbf{x}_1 | \mathbf{x}_2) p(\mathbf{x}_2), \tag{7.49}$$

where $p(\mathbf{x}_1|\mathbf{x}_2)$ is the conditional probability density of $\mathbf{x}_1$ given the probability of $\mathbf{x}_2$. We can now define the notion of independence. Thus, for the two vectors $\mathbf{x}_1$ and $\mathbf{x}_2$ to be independent the conditional probability density should not depend on the conditional vector $\mathbf{x}_1$. Thus, it follows that for the two vectors $\mathbf{x}_1$ and $\mathbf{x}_2$ to be independent the conditional and joint probability densities are, respectively, given by,

$$p(\mathbf{x}_1|\mathbf{x}_2) = p(\mathbf{x}_1), \tag{7.50}$$

$$p(\mathbf{x}_1, \mathbf{x}_2) = p(\mathbf{x}_1)p(\mathbf{x}_2). \tag{7.51}$$

Given that the covariance matrix of the vector $\mathbf{x}$ may be partitioned as,

$$\mathbf{P} = \begin{bmatrix} \mathbf{P}_{11} & \mathbf{P}_{12} \\ \mathbf{P}_{21} & \mathbf{P}_{22} \end{bmatrix}, \tag{7.52}$$

the two vectors $\mathbf{x}_1$ and $\mathbf{x}_2$ are said to be *uncorrelated* if $\mathbf{P}_{12} = \mathbf{P}_{21} = \mathbf{0}$. Thus, for the two vectors $\mathbf{x}_1$ and $\mathbf{x}_2$ to be independent, they should also be uncorrelated. The converse is only true when the density function of the vector process $\mathbf{x}$ is a Gaussian density function.

The Gaussian probability density function of the vector process $\mathbf{x}$ is defined as,

$$p(\mathbf{x}) = (2\pi)^{-1}\|\mathbf{P}\|^{-1/2} \exp\left(-\frac{1}{2}(\mathbf{x} - E\{\mathbf{x}\})^T \mathbf{P}^{-1}(\mathbf{x} - E\{\mathbf{x}\})\right), \tag{7.53}$$

where $\|\mathbf{P}\|$ is the determinant of the covariance matrix. Two key properties of the Gaussian density function are that any linear combination of the vector process $\mathbf{x}$ is also characterized by a Gaussian density function. More importantly, the product of two Gaussian density functions is also a Gaussian probability density function. Finally, the conditional probability density function $p(\mathbf{x}_1|\mathbf{x}_2)$ of $\mathbf{x}_1$ given the probability of $\mathbf{x}_2$ is also Gaussian, with the mean and covariance, respectively, given by,

$$E\{\mathbf{x}_1|\mathbf{x}_2\} = E\{\mathbf{x}_1\} + \mathbf{P}_{12}\mathbf{P}_{22}^{-1}(\mathbf{x}_2 - E\{\mathbf{x}_2\}), \tag{7.54}$$

$$\mathbf{P}(\mathbf{x}_1|\mathbf{x}_2) = \mathbf{P}_{1|2} = \mathbf{P}_{11} - \mathbf{P}_{12}\mathbf{P}_{22}^{-1}\mathbf{P}_{21}. \tag{7.55}$$

Moreover, if the two Gaussian distributed vectors $\mathbf{x}_1$ and $\mathbf{x}_2$ are uncorrelated they are also independent, and vice-versa.

A very important application is the computation of the mean and covariance of a sum of random variables. Let $\mathbf{x}$ be a vector random process and let us assume that another scalar random process is defined by the linear transformation,

$$y = \mathbf{H}\mathbf{x}. \tag{7.56}$$

The variance of the scalar output is,

$$P_y = \mathbf{H}\mathbf{P}\mathbf{H}^T = \text{trace}\left(\mathbf{H}^T\mathbf{H}\mathbf{P}\right)$$

If the elements of $\mathbf{x}$ are uncorrelated,

$$P_y = \sum_{i=1}^{n} h_{ii}^2 p_{ii}. \tag{7.57}$$

In particular, if all of the elements of $\mathbf{x}$ have the same covariance such that

$$p_{ii} = \sigma^2, \tag{7.58}$$

and if $h_{ii} = 1/n$, then,

$$P_y = \sigma^2/n. \tag{7.59}$$

Thus, the variance has reduced by a factor n when compared with each element of the set.

7.3 Random Processes

Now consider the case when all of the elements of **x** are uncorrelated and also have the same mean but different covariance. Assume further that the weights h_{ii} are strictly positive and unequal but add up to unity. Again the variance of the weighted sum of the elements of **x** is,

$$P_y = \sum_{i=1}^{n} h_{ii}^2 p_{ii}. \tag{7.60}$$

Suppose one is free to choose the weights so as to minimize the variance. It is natural to weight those elements with a lower variance, as they are definitely more reliable because the uncertainty in these elements is less. Thus, the optimum choice of the weights h_{ii} must be inversely proportional to p_{ii}. Hence, it follows that,

$$h_{ii}^o = 1/(cp_{ii}). \tag{7.61}$$

The constant c is found by applying the constraint,

$$\sum_{i=1}^{n} a_{ii}^o = 1, \tag{7.62}$$

so,

$$\sum_{i=1}^{n} a_{ii}^o = \frac{1}{c} \sum_{i=1}^{n} \frac{1}{p_{ii}} = 1. \tag{7.63}$$

Hence,

$$c = \sum_{i=1}^{n} 1/p_{ii}. \tag{7.64}$$

This formula may be used for localization by constructing a weighted average of the multiple range, differential bearing, or differential range (hyperbolic navigation) measurements.

To understand the basis of sensor fusion, one could consider two independent measurements of a certain physical quantity z_1 and z_2 at two different times t_1 and t_2, where $t_1 < t_2$ and they are almost equal to each other; i.e. $t_1 \cong t_2$. We wish to estimate the physical quantity as a linear combination of the two independent measurements, and hence it is assumed that the estimate has the form,

$$\hat{x}(t) = (1 - K)z_1 + Kz_2, \tag{7.65}$$

where the symbol "^" refers to the fact that the variable is an "estimate of" the original variable. It is possible to show that the statistical expectation or mean value of the estimate satisfies the equation,

$$E(\hat{x}(t)) = (1 - K)E(z_1) + KE(z_2). \tag{7.66}$$

The variance of the estimate denoted by p_x is equal to,

$$p_x = (1 - K)^2 p_{z1} + K^2 p_{z2}, \tag{7.67}$$

where p_{z1} and p_{z2} are the corresponding variances of the two measurements z_1 and z_2. The variance of the estimate has a minimum value when K is given by,

$$K = p_{z1}/(p_{z1} + p_{z2}). \qquad (7.68)$$

In the absence of z_2, z_1 may be considered the best estimate, $\hat{x}(t_1)$. The equation for the estimate at time $t = t_2$ may be written as,

$$\hat{x}(t_2) = \hat{x}(t_1) + K(t_2)(z_2 - \hat{x}(t_1)), \quad K = p_{z1}/(p_{z1} + p_{z2}). \qquad (7.69)$$

The equation for $\hat{x}(t_2)$ may be considered an update of the earlier estimate, $\hat{x}(t_1)$, when an additional measurement is available and $p_{z1} + p_{z2}$ may be interpreted, as the variance of the residual, $z_2 - \hat{x}(t_1)$. Not only is the equation defining the estimate $\hat{x}(t_2)$ the basis of the principle of sensor fusion but it is also the basis of the Kalman filter.

7.3.2 The Kalman Filter

It is now assumed that that the state of a system satisfies the discrete time update equations,

$$\mathbf{x}(k+1) = \mathbf{F}(k)\mathbf{x}(k) + \mathbf{G}(k)\mathbf{w}(k), \quad \mathbf{z}(k) = \mathbf{H}(k)\mathbf{x}(k) + \mathbf{v}(k), \qquad (7.70)$$

where $\mathbf{F}(k)$ is the state transition matrix of the state vector from the state at the time k to the time at $k+1$, and $\mathbf{z}(k)$ represents a measurement of a linear combination of states that is corrupted by additive discrete time white noise. It is assumed that $\mathbf{w}$ and $\mathbf{v}$ are discrete time white noise processes, independent of each other. It is assumed that,

$$E\{\mathbf{w}(k)\mathbf{w}^T(j)\} = \mathbf{Q}, \quad E\{\mathbf{v}(k)\mathbf{v}^T(j)\} = \mathbf{R}, \quad \text{when} \quad j = k, \qquad (7.71)$$

$$E\{\mathbf{w}(k)\mathbf{w}^T(j)\} = \mathbf{0}, \quad E\{\mathbf{v}(k)\mathbf{v}^T(j)\} = \mathbf{0}, \quad \text{when} \quad j \neq k, \qquad (7.72)$$

where $\mathbf{Q}$ and $\mathbf{R}$ is assumed to be constant.

The overall objective is to estimate $\mathbf{x}(k)$. The difference between the final estimate of $\mathbf{x}(k)$, denoted by $\hat{\mathbf{x}}(k)$ and $\mathbf{x}(k)$ itself, is termed the error and defined as, $\mathbf{e}(k) = \hat{\mathbf{x}}(k) - \mathbf{x}(k)$. The mean squared error (MSE) function is defined as,

$$\bar{\varepsilon}^2 = E\{\mathbf{e}(k)\mathbf{e}^T(k)\} = E\{(\hat{\mathbf{x}}(k) - \mathbf{x}(k))(\hat{\mathbf{x}}(k) - \mathbf{x}(k))^T\}. \qquad (7.73)$$

For the minimization of the MSE to yield the optimal filter it is essential, as will be seen later, to model the system measurement and the state error covariance matrices using Gaussian distributed random processes. The prior estimate of $\mathbf{x}(k)$, prior to the availability of the measurement, is denoted as $\hat{\mathbf{x}}^-(k)$. It should be possible to write an update equation for the final estimate after the availability of the measurement, combining the old estimate with measurement data. Thus,

$$\hat{\mathbf{x}}(k) = \hat{\mathbf{x}}^-(k) + \mathbf{K}(k)(\mathbf{z}(k) - \mathbf{H}(k)\hat{\mathbf{x}}^-(k)). \qquad (7.74)$$

In the above update equation, $\mathbf{K}(k)$ is the Kalman gain that we wish to specify to define the filter, and $\mathbf{z}(k) - \mathbf{H}(k)\hat{\mathbf{x}}^-(k)$ is the new information or *innovation* available in the measurement $\mathbf{z}(k)$. However, since,

$$\mathbf{z}(k) = \mathbf{H}(k)\mathbf{x}(k) + \mathbf{v}(k), \tag{7.75}$$

$$\hat{\mathbf{x}}(k) = \hat{\mathbf{x}}^-(k) + \mathbf{K}(k)(\mathbf{H}(k)\mathbf{x}(k) + \mathbf{v}(k) - \mathbf{H}(k)\hat{\mathbf{x}}^-(k)). \tag{7.76}$$

Thus,

$$\hat{\mathbf{x}}(k) = \hat{\mathbf{x}}^-(k) + \mathbf{K}(k)\mathbf{H}(k)(\mathbf{x}(k) - \hat{\mathbf{x}}^-(k)) + \mathbf{K}(k)\mathbf{v}(k). \tag{7.77}$$

It follows that the residual or estimation error is,

$$\mathbf{e}(k) = \hat{\mathbf{x}}(k) - \mathbf{x}(k) = (\mathbf{I} - \mathbf{K}(k)\mathbf{H}(k))(\hat{\mathbf{x}}^-(k) - \mathbf{x}(k)) + \mathbf{K}(k)\mathbf{v}(k). \tag{7.78}$$

Assuming that the a priori estimate error $\hat{\mathbf{x}}^-(k) - \mathbf{x}(k)$ and the measurement noise $\mathbf{v}(k)$ are uncorrelated and independent random processes,

$$\mathbf{P}_\varepsilon(k) = (\mathbf{I} - \mathbf{K}(k)\mathbf{H}(k))\mathbf{P}_\varepsilon^-(k)(\mathbf{I} - \mathbf{K}(k)\mathbf{H}(k))^T + \mathbf{K}(k)\mathbf{R}\mathbf{K}^T(k). \tag{7.79}$$

In this equation, $\mathbf{P}_\varepsilon^-(k)$ is the a priori estimate of the error covariance matrix $\mathbf{P}_\varepsilon(k)$. Suppressing the explicit dependence of terms on the time k, the equation may be expanded and expressed as,

$$\mathbf{P}_\varepsilon = \mathbf{P}_\varepsilon^- - \mathbf{K}\mathbf{H}\mathbf{P}_\varepsilon^- - \mathbf{P}_\varepsilon^-(\mathbf{K}\mathbf{H})^T + \mathbf{K}(\mathbf{H}\mathbf{P}_\varepsilon^-\mathbf{H}^T + \mathbf{R})\mathbf{K}^T. \tag{7.80}$$

In order find the optimum gain, we consider the trace of the matrix $\mathbf{P}_\varepsilon(k)$ and minimize. In doing so we have tacitly assumed that each of the elements of the error vector $\mathbf{e}(k)$ are uncorrelated with each other; this is only possible if the probability density function of the random error vector process $\mathbf{e}(k)$ is a Gaussian probability density function. Alternatively, we may minimize the mean square error. It can then be shown that the optimum Kalman gain is given by,

$$\mathbf{K} = \mathbf{P}_\varepsilon^- \mathbf{H}^T \left(\mathbf{H}\mathbf{P}_\varepsilon^- \mathbf{H}^T + \mathbf{R} \right)^{-1}. \tag{7.81}$$

and that,

$$\mathbf{P}_\varepsilon = (\mathbf{I} - \mathbf{K}\mathbf{H})\mathbf{P}_\varepsilon^-. \tag{7.82}$$

Now the a priori estimate of the error covariance matrix $\mathbf{P}_\varepsilon(k)$, $\mathbf{P}_\varepsilon^-(k)$ could be predicted from the error covariance estimate of the previous step. Thus, the priori estimate of the error covariance matrix $\mathbf{P}_\varepsilon^-(k+1)$ is related to the error covariance matrix $\mathbf{P}_\varepsilon(k)$ by the prediction equation,

$$\mathbf{P}_\varepsilon^-(k+1) = \mathbf{F}(k)\mathbf{P}_\varepsilon(k)\mathbf{F}(k)^T + \mathbf{G}(k)\mathbf{Q}\mathbf{G}(k)^T. \tag{7.83}$$

Thus, the Kalman filter could be implemented recursively. A simple example of its application in the continuous time domain will now be discussed.

Given the process dynamics in the continuous time domain, with a deterministic input vector $\mathbf{u}(t)$, in the form,

$$\dot{\mathbf{x}}(t) = \mathbf{A}(t)\mathbf{x}(t) + \mathbf{B}(t)\mathbf{u}(t) + \mathbf{G}(t)\mathbf{w}(t), \tag{7.84}$$

with the measurement vector $\mathbf{z}(t)$ linearly related to the state vector $\mathbf{x}(t)$ as,

$$\mathbf{z}(t) = \mathbf{C}(t)\mathbf{x}(t) + \mathbf{v}(t), \tag{7.85}$$

the optimal estimate satisfies the equation,

$$\dot{\hat{\mathbf{x}}}(t) = \mathbf{A}(t)\hat{\mathbf{x}}(t) + \mathbf{B}(t)\mathbf{u}(t) + \mathbf{P}\mathbf{C}^T\mathbf{R}^{-1}(\mathbf{z} - \mathbf{C}\hat{\mathbf{x}}(t)), \tag{7.86}$$

where **P** is the covariance matrix,

$$\mathbf{P} = E\{(\mathbf{x}-\hat{\mathbf{x}})(\mathbf{x}-\hat{\mathbf{x}})^T\}. \tag{7.87}$$

The matrix **P** satisfies the nonlinear matrix differential equation,

$$\dot{\mathbf{P}} = \mathbf{A}\mathbf{P} + \mathbf{P}\mathbf{A}^T - \mathbf{P}\mathbf{C}^T\mathbf{R}^{-1}\mathbf{C}\mathbf{P} + \mathbf{Q}. \tag{7.88}$$

In a simple application, it is assumed that one has noisy measurements of both the distance traveled $d(t)$ and the translational velocity $V(t)$ of a vehicle travelling along a straight path. Thus, the "process" and measurement models are, respectively, given by,

$$\dot{x}(t) = V(t) + w(t), \tag{7.89}$$

$$d(t) = x(t) + v(t). \tag{7.90}$$

Comparing with our general process model, and setting the number states to 1, $\mathbf{A}(t) = 0$, $\mathbf{B}(t) = \mathbf{G}(t) = 1$, $\mathbf{C}(t) = 1$, the variances of the two independent noise sources, $w(t)$ and $v(t)$, respectively, to the covariance of the noise signals, $\mathbf{w}(t)$ and $\mathbf{v}(t)$, so, $\mathbf{Q} = q^2$, $\mathbf{R} = r^2$, the equation for the matrix $\mathbf{P} = p$ reduces to,

$$\dot{p} = -(p^2/r^2) + q^2. \tag{7.91}$$

Under steady state conditions, $p = rq$. The optimal filter gain is p/r^2.

The near-steady state estimate of the distance traveled satisfies,

$$\dot{\hat{x}}(t) = V(t) + (q/r)(d - \hat{x}(t)). \tag{7.92}$$

It is seen from this differential equation for the estimate of the distance traveled that the distance measurement $d(t)$ is ignored when r, the standard deviation of the noise in the distance measurement, is large. On the other hand, when r, the standard deviation of the noise in the distance measurement, is small, the distance measurement $d(t)$ is given more importance and the velocity measurement $V(t)$ is practically ignored. It may be noted that the same filter could be implemented in the discrete time domain as well. To do this, one approximates the process dynamics in the discrete time domain as,

$$x(k+1) = x(k) + \Delta t V(k) + \Delta t w(k), \tag{7.93}$$

$$d(k) = x(k) + v(k). \tag{7.94}$$

In these equations, $t = k\Delta t$, where Δt is the sampling time step. The discrete time Kalman filter may now be applied to construct an estimate of the distance traveled by the vehicle.

Another important application is in estimating a constant such as a bias, assuming that a measurement of it is available. In this case, one obtains the same optimal filter as the one defined by Equation (7.92), with the velocity measurement $V(t)$ set to zero.

When the uncertainty associated with both the motion state vector and the measurement can be described by a Gaussian probability density function, and the uncertainty in the initial state is also specified by a Gaussian probability density function, then the probability density of the state estimate and the error in the estimate will remain Gaussian at all times. This corresponds to the case of the classical Kalman filter [2]. When both the state and measurement statistics are known with some accuracy, Kalman filter-based techniques have proven to be robust and reasonably accurate in continuously tracking the position and orientation of rover vehicles. As the filter can be implemented using just the mean and covariance matrix, which are also adequate for defining the probability density function, the Kalman filter algorithm can be implemented efficiently. Unfortunately, the assumption that the uncertainties in the state and measurement vectors are described by Gaussian density functions can be restrictive, and consequently the Kalman filter cannot fully accommodate either nonlinear or non-Gaussian motion and measurement uncertainties. Thus, the Kalman filter can sometimes exhibit a form of divergence when used for global localization applications. Although nonoptimal extensions of the Kalman filter exist, such as the *extended Kalman filter* (EKF) and the *unscented Kalman filter* (UKF), they are only known to perform accurately when the uncertainties can also be modeled by Gaussian probability density functions.

7.3.3 Probabilistic Methods and Essentials of Bayesian Inference

An alternative approach to Kalman filtering is to directly estimate the conditional probability density function of the current state vector, given a set of new measurements and the inputs to the system. The approach relies on Bayes' rule in probability theory (see for example [3]). Bayes' rule defines the probability of an event occurring based on the prior knowledge of the conditions that facilitate the occurrence of the event. Specifically when the conditional probability of the event A occurring conditioned to the event B having occurred, $p(A|B)$, is known, Bayes' rule allows one to estimate the conditional probability $p(B|A)$, using the fact that,

$$p(A, B) = p(A|B)p(B) = p(B|A)p(A). \tag{7.95}$$

Thus, it follows that, $p(B|A) = p(A|B)(p(B)/p(A))$. Of course, we have tacitly assumed that only two events A and B can occur. In general, if there is a possibility of multiple events A_i and another conditional event B,

$$p(A_i|B)p(B) = p(A_i|B)\left\{\sum_i p(B|A_i)p(A_i)\right\} = p(B|A_i)p(A_i); \tag{7.96}$$

an alternative form is,

$$p(A_i|B)\{p(B|A_i)p(A_i) + p(B|\text{not } A_i)p(\text{not } A_i)\} = p(B|A_i)p(A_i). \quad (7.97)$$

Here $p(B|A)$ is the posterior probability, $p(B)$, the prior probability, $p(A|B)$ is the likelihood and $p(A)$ the evidence. Thus Bayes' rule allows one to combine prior knowledge (prior statistical probability densities) and new information to estimate the a posteriori statistical probability densities. The "posterior" probability of the state given the new information may be considered to be an optimal combination of the prior statistics or probability density and new data presented by the information, weighted by their relative precision.

Consider a simple nonlinear dynamic system,

$$\mathbf{x}(k+1) = \mathbf{F}(\mathbf{x}(k), \mathbf{w}(k)), \quad \mathbf{z}(k) = \mathbf{H}(\mathbf{x}(k), \mathbf{v}(k)). \quad (7.98)$$

Since the process is Markovian, the probability density $p(\mathbf{x}(k+1)|\mathbf{x}(k))$ of the state $\mathbf{x}(k+1)$ is conditioned only on $\mathbf{x}(k)$, the measurements are conditionally independent given the states, $p(\mathbf{z}(k)|\mathbf{x}(k))$, and the probability of the initial state, $p(\mathbf{x}(0)|\mathbf{z}(0))$ is assumed to be known. The problem is to predict $p(\mathbf{x}(k+1)|\mathbf{z}_{1:k-1})$ from $p(\mathbf{x}(k)|\mathbf{z}_{1:k-1})$ when no additional information is available other than the measurements from the initial time $k=1$ up to the current time $k-1$; i.e.

$$p(\mathbf{x}(k)|\mathbf{z}_{1:k-1}) \leftarrow p(\mathbf{x}(k-1)|\mathbf{z}_{1:k-1}). \quad (7.99)$$

At each point in time, the probability distribution $p(\mathbf{x}(k-1)|\mathbf{z}_{1:k-1})$ is also called the belief, $Bel(\mathbf{x}(k-1))$, and represents the uncertainty. The prediction is followed by a measurement update, to update the probability in the light of new data,

$$p(\mathbf{x}(k)|\mathbf{z}_{1:k}) \leftarrow p(\mathbf{x}(k)|\mathbf{z}_{1:k-1})$$

The prediction is based on the system model $p(x(k)|x(k-1))$ and it follows that,

$$p(\mathbf{x}(k)|\mathbf{z}_{1:k-1}) = \int p(x(k)|x(k-1))p(\mathbf{x}(k-1)|\mathbf{z}_{1:k-1})dx(k-1). \quad (7.100)$$

To update the probability in the light of new data, we adopt the Bayesian approach. Thus,

$$p(\mathbf{x}(k)|\mathbf{z}_{1:k}) = p(\mathbf{x}(k)|\mathbf{z}_{1:k-1}, \mathbf{z}(k)). \quad (7.101)$$

Using Bayes' rule, the measurement model or likelihood, the prior state probability and the evidence, the update of the probability density is,

$$p(\mathbf{x}(k)|\mathbf{z}_{1:k-1}, \mathbf{z}(k)) = p(\mathbf{z}_{1:k-1}, \mathbf{z}(k)|\mathbf{x}(k))p(\mathbf{x}(k)|\mathbf{z}_{1:k-1})/p(\mathbf{z}(k)|\mathbf{z}_{1:k-1}). \quad (7.102)$$

The evidence is expressed as,

$$p(\mathbf{z}(k)|\mathbf{z}_{1:k-1}) = \int p(\mathbf{z}(k)|x(k))p(\mathbf{x}(k)|\mathbf{z}_{1:k-1})dx(k). \quad (7.103)$$

The problem now reduces to estimating the actual probability density functions. This requires the evaluation of integrals. Moreover, the Kalman filter is seen to be a special case of the general filtering approach described previously.

7.4 Probabilistic Representation of Uncertain Motion Using Particles

In one approach to Bayesian filtering, the joint posterior probability density function at the time k is approximated by a weighted summation of Dirac Delta functions at a finite set of sample points. These sample points are referred to as "particles," and consequently the joint posterior probability density function at the time k is characterized by a set of particles and importance weights. Particle filters are an efficient way to represent non-Gaussian distributions. Two techniques, one known as *Monte Carlo integration* and the other known as *Importance sampling*, form the basis of this approach. The principle of Monte Carlo integration is based on evaluating the expected value of a nonlinear function of a set of independent, identically distributed random variables. In computing the expected value of a nonlinear function of a set of independent, identically distributed random variables, one is not constrained to using a particular distribution of random variables. One could use any distribution of random variables. However, since such a distribution would differ from the joint prior probability density function, one multiplies the integrand, at each sample point, by the corresponding importance weight. The importance weights therefore represent the ratio of the target to the sampling distribution. Another important operation in the execution of the particle-based Bayesian filter is *resampling*, where unlikely samples are replaced by more likely samples. Importance sampling for Bayesian estimation involves two steps: (i) weight update and (ii) particle update. After introducing key ideas underpinning the particle filter, the implementation and operation of the particle filter algorithm will be briefly discussed.

7.4.1 Monte Carlo Integration, Normalization, and Resampling

To fix ideas, Monte Carlo integration is introduced. Consider the weighted integral,

$$I = \int_a^b f(x)t(x)dx, \qquad (7.104)$$

where the weighting function $t(x)$ has the properties of a probability function and is referred to as a target probability density satisfying,

$$1 = \int_a^b t(x)dx. \qquad (7.105)$$

The integral I may be defined as an expectation and evaluated by drawing a large number of samples x_i, which are independent and identically distributed. Thus,

$$I = E\{f(x)\} = \int_a^b f(x)t(x)dx = \frac{1}{N}\sum_{i=1}^{N} f(x_i), \quad N \to \infty. \qquad (7.106)$$

It may not always be possible to find a target probability density function $t(x)$ suitable for a particular integral and so we consider the use of an alternative probability density function $q(x)$ defined over an alternative set of limits, $[-\infty \ \ \infty]$. Thus, the integral I may be expressed as,

$$I = \int_{-\infty}^{\infty} \frac{t(x)h(x)}{q(x)} f(x)q(x)dx \equiv \int_{-\infty}^{\infty} w(x)f(x)q(x)dx, w(x) \equiv \frac{t(x)h(x)}{q(x)}, \quad (7.107)$$

where $h(x)$ is equal to 1 when $a \leq x \leq b$ and zero otherwise. Again, the integral I may be defined as an expectation and evaluated by drawing a large number of samples x_i, which are independent and identically distributed with probability density function $q(x)$ defined over an alternative set of limits, $[-\infty \ \ \infty]$. Thus,

$$I = E\{f(x)\} = \int_{-\infty}^{\infty} w(x)f(x)q(x)dx = \sum_{i=1}^{N} w_i f(x_i). \quad (7.108)$$

The weights w_i are known as *importance weights* and are the principle behind *importance sampling*. Moreover, when $f(x) = 1$, $q(x)$ is a normalized probability density function, the weights must satisfy,

$$1 = \sum_{i=1}^{N} w_i. \quad (7.109)$$

When the weights do not satisfy the normalizing condition, they are normalized. Thus,

$$I = \sum_{i=1}^{N} \left(w_i \bigg/ \sum_i w_i \right) f(x_i) = \sum_{i=1}^{N} \bar{w}_i f(x_i), \quad \bar{w}_i \equiv w_i \bigg/ \sum_i w_i. \quad (7.110)$$

Now the normalized weights $\bar{w}_i$ may themselves be considered as probabilities, and therefore it must be possible to resample with a set of samples $\bar{x}_i$ with probabilities, $P(\bar{x}_i = x_i) = \bar{w}_i$. This is the principle behind *importance resampling*. There are indeed a number of algorithms both for importance sampling and for importance resampling, such as *sequential importance sampling* and *sampling importance resampling*, and the actual algorithm used can depend on the nature of the application. Thus, following importance resampling,

$$I = \frac{1}{N} \sum_{i=1}^{N} f(\bar{x}_i). \quad (7.111)$$

7.4.2 The Particle Filter

It is possible to introduce the particle filter. The primary assumption in the development of the particle filter is to assume that the conditional probability density function $p(\mathbf{x}(k)|\mathbf{z}_{1:k})$ is approximated by a weighted summation of Dirac delta functions. Thus, given the state vector at time k, and a set of measurements made prior to this time, the conditional probability density function $p(\mathbf{x}(k)|\mathbf{z}_{1:k})$ is approximated as,

$$p(\mathbf{x}(k)|\mathbf{z}_{1:k}) = \sum_{i=1}^{N} \mathbf{w}_i(k)\delta(\mathbf{x}(k) - \mathbf{x}_i(k)), \quad \mathbf{w}_i(k) \geq 0, \quad \sum_{i=1}^{N} \mathbf{w}_i(k) = 1. \quad (7.112)$$

In this representation, $\mathbf{x}_i(k)$ are the sample vectors, which are known as "*particles*," while $\mathbf{w}_i(k)$ are the weights. The samples are drawn from a certain target density approximation function, which is given by,

$$t(\mathbf{x}) = \frac{1}{N}\sum_{i=1}^{N} \delta(\mathbf{x}(k) - \mathbf{x}_i(k)). \quad (7.113)$$

Thus, given a function $g(\mathbf{x})$, the expected value is given by,

$$E\{g(\mathbf{x})\} = \frac{1}{N}\int_{\mathbf{x}} g(\mathbf{x}) \sum_{i=1}^{N} \delta(\mathbf{x}(k) - \mathbf{x}_i(k)) d\mathbf{x} = \frac{1}{N}\sum_{i=1}^{N} g(\mathbf{x}_i(k)). \quad (7.114)$$

On the other hand, given that the samples are drawn from the alternative of the *importance probability density function*, $q(\mathbf{x})$, the target and importance density approximation functions may be, respectively, represented in the form,

$$t(\mathbf{x}) = \sum_{i=1}^{N} \mathbf{w}_i(k)\delta(\mathbf{x}(k) - \mathbf{x}_i(k)), \quad q(\mathbf{x}) = \frac{1}{N}\sum_{i=1}^{N} \delta(\mathbf{x}(k) - \mathbf{x}_i(k)). \quad (7.115)$$

Thus, given the function $g(\mathbf{x})$, the expected value is given by,

$$E\{g(\mathbf{x})\} = \int_{\mathbf{x}} g(\mathbf{x}) \sum_{i=1}^{N} \mathbf{w}_i(k)\delta(\mathbf{x}(k) - \mathbf{x}_i(k)) d\mathbf{x} = \sum_{i=1}^{N} \mathbf{w}_i(k)g(\mathbf{x}_i(k)). \quad (7.116)$$

Based on the representation of $p(\mathbf{x}(k)|\mathbf{z}_{1:k-1})$, the conditional expected value is given by,

$$E\{g(\mathbf{x})\} = \int_{\mathbf{x}} g(\mathbf{x}) \sum_{i=1}^{N} \mathbf{w}_i(k)\delta(\mathbf{x}(k) - \mathbf{x}_i(k)) d\mathbf{x} = \sum_{i=1}^{N} \mathbf{w}_i(k)g(\mathbf{x}_i(k)). \quad (7.117)$$

It is clear that the preceding two expressions are the same; the conditional probability density function $p(\mathbf{x}(k)|\mathbf{z}_{1:k-1})$ could be used as the importance probability density function. For the importance distribution, though, one has,

$$p(\mathbf{x}(k)|\mathbf{z}_{1:k-1}) = \frac{1}{N}\sum_{i=1}^{N} \delta(\mathbf{x}(k) - \mathbf{x}_i(k)). \quad (7.118)$$

Furthermore,

$$p(\mathbf{x}(k-1)|\mathbf{z}_{1:k-1}) = \frac{1}{N}\sum_{i=1}^{N} \delta(\mathbf{x}(k-1) - \mathbf{x}_i(k-1)). \quad (7.119)$$

Considering, the Bayesian prediction density,

$$p(\mathbf{x}(k)|\mathbf{z}_{1:k-1}) = \int p(\mathbf{x}(k)|\mathbf{x}(k-1))p(\mathbf{x}(k-1)|\mathbf{z}_{1:k-1})dx(k-1), \quad (7.120)$$

it follows that,

$$p(\mathbf{x}(k)|\mathbf{z}_{1:k-1}) = \frac{1}{N}\int p(\mathbf{x}(k)|\mathbf{x}(k-1))\sum_{i=1}^{N}\delta(\mathbf{x}(k-1)-\mathbf{x}_i(k-1))dx(k-1). \quad (7.121)$$

Thus,

$$p(\mathbf{x}(k)|\mathbf{z}_{1:k-1}) = \frac{1}{N}\sum_{i=1}^{N}p(\mathbf{x}(k)|\mathbf{x}(k-1)). \quad (7.122)$$

It is quite feasible to use the conditional density $q(\mathbf{x}(k)|\mathbf{x}_i(k-1), \mathbf{z}_{1:k})$ as the importance distribution and $p(\mathbf{x}(k)|\mathbf{z}_{1:k-1})$ as the target distribution, so one has,

$$p(\mathbf{x}(k)|\mathbf{z}_{1:k-1}) = \sum_{i=1}^{N}\mathbf{w}_i(k)\delta(\mathbf{x}(k)-\mathbf{x}_i(k)). \quad (7.123)$$

The importance density is approximated as,

$$q(\mathbf{x}(k)|\mathbf{x}_i(k-1), \mathbf{z}_{1:k}) = \frac{1}{N}\sum_{i=1}^{N}\delta(\mathbf{x}(k)-\mathbf{x}_i(k)). \quad (7.124)$$

Hence, weights must satisfy,

$$\mathbf{w}_i(k) = p(\mathbf{x}_i(k)|\mathbf{z}_{1:k-1})/q(\mathbf{x}(k)|\mathbf{x}_i(k-1), \mathbf{z}_{1:k}). \quad (7.125)$$

Furthermore,

$$p(\mathbf{x}(k-1)|\mathbf{z}_{1:k-1}) = \sum_{i=1}^{N}\mathbf{w}_i(k-1)\delta(\mathbf{x}(k-1)-\mathbf{x}_i(k-1)). \quad (7.126)$$

Re-considering the Bayesian prediction density,

$$p(\mathbf{x}(k)|\mathbf{z}_{1:k-1}) = \int p(\mathbf{x}(k)|\mathbf{x}(k-1))p(\mathbf{x}(k-1)|\mathbf{z}_{1:k-1})dx(k-1), \quad (7.127)$$

it follows that,

$$p(\mathbf{x}(k)|\mathbf{z}_{1:k-1}) = \int p(\mathbf{x}(k)|\mathbf{x}(k-1))\sum_{i=1}^{N}\mathbf{w}_i(k-1)\delta(\mathbf{x}(k-1)-\mathbf{x}_i(k-1))dx(k-1). \quad (7.128)$$

It follows that,

$$p(\mathbf{x}(k)|\mathbf{z}_{1:k-1}) = \sum_{i=1}^{N}\mathbf{w}_i(k-1)p(\mathbf{x}(k)|\mathbf{x}(k-1)). \quad (7.129)$$

7.4 Probabilistic Representation of Uncertain Motion Using Particles

Thus, one has the weights time-update equation,

$$\mathbf{w}_i(k) = \sum_{i=1}^{N} \frac{p(\mathbf{x}(k)|\mathbf{x}(k-1))}{q(\mathbf{x}(k)|\mathbf{x}_i(k-1),\mathbf{z}_{1:k})} \mathbf{w}_i(k-1). \tag{7.130}$$

The Bayesian measurement probability update was expressed as,

$$p(\mathbf{x}(k)|\mathbf{z}_{1:k-1},\mathbf{z}(k)) = p(\mathbf{z}_{1:k-1},\mathbf{z}(k)|\mathbf{x}(k))p(\mathbf{x}(k)|\mathbf{z}_{1:k-1})/p(\mathbf{z}(k)|\mathbf{z}_{1:k-1}), \tag{7.131}$$

with,

$$p(\mathbf{z}(k)|\mathbf{z}_{1:k-1}) = \int p(\mathbf{z}(k)|\mathbf{x}(k))p(\mathbf{x}(k)|\mathbf{z}_{1:k-1})dx(k). \tag{7.132}$$

From the preceding equation,

$$p(\mathbf{z}(k)|\mathbf{z}_{1:k-1}) = \int p(\mathbf{z}(k)|\mathbf{x}(k)) \sum_{i=1}^{N} \mathbf{w}_i(k)\delta(\mathbf{x}(k) - \mathbf{x}_i(k))dx(k). \tag{7.133}$$

It follows that,

$$p(\mathbf{z}(k)|\mathbf{z}_{1:k-1}) = \sum_{i=1}^{N} \mathbf{w}_i(k)p(\mathbf{z}(k)|\mathbf{x}_i(k)). \tag{7.134}$$

Thus, we have the following discrete approximation to the measurement probability update,

$$p(\mathbf{x}(k)|\mathbf{z}_{1:k}) = \sum_{i=1}^{N} \frac{p(\mathbf{z}_{1:k-1},\mathbf{z}(k)|\mathbf{x}_i(k))\mathbf{w}_i(k)}{\sum_{i=1}^{N} \mathbf{w}_i(k)p(\mathbf{z}(k)|\mathbf{x}_i(k))} \delta(\mathbf{x}(k) - \mathbf{x}_i(k)). \tag{7.135}$$

If one lets,

$$p(\mathbf{x}(k)|\mathbf{z}_{1:k}) = \sum_{i=1}^{N} \bar{\mathbf{w}}_i(k)\delta(\mathbf{x}(k) - \mathbf{x}_i(k)), \tag{7.136}$$

the weights probability update is,

$$\bar{\mathbf{w}}_i(k) = \frac{p(\mathbf{z}_{1:k-1},\mathbf{z}(k)|\mathbf{x}_i(k))\mathbf{w}_i(k)}{\sum_{i=1}^{N} \mathbf{w}_i(k)p(\mathbf{z}(k)|\mathbf{x}_i(k))}. \tag{7.137}$$

One of the problems that one faces with the use of the particle filter for constructing motion estimates is changes in the samples after only a few iterations. Most of the samples drift quite far from the current position, so their weights lose their significance in terms of their contribution to the probability density function. There is a need to continuously update the samples and not just their weights, and this establishes the need for resampling. Generally, the resampling is done to pick up weights of a specified probability. Once resampled, it becomes necessary to re-compute the corresponding weights which could be set to equal values when certain resampling methods are used. In general, it is essential that they are re-computed. The basis of all resampling methods is eliminate the weights that are near zero are eliminated in favor of weights which relative large. Thus, first cumulative sum of the particle weights is calculated and the N sorted random numbers uniformly distributed in [0 1] are selected. Next the number of the sorted random number that appear in a particular interval of the cumulative sum gives the number of copies of this particular particle (or sample) which is propagated to the next stage. Thus if a particle has a weight of small magnitude, the corresponding interval of the cumulative sum is small, thus reducing the number of random numbers appearing in that interval. On the contrary, when the weight is large in magnitude, the corresponding interval of the cumulative sum is large, and so are the number of random numbers appearing in that interval. This is the principle of the resampling algorithm. A number of other variations in resampling have been proposed. In our work only the simplest will be used.

To recursively implement the algorithm, with a particular dynamic model defining updates of the state vector and the measurement model, simplifications of the algorithm are possible. For the sate update, the ith sample vector is updated using the process dynamic model,

$$\mathbf{x}_i(k+1) = \mathbf{F}(\mathbf{x}_i(k), \mathbf{w}(k)). \tag{7.138}$$

From the measurement dynamic model and the actual measurement,

$$\mathbf{z}(k) = \mathbf{H}(\mathbf{x}_i(k), \mathbf{v}(k)). \tag{7.139}$$

It is not only possible to construct a typical estimate of the error but also to build a model for the probability density function of the error. Thus, an estimate of the weight is obtained, which is used to update the estimate of the state vector and to resample the particles, before proceeding to next time step. For a detailed view of particle filters and their application to Monte Carlo localization, the reader is referred to Thrun [4] and Thrun, Fox, Burgard, and Dellaert [5]. Application examples are discussed in the upcoming sections.

7.4.3 Application to Rover Localization

The application of the particle filter to a typical mobile rover localization problem is demonstrated. It is assumed that the mobile rover is moving on a horizontal plane. At any particular instant its location is defined by a pair of Cartesian coordinates (x, y) with reference to a space-fixed Cartesian frame, and its orientation to the x axis is assumed to be defined by θ. It is also assumed that it is moving forward with a velocity v, and is supported by four or more wheels, with two driven wheels at the rear of the rover separated by a distance d and with each wheel driven by an independent motor.

The equations of motion may be expressed as,

$$\dot{x}(t) = v \cos(\theta(t)), \quad \dot{y}(t) = v \sin(\theta(t)), \quad \dot{\theta}(t) = \omega. \tag{7.140}$$

7.4 Probabilistic Representation of Uncertain Motion Using Particles

If the velocity of the rover is measured at two points just above each of the two rear wheels to the left and to the right of the rover longitudinal axis, the forward velocity of the rover v and the angular velocity ω of the rover are, respectively, related to the two velocity measurements by,

$$v = 0.5(v_L + v_R) + w_1, \quad \omega = 0.5(v_R - v_L)/d + w_2, \qquad (7.141)$$

where w_1 and w_2 are assumed to be independent white noise disturbance processes with known variances q_1^2 and q_2^2, respectively. Denoting v_m and ω_m as,

$$v_m = 0.5(v_L + v_R), \quad \omega_m = 0.5(v_R - v_L)/d, \qquad (7.142)$$

the forward velocity of the rover v and the angular velocity ω of the rover are, respectively, expressed as,

$$v = v_m + w_1, \quad \omega = \omega_m + w_2. \qquad (7.143)$$

The equations of motion are expressed in matrix form as,

$$\begin{bmatrix} \dot{x}(t) \\ \dot{y}(t) \\ \dot{\theta}(t) \end{bmatrix} = \begin{bmatrix} \cos(\theta(t)) & 0 \\ \sin(\theta(t)) & 0 \\ 0 & 1 \end{bmatrix} \begin{bmatrix} v_m \\ \omega_m \end{bmatrix} + \begin{bmatrix} \cos(\theta(t)) & 0 \\ \sin(\theta(t)) & 0 \\ 0 & 1 \end{bmatrix} \begin{bmatrix} w_1 \\ w_2 \end{bmatrix}. \qquad (7.144)$$

Thus, the process model is established in the continuous time domain. Given a sampling time interval, Δt, the process model in the discrete time domain is,

$$\begin{bmatrix} x(k+1) \\ y(k+1) \\ \theta(k+1) \end{bmatrix} = \begin{bmatrix} x(k) \\ y(k) \\ \theta(k) \end{bmatrix} + \begin{bmatrix} \cos(\theta(k)) & 0 \\ \sin(\theta(k)) & 0 \\ 0 & 1 \end{bmatrix} \begin{bmatrix} \bar{v}_m \\ \bar{\omega}_m \end{bmatrix} + \begin{bmatrix} \cos(\theta(k)) & 0 \\ \sin(\theta(k)) & 0 \\ 0 & 1 \end{bmatrix} \begin{bmatrix} \bar{w}_1 \\ \bar{w}_2 \end{bmatrix}, \qquad (7.145)$$

with $t = k\Delta t$, $\bar{v}_m = v_m \Delta t$, $\bar{\omega}_m = \omega_m \Delta t$ and the discrete process noise sources, $\bar{w}_i = w_i \Delta t$.

It is also assumed that a landmark can be established at the start of the rover's motion and that the distance traveled by the rover and its orientation to the Cartesian frame relative to its initial position can be measured. Thus, the measurement model is defined by,

$$\begin{bmatrix} z_1 \\ z_2 \end{bmatrix} = \begin{bmatrix} \sqrt{x^2 + y^2} \\ \tan^{-1}(y/x) \end{bmatrix} + \begin{bmatrix} v_1 \\ v_2 \end{bmatrix}, \qquad (7.146)$$

where v_1 and v_2 are assumed to be independent white noise disturbance processes with known variances r_1^2 and r_2^2, respectively.

The rover localization problem reduces to one of estimation of its position and orientation, which is done by adopting the particle filter, with resampling as discussed in the preceding section. The estimates of the position coordinates $(\hat{x}, \hat{y})$ and of the orientation $\hat{\theta}$ obtained by a typical particle filter implementation are compared with the simulated position coordinates (x, y) and the orientation θ in Figure 7.8. It may be noted from Figure 7.8 that there appears to be a constant bias in the orientation of the rover.

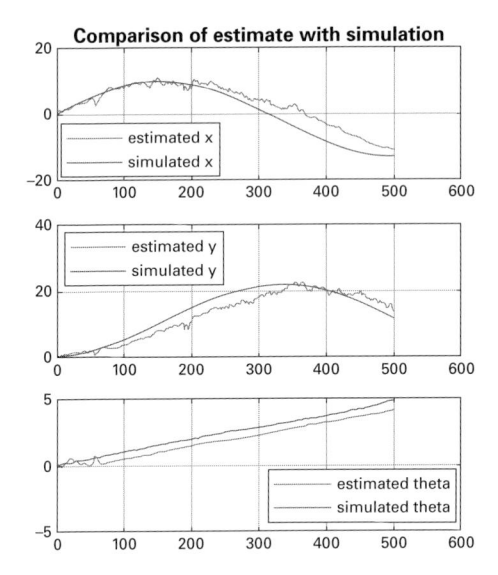

Figure 7.8 Comparison of the estimated positions and orientation of a rover with the corresponding simulated positions and orientation.

It is possible in principle to estimate and eliminate the bias from the orientation. To do this, the process model is modified to,

$$\begin{bmatrix} x(k+1) \\ y(k+1) \\ \theta(k+1) \\ b(k+1) \end{bmatrix} = \begin{bmatrix} x(k) \\ y(k) \\ \theta(k) \\ b(k) \end{bmatrix} + \begin{bmatrix} \cos(\theta(k)+b(k)) & 0 \\ \sin(\theta(k)+b(k)) & 0 \\ 0 & 1 \\ 0 & 0 \end{bmatrix} \begin{bmatrix} \bar{v}_m \\ \bar{\omega}_m \end{bmatrix}$$
$$+ \begin{bmatrix} \cos(\theta(k)+b(k)) & 0 & 0 \\ \sin(\theta(k)+b(k)) & 0 & 0 \\ 0 & 1 & 0 \\ 0 & 0 & 1 \end{bmatrix} \begin{bmatrix} \bar{w}_1 \\ \bar{w}_2 \\ \bar{w}_3 \end{bmatrix}. \quad (7.147)$$

The process model includes one more discrete process noise sources $\bar{w}_3$. The measurement model is unchanged and continues to be defined by Equation (7.146). The particle filter is now implemented with the additional bias state $b(k)$ in the model. The new estimates of the position coordinates $(\hat{x}, \hat{y})$ and of the orientation $\hat{\theta}$ obtained by a typical particle filter implementation are compared with the simulated position coordinates (x, y) and the orientation θ in Figure 7.9. The reduction in the bias in the estimate of the orientation is quite apparent when one compares Figures 7.8 and 7.9.

7.4.4 Monte Carlo Localization

Monte Carlo localization (MCL), which is also known as particle filter localization, is an algorithm to localize a mobile rover using a particle filter. The principle of operation

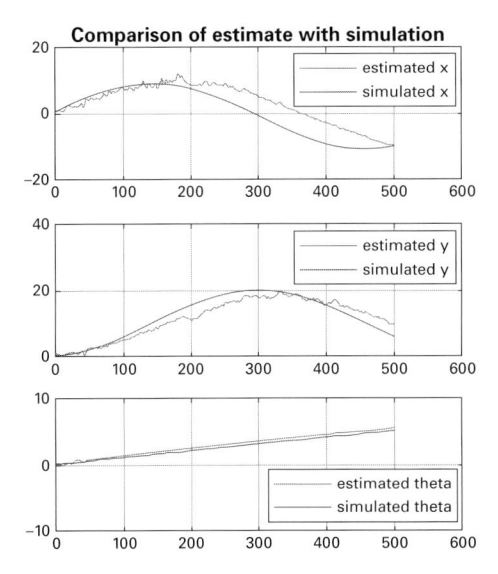

Figure 7.9 Comparison of the estimated positions and orientation of a rover with the corresponding simulated positions and orientation, with the bias in the orientation eliminated.

of Monte Carlo localization is based on the particle filter. An initial set of particles are generated based on the knowledge of the uncertainty statistics of the initial conditions. Each particle provides a hypothesis for the current pose. The proposal probability density function and the corresponding particles,

$$\mathbf{x}_t^i \sim p(\mathbf{x}_t | \mathbf{x}_{t-1}, \mathbf{u}_t), \quad (7.148)$$

are obtained from the motion model as demonstrated in the preceding example. Corrections are established by applying the observation model. This involves the updating of the weights as,

$$\mathbf{w}_t^i = \frac{\text{target}}{\text{proposal}} \alpha \, p(\mathbf{z}_t | \mathbf{x}_t, \mathbf{z}_{1:t-1}). \quad (7.149)$$

Resampling, to replace unlikely samples by more likely or equally likely ones, allows the weight updates to be transformed into particle updates. The whole process is then recursively applied in the next step.

7.4.5 Probabilistic Localization within a Map, Using Odometry and Range Measurements

A primary advantage of using Cartesian coordinates for localization is that one could plot the pose information on a map or rectangular grid to indicate the position of the rover by a marker. The orientation of the marker could also be depicted on the map or grid. Thus, it is possible in principle to localize the rover on a *standard* or *global* map, which could also serve as a *reference* map. A map, known as an *occupancy map*, provides a probabilistic representation of the estimated coordinates on a Cartesian grid.

A probability of 1 means that the cell is definitely occupied and the robot cannot pass through it. If it is 0, the cell is definitely vacant, and the rover can traverse it. When the expected value is indicated on the map, the variance provides a measure of the probability. Topological maps and landmark maps are two other map representations that are necessary for mobile rover navigation. Landmarks may also be overlaid on a typical occupancy map. The advantages of such an approach, which permits the overlay of obstacles, are briefly discussed in the next section.

A rover needs to continuously keep track of its position coordinates (x, y) and its orientation (θ). In most real applications this is done by odometry or the measurement of the left and right wheel speeds rather than the velocities of the rover where the wheels are located. Positioning of a mobile rover using odometry is inexpensive but inaccurate due to slipping at the wheel-ground interface and other disturbances. In terms of odometry measurements, the forward velocity of the rover v and the angular velocity ω of the rover are, respectively, related to the two-wheel angular velocity, ω_L and ω_R, measurements by,

$$v = 0.5 r_w (\omega_L + \omega_R) + w_{w1}, \quad \omega = 0.5 r_w (\omega_R - \omega_L)/d + w_{w2}, \quad (7.150)$$

where w_{1w} and w_{2w} are assumed to be independent noise disturbance processes with known statistics such as the probability density functions. Again, the particle filter approach may be used to estimate the rover's position and orientation. When both the velocity measurements and the odometry measurements of the wheel speeds are available, it is possible to estimate the slip at each wheel by constructing filtered estimates of the wheel speeds.

7.5 Place Recognition and Occupancy Mapping: Advanced Sensing Techniques and Ranging

It is now appropriate to discuss the role of advanced sensing and ranging techniques. It is now assumed that several landmarks are monitored and ranges of the rover are measured. Consider a landmark with Cartesian coordinates, (x_{lm}^r, y_{lm}^r), from the rover. In terms of the base or initial coordinates, the Cartesian coordinates of the landmark are, $(x_{lm}, y_{lm}) = (x + x_{lm}^r, y + y_{lm}^r)$. If the landmark is a fixed landmark, one has the process model,

$$\dot{x}_{lm} = w_{1lm}, \quad \dot{y}_{lm} = w_{2lm}.$$

Since the range from the landmark to the rover is measured, the measurement model is,

$$z_{lm} = \sqrt{(x_{lm} - x)^2 + (y_{lm} - y)^2} + v_{lm}, \quad (7.151)$$

where v_{lm} is assumed to be an independent white noise disturbance process with known variance r_{lm}^2.

When this model is appended to the rover model, one could, in principle, also jointly estimate the landmark's Cartesian position coordinates (x_{lm}, y_{lm}) and indicate the

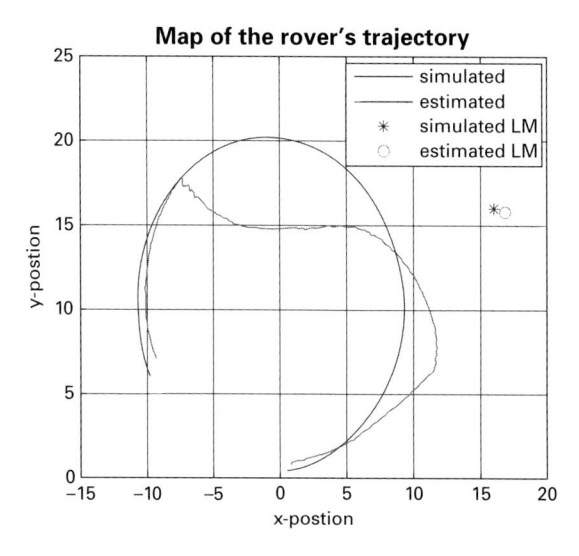

Figure 7.10 A typical map of the rover's trajectory starting near (0,0).

landmark on the same grid indicating the position and orientation of the rover. When the landmark is also slowly moving and indication of its motion can also be obtained from the estimates of the drift in the landmark's Cartesian position coordinates (x_{lm}, y_{lm}). In fact, this can be done for several landmarks, and all of the landmarks could be indicated on the same map or grid. Thus, one could construct a probabilistic *occupancy map* of all of the landmarks on the grid. In Figure 7.10 a typical map of the rover's trajectory starting near (0,0) is shown. Also shown are the estimated and simulated positions of a landmark (LM). A reduction in the error of the estimated position coordinates can be affected by reducing the initial uncertainties, the process noise, and the measurement uncertainties.

7.5.1 Place Recognition Using Ranging Signatures: Occupancy Mapping of Free Space and Obstacles

Let one suppose that it is now possible to conduct patterned ranging of several landmarks from the rover. One example of such a patterned ranging is a continuous sweep using a LASER ranger over an azimuth covering 360 degrees. The returns from such a sweep collectively constitute a "signature" of the particular location, when all the returns from any moving landmarks are detected and eliminated from the totality of the returns over a single sweep. This may be recorded and used as a unique signature of the particular location.

7.6 The Extended Kalman Filter

The Kalman filter, discussed in Section 7.3.2, is a recursive algorithm to optimally estimate and obtain a linear, unbiased, and minimum error variance response for the unknown state of a linear dynamic system from noisy data taken at discrete real-time intervals. If the process to be estimated and (or) the measurement relationship to the process is nonlinear, the Kalman filter can be applied by linearizing the dynamics and the measurement's relationship to the states about the current mean and covariance and is known as the EKF. It is an empirical extension of the classical Kalman filter as applied to a linear model of a plant. There is, however, a flaw in the basis for the EKF in that the distributions (or densities in the continuous case) of the various random variables are no longer normal after undergoing their respective nonlinear transformations. The EKF is simply an improvised state estimator that approximates the optimality of Bayes' rule by linearization. Julier and Uhlmann [6] have developed a variation to the EKF, the UKF, using methods that preserve the normal distributions up to a certain order, provided the nonlinear transformation can be suitably approximated.

The plant disturbance and measurement noise statistics ought to be known a priori to be able to implement the optimum EKF. Moreover, the use of incorrect variance estimates for the plant disturbances could render the filter unstable, leading to the classic problem of filter divergence. The solution to this problem is to adopt one of three strategies:

(i) Error sensitivity analysis to determine the performance sensitivity to uncertainties in the noise statistics to ensure that any performance degradation is within acceptable bounds;
(ii) A min-max approach such as a nonlinear H^∞ type filter to determine the filter parameters to minimize the worst case errors;
(iii) An adaptive approach to evaluate the noise statistics in real-time.

Four classes of methods have evolved to obtain the noise statistics in real-time, which are (a) Bayesian methods, (b) Maximum likelihood based methods, (c) correlation methods, and (d) the covariance matching method. The last two classes of methods are particularly suited to alleviate the problem of filter divergence due to improper knowledge of the variance of the process disturbance noise.

Most dynamic models employed for purposes of estimation or filtering of ranging errors are generally not linear. To extend and overcome the limitations of linear models, a number of approaches such as the EKF have been proposed in the literature for nonlinear estimation using a variety of approaches. Unlike the KF, the EKF may diverge if the consecutive linearizations are not a good approximation of the linear model over the entire uncertainty domain. Yet the EKF provides a simple and practical approach to dealing with essential nonlinear dynamics. To simplify the notation, the process and the measurement models are, respectively, written in the form,

$$\mathbf{x}_k = \mathbf{f}_{k-1}(\mathbf{x}_{k-1}) + \mathbf{w}_{k-1}, \quad \mathbf{z}_k = \mathbf{h}_k(\mathbf{x}_k) + \mathbf{v}_k. \quad (7.152)$$

Given the Jacobians, the state and measurement dynamics, respectively, are,

$$\Phi_{k-1} = \nabla \mathbf{f}_{k-1}(\hat{\mathbf{x}}_{k-1})|_{k-1}, \quad \mathbf{H}_k = \nabla \mathbf{h}_k(\hat{\mathbf{x}}_k^-)|_k, \quad (7.153)$$

the state prediction equation defining the EKF is,

$$\hat{\mathbf{x}}_k^- = \mathbf{f}_{k-1}(\hat{\mathbf{x}}_{k-1}), \tag{7.154}$$

while the covariance prediction equation is,

$$\hat{\mathbf{P}}_k^- = \Phi_{k-1}\mathbf{P}_{k-1}\Phi_{k-1}^T + \mathbf{Q}_{k-1}. \tag{7.155}$$

The measurement correction and filter gain equations defining the EKF are,

$$\hat{\mathbf{x}}_k = \hat{\mathbf{x}}_k^- + \mathbf{K}_k\left[\mathbf{z}_k - \mathbf{h}_k\left(\hat{\mathbf{x}}_k^-\right)\right] \tag{7.156}$$

$$\mathbf{K}_k = \hat{\mathbf{P}}_k^- \mathbf{H}_k^T \left(\mathbf{H}_k \hat{\mathbf{P}}_k^- \mathbf{H}_k^T + \mathbf{R}_k\right)^{-1} \tag{7.157}$$

$$\hat{\mathbf{P}}_k = (\mathbf{I} - \mathbf{K}_k \mathbf{H}_k)\hat{\mathbf{P}}_k^-. \tag{7.158}$$

Equations (7.156), (7.157), and (7.158) are identical to Equations (7.74), (7.81), and (7.82), respectively. The main difficulty in applying the algorithm to problems related to localization is in determining the proper Jacobian matrices.

The EKF is applied to the rover localization problem considered in Section 7.4.3. In Figure 7.11 the estimates of the components of the pose of the rover, obtained by using the EKF, are compared with the corresponding simulated components. In Figure 7.12

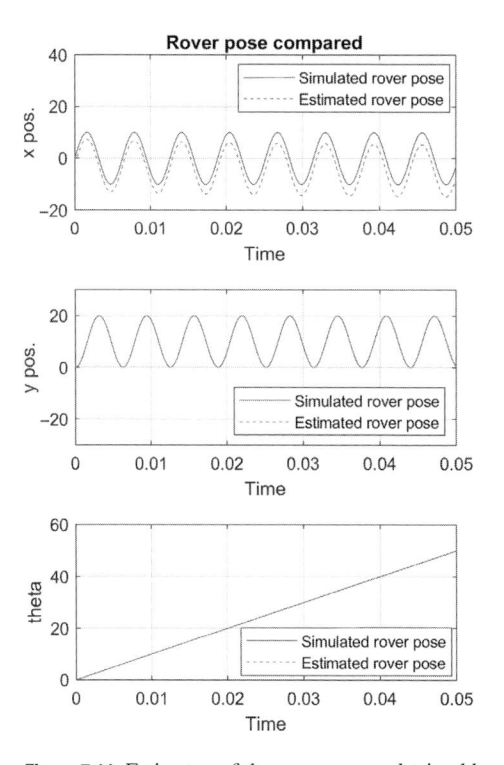

Figure 7.11 Estimates of the rover pose, obtained by using the EKF, compared with corresponding simulations.

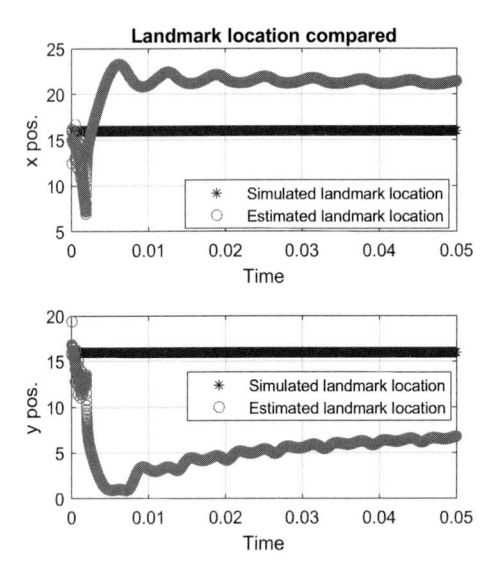

Figure 7.12 Estimates of a landmark location, obtained by using the EKF, compared with corresponding simulations.

the estimates of a landmark location, obtained by using the EKF and with measurements of its range only, are compared with corresponding simulations.

7.6.1 The Unscented Kalman Filter (UKF)

The UKF is a feasible alternative that has been proposed, to overcome this difficulty, by Julier and Uhlman [4], as an effective way of applying the Kalman filter to nonlinear systems. The UKF gets its name from the unscented transformation, which is a method of calculating the mean and covariance of a random variable undergoing nonlinear transformation $\mathbf{y} = \mathbf{f}(\mathbf{w})$. Although it is a derivative-free approach, it does not really address the divergence problem. In essence, the method constructs a set of *sigma vectors* and propagates them through the same nonlinear function. The mean and covariance of the transformed vector are approximated as a weighted sum of the transformed *sigma vectors* and their covariance matrices. In the nonlinear estimator, the complete nonlinear functional $\mathbf{f}(\mathbf{x})$ is used rather than its approximation so as to generate a reasonably accurate estimate of the state.

Consider a random variable $\mathbf{w}$ with dimension L that is going through the nonlinear transformation, $\mathbf{y} = \mathbf{f}(\mathbf{w})$. The initial conditions are that $\mathbf{w}$ has a mean $\bar{\mathbf{w}}$ and a covariance $\mathbf{P}_{ww}$. To calculate the statistics of $\mathbf{y}$, a matrix χ of $2L+1$ sigma vectors is formed. Sigma vector points are calculated according to the following conditions,

$$\chi_0 = \bar{\mathbf{w}} \qquad (7.159)$$

$$\chi_i = \bar{\mathbf{w}} + \left(\sqrt{(L+\lambda)\mathbf{P}_{ww}}\right)_i, \quad i = 1, 2, \ldots, L, \qquad (7.160)$$

$$\chi_i = \bar{\mathbf{w}} - \left(\sqrt{(L+\lambda)\mathbf{P}_{ww}}\right)_i, \quad i = L+1, L+2, \ldots, 2L, \tag{7.161}$$

where $\lambda = \alpha^2(L+\kappa) - L$, α is a scaling parameter between 0 and 1, and κ is a secondary scaling parameter. $\left(\sqrt{(L+\lambda)\mathbf{P}_{ww}}\right)_i$ is the ith column of the matrix square root. This matrix square root can be obtained by Cholesky factorization. The weights associated with the sigma vectors are calculated from the following,

$$W_0^{(m)} = \lambda/(L+\lambda) \tag{7.162}$$

$$W_0^{(c)} = (\lambda/(L+\lambda)) + 1 - \alpha^2 + \beta \tag{7.163}$$

$$W_i^{(m)} = W_i^{(c)} = 1/2\,(L+\lambda), \quad i = 1,\,2,\,\ldots,2L, \tag{7.164}$$

where β is chosen as 2 for Gaussian distributed variables. The mean, covariance, and cross-covariance of $\mathbf{y}$ calculated using the unscented transformations are given by,

$$\mathbf{y}_i = \mathbf{f}(\chi_i) \tag{7.165}$$

$$\bar{\mathbf{y}} \approx \sum_{i=0}^{2L} W_i^{(m)} \mathbf{y}_i \tag{7.166}$$

$$\mathbf{P}_{yy} \approx \sum_{i=0}^{2L} W_i^{(c)} (\mathbf{y}_i - \bar{\mathbf{y}})(\mathbf{y}_i - \bar{\mathbf{y}})^T \tag{7.167}$$

$$\mathbf{P}_{xy} \approx \sum_{i=0}^{2L} W_i^{(c)} (\chi_i - \bar{\chi})(\mathbf{y}_i - \bar{\mathbf{y}})^T, \tag{7.168}$$

where $W_i^{(m)}$ and $W_i^{(c)}$ are the set of weights defined in a manner so that approximations of the mean and covariance are accurate up to third order for Gaussian inputs for all nonlinearities, and to at least second order for non-Gaussian inputs. The sigma points in the sigma vectors are updated using the nonlinear model equations without any linearization.

Given a general discrete nonlinear dynamic system in the form,

$$\mathbf{x}_{k+1} = \mathbf{f}(\mathbf{x}_k, \mathbf{u}_k) + \mathbf{w}_k, \quad \mathbf{y}_k = \mathbf{h}(\mathbf{x}_k) + \mathbf{v}_k, \tag{7.169}$$

where $\mathbf{x}_k \in R^n$ is the state vector, $\mathbf{u}_k \in R^r$ is the known input vector, and $\mathbf{y}_k \in R^m$ is the output vector at the end of the time kth time step. The vectors $\mathbf{w}_k$ and $\mathbf{v}_k$ are, respectively, the disturbance or process noise and sensor noise vectors, which are assumed to be Gaussian white noise processes with zero mean. Furthermore, $\mathbf{Q}_k$ and $\mathbf{R}_k$ are assumed to be the covariance matrices of the process noise sequence $\mathbf{w_k}$ and the measurement noise sequence $\mathbf{v_k}$, respectively. The unscented transformations of the state and measurement defining functions are denoted as,

$$\mathbf{f}^{UT} = \mathbf{f}^{UT}(\mathbf{x}_k, \mathbf{u}_k), \quad \mathbf{h}^{UT} = \mathbf{h}^{UT}(\mathbf{x}_k), \tag{7.170}$$

while the transformed covariance matrices and cross-covariance are, respectively, denoted as,

$$\mathbf{P}_k^f = \mathbf{P}_k^f(\hat{\mathbf{x}}_k, \mathbf{u}_k), \quad \mathbf{P}_k^h = \mathbf{P}_k^h(\hat{\mathbf{x}}_k), \quad \mathbf{P}_k^{fh} = \mathbf{P}_k^{fh}(\hat{\mathbf{x}}_k, \mathbf{u}_k). \quad (7.171)$$

The state time-update equation, the predicted covariance, the Kalman gain, the state estimate, and the corrected covariance are, respectively, given by,

$$\hat{\mathbf{x}}_k^- = \mathbf{f}_{k-1}^{UT}(\hat{\mathbf{x}}_{k-1}), \quad (7.172)$$

$$\hat{\mathbf{P}}_k^- = \mathbf{P}_{k-1}^{ff} + \mathbf{Q}_{k-1}, \quad (7.173)$$

$$\mathbf{K}_k = \hat{\mathbf{P}}_k^{xh-}\left(\hat{\mathbf{P}}_k^{hh-} + \mathbf{R}_k\right)^{-1}, \quad (7.174)$$

$$\hat{\mathbf{x}}_k = \hat{\mathbf{x}}_k^- + \mathbf{K}_k\left[\mathbf{y}_k - \mathbf{h}_k^{UT}(\hat{\mathbf{x}}_k^-)\right], \quad (7.175)$$

$$\hat{\mathbf{P}}_\mathbf{k} = \hat{\mathbf{P}}_k^- - \mathbf{K}_k\left(\hat{\mathbf{P}}_k^{hh-} + \mathbf{R}_k\right)\mathbf{K}_k^T. \quad (7.176)$$

Equations (7.172)–(7.176), are referred to as UKF equations. The UKF is applied to the rover localization problem considered in Section 7.4.3. In Figure 7.13 the estimates of the components of the pose of the rover, obtained by using the UKF, are compared with the corresponding simulated components.

In Figure 7.14 the estimates of a landmark location, obtained by using the UKF and with measurements of its range only, are compared with corresponding simulations. It is clear that the UKF estimator is significantly better than the estimation obtained by the EKF.

7.7 Nonlinear Least Squares, Maximum Likelihood (ML), Maximum A Posteriori (MAP) Estimation

Before considering the problems of simultaneous localization or position estimation and map building, it is essential to briefly consider some alternative approaches to estimation. Thus far, only the cases of Bayesian estimation and Kalman filtering have been considered. The latter can be viewed as a special case of linear least squares filtering. However, it is interesting to consider the more general setting and briefly review other estimation approaches, which could be significantly different when considering nonlinear estimation problems.

The fundamental of estimation to obtain our best estimate $\hat{\mathbf{x}}$ for a state vector $\mathbf{x}$, given a set of k measurements, is $\mathbf{z}_{1:k} = [\mathbf{z}_1 \, \mathbf{z}_2 \, \mathbf{z}_3 \cdots \mathbf{z}_k]$. Generally, one also has a set of constraints imposed on the state vector in that it is also required to satisfy some differential equations. However, in the first instance, it is assumed that there are no constraints on the state vector $\mathbf{x}$. It is assumed that the measurements at any time t are linearly related to the state vector by a linear relation,

$$\mathbf{z}(t) = \mathbf{H}(t)\mathbf{x}(t), \quad (7.177)$$

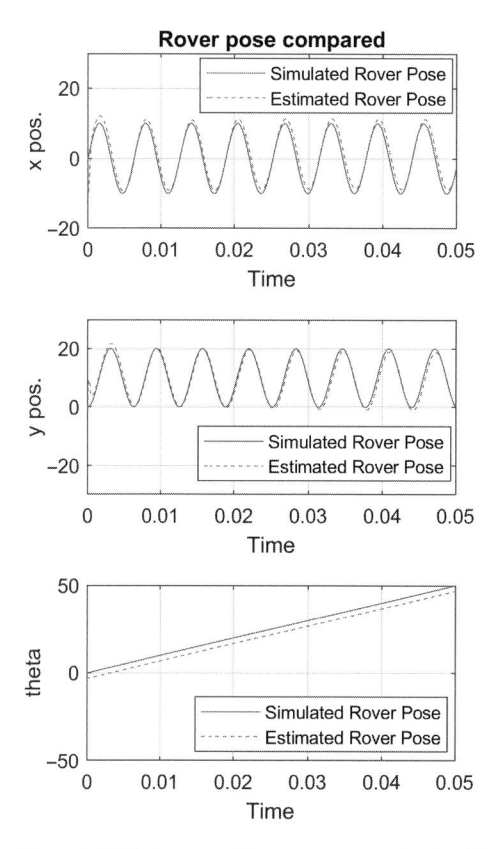

Figure 7.13 Estimates of the rover pose, obtained by using the UKF, compared with corresponding simulations.

where the matrix $\mathbf{H}(t)$ is generally not square or invertible. The best estimate of $\mathbf{x}(t)$, given by $\hat{\mathbf{x}}(t)$, is obtained by minimizing a performance index,

$$J = \left\| (\mathbf{z}(t) - \mathbf{H}(t)\mathbf{x}(t))^T (\mathbf{z}(t) - \mathbf{H}(t)\mathbf{x}(t)) \right\|, \tag{7.178}$$

where $\|.\|$ is some norm of the argument. Thus,

$$\hat{\mathbf{x}}(t) = \min_{\mathbf{x}} J = \min_{\mathbf{x}} \left\| (\mathbf{z}(t) - \mathbf{H}(t)\mathbf{x}(t))^T (\mathbf{z}(t) - \mathbf{H}(t)\mathbf{x}(t)) \right\|. \tag{7.179}$$

In the case of linear least squares, one obtains the solution,

$$\hat{\mathbf{x}}(t) = \left(\mathbf{H}^T(t)\mathbf{H}(t)\right)^{-1} \mathbf{H}^T(t)\mathbf{z}. \tag{7.180}$$

One could introduce a weighted performance index,

$$J = \left\| (\mathbf{z}(t) - \mathbf{H}(t)\mathbf{x}(t))^T \mathbf{R}^{-1} (\mathbf{z}(t) - \mathbf{H}(t)\mathbf{x}(t)) \right\|. \tag{7.181}$$

In this case, one has the solution,

$$\hat{\mathbf{x}}(t) = \left(\mathbf{H}^T(t)\mathbf{R}^{-1}\mathbf{H}(t)\right)^{-1} \mathbf{H}^T(t)\mathbf{R}^{-1}\mathbf{z}(t). \tag{7.182}$$

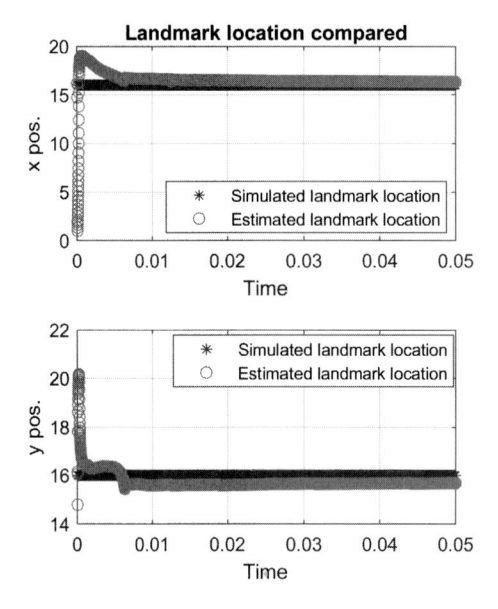

Figure 7.14 Estimates of a landmark location, obtained by using the UKF, compared with corresponding simulations.

Two important generalizations follow,

(i) First the linear least squares method can be generalized to a nonlinear estimation problem;
(ii) Second, the solution may be recursively improved when the measurements that were initially available only at the time $t = k\Delta t$ are now also additionally available at time $t + \Delta t = (k + 1)\Delta t$.

Considering the recursive generalization first, given the set of measurements at time $t = k\Delta t$, in the form,

$$\mathbf{z}(k) = \mathbf{H}(k)\mathbf{x}(k) + \mathbf{v}(k), \qquad (7.183)$$

$\mathbf{v}(k)$ is a disturbance vector of some form. However, in order that the solution is physically meaningful, it is necessary to assume that $\mathbf{v}(k)$ is a zero-mean white noise process with covariance $\mathbf{R}$. The solution for the estimate is then written as,

$$\hat{\mathbf{x}}(k) = \left(\mathbf{H}^T(k)\mathbf{R}^{-1}\mathbf{H}(k)\right)^{-1}\mathbf{H}^T(k)\mathbf{R}^{-1}\mathbf{z}(k). \qquad (7.184)$$

When a new measurement becomes available, the new estimate is expressed as,

$$\hat{\mathbf{x}}(k+1) = \hat{\mathbf{x}}(k) + \mathbf{K}(k+1)\mathbf{i}(k+1), \qquad (7.185)$$

with,

$$\mathbf{i}(k+1) = \mathbf{z}(k+1) - \mathbf{H}(k+1)\hat{\mathbf{x}}(k). \qquad (7.186)$$

In this equation, $\mathbf{i}(k+1)$ is the *innovation* or the new information present in the additional measurement $\mathbf{z}(k+1)$. While the mean of the estimation error can be shown to be zero, estimator gain $\mathbf{K}(k+1)$,

$$\mathbf{K}(k+1) = \mathbf{P}(k)\mathbf{H}^T(k+1)\left(\mathbf{H}(k+1)\mathbf{P}(k)\mathbf{H}^T(k+1) + \mathbf{R}(k+1)\right)^{-1}, \quad (7.187)$$

$$\mathbf{P}(k+1) = (\mathbf{I} - \mathbf{K}(k+1)\mathbf{H}(k+1))\mathbf{P}(k). \quad (7.188)$$

The solution is in the same form as the measurement updates of the Kalman filter.

Considering the nonlinear least squares problem, it is assumed that the measurements at any time t are linearly related to the state vector by a nonlinear relation,

$$\mathbf{z}(t) = \mathbf{h}(\mathbf{x}(t), t). \quad (7.189)$$

The best estimate of $\mathbf{x}(t)$, given by $\hat{\mathbf{x}}(t)$, is obtained by minimizing the performance index of the form,

$$J = \left\| (\mathbf{z}(t) - \mathbf{h}(\mathbf{x}(t), t))^T (\mathbf{z}(t) - \mathbf{h}(\mathbf{x}(t), t)) \right\|. \quad (7.190)$$

The solution is given by,

$$\hat{\mathbf{x}}(t) = \min_{\mathbf{x}} J = \min_{\mathbf{x}} \left\| (\mathbf{z}(t) - \mathbf{h}(\mathbf{x}(t), t))^T (\mathbf{z}(t) - \mathbf{h}(\mathbf{x}(t), t)) \right\|. \quad (7.191)$$

To start the minimization process, one sets an initial estimate as $\hat{\mathbf{x}}(t) = \mathbf{x}_0(t)$ and linearizes the problem about $\mathbf{x}(t) = \mathbf{x}_0(t)$. The linear estimation problem is solved and the perturbation estimate $\delta\hat{\mathbf{x}}_0(t)$ is obtained. Linearization involves finding the Jacobian matrix,

$$\mathbf{H}(t) = \left. \frac{\partial \mathbf{h}(\mathbf{x}(t), t)}{\partial \mathbf{x}(t)} \right|_{\mathbf{x}(t) = \mathbf{x}_0(t)}. \quad (7.192)$$

A new estimate is obtained as $\hat{\mathbf{x}}(t) = \mathbf{x}_0(t) + \delta\hat{\mathbf{x}}_0(t)$ and the problem is again linearized about $\mathbf{x}(t) = \mathbf{x}_0(t) + \delta\hat{\mathbf{x}}_0(t)$. The process is repeated until the estimate converges to a limit within a certain tolerance. One application of nonlinear estimation is iterative solution of the trilateration equations introduced in Section 7.2.1.

It is natural to expect that the measurement $\mathbf{z}(t)$ one obtains from observations of the state vector $\mathbf{x}(t)$ is in some way related to the state vector. Thus, one cost function that could be minimized is the conditional probability density function of the measurement, given the state vector. Such a cost function is often known as the *likelihood function*. Assuming the conditional probability density function of the measurement to be Gaussian, it can be expressed as,

$$p(\mathbf{z}|\mathbf{x}) = (2\pi)^{-1} \|\mathbf{R}\|^{-1/2} \exp\left(-\frac{1}{2}(\mathbf{z} - \mathbf{H}\mathbf{x})^T \mathbf{R}^{-1}(\mathbf{z} - \mathbf{H}\mathbf{x})\right). \quad (7.193)$$

This is the *likelihood function*, which is minimized with respect to $\mathbf{x}(t)$. The resulting estimate is known as the *Maximum Likelihood* (ML) *estimate*. It may be observed that the earlier least squares solution may also be interpreted as the *Maximum Log-likelihood estimate*.

7.7.1 Nonlinear Least Squares Problems Solution Using Gauss-Newton and Levenberg Marquardt Optimization Algorithms

A very popular approach to localization and mapping is to directly minimize a chosen cost function. The problem is formulated as one of fitting a generalized curve through a set of known data points. Thus, given a set of data points z_i, $i = 1, 2 \cdots N$, and a set of basic functions $\Psi_j(\mathbf{x})$ of an independent vector variable $\mathbf{x}$, the problem can be reduced to finding a set of coefficients p_j, $j = 1, 2 \cdots M$, such that the estimate.,

$$\hat{z}_i = \sum_{j=1}^{M} p_j \Psi_j(\mathbf{x}_i), \tag{7.194}$$

is optimal in the sense that it minimizes a performance index,

$$J = \frac{1}{2} \sum_{i=1}^{N} \frac{\|z_i - \hat{z}_i\|^2}{\sigma_i^2} = \frac{1}{2} (\mathbf{z} - \hat{\mathbf{z}})^T \mathbf{W} (\mathbf{z} - \hat{\mathbf{z}}). \tag{7.195}$$

In this definition of the performance index or objective function J, $\mathbf{W}$ is a diagonal weighting matrix with entries, $1/\sigma_i^2$, $\mathbf{z}$ is the vector of data points z_i, and $\hat{\mathbf{z}}$ is the vector of estimates. Expanding the function J,

$$J = \frac{1}{2} (\mathbf{z} - \hat{\mathbf{z}})^T \mathbf{W} (\mathbf{z} - \hat{\mathbf{z}}) = \frac{1}{2} \mathbf{z}^T \mathbf{W} \mathbf{z} - \mathbf{z} \mathbf{W} \hat{\mathbf{z}} + \frac{1}{2} \hat{\mathbf{z}}^T \mathbf{W} \hat{\mathbf{z}}. \tag{7.196}$$

The gradient of the function J with respect to the parameters p_j, which are yet to be determined, is,

$$\frac{\partial J}{\partial p_j} = \left(\hat{\mathbf{z}}^T - \mathbf{z}^T\right) \mathbf{W} \frac{\partial \hat{\mathbf{z}}}{\partial p_j} \equiv \left(\hat{\mathbf{z}}^T - \mathbf{z}^T\right) \mathbf{W} \Psi_j. \tag{7.197}$$

When we are searching for an optimal set of parameters, starting from an initial parameter set, $p_{j,k}$, an appropriate manner of updating the parameter for conducting the search is to increment that parameter in the direction of steepest descent from the current parameter set as,

$$p_{j,k+1} = p_{j,k} - \Delta\alpha \left(\hat{\mathbf{z}}^T - \mathbf{z}^T\right) \mathbf{W} \Psi_j. \tag{7.198}$$

In this update equation, $\Delta\alpha$ specifies the size of the step change in the direction of the steepest descent.

An alternative approach is based on expanding the estimate $\hat{\mathbf{z}}$ as a Taylor's series in the parameter set, p_j. Thus, $\hat{\mathbf{z}}$ is approximated as,

$$\hat{\mathbf{z}}(p_j + \delta) \approx \hat{\mathbf{z}}(p_j) + \delta \frac{\partial \hat{\mathbf{z}}}{\partial p_j}. \tag{7.199}$$

In vector form,

$$\hat{\mathbf{z}}(\mathbf{p} + \boldsymbol{\delta}) \approx \hat{\mathbf{z}}(\mathbf{p}) + \frac{\partial \hat{\mathbf{z}}}{\partial \mathbf{p}} \boldsymbol{\delta} \equiv \hat{\mathbf{z}}(\mathbf{p}) + \mathbf{J}\boldsymbol{\delta}, \tag{7.200}$$

where **J** is a Jacobian matrix. The Gauss-Newton approach assumes that one is relatively close to the minimum, so one could use the Taylor's series expansion to approximate the performance index or objective function J in the vicinity of the minimum. Thus, expanding the performance index or objective function J up to second order in $\boldsymbol{\delta}$,

$$J = \frac{1}{2}\mathbf{z}^T\mathbf{W}\mathbf{z} - \mathbf{z}^T\mathbf{W}\hat{\mathbf{z}}(\mathbf{p}) + \frac{1}{2}\hat{\mathbf{z}}^T(\mathbf{p})\mathbf{W}\hat{\mathbf{z}}(\mathbf{p}) + (\hat{\mathbf{z}}(\mathbf{p}) - \mathbf{z})^T\mathbf{W}\mathbf{J}\boldsymbol{\delta} + \frac{1}{2}\boldsymbol{\delta}^T\mathbf{J}^T\mathbf{W}\mathbf{J}\boldsymbol{\delta}. \tag{7.201}$$

The Hessian of the performance index or objective function J is given by,

$$\mathbf{H} = \mathbf{J}^T\mathbf{W}\mathbf{J}. \tag{7.202}$$

Minimizing J with respect to $\boldsymbol{\delta}$ gives,

$$\frac{\partial J}{\partial \boldsymbol{\delta}} = -(\hat{\mathbf{z}}(\mathbf{p}) - \mathbf{z})^T\mathbf{W}\mathbf{J} + \boldsymbol{\delta}^T\mathbf{J}^T\mathbf{W}\mathbf{J} = 0. \tag{7.203}$$

Hence,

$$\boldsymbol{\delta} = -\left(\mathbf{J}^T\mathbf{W}\mathbf{J}\right)^{-1}\mathbf{J}^T\mathbf{W}(\hat{\mathbf{z}}(\mathbf{p}) - \mathbf{z}). \tag{7.204}$$

Thus, the parameter update equation, known as the Gauss-Newton update law, may be expressed as,

$$\mathbf{p}_{k+1} = \mathbf{p}_k - \Delta\beta\left(\mathbf{J}^T\mathbf{W}\mathbf{J}\right)^{-1}\mathbf{J}^T\mathbf{W}(\hat{\mathbf{z}} - \mathbf{z}), \tag{7.205}$$

where $\Delta\beta$ specifies the size of the step change in the direction of the minimum J. One can express the steepest descent update formula in the same format, which is,

$$\mathbf{p}_{k+1} = \mathbf{p}_k - \Delta\alpha\mathbf{J}^T\mathbf{W}(\hat{\mathbf{z}} - \mathbf{z}). \tag{7.206}$$

In the Levenberg-Marquardt algorithm, one combines the two update formulae, obtained by the steepest descent approach and by the Gauss-Newton approach. The parameter update equation is expressed as,

$$\mathbf{p}_{k+1} = \mathbf{p}_k - \Delta\beta\left(\left(\mathbf{J}^T\mathbf{W}\mathbf{J}\right)^{-1} + \lambda\mathbf{I}\right)\mathbf{J}^T\mathbf{W}(\hat{\mathbf{z}} - \mathbf{z}), \tag{7.207}$$

where the parameter λ is adaptively chosen. When λ is very large, the parameter update law resembles the steepest descent approach, while in the case when λ is very small, the parameter update law resembles the Gauss-Newton update law. Initially, one can start the search for a minimum with the steepest descent approach with high value of λ, and as the minimum is approached, one gradually reduces λ, so the solution converges to a minimum. This is the basis of the Levenberg-Marquardt algorithm, which is the most widely used optimization algorithm, as it performs better than most other algorithms, such as simple gradient descent and conjugate gradient-based methods.

7.8 Simultaneous Localization and Mapping (SLAM)

7.8.1 Introduction to the Essential Principles and Method of SLAM

The concept of simultaneous localization and mapping has already been introduced in an earlier section. It was first introduced by Cheeseman and Smith [7]. Thrun [8] and Bailey and Durrant-Whyte [9] have provided extensive surveys of the development of SLAM. Dissanayake, Durrant-Whyte, and Bailey [10] and Clark, Dissanayake, Newman, and Durrant-Whyte [11] have suggested improvements to the basic SLAM algorithms. SLAM was first implemented based on the use of the EKF and UKF. However, the real issue in SLAM is not only about localization but also about simultaneously mapping a large number of locations and/or landmarks. In the case of the approaches based on the EKF and UKF, although still feasible, require large amounts of memory and could be slow in converging to the correct solutions. For this reason alternate approaches to SLAM have been developed which are now considered to be the very basis of SLAM. Moreover, when either the EKF or UKF is applied for landmark localization, it is important that the landmark's location is measured in such a way that it is quite accurately observable. Thus, it is important that both the range and bearing measurements of the landmark are made.

Reconsider the example of the localization of a rover's pose, introduced in Section 7.4.3. The rover's discrete time dynamics were given by,

$$\begin{bmatrix} x(k+1) \\ y(k+1) \\ \theta(k+1) \\ b(k+1) \end{bmatrix} = \begin{bmatrix} x(k) \\ y(k) \\ \theta(k) \\ b(k) \end{bmatrix} + \begin{bmatrix} \cos(\theta(k)+b(k)) & 0 \\ \sin(\theta(k)+b(k)) & 0 \\ 0 & 1 \\ 0 & 0 \end{bmatrix} \begin{bmatrix} \bar{v}_m \\ \bar{\omega}_m \end{bmatrix}$$
$$+ \begin{bmatrix} \cos(\theta(k)+b(k)) & 0 & 0 \\ \sin(\theta(k)+b(k)) & 0 & 0 \\ 0 & 1 & 0 \\ 0 & 0 & 1 \end{bmatrix} \begin{bmatrix} \bar{w}_1 \\ \bar{w}_2 \\ \bar{w}_3 \end{bmatrix}. \quad (7.208)$$

The two measurements related to the rover's location are,

$$\begin{bmatrix} z_1 \\ z_2 \end{bmatrix} = \begin{bmatrix} \sqrt{x^2+y^2} \\ \tan^{-1}(y/x) \end{bmatrix} + \begin{bmatrix} v_1 \\ v_2 \end{bmatrix}. \quad (7.209)$$

The landmark's location is assumed to be fixed and modeled as,

$$\dot{x}_{lm} = w_{1lm}, \quad \dot{y}_{lm} = w_{2lm}. \quad (7.210)$$

In discrete time, it is expressed as,

$$x_{lm}(k+1) = x_{lm}(k) + \bar{w}_{1lm}, \quad y_{lm}(k+1) = y_{lm}(k) + \bar{w}_{2lm}, \quad (7.211)$$

where w_{1lm} and w_{2lm} are assumed to be independent white noise disturbance processes with known variances q_{lm1}^2 and q_{lm2}^2, respectively.

7.8 Simultaneous Localization and Mapping (SLAM)

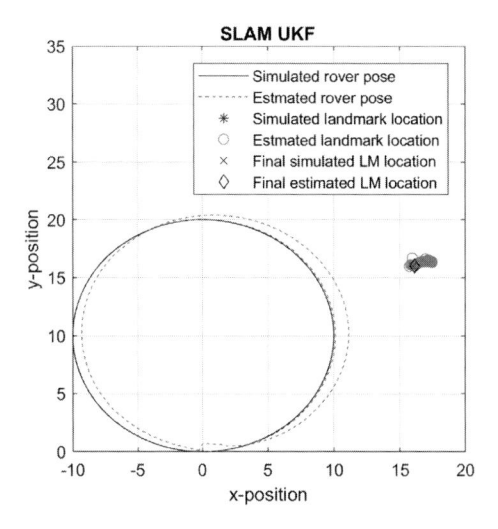

Figure 7.15 Demonstration of SLAM-UKF.

With the landmark's range measurement only,

$$z_{lm} = \sqrt{(x_{lm} - x)^2 + (y_{lm} - y)^2} + v_{lm}. \tag{7.212}$$

With the landmark's range and bearing measurement, the measurement equation is modified to,

$$\begin{bmatrix} z_{lm1} \\ z_{lm2} \end{bmatrix} = \begin{bmatrix} \sqrt{(x_{lm} - x)^2 + (y_{lm} - y)^2} \\ \tan^{-1}((y_{lm} - y)/(x_{lm} - x)) \end{bmatrix} + \begin{bmatrix} v_{lm1} \\ v_{lm2} \end{bmatrix}, \tag{7.213}$$

where v_{lm1} and v_{lm2} are assumed to be independent white noise disturbance processes with known variances r_{lm1}^2 and r_{lm2}^2, respectively.

It may be observed that, in this case, it is possible to solve for (x, y) in terms of z_{lm1} and z_{lm2}. In Figure 7.15 a plot of the rover's pose evolution is shown, as well as the evolution of the estimated position of the landmark on a two-dimensional grid, based on the UKF. Once the estimated position of the landmark converges, it can be used to measure the rover's position relative to it and used to improve the estimate the rover's position. Thus, the joint robot pose and landmark localization comprises the dual task of including detected features to the map using the current robot pose as a reference, while employing the existing landmarks to estimate the robot pose itself. Figure 7.16 illustrates the simulated and estimated rover pose, when an additional known landmark is included to increase the number of measurements and improve the rover's pose estimate. Also shown on the graph are tiny shaded ellipses overlaid on the estimates, representing the 3-sigma limits of the (x, y) coordinates of the rover's position. The ellipses represent the extent of the uncertainty associated with the position estimates of the rover and given indication of the rover's probabilistic location at any particular instant of time.

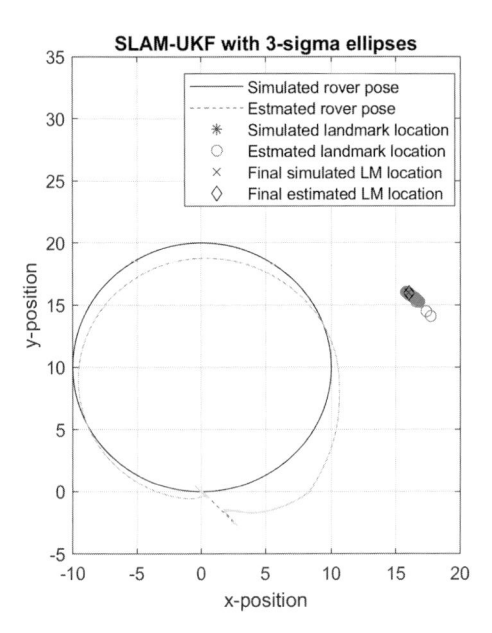

Figure 7.16 Improved position estimation of the rover with uncertainty indication.

The SLAM-UKF or SLAM-EKF algorithms are also the basis of a fast-slam algorithm when multiple landmarks are present, where the assumption is made that landmarks do not need to be correlated with each other, and are represented by employing an individual UKF or EKF for each landmark. Thus, for each landmark only a single covariance matrix of constant size is maintained.

An alternative approach to SLAM is based on a graph representation of the evolution of the rover's pose. The concept of "*graph SLAM*" was introduced by Lu & Milos [12]. Grisetti, Kummerle, Stachniss, and Burgard [13] have presented a tutorial paper on "*graph SLAM*." Thrun, Burgard, and Fox [14] and Thrun and Montemerlo [15] made significant improvements to the algorithm. The SLAM problem may be depicted as a Bayes network as shown in Figure 7.17 and it provides for an understanding of the dependencies in the SLAM problem. The figure shows the changes of the rover's poses as it moves from a state vector $\mathbf{x}(k-1)$ to a new state vector $\mathbf{x}(k)$ following the application of a control signal $\mathbf{u}(k)$ and making corrections from the observation $\mathbf{z}(k-1)$ to the rover's pose employing the known landmarks $\mathbf{m}(j)$. The arrows show direct dependencies, while there is no direct relationship between the rover's poses and landmarks. Shaded nodes represent data directly measured by the rover. Following SLAM, the rover estimates all of the state variables and landmarks to complete the localization and mapping. Landmarks that are continuously monitored and tracked provide the necessary information to reduce the uncertainty in the rover's poses. When a rover traverses a loop or a near loop as it detects a location in the vicinity of another that it has been to before, there is then another opportunity for reducing the uncertainty in the rover's pose. When the observed pose at the start of the motion has low uncertainty, revisiting this location reduces the uncertainty of the predicted robot pose

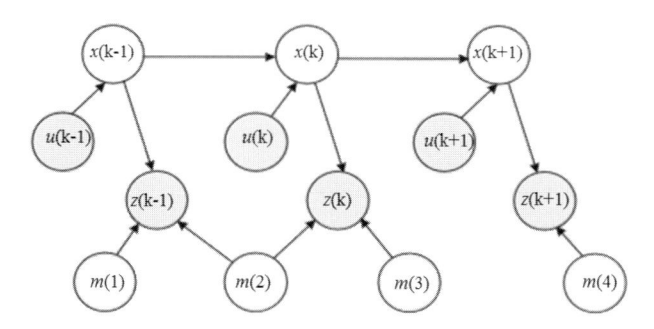

Figure 7.17 A partial graph representation of the SLAM problem.

at this point in time. This is known as *loop-closure*. Thus, loop-closure constraints are used when the rover revisits a previously explored part of the environment.

To formulate the problem in mathematical terms, the rover's path is defined by a series of robot poses. The robot pose at time k is defined by $\mathbf{x}(k)$, and the rover's path $\mathbf{x}_{1:T}$ is written as,

$$\mathbf{x}_{1:T} = \begin{bmatrix} \mathbf{x}(1) & \mathbf{x}(2) & \cdots & \mathbf{x}(T) \end{bmatrix}. \tag{7.214}$$

In SLAM, the rover's initial position $\mathbf{x}(1)$ is known. It is assumed that the rover's control input is defined by $\mathbf{u}(k)$ and it takes the rover from the state $\mathbf{x}(k)$ to the state $\mathbf{x}(k-1)$. The rover's control sequence is then defined as,

$$\mathbf{u}_{1:T} = \begin{bmatrix} \mathbf{u}(1) & \mathbf{u}(2) & \cdots & \mathbf{u}(T) \end{bmatrix}. \tag{7.215}$$

The landmarks employed are denoted by,

$$\mathbf{m}_{1:M} = \begin{bmatrix} m(1) & m(2) & \cdots & m(M) \end{bmatrix}, \tag{7.216}$$

which are denoted simply by $\mathbf{m}$. The observations are denoted by,

$$\mathbf{z}_{1:T} = \begin{bmatrix} \mathbf{z}(1) & \mathbf{z}(2) & \cdots & \mathbf{z}(T) \end{bmatrix}. \tag{7.217}$$

In order to solve the SLAM problem, the probabilistic state transition model and the probability of measurement or the observation model are necessary. The probabilistic state transition model is derived from rover's motion kinematics. The rover's current state $\mathbf{x}(k)$ is obtained from the previous state $\mathbf{x}(k-1)$ and the control input $\mathbf{u}(k)$.

The conditional state transition probability of the current state, given the previous state, is, $p(\mathbf{x}(k)|\mathbf{x}(k-1), \mathbf{u}(k))$, while the probability of measurement, given the location of the landmarks $\mathbf{m}(k)$, at the current time instant k, is defined by, $p(\mathbf{z}(k)|\mathbf{x}(k), \mathbf{m}(k))$. The measurement model is assumed to be of the form,

$$\mathbf{z}(k) = \mathbf{h}_k(\mathbf{x}(k)) + \mathbf{v}(k). \tag{7.218}$$

Consequently, the probability of measurement takes the form,

$$p(\mathbf{z}(k)|\mathbf{x}(k), \mathbf{m}(k)) = p_{m0} \exp\left(-(\mathbf{z}(k) - \mathbf{h}_k(\mathbf{x}(k)))^T \mathbf{R}^{-1} (\mathbf{z}(k) - \mathbf{h}_k(\mathbf{x}(k)))\right). \tag{7.219}$$

Similarly, the state transition is assumed to be governed by,

$$\mathbf{x}(k) = \mathbf{f}_{k-1}(\mathbf{x}(k-1), \mathbf{u}(k)) + \mathbf{G}_{w,k-1}\bar{\mathbf{w}}(k-1). \quad (7.220)$$

Consequently, state transition probability takes the form,

$$p(\mathbf{x}(k)|\mathbf{x}(k-1), \mathbf{u}(k)) =$$
$$p_0 \exp\left(-(\mathbf{x}(k) - \mathbf{f}_{k-1}(\mathbf{x}(k-1), \mathbf{u}(k)))^T \mathbf{Q}^{-1}(\mathbf{x}(k) - \mathbf{f}_{k-1}(\mathbf{x}(k-1), \mathbf{u}(k))))\right). \quad (7.221)$$

In graph-based SLAM, as was seen earlier, the poses of the robot are modeled by nodes in a graph and labeled with their state in a map. Spatial constraints between poses and the data from the observations $\mathbf{z}(k-1)$ and from the control input $\mathbf{u}(k)$ are represented by the edges between the nodes. A graph-based SLAM algorithm then constructs a graph out of the raw measurements. Each node in the graph represents a rover state or a measurement acquired at that state. An edge between two nodes represents a spatial probabilistic performance constraint relating the earlier rover pose to the latter rover pose. A probabilistic performance constraint is obtained from the probability distribution over the transformations between the earlier rover pose and the latter rover pose. The constraint cost takes the form,

$$J_s(k) = (\mathbf{x}(k) - \mathbf{f}_{k-1}(\mathbf{x}(k-1), \mathbf{u}(k)))^T \mathbf{Q}^{-1}(\mathbf{x}(k) - \mathbf{f}_{k-1}(\mathbf{x}(k-1), \mathbf{u}(k))). \quad (7.222)$$

These transformations are either control inputs relating to sequential robot positions or are determined by relating the observations acquired at the two robot locations. The observational constraint cost takes the form,

$$J_m(k) = (\mathbf{z}(k) - \mathbf{h}_k(\mathbf{x}(k)))^T \mathbf{R}^{-1}(\mathbf{z}(k) - \mathbf{h}_k(\mathbf{x}(k))). \quad (7.223)$$

An initial anchoring constraint cost of the form $J_i = \mathbf{x}(1)^T \mathbf{Q}_0^{-1}\mathbf{x}(1)$ is also included.

Once the graph is constructed, one seeks to find the configuration of the robot poses that best satisfies the total constraint cost. Thus, in graph-based SLAM the problem is decoupled in two tasks: the first involves graph construction or constructing of the graph from the raw measurements only, while the second involves graph optimization or determining the optimal configuration of the poses given the edges of the graph. The graph construction is known as the front-end problem, as it is heavily observation dependent, while the graph optimization part is known as the back-end problem, as it relies on the representation of the state and is independent of the observations. To develop a consistent map, the back end must necessarily store data about the vehicle's trajectory in a data structure that represents the graph. The graphical representation of a vehicle's path that encodes the vehicle's global pose estimates at graph vertices and the relative transformation between two poses as graph edges must be effectively stored in the data structure. The graph representation is then effectively optimized by efficient optimization algorithms that can be applied to refine the pose estimates. The graph optimization involves optimization of the total constraint cost,

$$J = \sum_j J_j. \quad (7.224)$$

It is a function defined over the pose states $\mathbf{x}_{1:T}$ and all landmark locations in the map. Additionally, it could include any costs incurred due to loop closure. Now, reconsidering the state transition model and examining it closely, it takes the form,

$$\mathbf{x}(k) = \mathbf{f}_{k-1}(\mathbf{x}(k-1), \mathbf{u}(k)) + \mathbf{G}_{w,k-1}\bar{\mathbf{w}}(k-1), \quad (7.225)$$

which takes the form,

$$\mathbf{x}(k) = \mathbf{x}(k-1) + \mathbf{G}_{u,k}\mathbf{u}(k) + \mathbf{G}_{w,k-1}\bar{\mathbf{w}}(k-1). \quad (7.226)$$

Since $\mathbf{G}_{u,k}\mathbf{u}(k)$ is a rotational transformation of $\mathbf{u}(k)$, it may be expressed as,

$$\tilde{\mathbf{u}}(k) = \mathbf{G}_{u,k}\mathbf{u}(k). \quad (7.227)$$

The state transition error is,

$$\varepsilon(k) = \mathbf{x}(k) - \mathbf{x}(k-1) - \tilde{\mathbf{u}}(k). \quad (7.228)$$

The corresponding state transition error constraint cost is,

$$J_s(k) = (\mathbf{x}(k) - \mathbf{x}(k-1) - \tilde{\mathbf{u}}(k))^T \mathbf{Q}^{-1} (\mathbf{x}(k) - \mathbf{x}(k-1) - \tilde{\mathbf{u}}(k)). \quad (7.229)$$

When a loop closure is encountered between an earlier node j and node k, the state transition error takes the form,

$$\varepsilon_{lc}(j) = \mathbf{x}(j) - \mathbf{x}(k) - \mathbf{z}_{jk}, \quad (7.230)$$

where $\mathbf{z}_{jk}$ is a fixed known vector between the current node k and the previously visited node j. Thus, the loop closure constraint cost takes the form,

$$J_{lc}(j) = (\mathbf{x}(j) - \mathbf{x}(k) - \mathbf{z}_{jk})^T \mathbf{Q}_{lc}^{-1} (\mathbf{x}(j) - \mathbf{x}(k) - \mathbf{z}_{jk}). \quad (7.231)$$

Thus, one may combine the anchoring, state transition, measurement constraint, and loop closure constraint costs. The total cost is then minimized subject to the state transition and loop closure constraints non-recursively to produce an update of the rover's trajectory. A simple example of a graph SLAM with a loop closure constraint is illustrated in Figure 7.18.

The solution is based on the "back end," which optimizes the positions for a given landmark based on the information stored in the "front end" of the graph. This is done by finding the minimum of the least-squares objective, typically by the Gauss-Newton or the Levenberg-Marquardt algorithms. The errors are the differences between the initial positions of the rover that define the state vector and the estimated positions of the rover. The loop closure constraint clearly helps in reducing the location error as the rover approaches a previously visited location.

7.8.2 Multi-Sensor Fusion and SLAM

Multi-sensor fusion, which involves the use of multiple sensors for sensing features related to the same set of landmarks, provides partial redundancy, due to the large number of measurements that are made for localizing the robot and mapping the

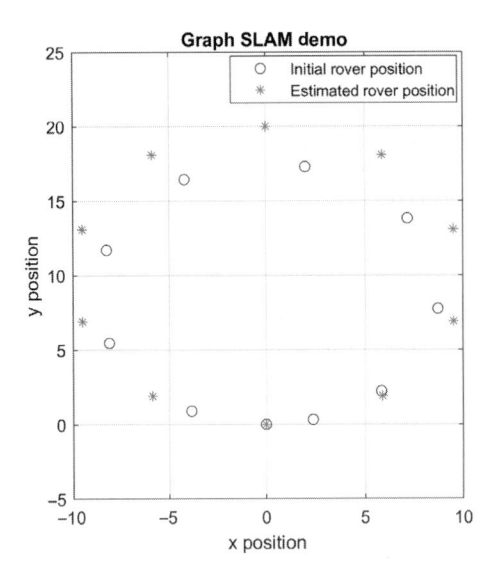

Figure 7.18 Demonstration of simple graph SLAM with a loop closure constraint.

navigation area, particularly if the sensors are dissimilar but complementary in nature. As a consequence, both the reliability and precision of the estimates are increased from the initial stages of the processing. The real difficulty of fusion strategies arises when one has to combine potentially conflicting information coming from an array of heterogeneous sensors in order to accurately localize a rover and estimate the evolving pose relative to the surrounding dynamic environment. Due to the existence of features associated with a given landmark, which could be correlated, it is essential that the identified features are compatible with each other. Consequently, the compatibility of the correlated features must be accounted for via probabilistic techniques, and for this purpose both the Kalman filter-based or the graph-based approaches may be used to mechanize the SLAM algorithm. Compatibility constraints may be included as auxiliary "measurements" and used routinely to improve the accuracy of the estimates in both the EKF and UKF algorithms. They may also be treated as "parity constraints," as has been done in the field of sensor fault detection and diagnosis by the application of Kalman filtering techniques. Yet the particle filter- based approaches, which allow for fewer constraining assumptions to be made relating to the probability distribution of the uncertainties associated with the measurements, may be more appropriate to obtain accurate pose or position estimates for SLAM applications.

7.8.3 Large-Scale Map Building via Sub-Maps

When multiple rovers are present, each rover will produce an individual map. The obvious operation to perform is to merge the individual maps into one single global map by the process referred to as map merging. As different maps may use different coordinate systems, the primary issue in map merging is to find a map transformation matrix that transforms locations on an individual map into the corresponding locations on the global map. To be able to do this, it is essential that certain features are covered by two or more of the maps. Feature-matching techniques or point-matching methods based on statistical techniques play a key role in the ability to merge maps. Each rover generally uses its own custom coordinate system for performing SLAM. So the different maps produced by different rovers generally cannot be merged directly. The individual maps usually need to be rotated or translated prior to merging. Thus, the pose of one of the two rovers relative to the other must be known or estimated, so the maps constructed by each individual rover can be transformed to the global frame.

Merging of map features and data is closely related to the problem of image registration, which arises when a camera is being used to monitor a landmark. Image registration is generally used when sensors such as cameras are employed to capture images of landmarks or other features. In image registration, different sets of data are not only transformed into a common coordinate frame by matching two or more images. An alternative approach is based on creating a landscape image from separate but overlapping images, which are then "stitched" together to form a composite image. A rigid body transformation of the image coordinates, such as a linear transformation, which include rotation, scaling, translation, and other affine transforms, is generally used to fit a segment of an image of a specific geometric feature with another view of the same specific geometric feature. Comparisons of several of the applicable transformations have been made by Eggert, Lorusso, and Fisher [16].

7.8.4 Vision-Based SLAM

Visual motion estimation techniques from sequences of measurements made by either monocular or stereo cameras that can provide precise measurements that could be used in traditional motion estimation algorithms are widely accepted for SLAM applications.

Lemaire, Berger, Jung, and Lacroix [17] present two approaches to the SLAM problem using vision: one with stereo vision, and one with monocular images. Hrabar and Sukhatme [18] have successfully applied stereo cameras to the problem of road recognition. Camera measurements provide images which are particularly suitable for feature-based matching, which requires the extraction of the primary features from images. There are many feature detection algorithms associated with the field of image processing. Algorithms such as the Hough transform can be used to detect simple geometric shapes such as straight lines, circles, and squares as outlined in Ballard [19]. The Harris and SUSAN algorithms may be used to extract the corners. The Canny and Sobel edge detection algorithms may be used to detect the edges in images. These image-processing algorithms can be incorporated into a typical EKF, UKF, or particle filter. The use and application of cameras and other related sensors for motion estimation are discussed in the next chapter.

7.9 Localization in Space and Mobile Robotics

The application of the SLAM techniques need not be restricted to a mobile rover. The dimension of the pose vector for the spacecraft would now include all six degrees of freedom. These could include the position coordinates of the spacecraft as well as its attitude, specified in terms of an attitude quaternion. All of the techniques discussed in this chapter are applicable to six degree of freedom spacecraft navigating in three-dimensional space. A range of applications are currently being developed for future space missions. They may be applied to a spacecraft navigating in the vicinity of a comet or an asteroid in three-dimensional space.

References

[1] Krause, L. O. (1987) A direct solution to GPS Type Navigation equations. *IEEE Transactions on Aerospace and Electronic Systems*, AES-23(2): 225–232.

[2] Brown, R. G. and Hwang, P. Y. C. (1992) *Introduction to Random Signals and Applied Kalman Filtering*, Hoboken, NJ: John Wiley.

[3] Papoulis, A. (1984) *Probability, Random Variables, and Stochastic Processes*, 2nd edn. New York: McGraw-Hill.

[4] Thrun, S. (2002) Particle Filters in Robotics, *In Proceedings of Uncertainty in AI* (UAI).

[5] Thrun, S., Fox, D., Burgard, W., and Dellaert, F. (2001) Robust Monte Carlo localization for mobile robots. *Artificial Intelligence*, 128(1–2): 99–141.

[6] Julier, S. J. and Uhlmann, J. (2000) Unscented filtering and nonlinear estimation. *Proceedings of the IEEE*, 92(3): 401–422.

[7] Cheeseman, P. and Smith, P. (1986) On the representation and estimation of spatial uncertainty. *International Journal of Robotics*, 5: 56–68.

[8] Thrun, S. (2008) Simultaneous localization and mapping. In *Robotics and Cognitive Approaches to Spatial Mapping*, M. E. Jefferies, ed., Berlin/Heidelberg: Springer, 13–41.

[9] Bailey, T. and Durrant-Whyte, H. (2016) Simultaneous localization and mapping (SLAM): Part II the state of the art. *IEEE Robotics And Automation Magazine*, 13: 108–117.

[10] Dissanayake, G., Durrant-Whyte, H., and Bailey, T. (2000) A computationally efficient solution to the simultaneous localisation and map building (SLAM) problem, in *Proceedings of the 2000 IEEE International Conference on Robotics & Automation*, San Francisco, CA, 1009–1014.

[11] Clark, S., Dissanayake, G., Newman, P., and Durrant-Whyte, H. (2001). A solution to simultaneous localization and map building (SLAM) problem. *IEEE Journal of Robotics and Automation*, 17(3): 229–241.

[12] Lu, F. and Milios, E. (1997). Globally consistent range scan alignment for environment mapping. *Autonomous Robots* 4: 333–349.

[13] Grisetti, G., Kummerle, R., Stachniss, C., and Burgard, W. (2010) A Tutorial on graph-based SLAM. *IEEE Intelligent Transport Systems Magazine*, 2(4): 31–43.

[14] Thrun, S., Burgard, W., and Fox, D. (2005) *Probabilistic Robotics: Intelligent Robotics and Autonomous Agents*, Cambridge, MA: MIT Press.

[15] Thrun, S. and Montemerlo, M. (2006) The graph SLAM algorithm with applications to large-scale mapping of urban structures. *The International Journal of Robotics Research*, 25(5–6): 403–429.

[16] Eggert, D. W., Lorusso, A., and Fisher, R. B. (1997) Estimating 3-D rigid body transformations: A comparison of four major algorithms. *Machine Vision and Applications*, 9(5–6): 272–290.

[17] Lemaire, T., Berger, C., Jung, I.-K., and Lacroix, S. (2007) Vision-based SLAM: Stereo and monocular approaches. *International Journal of Computer Vision*, 74(3): 343–364.

[18] Hrabar, S. and Sukhatme, G. (2009). Vision-based navigation through urban canyons. *Journal of Field Robotics*, 26(5): 431–452.

[19] Ballard, D. H. (1981) Generalizing the Hough transform to detect arbitrary shapes. *Pattern Recognition*, 13(2): 111–122.

8 Sensing and Estimation of Spacecraft Dynamics

8.1 Introduction

A variety of sensors are used for the sensing and estimation of spacecraft dynamics. These include classical mechanical motion sensors such as accelerometers and gyroscopes, and mechanical encoders, as well as a number of dedicated sensors used in space flight such as sun sensors, Earth horizon sensors, and star sensors. A whole class of secondary sensors are also emerging due to recent developments in radio detection and ranging (RADAR), sound navigation and ranging (SONAR), ultrasonics, optics, light amplification due to stimulated emission of radiation (LASER), LASER detection and ranging (LADAR), light detection and ranging (LIDAR), and other similar systems involving the use, propagation, and detection of various bands of the electro-magnetic spectrum. These new sensor designs have also been facilitated in part due to the developments of new semiconductor and magnetic materials. Thus, this chapter is dedicated to the application of modern attitude, localization, and navigation sensors and the associated computational algorithms.

8.2 Spacecraft Attitude Sensors

The primary sensors that are used in the determination of a spacecraft's motion are mechanical motion sensors such as accelerometers and gyroscopes, as these are generally inexpensive. The outputs of typical accelerometers and gyroscopes are used to primarily measure the motion of a rigid body, such as a spacecraft, at a particular location relative to a body-fixed reference frame, and the outputs are generally processed in a computer to generate the position of the spacecraft. This is the principle behind strapped-down navigation. Rate-integrating gyroscopes mounted alongside the accelerometers may be used to generate outputs, which are then processed in a computer to generate the attitude of the spacecraft. The principle of operation of the accelerometers and gyroscopes is first discussed in the following section.

8.2.1 The Principle of Operation of Accelerometers and Gyroscopes

The principle of operation of an accelerometer can be illustrated by modeling it using a mass supported by a spring and a damper. For detection of acceleration, a motion sensor

8.2 Spacecraft Attitude Sensors

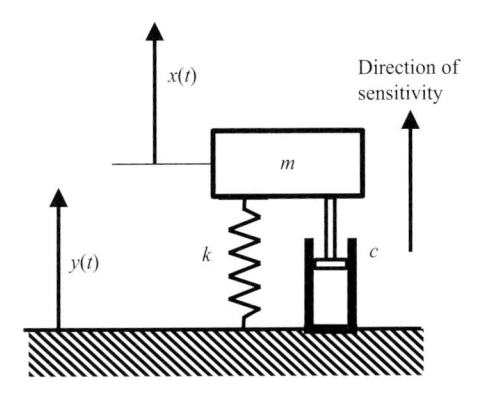

Figure 8.1 Schematic diagram of an accelerometer to indicate the principles of operation.

based on the piezoelectric effect is often employed in practice. A piezoelectric crystal or ceramic wafer is employed to detect acceleration, and it can be modeled as a mass supported on a spring and damper. The inertial force acting on the piezoelectric material by a heavy mass attached to one side of the piezoelectric wafer or a piezoelectric crystal generates a stress across two faces on the material, which in turn generates an output charge. The charge is generally sensed to generate an output that is directly proportional to the acceleration of the attached mass.

Consider a schematic diagram of an accelerometer as shown in Figure 8.1. The direction of sensitivity is shown on the right of the figure. The ground motion is given as $y(t)$. The equation of motion in terms of $x(t)$ is,

$$m\ddot{x} = -k(x-y) - c(\dot{x}-\dot{y}) - mg. \qquad (8.1)$$

Let the relative displacement of the mass relative to the base be,

$$x_r(t) = x(t) - y(t). \qquad (8.2)$$

Then it follows that,

$$m\ddot{x}_r + c\dot{x}_r + kx_r = -m\ddot{y} - mg. \qquad (8.3)$$

As the stiffness constant is usually very high, neglecting the relative inertial term in comparison with the spring force,

$$F = kx_r + c\dot{x}_r = -m\ddot{y} - mg = -m(\ddot{y}+g) = -m(\ddot{y}-G), \qquad (8.4)$$

where F is the force acting on the piezoelectric material surface and $G = -g$ is the component of the acceleration due to gravity in the direction of measurement. Thus, the relative acceleration of the mass attached to one side of the piezoelectric wafer or a piezoelectric crystal is $a_{piezo} \propto A - G$, where A is the ground acceleration given by $A = \ddot{y}$. This is the principle of acceleration measurement.

To explain the principle of the rate gyroscope and the rate-integrating gyroscope, consider the Euler equations of motion of a rigid body,

$$I_{xx}\dot{p} + (I_{zz} - I_{yy})qr = L, \qquad (8.5)$$

$$I_{yy}\dot{q} + (I_{xx} - I_{zz})rp = M, \tag{8.6}$$

$$I_{zz}\dot{r} + (I_{yy} - I_{xx})pq = N. \tag{8.7}$$

The most interesting application of the Euler equations is the gyroscope. In an ideal gyroscope a rotor or wheel is kept spinning at constant angular velocity. The axis of the wheel is assumed to be coincident with the z axis. It is assumed that the Oz axis is such that $\dot{r} = 0$, i.e. such that,

$$h_z = I_{zz}r = I_{zz}\dot{\psi} = \text{constant}, \tag{8.8}$$

where I_{zz} is the polar moment of inertia of the wheel. The other two moments of inertia of the wheel are assumed to be equal to each other and equal to I_d, and hence the first two of the Euler equations may be written as,

$$I_d\dot{p} + Hq = L, \tag{8.9}$$

$$I_d\dot{q} - Hp = M, \tag{8.10}$$

where $H = (I_{zz} - I_d)r$.

These are the general equations of motion for two degrees of freedom gyroscopes. If one assumes the gyroscope to be gimballed with a single gimbal, as in Figure 8.2, with a prescribed input angular velocity in the Ox direction, then the equation of motion in the Oy axis is,

$$I_d\dot{q} - Hp = M, \tag{8.11}$$

where p is the prescribed input angular velocity and M is the net torque acting on the rotor-gimbal assembly.

If the restoring torque is given by a lightly damped torsional spring force, the net torque acting on the rotor-gimbal assembly is given by,

$$M = -Bq - K\theta = -B\dot{\theta} - K\theta, \quad q = \dot{\theta}. \tag{8.12}$$

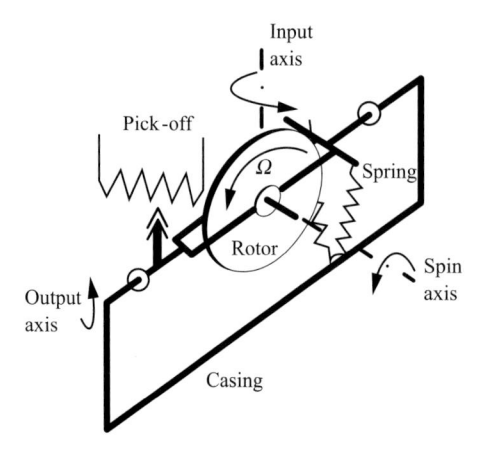

Figure 8.2 Schematic diagram illustrating the principle of operation of a gyroscope.

The equation of motion in the Oy axis may be expressed as,

$$I_d\ddot{\theta} + B\dot{\theta} + K\theta = Hp. \tag{8.13}$$

Under steady state conditions, the output angular displacement is proportional to the input angular velocity,

$$\theta \approx (H/K)p. \tag{8.14}$$

The gyroscope behaves like a rate gyroscope.

When the restoring torque is given by a torsional damper force, the net torque acting on the rotor-gimbal assembly is given by,

$$M = -Bq = -B\dot{\theta}, \quad q = \dot{\theta}, \tag{8.15}$$

and the equation of motion in the Oy axis can be expressed as,

$$I_d\ddot{\theta} + B\dot{\theta} = Hp. \tag{8.16}$$

Under steady state conditions, the output angular displacement is proportional to the input angular velocity,

$$\dot{\theta} = q \approx (H/B)p. \tag{8.17}$$

Integrating both sides, it follows that the gyroscope output is directly proportional to the integral of the prescribed input angular velocity p. Thus, the gyroscope now behaves like a rate-integrating gyroscope or an attitude-sensing gyroscope.

8.2.2 Magnetic Field Sensor

The Earth's magnetic field is a vector quantity with both magnitude and direction at any point in space. A magnetic field sensor measures the field's total magnitude in a particular direction. The sensor measures the component of a field that lies along its axis of sensitivity. A multi-directional sensor is capable of making magnetic field measurements in several directions. If the axes of sensitivity are aligned in three mutually perpendicular directions, one has a vector magnetic sensor that is capable of measuring the three components of the field at a particular point in space. Most current magnetic sensors give a digital output and have a built-in threshold, and consequently only produce an output when the threshold is exceeded.

While complex semiconductor-based magnetic field sensors based on the Josephson junction have been built, the magnetic field sensors in their simplest form are induction-coil magnetometers, which function according to Faraday's law of induction, which states that the voltage induced in a coil is proportional to the change in the magnetic field in the coil. This induced voltage creates a current that is proportional to the rate of change of the field. The sensitivity of the induction coil is dependent on the permeability of the core and the coil's cross-sectional area and number of turns of the coil. As the induction coil is only sensitive to a varying magnetic field, they cannot detect static or slowly changing fields. For this reason, one employs two coils, a primary and a secondary, wrapped around a long, common high-permeability ferromagnetic core. The core's magnetic induction changes in the presence of an external magnetic field. A typical square or sine wave drive signal applied to the primary winding at a particular carrier frequency causes the magnetic field in the core to oscillate between saturation points. The secondary winding outputs a signal that is coupled through the core from the primary winding, and the signal is essentially amplitude modulated by the Earth's magnetic field in the direction of measurement. While there are several modes in which such a flux-gate magnetometer can operate, the output of the secondary coil is generally demodulated to obtain a signal directly proportional to the Earth's magnetic field in the direction of measurement.

8.2.3 Sun Sensors

Sun sensors measure the direction of the Sun relative to a spacecraft and provide attitude measurement vectors for attitude determination. The sensors measure the incident solar radiation on a photovoltaic-sensitive surface. An electronic computation unit is used to process the measurements and determine the direction of the Sun with reference to a frame fixed in the body of the satellite. Most modern Sun sensors are digital, so one has to account for the quantization error in addition to the sensor noise generated within the photovoltaic-sensitive surface. Both these errors are assume to be Gaussian random noise processes and are usually combined and modeled by a single Gaussian white noise process of known variance. Furthermore, the sensors are active only during the Sun-lit phase of the orbit and cannot be used during an eclipse phase, when the Sun is not visible from the satellite. Generally, Sun sensors are mounted on the faces of a cube, so they are capable of taking six measurements. However, only three of the faces of the cube may be exposed to the Sun at a particular point on an orbit, so only a minimum of three measurements are generally available at any instant. Ideally, each measurement is equal to the dot product of the Sun vector with the local outward normal of the face of the cube, which is equivalent to the direction cosine of the face of the cube with respect to the Sun vector. There is also the additive white noise to contend with. Since the local outward normals of the faces of the cube are known in the body-fixed frame of the spacecraft, one has a set of three equations for the three unknown components of the Sun vector in the body-fixed frame of the satellite. The Sun vector measurements can then be used for attitude determination, which will be discussed in a subsequent section.

8.2.4 Earth Horizon Sensors

Horizon sensors, essentially, measure the direction of the Earth by observing the shape of the Earth's surface as observed from the spacecraft and comparing it with a modeled shape, to arrive at the spacecraft attitude. They are particularly useful for nadir-pointing satellites. A nadir-pointing satellite points toward a geodetic sub-satellite point. The geodetic sub-satellite point is defined as a point on the surface of the Earth's reference ellipsoid that is nearest to the satellite. This is also the point on the Earth's ellipsoid where the satellite is at zenith or vertically above. A vector pointing to the center of the

Earth will intersect the Earth at a point (geocentric sub-satellite point) that is up to 21 km away from the geodetic sub-satellite point for a satellite in a sun-synchronous orbit. Horizon sensors are generally imaging sensors where an infrared image is post-processed to detect the horizon from the Earth and its surroundings. The frequency band of the radiation that is detected is chosen so as to maximize the radiation detected from the Earth and minimize the radiation detected from the surrounding atmosphere. Like other imaging or ranging sensors, discussed later, horizon sensors can be either static or scanning types. The horizon detection algorithm is typically a curve detection type, such as the Hough transform.

8.2.5 Star Sensors

Star sensors are similar to Sun sensors and generally use an imaging camera to detect stars in the sky that are emitting light beyond a certain threshold. Once the stars are identified and mapped on a celestial map, the map and the star locations are compared with a star map from a catalogue of the stars that are expected to be visible at that point in the orbit. Using pattern recognition techniques, the attitude transformation relating the measured star map to the star map constructed in real time from the star catalogue is constructed. Thus, one is able to construct the star vector pointing in the direction of the brightest star in the map. The star vector can then be used, along with other directional measurements, for attitude determination of a spacecraft, which will be discussed in a subsequent section.

8.2.6 Use of Navigation Satellite as a Sensor for Attitude Determination

While most of the early applications were based on using gyroscopes, star trackers, magnetometers, and gravity sensors, the recent availability of interferometric measurements of carrier phase differences arising from navigation satellite transmissions has led to cheaper and more reliable alternative sensors for spacecraft attitude determination. Because these sensors operate without the need to rely on pseudo-random number-based codes, they are often referred to as codeless satellite navigation sensors. They operate by measuring the phase difference in the carrier signals received at two antennae located at opposite ends of a baseline vector. Given the baseline vector **d** and the navigation satellite's sight line unit vector **r**, as well as the carrier signal wavelength λ, the differential phase measurement is given by,

$$\Delta\phi = \frac{2\pi}{\lambda}(\mathbf{d}\cdot\mathbf{r}) - 2\pi N, \qquad (8.18)$$

where N is the number of integer wavelengths occurring between the two antennae. Since the antenna's baseline vector is a function of the spacecraft's attitude, the differential phase measurement $\Delta\phi$ may also be construed to be a nonlinear function of the spacecraft's attitude. A number of studies of the spacecraft attitude determination problem have appeared based on the differential measurement idea. The use of these codeless satellite navigation sensors, coupled with the quaternion representations of the attitude, opens up the possibility of determining large attitudes without any restriction on their magnitudes, over a large time frame.

The application of a typical codeless satellite navigation sensor to attitude estimation has been described by Vepa [1]. It is assumed that one is able to construct a codeless satellite navigation receiver antenna pair located at the two ends of a graphite-epoxy bar of a length less than the wavelength λ of the carrier signal. When the length of the graphite-epoxy bar is greater than the wavelength λ of the carrier signal, the ambiguity in N must first be resolved. Satellite navigation ambiguity resolution is the process of estimating the correct carrier phase integer ambiguity. The ambiguity term must be determined correctly if one intends to achieve centimeter-level positioning accuracies. One method of estimating N, when multiple satellites are available, is the residual sensitivity matrix approach proposed by Hatch and Sharpe [2]. The most successful methods of ambiguity resolution are based on using pseudo-satellite (*pseudolite*) carrier wave transmitters to complement measurements from available satellites and were proposed by Pervan, Cohen, and Parkinson [3]. When it can be assumed that one is able to construct a codeless satellite navigation receiver antenna pair located at the two ends of a graphite-epoxy bar that is less than the wavelength λ of the carrier signal, it is possible in principle to assume that N equals a known integer or zero in Equation (8.18), which may be expressed as,

$$\Delta\phi = \frac{2\pi}{\lambda}(\mathbf{d}\cdot\mathbf{r}). \qquad (8.19)$$

The primary GPS signals are known to be transmitted on two radio frequencies in the UHF band. These frequencies are referred to as L1 and L2, with

$$f_{L1} = 154 \cdot f_0 = 1{,}575.42 \text{ MHz}, \quad f_{L2} = 120 \cdot f_0 = 1{,}227.6 \text{ MHz}, \qquad (8.20)$$

where $f_0 = 10.23$ MHz is a common frequency. Recently, another frequency was introduced, referred to as L5, with,

$$f_{L5} = 115 \cdot f_0 = 1{,}176.45 \text{ MHz}. \qquad (8.21)$$

The primary GPS signals are phase-modulated signals using the modulation method known as binary-phase shift keying (BPSK). A change in code state (± 1) causes a 180 degree phase shift in the carrier. The phase change rate is often referred to as the chip rate. In the case of the newer L5 signal, the modulation method used is known as quadrature phase shift keying. In this case, binary bit combinations corresponding to $m = 0, 1, 2,$ and 3 are sensed and phase changes are equal to $45° + m \times 90°$, for $m = 0, 1, 2,$ and 3. For the case of the highest frequency carrier, which is the L1 carrier signal (1,575.42 MHz), the distance between the two antennae located at the two ends of the graphite-epoxy bar should be less than 0.19 m. The distance between the two antennae would then be sufficient to operate even at the L2 carrier frequency (1,227.6 MHz). In practice, these frequencies ought to be Doppler shifted. It is also feasible to construct such an antenna by locating two micro-strip patch antennae at the two ends of the bar, although the field of view is considerably limited. However, in principle, the two-channel receiver can be constructed and the differential phase angle extracted by employing the fast Fourier Transform (FFT) algorithm implemented on suitable digital signal processing hardware.

The GLONASS satellite navigation system offers many features in common with the GPS satellite navigation system. It is also based on 24 orbiting satellites forming the space segment (21 operational satellites with 3 in-orbit spares) but will use only 3 orbital planes separated by 120 degrees of longitude and with equal spacing between satellites of 45 degrees within the plane. The GLONASS orbits are roughly circular orbits with an inclination of about 64.8 degrees, a semi-major axis of 25,440 km, and a period of 11 h 15 m 44 s. Similar to GPS, GLONASS also transmits two spread-spectrum signals in the L-band at around the same power levels (−160 dBW at L1, −163 dBW at L2). However, individual GLONASS satellites are distinguished by a dedicated radio frequency channel rather than spread spectrum code. In GLONASS, a single code of length 511 bits repeating every 1 ms is used. Information is encoded differentially in an RZ (return to zero) format with a final data rate of 50 baud.

Radio-frequency carriers used by GLONASS occupy channels within the L-band ranging from 1,240–1,260 MHz and 1,597–1,617 MHz, the channel spacing being 9/16 (or 0.5625) MHz at the higher frequencies and 7/16 (or 0.4375) MHz at the lower frequencies. The carrier frequencies themselves are also multiples of channel spacing, and the number of planned channels is 24. GLONASS L1 transmission carrier frequencies (f_n, in megahertz) and channel numbers (C_n) are related by the expression,

$$f_n = 1,602 + 0.5625 C_n = 0.5625(2,848 + C_n). \tag{8.22}$$

A similar formula relates the GLONASS L2 transmission carrier frequencies to the channel numbers, where the frequencies are in the ratio 7/9.

This is quite different from the GPS system, which uses the same frequency for all satellites and differentiates one satellite from another by individual Gold codes, a form of CDM (code division multiplex). The GLONASS system uses FDMA (frequency division multiple access) to distinguish between satellites. This difference between the two systems has a major impact in designing codeless receivers capable of joint operation. However, receivers capable of measuring carrier phase differential have been designed and built for both the GLONASS and GPS systems. In fact, a receiver with an antenna separation distance of 0.18 m would be adequate for the relation (Equation 8.19) to hold. Thus, codeless satellite navigation sensors could be constructed based on either the GPS or GLONASS carrier signals or even the more recent GALLILEO satellite system-based carrier signals.

8.3 Attitude Determination

The objective of attitude determination is to obtain accurately the orientation of a satellite in space, at any given instant, in terms of known reference axes, using measurements obtained from sensors strapped on to the satellite's body, by solving a set of nonlinear mathematical relations between the attitude matrix or attitude parameters and the measurements. When a spacecraft has a preferred axis, such as the direction of the angular momentum vector of a spinning satellite, the actual three axis attitude is the computed attitude about the preferred axis plus a phase angle about it. Such an approach is referred to as the geometric approach because the additional phase angle is computed by resorting to spherical trigonometry. In the algebraic method, the attitude matrix is computed from two vector observations of the satellite's attitude relative to two corresponding reference axes. This is the basis of the TRIAD algorithm first proposed by Black [4]. Finally, in the **q**-method as well as in the QUEST algorithm, the three-axis attitude matrix is computed from several vector observations relative to corresponding reference directions by minimizing a cost function. In the **q**-method, the minimization problem is transformed to an equivalent eigenvalue problem, while in the QUEST algorithm is directly solved by an alternate algorithm rather than by the **q**-method. It can also be minimized by a method such as the Levenberg-Marquardt algorithm.

An alternative class of methods is based on the statistical approaches to estimation that are the foundation of recursive filters such as Kalman filters and particle filters.

Given two nonparallel reference vectors in an inertial frame and the respective observation vectors in a common spacecraft fixed-body frame, it is possible to determine the orthogonal matrix transformation that relates each of the observation vectors to the corresponding reference vector. This is the basis of the TRIAD algorithm. Thus, given the reference vectors v_1 and v_2 and the measurement vectors w_1 and w_2, the attitude transformation matrix $\mathbf{R}_{BI}$, from the reference inertial to the body frames, satisfies the relations,

$$\mathbf{R}_{BI} v_1 = w_1, \quad \mathbf{R}_{BI} v_2 = w_2. \tag{8.23}$$

A reference triad is set up from the relations,

$$r_1 = v_1, \quad r_2 = (v_1 \times v_2)/|v_1 \times v_2|, \quad r_3 = v_1 \times r_2. \tag{8.24}$$

The corresponding measurement triad is defined from the relations,

$$s_1 = w_1, \quad s_2 = (w_1 \times w_2)/|w_1 \times w_2|, \quad s_3 = w_1 \times s_2. \tag{8.25}$$

Thus the two triads are related by,

$$\mathbf{R}_{BI} r_i = s_i, \quad i = 1, 2, 3. \tag{8.26}$$

Only six of these equations are independent. The matrix $\mathbf{R}_{BI}$ has six independent parameters and they can be obtained by solving the six independent equations. The general solution is given by,

$$\mathbf{R}_{BI} = \begin{bmatrix} s_j \end{bmatrix} \begin{bmatrix} r_j \end{bmatrix}^T = \begin{bmatrix} s_1 & s_2 & s_3 \end{bmatrix} \begin{bmatrix} r_1 & r_2 & r_3 \end{bmatrix}^T. \tag{8.27}$$

As an example of the application of the TRIAD algorithm to determine the attitude transformation for a vehicle moving on the surface of the Earth, the reference vectors v_1 and v_2 could be the north-pole pointing vector and the down-pointing gravity vector at the vehicle's location. The corresponding accurate measurement vectors are obtained by using a high-precision compass to measure the vehicle's heading and an inclinometer to measure the inclination of the local surface normal to the gravity vector. Provided these

8.3 Attitude Determination

measurements are sufficiently accurate, the attitude of the ground vehicle may be estimated using the TRIAD algorithm.

The **q**-method devised by Davenport [5, 6] uses several observation vectors, and it can be defined by considering n unitary vectors w_i, $i = 1, 2, 3 \cdots n$, where n corresponds to the number of sensors installed in the satellite body that are assumed to each make independent observations of a landmark, and w_i the observation vectors. For each observed vector, a reference vector v_i is necessary for the attitude computation. Therefore, the attitude matrix $\mathbf{R}_{BI}$ associates the observed and reference vectors by,

$$w_i = \mathbf{R}_{BI} v_i, \quad i = 1, 2, 3 \cdots n. \tag{8.28}$$

A solution to this problem can be found using optimization methods. The matrix $\mathbf{R}_{BI}$ is obtained when minimizing the error defined by the weighted sum of the norms $\|w_i - \mathbf{R}_{BI} v_i\|$. The **q**-method searches a solution that maximizes the *gain function* $g(\mathbf{R}_{BI})$ given by,

$$g(\mathbf{R}_{BI}) = 1 - L(\mathbf{R}_{BI}) = \sum_{j=1}^{n} a_j \left(w_j^T \mathbf{R}_{BI} v_j \right), \tag{8.29}$$

where a_j are the weights associated with the jth vector pair and $\sum_{j=1}^{n} a_j = 1$.

To simply the *gain function*, it is best to express the attitude matrix in terms of the quaternion attitude components,

$$\mathbf{q} = \begin{bmatrix} q_1 & q_2 & q_3 & q_4 \end{bmatrix}^T. \tag{8.30}$$

Thus, the attitude matrix is written as,

$$\mathbf{R}_{BI} = \mathbf{I} - 2q_4 \mathbf{S}(\vec{q}) + 2\mathbf{S}^2(\vec{q}), \quad \vec{q} = \begin{bmatrix} q_1 & q_2 & q_3 \end{bmatrix}^T, \tag{8.31}$$

where $\mathbf{I}$ is the identity matrix and $\mathbf{S}(\vec{q})$ is the cross-product operator and is defined by,

$$\mathbf{S}(\vec{q}) = -\mathbf{S}(\vec{q})^T = \begin{bmatrix} 0 & -q_3 & q_2 \\ q_3 & 0 & -q_1 \\ -q_2 & q_1 & 0 \end{bmatrix}. \tag{8.32}$$

The cross-product operator is also denoted as, $\mathbf{S}(\vec{q}) = \vec{q} \times$.

The attitude transformation matrix is also expressed as,

$$\mathbf{R}_{BI} = (\mathbf{q} \cdot \mathbf{q} - 2\vec{q} \cdot \vec{q})\mathbf{I} - 2q_4 \mathbf{S}(\vec{q}) + 2\vec{q} \cdot \vec{q}^T. \tag{8.33}$$

The gain function, in terms of quaternions, reduces to,

$$g(\mathbf{R}_{BI}) = (\mathbf{q} \cdot \mathbf{q} - 2\vec{q} \cdot \vec{q}) tr(\mathbf{B}^T) - 2q_4 tr(\mathbf{S}(\vec{q})\mathbf{B}^T) + 2tr(\vec{q} \cdot \vec{q}^T \mathbf{B}^T), \tag{8.34}$$

which may be expressed as,

$$g(\mathbf{R}_{BI}) = \mathbf{q}^T \mathbf{K} \mathbf{q}, \tag{8.35}$$

where **K** is a 4×4 matrix of the form,

$$\mathbf{K} = \begin{bmatrix} \mathbf{S} - \sigma \mathbf{I} & \mathbf{Z} \\ \mathbf{Z}^T & \sigma \end{bmatrix}, \tag{8.36}$$

and σ, **S**, and **Z** are, respectively, given by,

$$\sigma = tr(\mathbf{B}) = \sum_{j=1}^{n} a_j(w_j \cdot v_j), \mathbf{S} = \mathbf{B} + \mathbf{B}^T = \sum_{j=1}^{n} a_j\left(w_j v_j^T + v_j w_j^T\right)$$
$$\mathbf{Z} = \sum_{j=1}^{n} a_j(w_j \times v_j).$$
(8.37)

The problem of determining the attitude is reduced to finding the quaternion that maximizes the gain function. It can be expressed by,

$$\tilde{g}(\mathbf{R}_{BI}) = \mathbf{q}^T \mathbf{K} \mathbf{q} - \lambda \mathbf{q}^T \mathbf{q},$$
(8.38)

where λ is chosen in order to satisfy an eigenvalue constraint. It is the maximum eigenvalue and the corresponding optimal eigenvector of the eigenvalue problem defined by,

$$\mathbf{K}\mathbf{q}_{opt} = \lambda \mathbf{q}_{opt}.$$
(8.39)

The QUEST algorithm (Quaternion Estimator) estimates the optimal eigenvalue and eigenvector for the problem formulated in the **q**-method, without the necessity of eigenvalue analysis. It was proposed by Shuster and Oh [7], and it can be seen as an improvement of the **q**-method, searching the minimization of a cost function,

$$L(\mathbf{R}_{BI}) = \frac{1}{2}\sum_{j=1}^{n} a_j(w_j - \mathbf{R}_{BI} v_j)^2,$$
(8.40)

or to maximize the gain function,

$$g(\mathbf{R}_{BI}) = \sum_{j=1}^{n} a_j\left(w_j^T \mathbf{R}_{BI} v_j\right),$$
(8.41)

where the maximum is given by λ_{opt}. The solution for the optimal value of λ_{opt} is,

$$\lambda_{opt} \approx \sum_{j=1}^{n} a_j - L(\mathbf{R}_{BI}).$$
(8.42)

When $L(\mathbf{R}_{BI})$ is relatively very small,

$$\lambda_{opt} \approx \sum_{j=1}^{n} a_j.$$
(8.43)

This provides a first estimate of the eigenvalue that may be used to estimate the corresponding eigenvector, which is in fact the solution for the quaternion attitude.

A simplified solution for the eigenvector problem may be obtained by defining the vector,

$$\mathbf{p} = \vec{q}/q_4 = [q_1 \quad q_2 \quad q_3]^T/q_4.$$
(8.44)

The vector **p** satisfies the equation,

$$\mathbf{p} = (\lambda_{opt}\mathbf{I} - \mathbf{S})^{-1}\mathbf{Z}.$$
(8.45)

The quaternion attitude is then defined by,

$$\mathbf{q} = \frac{1}{\sqrt{1 + \mathbf{p}^T \mathbf{p}}} \begin{bmatrix} \mathbf{p} \\ 1 \end{bmatrix}. \tag{8.46}$$

There are certain special cases when the algorithm fails, but in these cases the quaternion attitude can be easily obtained by applying physical arguments.

8.4 Spacecraft Large Attitude Estimation

The Wahba problem [8] arose in the estimation of the attitude of a satellite by using direction cosines of bodies observed in a satellite-fixed frame of reference and direction cosines of the same bodies in a known frame of reference. The problem was to find a least squares estimate of the rotation matrix that transforms the known frame of reference into the satellite-fixed frame of reference. Since the publication of Wahba's paper [8], a number of improvements to the attitude estimation method have been reported. This problem was reformulated by Farrell [9] as a Kalman filtering problem by using Sun sensor and magnetometer sensor data to achieve a 10-fold increase in accuracy of the estimate over the accuracy of the measurement. While the problem of attitude determination may be posed as a purely kinematic problem, the problem reduces to the case of optimum mixing of rate and attitude measurements if one uses a set of rate gyroscopes to measure the components of the angular velocity vector and attitude gyroscopes to measure the attitudes. Potter and Vander Velde [10] adopted such an approach to achieve significant increases in accuracy of the estimate over the accuracy of the measurements. An alternative possibility is to determine the angular velocity components from the spacecraft's Eulerian dynamics, but such an approach no longer remains a purely kinematic approach as in Fujikawa and Zimbelman [11] and Huang and Juang [12]. The representation of the attitudes either by the three-parameter varieties of Euler angles or equivalently by the four-parameter representations such as quaternions leads to several different implementations of the attitude estimation problem. In spite of the need for quaternion normalization as discussed in Lefferts, Markley, and Schuster [13], quaternions have emerged as the basis for the standard method described by Lerner [14] for representing the attitude of a space vehicle.

The problem of quaternion- based attitude estimation is patently nonlinear, and most early efforts focused on the Extended Kalman filter (EKF). The EKF-based approaches are still exceptionally popular, and some applications of it are illustrated by Marins et al. [15] and de Ruiter et al. [16]. However, the need for including higher order nonlinear effects in the quaternion prediction and the state update phases as well as in the error covariance prediction and update was soon recognized, and a number of extensions have appeared in the literature. The first class of these extensions account for the second-order effects as in Vathsal, [17] and these employ the mean and covariance updating methods outlined in Sage and Melsa [18] and Jazwinski [19]. The EKF employs the Jacobians of the nonlinear transformations of the process states, and this has led to the development of two other extensions to it. The first of these has been applied to the spacecraft attitude problem as outlined by Crassidis and Markley [20, 21]. The need to avoid the determination of the Jacobian linearizations as in the EKF has led to the development of the unscented Kalman filter (UKF) based on the unscented transformation by Julier and Uhlmann [22]. A recent application of the UKF to the attitude estimation problem is discussed by Vandyke et al. [23] and Vepa [1].

Kingston and Beard [24] have also employed codeless satellite navigation sensors for attitude determination but their formulations are not exclusively based on the quaternion. Thus, codeless satellite navigation sensors are the only sensors used with no prospect of mixing and the attitudes are not estimated purely from the kinematic considerations. A mixing-type attitude determination filter purely from kinematics is proposed by Creamer [25], but experience with using this approach did not indicate a long enough prediction window for long-term autonomous operations. This was identified to be due to the use of the EKF approach, which is known to diverge under certain conditions. Crassidis and Markley [26] have proposed a method of applying the UKF to the problem of spacecraft estimation by extending the multiplicative EKF, using equivalent quaternion multiplications leading to rotation additions rather than component vector additions. They transform the quaternion to modified Rodrigues parameters, which need not be explicitly constrained and define the mean and covariance in this space. The transformation and inverse involve the application of trigonometric and inverse trigonometric functions that cannot be adequately approximated by second- and third-order functions uniformly for all rotations. Although the method is novel, it is not always applicable, particularly when disturbance models are naturally additive. A similar method was also proposed by Kraft [27]. With a view to improving the accuracy and speed of estimation, Cheon and Kim [28] have proposed a novel unscented filter in a unit quaternion space, where the sigma vectors maintain normalization during the update stage. However, they implicitly assume that the entire quaternion normalization error is due to the magnitude of the rotation, which could be restrictive. Moreover, in many satellite applications the speed of prediction is not as important as the ability to accurately predict the attitude over a long timeframe. For a recent discussion of the method of attitude determination and estimation, the reader is referred to Markley and Crassidis [29].

In the following sections the method discussed by Vepa [1] is briefly presented and analyzed in detail.

8.4.1 Attitude Kinematics Process Modeling

The quaternion vector update equation relating the quaternion rate to the angular velocity vector ω and the quaternion normalization relation are given by well-known differential equations. When the angular velocity vector is measured by rate gyros and corrupted by biases and noise, the dynamics may be expressed in terms of the attitude quaternion $\mathbf{q}$ as (see, for example, Vathsal [17]),

$$\dot{\mathbf{x}} = \begin{bmatrix} \dot{\mathbf{q}} \\ \dot{\mathbf{b}} \end{bmatrix} = \begin{bmatrix} \frac{1}{2}\Omega(\mathbf{u}) & -\frac{1}{2}\Gamma(\mathbf{q}) \\ \mathbf{0} & \mathbf{0} \end{bmatrix} \begin{bmatrix} \mathbf{q} \\ \mathbf{b} \end{bmatrix} + \begin{bmatrix} -\frac{1}{2}\Gamma(\mathbf{q}) & \mathbf{0} \\ \mathbf{0} & \mathbf{I} \end{bmatrix} \begin{bmatrix} \boldsymbol{\eta}_1 \\ \boldsymbol{\eta}_2 \end{bmatrix}, \qquad (8.47)$$

which is expressed in the form as,

$$\dot{\mathbf{x}} \equiv \mathbf{A}(\mathbf{q})\mathbf{x} + \mathbf{B}(\mathbf{q})\mathbf{\eta}, \tag{8.48}$$

where,

$$\mathbf{\Omega}(\mathbf{\omega}) = \begin{bmatrix} 0 & \omega_3 & -\omega_2 & \omega_1 \\ -\omega_3 & 0 & \omega_1 & \omega_2 \\ \omega_2 & -\omega_1 & 0 & \omega_3 \\ -\omega_1 & -\omega_2 & -\omega_3 & 0 \end{bmatrix}, \quad \mathbf{\omega} = \begin{bmatrix} \omega_1 \\ \omega_2 \\ \omega_3 \end{bmatrix}, \quad \mathbf{\Gamma}(\mathbf{q}) = \begin{bmatrix} q_4 & -q_3 & q_2 \\ q_3 & q_4 & -q_1 \\ -q_2 & q_1 & q_4 \\ -q_1 & -q_2 & -q_3 \end{bmatrix}, \tag{8.49}$$

$$\mathbf{q}^T \cdot \mathbf{q} = 1, \tag{8.50}$$

and $\mathbf{b}$ is a rate-gyro drift rate bias, $\mathbf{\eta}_1$ is a Gaussian white noise process corrupting the angular velocity measurements, while the bias rate is assumed to be driven by another independent Gaussian white noise process, $\mathbf{\eta}_2$.

We also assume that the quaternion rate satisfies the constraint,

$$\mathbf{q}^T \cdot \dot{\mathbf{q}} = 0. \tag{8.51}$$

To satisfy the constraints (8.50) and (8.51), we assume a time step equal to Δt and discretize Equation (8.48) as,

$$\mathbf{x}(k+1) \approx \mathbf{F}\mathbf{x}(k) + \mathbf{G}\mathbf{\eta}(k), \quad \mathbf{F} = \exp(\mathbf{A}\Delta t). \quad \mathbf{G} = \mathbf{B}\Delta t. \tag{8.52}$$

The evaluation of $\mathbf{F}$ is done in accordance with the approach recommended by Markley [30], which is equivalent to Equation (8.45) and given by the pair of equations,

$$\mathbf{q}(k+1) = \mathbf{F}_1(\mathbf{\omega}_k, \mathbf{b}_k + \mathbf{\eta}(k))\mathbf{q}(k) \approx \mathbf{F}_1(\mathbf{\omega}_k, \mathbf{b}_k)\mathbf{q}(k) + \mathbf{G}_1(\mathbf{q}(k))\mathbf{\eta}(k), \tag{8.53}$$

$$\mathbf{b}(k+1) = \mathbf{b}(k) + \mathbf{G}_2 \mathbf{\eta}(k), \tag{8.54}$$

where,

$$\mathbf{F}_1(\mathbf{\omega}, \mathbf{b}) = \begin{bmatrix} \alpha_k \mathbf{I} & \beta_k(\mathbf{\omega} + \mathbf{b}) \\ -\beta_k(\mathbf{\omega} + \mathbf{b})^T & \alpha_k \end{bmatrix}, \quad \mathbf{G}_1(\mathbf{q}) = -\frac{\Delta t}{2}\mathbf{\Gamma}(\mathbf{q})[\mathbf{I} \quad \mathbf{0}],$$

$$\mathbf{G}_2 = \Delta t[\mathbf{0} \quad \mathbf{I}], \tag{8.55}$$

$$\alpha_k = \cos\left(\frac{\Delta t}{2}\|\mathbf{\omega} + \mathbf{b}\|\right), \quad \beta_k = \sin\left(\frac{\Delta t}{2}\|\mathbf{\omega} + \mathbf{b}\|\right)/\|\mathbf{\omega} + \mathbf{b}\|.$$

The last column of the matrix $\mathbf{F}_1$ may be interpreted as a change in the quaternion components due to the angular velocity vector, giving the updates to the quaternion given an initial rotation magnitude or equivalently the last component of the quaternion. Thus, if the quaternion normalization error is assumed to be entirely due to the last component, as by Cheon and Kim [28], the corrections to the quaternion components due to an angular rate error correction or a multiplicative error quaternion $\mathbf{q}_e^+$ can be computed from the last column of the matrix $\mathbf{F}_1$. The quaternion update can then be done by quaternion multiplication defined by,

$$\mathbf{q}^+(k+1) = \mathbf{q}_e^+ \otimes \mathbf{q}^-(k+1). \tag{8.56}$$

The use of Equation (8.56) involves a nonlinear transformation of the random variable representing the noise vector. The noise in the rate gyro measurements could have also been transformed to the modified Rodrigues parameters, to maintain the post-update quaternion normalization. This would lead to a filter similar to that of Crassidis and Markley [26], provided the quaternion measurement error is also modeled as a multiplicative quaternion. A Gaussian distributed vector random variable, when transformed to an equivalent quaternion or a rotation, need not remain Gaussian distributed. Hence, we choose to include the rate gyro noise vector as an additive Gaussian-distributed vector to demonstrate an alternative approach to unscented filtering. An interesting feature of this formulation is that mean quaternion update continues to be normalized if it is initially normalized. The matrices $\mathbf{G_1}(\mathbf{q})$ and $\mathbf{G_2}$ are only needed for purposes of updating the covariance and for evaluating the sigma points. The mean quaternion at the sigma points therefore continues to be normalized if the initial sigma point vector is normalized. However, the state estimate does not preserve the normalization, as these are evaluated as weighted linear combinations of the means at the sigma points. For this reason, we propose a predictor-corrector implementation of the filter (this is in addition to the already existing propagate and update structure of the Kalman filter), which is based on a process and measurement predictor model given by,

$$\mathbf{q}^-(k+1) = \mathbf{F_1}(\boldsymbol{\omega}_k, \mathbf{b}_k)\mathbf{q}(k) + \mathbf{G_1}(\mathbf{q}(k))\boldsymbol{\eta}(k), \quad \mathbf{b}^-(k+1) = \mathbf{b}(k) + \mathbf{G_2}\boldsymbol{\eta}(k) \quad (8.57)$$

$$z_{mi} = h_i(\mathbf{q}^-(k)) + v_i, \quad i = 1, 2, 3. \quad (8.58)$$

The post-estimate process and output corrector model is given by,

$$\mathbf{q}^*(k+1) = \mathbf{q}^-(k+1), n^*(k+1) = (\mathbf{q}^-(k+1)\cdot\mathbf{q}^-(k+1))^{1/2}, \quad (8.59)$$

$$\mathbf{q}(k) = \mathbf{q}^*(k)/n^*(k). \quad (8.60)$$

Equations (8.57) and (8.58) are used to implement a UKF, while Equations (8.59) and (8.60) are used to define a second unscented transformation to renormalize the estimated mean and modify the corresponding covariance of the estimates.

8.4.2 Codeless Satellite Navigation Attitude Sensor Model

The change in attitude of the spacecraft over a period of time can be observed by comparing the current measured phase differential with the initial phase differential measured at some initial reference time. Thus, this difference in the measured phase differential can be expressed as,

$$\Delta\phi_m = \frac{2\pi}{\lambda}(\mathbf{d}\cdot(\mathbf{r}_B - \mathbf{r_0})), \quad (8.61)$$

where $\mathbf{r}_B$ is the navigation satellite's sight line vector at the current time and $\mathbf{r}_0$ is the navigation satellite's sight line vector at the initial reference time. The navigation satellite's sight line vector $\mathbf{r}_B$ could be expressed in terms of the satellite's body coordinates. However, since the body attitude may be defined in terms of the quaternion, the transformation relating the estimate of current sight line vector $\hat{\mathbf{r}}$ in

8.4 Spacecraft Large Attitude Estimation

the orbiting coordinates to the current sight line vector $\mathbf{r}_B$ in body coordinates may be expressed in terms of the quaternion components. Hence,

$$\mathbf{r}_B = \mathbf{T}(\mathbf{q})\hat{\mathbf{r}}, \tag{8.62}$$

where,

$$\mathbf{T}(\mathbf{q}) = \begin{bmatrix} q_4^2 + q_1^2 - q_2^2 - q_3^2 & 2(q_1 q_2 + q_3 q_4) & 2(q_1 q_3 - q_2 q_4) \\ 2(q_1 q_2 - q_3 q_4) & q_4^2 - q_1^2 + q_2^2 - q_3^2 & 2(q_2 q_3 + q_1 q_4) \\ 2(q_1 q_3 + q_2 q_4) & 2(q_2 q_3 - q_1 q_4) & q_4^2 - q_1^2 - q_2^2 + q_3^2 \end{bmatrix}. \tag{8.63}$$

An estimate of current sight line vector $\hat{\mathbf{r}}$ in the orbiting coordinates can generally be obtained by an independent Kalman filter or by employing a classical algorithm for estimating the orbital position of the satellite, as by Hoots et al. [31]. It therefore follows that the difference in the measured phase differential could be expressed as,

$$\Delta \phi_m = \frac{2\pi}{\lambda} (\mathbf{d} \cdot (\mathbf{T}(\mathbf{q}) - \mathbf{I})\hat{\mathbf{r}}) + \frac{2\pi}{\lambda} (\mathbf{d} \cdot (\hat{\mathbf{r}} - \mathbf{r_0})), \tag{8.64}$$

And, using Equation (8.63), we may write $\mathbf{T}(\mathbf{q}) - \mathbf{I} \equiv \Delta \mathbf{T}(\mathbf{q})$ as,

$$\Delta \mathbf{T}(\mathbf{q}) = -2 \begin{bmatrix} (q_2^2 + q_3^2) & -(q_1 q_2 + q_3 q_4) & -(q_1 q_3 - q_2 q_4) \\ -(q_1 q_2 - q_3 q_4) & (q_1^2 + q_3^2) & -(q_2 q_3 + q_1 q_4) \\ -(q_1 q_3 + q_2 q_4) & -(q_2 q_3 - q_1 q_4) & (q_1^2 + q_2^2) \end{bmatrix}, \tag{8.65}$$

which is a homogeneous quadratic function of the components of the quaternion. Thus, a discrete measurement of the error in the difference of the phase differentials due to changes in the attitude can be expressed as,

$$z_m \equiv \Delta \phi_m - \frac{2\pi}{\lambda}(\mathbf{d} \cdot (\hat{\mathbf{r}} - \mathbf{r_0})) = \frac{2\pi}{\lambda}(\mathbf{d} \cdot \Delta \mathbf{T}(\mathbf{q}(k))\hat{\mathbf{r}}) + v, \tag{8.66}$$

where v is an additive Gaussian random variable representing a white noise or delta-correlated stochastic process. For three-axis measurement of the attitude, one would require three independent measurements, which may be expressed as,

$$z_{mi} \equiv \Delta \phi_{mi} - \frac{2\pi}{\lambda}(\mathbf{d}_i \cdot (\hat{\mathbf{r}} - \mathbf{r_0})) = \frac{2\pi}{\lambda}(\mathbf{d}_i \cdot \Delta \mathbf{T}(\mathbf{q}(k))\hat{\mathbf{r}}) + v_i, \quad i = 1, 2, 3. \tag{8.67}$$

It may be observed that the process model given by Equations (8.53) and (8.54), and the measurement model given by Equation (8.67) are both nonlinear in the components of the quaternion. Generally, the measurements cannot be cast as multiplicative quaternion error models. The estimation problem may now be stated: Given a sequence of noisy observations by Equation (8.67), we need to estimate the sequence of state vectors of the nonlinear system driven by noise, defined by Equations (8.53) and (8.54). This problem is now dealt with by using a UKF after briefly discussing the reasons for its choice in the next section.

8.4.3 Application of Nonlinear Kalman Filtering to Attitude Estimation

Most dynamic models employed for purposes of estimation or filtering of pseudo range errors or orbit ephemeris errors are generally not linear. To extend and overcome the limitations of linear models, a number of approaches such as the EKF have been proposed in the literature for nonlinear estimation using a variety of approaches. Unlike the Kalman filter, the EKF may diverge if the consecutive linearizations are not a good approximation of the linear model over the entire uncertainty domain. Yet the EKF provides a simple and practical approach to dealing with essential nonlinear dynamics. The main difficulty in applying the algorithm to problems related to the estimation of a spacecraft's orbit and attitude is in determining the proper Jacobian matrices. The UKF is a feasible alternative that has been proposed to overcome this difficulty by Julier, Uhlmann, and Durrant-Whyte [32] as an effective way of applying the Kalman filter to nonlinear systems. It is based on the intuitive concept that it is easier to approximate a probability distribution than it is to approximate an arbitrary nonlinear function or transformation of a random variable. The complete UKF algorithm, first presented by Julier [33], is presented in Section 7.6.1 and will not be repeated here. Only the results of the application are discussed here.

The success of the application of the UKF depends largely on the approximation to the covariance that is estimated as a weighted linear sum of the covariance at the sigma points. When this approximation is such that the covariance is not positive definite, the UKF algorithm fails, as the Cholesky decomposition is not possible. To ensure this covariance, it is essential to adjust the scaling parameter α, if and when necessary. In the following example, α was chosen to be a very small positive number in the case of both the transformations. Alternatively, one could start the first few steps by using an EKF and then switch to the UKF.

In the first instance, it was attempted to estimate the attitude of the spacecraft by employing a single carrier phase sensor. As expected, a single sensor was not able to produces a meaningful estimate of all the components of the attitude quaternion over a relatively large time frame (1,440 minutes). This was due to the need for three vector measurements of nonzero, nonco-linear attitude vectors for purposes of the attitude estimation from kinematic considerations.

The three-receiver based carrier phase sensor was then simulated. The spacecraft, a navigation satellite in a typical GLONASS orbit, was assumed to be performing sustained three-axis rotations with a body axes angular velocity vector given by, $\omega = \begin{bmatrix} 0.1 & 0.2 & 0.3 \end{bmatrix}$ rads/hr. These are measured by three rate gyros and the measurements are assumed to be biased, where the bias rate is driven by white noise. To establish the parameters for implementing the UKF, the standard deviations were assumed to be the same order of magnitude as the rate gyros (0.002 arc-s/s) considered by Farrenkopf [34], where the single-axis attitude estimation problem was considered with attitude measurements provided by a star tracker ($\sigma = 20$ arc-s) and an update rate of 0.1 per minute. The standard deviations of the carrier phase measurements were assumed to be between $\sigma = 50 - 70$ arc-s. Independent simulations of the process and the measurements, including the rate-gyros, were performed, and standard deviations of

8.4 Spacecraft Large Attitude Estimation

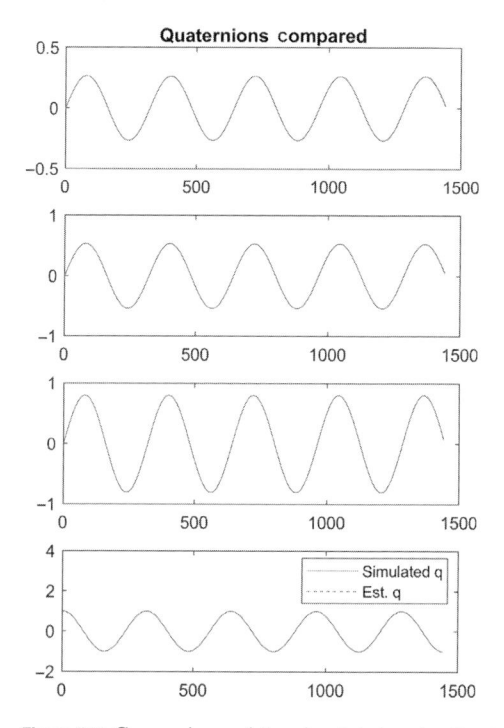

Figure 8.3 Comparison of the simulated and estimated attitude quaternion over a time frame of 1,440 minutes.

the white noise sources were assumed to be 30 times larger than those assumed for the design calculations. Figure 8.3 illustrates the simulated and estimated attitude quaternion plotted to the same scale.

In Figure 8.4 the corresponding errors in the simulated quaternion and estimated quaternion over the same timeframe are shown. Figure 8.5 illustrates the errors in the corresponding 3, 2, 1 sequence Euler angles, ψ, θ, and ϕ. Attitude accuracies of the order of $\pm 1°$–$\pm 2°$ may be obtained by using antenna patches separated by distances in the order of 0.2 m. A tenfold decrease in the order of the accuracies (errors) is possible if the antenna patch distance is increased fivefold. In Figure 8.6 a) the growth of the simulated and estimated bias states, related to the gyro drift rate with the UKF are shown. In Figure 8.6 b) the growth of the simulated and estimated bias states, related to the gyro drift rate without the second unscented transformation are shown. It is clear that in the long term the bias states slowly tend to diverge from the corresponding simulated states.

Comparing our implementation of the UKF with the standard UKF, which did not involve the use of the unscented transformation due to normalization of the quaternion given by Equations (8.59), although there were practically no differences between the predicted mean and covariance of the estimate, it was observed that the bias states slowly began to diverge from the simulations in the case of the standard UKF. Thus, the performance of the UKF was relatively more stable over longer timeframe.

Figure 8.7 shows the simulated measurement error of a typical carrier phase sensor. The carrier frequency was assumed to be 1,602 MHz. It was assumed that three pairs of

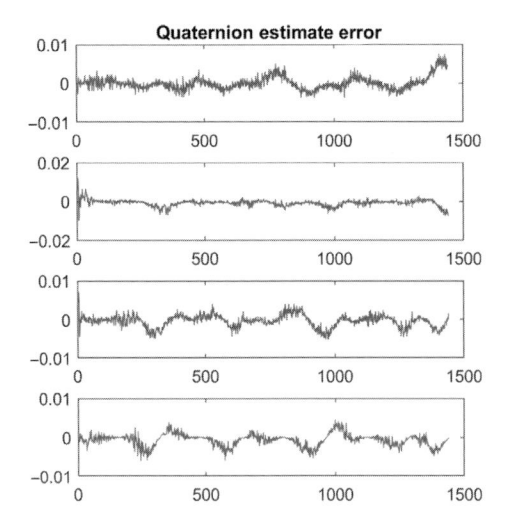

Figure 8.4 Errors in the components of the simulated and estimated attitude quaternion over a timeframe of 1,440 minutes.

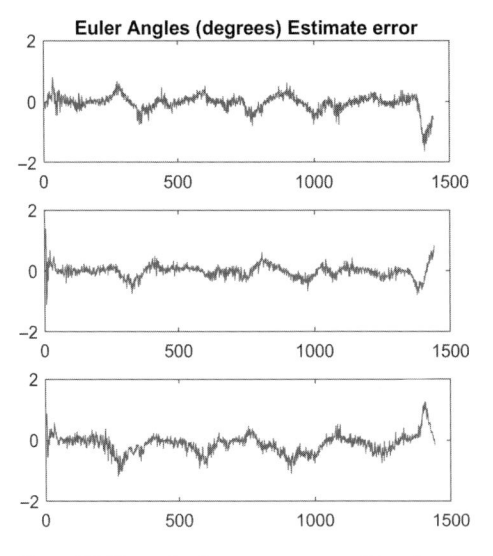

Figure 8.5 Errors in the components of the simulated and estimated attitude Euler angles φ, θ, and ψ over a timeframe of 1,440 minutes.

patch antennae are located along mutually perpendicular three-axes and that each pair was separated by 18 cm. Satellite Navigation receiver-based carrier phase measurements are subject to additional measurement errors, such as tropospheric and ionospheric errors, orbital ephemeris errors, multipath, frequent cycle slip occurrences, antenna phase center variation, and noise. In addition to measuring the arrival time of the satellite navigation signal using code modulation, receivers also measure the phase of the carrier frequency and the total phase change (both whole and partial cycles), since

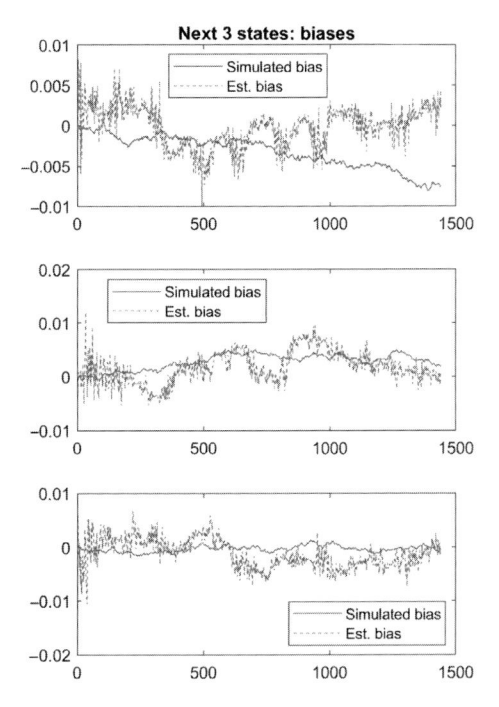

Figure 8.6 a) Growth and comparison of the simulated and estimated bias states in gyros with the second unscented transformation.

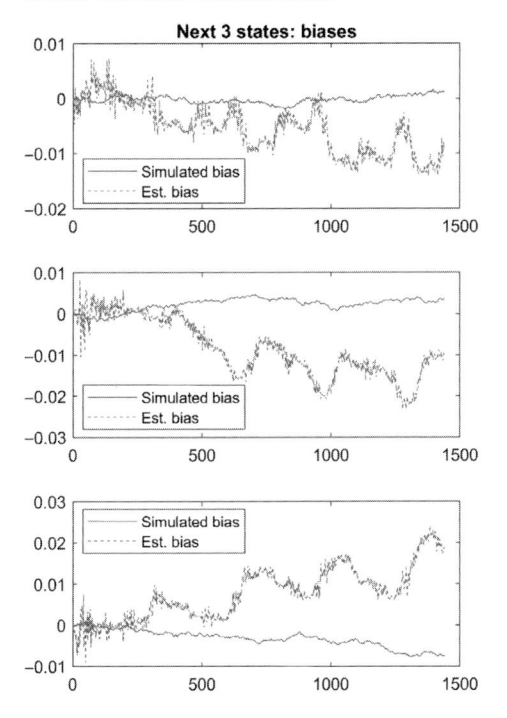

Figure 8.6 b) Growth and comparison of the simulated and estimated bias states in gyros without the second unscented transformation.

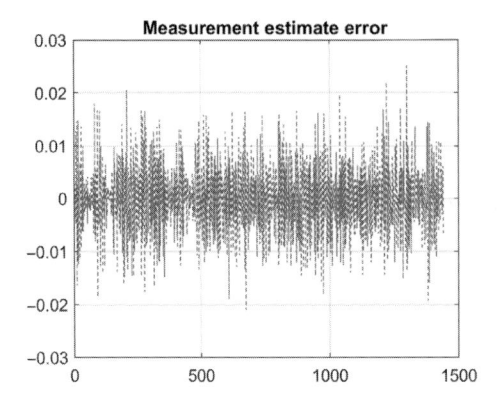

Figure 8.7 Simulated measured carrier phase error (in radians) of a typical carrier phase sensor.

the initial measurements are currently available. This measure, referred to as the integrated carrier phase, is statistically independent of the code measurement and is also about two orders of magnitude less noisy. Yet these error sources severely deteriorate attitude determination availability. Moreover, these errors to the carrier phase cannot be modeled as multiplicative quaternion errors because of the nature of the probability density functions. To model these errors adequately, the covariance of the errors in the measurements was assumed to be much higher than was assumed at the design stage. This fact is adequately reflected in Figure 8.7.

Finally, it must be said that the filter was run over a much longer timeframe (over 3,600 minutes) and the performance of the filter did not deteriorate in spite of this long term operation. Thus, the implementation of an attitude determination system over a relatively long timeframe is successfully demonstrated.

8.5 Nonlinear State Estimation for Spacecraft Rotation Rate Synchronization with an Orbiting Body

Noncooperative target bodies in space are those bodies in orbit that cannot convey any information about their position, attitude, or velocities or facilitate rendezvous and docking or berthing. In order for a spacecraft to rendezvous with such an object, a multi-phase approach is often adopted. In the first phase, the primary chaser spacecraft often tracks the secondary orbiting body at a safe but short distance from it. In the next phase, it is essential that the chaser spacecraft completely align its attitude with that of the secondary orbiting body or target body. In the final phase, rendezvous and docking is initiated and completed. This requires the chaser spacecraft to continuously and dynamically synchronize its attitude with that of the target body.

In an early paper, Wen and Kreutz-Delgado [35] presented a general theory for the synthesis of the attitude tracking control laws. They considered several control schemes based on quaternion feedback, and developed a methodology for establishing the stability of the closed loop system rigorously using Lyapunov theory. Bhat and

8.5 Nonlinear State Estimation for Spacecraft Rotation Rate Synchronization

Bernstein [36] showed that attitude kinematics cannot be globally stabilized using continuous feedback from within the three-dimensional rotation group characterized by special orthogonal transformations, *SO* (3). This led to subsequent development of discontinuous feedback control laws using quaternion feedback by Fjellstad and Fossen [37] and later by Fragopoulos and Innocenti [38]. In the literature over the last two decades, attitude synchronization control has been studied with reference to spacecraft formation control, and attitude synchronization controllers were designed for cooperating spacecraft using a leader-follower model by Dimarogonas, Tsiotras, and Kyriakopoulos [39], Kang and Yeh [40], and Wang, Hadaegh, and Lau [41]. Pan and Kapila [42] also adopted a similar approach and developed an adaptive nonlinear controller, while Nijmeijer and Rodriguez-Angeles [43] presented a generic synchronization theory for mechanical systems. Lawton and Beard [44] adopted a behavior-based approach for synchronizing multiple spacecraft rotations. Ren and Beard [45] have developed a decentralized scheme based on the virtual structure approach, where the spacecraft formation is treated as a virtual rigid body. Egeland and Godhavn [46] developed an attitude-tracking controller based on the passivity approach, while Bai, Arcak, and Wen [47] have developed a passivity-based method, which relies only on relative attitude information.

A range of different control law synthesis techniques have been employed for designing a rotation rate synchronization or an attitude synchronization controller. Terui [48] used sliding mode control for position and attitude tracking for capturing and removing large space debris, while Stansbery and Cloutier [49] adopted a state-dependent nonlinear, suboptimal Riccati equation technique to design both the position and attitude control laws. Singla, Subbarao, and Junkins [50] adopted a model reference adaptive control approach to design an output feedback-based control law for spacecraft rendezvous and docking problems. Subbarao and Welsh [51] solved the coupled problem of relative position tracking with attitude reorientation for robotic spacecraft using a methodology for synthesizing synchronization maneuvers. Xin and Pan [52] proposed a nonlinear optimal control technique to design a closed-form feedback control law for the regulation of both translational and rotational motion. Their approach also dealt with vibrations due to flexibility.

Several authors have dealt with the problems associated with a lack of full information or uncertainties in the measurements. Lizarralde and Wen [53] developed control laws for rigid-body attitude stabilization when the angular velocity vector is not available for feedback. They used a velocity filter to reconstruct the desired angular velocity vector. Costic, Dawson, De Queiroz, and Kapila [54] designed an adaptive quaternion- based controller based on an angular velocity filter and were able to synthesis a converging and asymptotically stable control law. Singla, Subbarao, and Junkins [50] and Akella [55] also designed control laws without the need for angular velocity measurement, and show convergence and asymptotic stability of the tracking error.

Recently, Sun and Huo [56] studied relative position tracking and attitude synchronization of noncooperative spacecraft seeking to rendezvous. Lyapunov theory is used to show that closed-loop system errors asymptotically converge to zero. Lu, Geng, Chen, and Zhang [57] have developed a robust sliding mode controller for a spacecraft autonomously approaching and docking with a freely tumbling target. They used the notion of the relative attitude quaternion and employed an extension of Brown's method [58] to design the control law. Control constraints were incorporated using Wie and Lu's method [59]. Liu, Shi, Bi, and Zhang [60] designed a backstepping-based terminal sliding mode chatter controller for rendezvous and docking with a tumbling spacecraft. Stability of the system was established based on Lyapunov's theory.

In this section, we consider the state estimation for the design of a rotation rate synchronization controller to align a chaser spacecraft's angular velocity with that of a target orbiting body. Our aim is to control the chaser spacecraft so it synchronizes its attitude rate with that of the target. It is assumed that the target body's angular velocity can be measured in the frame of reference attached to the chaser spacecraft. The chaser spacecraft's angular velocity components are assumed to be measured as well. Yet all of the measurements are assumed to be noisy, with the noise statistics well known. Thus, we could use an UKF to filter the measurements and estimate the states of the dynamic model. Initially, a discrete time, finite horizon optimal controller is synthesized. When the feedbacks are implemented using the estimated states, the controller is not stable in the long term. For this reason, after the errors are within manageable limits, a nonlinear controller that resembles a proportional-derivative controller is synthesized using a BLF. The feedback is implemented using the estimated states. The results, presented in Chapter 4, indicate that the synchronization error between the target body and the chaser's angular velocity components can be reduced to within 1%, and it is shown that the chaser spacecraft can track the target's rotation rate autonomously. This section is organized as follows. In the next sub-section, we review the chaser spacecraft's nonlinear attitude dynamics, including the gravity-gradient torques and the quaternion attitude dynamics. This is followed by a sub-section on the attitude dynamics of the chaser spacecraft, relative to an orbiting target body. The following sub-section considers the nonlinear state estimation problem followed by the measurements that must be made to successfully estimate the states of the model are enunciated. The control law synthesis for attitude synchronization is then briefly discussed.

8.5.1 Chaser Spacecraft's Attitude Dynamics

In matrix form, when the inertia matrix is not diagonal, the equations of attitude motion of chaser spacecraft are,

$$\mathbf{I}\dot{\boldsymbol{\omega}} + \boldsymbol{\Omega}\mathbf{I}\boldsymbol{\omega} = \mathbf{M} + \mathbf{M}_{gg} + \mathbf{M}_d, \tag{8.68}$$

where $\mathbf{I}$ is the moment of inertia matrix, which is assumed to be,

$$\mathbf{I} = \begin{bmatrix} I_{11} & I_{12} & I_{13} \\ I_{12} & I_{22} & I_{23} \\ I_{13} & I_{23} & I_{33} \end{bmatrix}, \quad \boldsymbol{\omega} \equiv \begin{bmatrix} \omega_1 \\ \omega_2 \\ \omega_3 \end{bmatrix}, \quad \boldsymbol{\Omega} = \begin{bmatrix} 0 & -\omega_3 & \omega_2 \\ \omega_3 & 0 & -\omega_1 \\ -\omega_2 & \omega_1 & 0 \end{bmatrix} \tag{8.69}$$

$\mathbf{M}_d$ are the disturbance torques and $\mathbf{M}_{gg}$ are the gravity-gradient torques.

8.5 Nonlinear State Estimation for Spacecraft Rotation Rate Synchronization

It is important to emphasize that the target's dynamics are irrelevant to us, as there is little or no chance of acquiring the target's inertia properties. However, the target's angular velocity vector is assumed to be given by $\boldsymbol{\omega}_d$, its attitude quaternion relative to the chaser's body frame is assumed to be $\mathbf{q}_d$, or its relative attitude quaternion $\Delta \mathbf{q}$, relative to the chaser's body frame, can in principle be measured from within the chaser spacecraft.

Expressions for the gravity-gradient moment are obtained, assuming that z axis of the satellite body is nominally pointing to the Earth. The direction vector of the center of gravity of the satellite pointing to the Earth is given by the last column of $\mathbf{T}_{BR}$, the transformation from the Earth-orbiting frame to the body-fixed frame of the satellite as,

$$\mathbf{c} = \begin{bmatrix} c_1 & c_2 & c_3 \end{bmatrix}^T. \tag{8.70}$$

The corresponding cross-product operator $\mathbf{c}_\times$ is defined as,

$$\mathbf{c}_\times = \begin{bmatrix} 0 & -c_3 & c_2 \\ c_3 & 0 & -c_1 \\ -c_2 & c_1 & 0 \end{bmatrix}. \tag{8.71}$$

Hence, the gravity-gradient moments acting on the satellite and manipulator body are,

$$\mathbf{M}_{gg} = 3n^2 \mathbf{c}_\times \mathbf{I}\mathbf{c} \equiv 3n^2 \mathbf{c}_\times \mathbf{I}\mathbf{c} = \begin{bmatrix} L_{gg} & M_{gg} & N_{gg} \end{bmatrix}^T. \tag{8.72}$$

Thus,

$$\mathbf{M}_{gg} = \begin{bmatrix} L_{gg} \\ M_{gg} \\ N_{gg} \end{bmatrix} = 3n^2 \begin{bmatrix} c_2 c_1 I_{31} + c_2^2 I_{32} + c_2 c_3 (I_{33} - I_{22}) - c_3 c_1 I_{21} - c_3^2 I_{23} \\ c_3 c_1 (I_{11} - I_{33}) + c_2 c_3 I_{12} + c_3^2 I_{13} - c_1^2 I_{31} - c_1 c_2 I_{32} \\ c_1^2 I_{21} + c_1 c_2 (I_{22} - I_{11}) + c_1 c_3 I_{23} - c_2^2 I_{12} - c_2 c_3 I_{13} \end{bmatrix}. \tag{8.73}$$

If we express the transformation from the orbiting frame to the body coordinates in terms of an attitude quaternion of the chaser spacecraft with components $\varepsilon_1, \varepsilon_2, \varepsilon_3$, and η as,

$$\mathbf{T}_{BR}(\mathbf{q}) = \begin{bmatrix} \eta^2 + \varepsilon_1^2 - \varepsilon_2^2 - \varepsilon_3^2 & 2(\varepsilon_1 \varepsilon_2 + \varepsilon_3 \eta) & 2(\varepsilon_1 \varepsilon_3 - \varepsilon_2 \eta) \\ 2(\varepsilon_1 \varepsilon_2 - \varepsilon_3 \eta) & \eta^2 - \varepsilon_1^2 + \varepsilon_2^2 - \varepsilon_3^2 & 2(\varepsilon_2 \varepsilon_3 + \varepsilon_1 \eta) \\ 2(q_1 q_3 + q_2 \eta) & 2(q_2 q_3 - q_1 \eta) & \eta^2 - \varepsilon_1^2 - \varepsilon_2^2 + \varepsilon_3^2 \end{bmatrix}, \tag{8.74}$$

then from the last column, the Earth-pointing direction vector is,

$$\mathbf{c} = \begin{bmatrix} c_1 \\ c_2 \\ c_3 \end{bmatrix} = \begin{bmatrix} 2(\varepsilon_1 \varepsilon_3 - \varepsilon_2 \eta) \\ 2(\varepsilon_2 \varepsilon_3 + \varepsilon_1 \eta) \\ \eta^2 - \varepsilon_1^2 - \varepsilon_2^2 + \varepsilon_3^2 \end{bmatrix}. \tag{8.75}$$

The quaternion kinematics satisfy,

$$\frac{d\mathbf{q}}{dt} = \frac{1}{2} \mathbf{A}_\omega(\boldsymbol{\omega}) \mathbf{q}, \tag{8.76}$$

where the quaternion $\mathbf{q} = [\varepsilon_1 \ \varepsilon_2 \ \varepsilon_3 \ \eta]^T$, consists of a vector part, $\boldsymbol{\varepsilon} = [\varepsilon_1 \ \varepsilon_2 \ \varepsilon_3]^T$ and the scalar η so,

$$\mathbf{q} = \begin{bmatrix} \boldsymbol{\varepsilon} \\ \eta \end{bmatrix} \text{ and } \mathbf{A}_\omega = \begin{bmatrix} -\boldsymbol{\Omega}(\boldsymbol{\omega}) & \boldsymbol{\omega} \\ -\boldsymbol{\omega}^T & 0 \end{bmatrix}, \ \boldsymbol{\Omega}(\boldsymbol{\omega}) = \begin{bmatrix} 0 & -\omega_3 & \omega_2 \\ \omega_3 & 0 & -\omega_1 \\ -\omega_2 & \omega_1 & 0 \end{bmatrix}. \tag{8.77}$$

The quaternion kinematics may also be compactly expressed as,

$$\frac{d\mathbf{q}}{dt} = \frac{d}{dt}\begin{bmatrix} \boldsymbol{\varepsilon} \\ \eta \end{bmatrix} = \frac{1}{2}\boldsymbol{\Gamma}(\mathbf{q})\boldsymbol{\omega}, \ \boldsymbol{\Gamma}(\mathbf{q}) = \begin{bmatrix} \eta\mathbf{I}_{3\times3} + \mathbf{S}(\boldsymbol{\varepsilon}) \\ -\boldsymbol{\varepsilon}^T \end{bmatrix}, \ \mathbf{S}(\boldsymbol{\varepsilon}) = \begin{bmatrix} 0 & -\varepsilon_3 & \varepsilon_2 \\ \varepsilon_3 & 0 & -\varepsilon_1 \\ -\varepsilon_2 & \varepsilon_1 & 0 \end{bmatrix} \tag{8.78}$$

where $\mathbf{I}_{3\times3}$ is the 3×3 unit matrix. These relations may be inverted as,

$$\boldsymbol{\omega} = 2[\eta\mathbf{I}_{3\times3} + \mathbf{S}^T(\boldsymbol{\varepsilon}) \ \ -\boldsymbol{\varepsilon}][\dot{\boldsymbol{\varepsilon}} \ \dot{\eta}]^T = 2\boldsymbol{\Gamma}^{-1}(\mathbf{q})[\dot{\boldsymbol{\varepsilon}} \ \dot{\eta}]^T. \tag{8.79}$$

The desired attitude quaternion relative to the chaser's body frame, which is assumed to be $\mathbf{q}_d$ and the relative attitude quaternion $\Delta\mathbf{q}$, relative to the chaser's body frame are related to the chaser's attitude by,

$$\mathbf{q}_d = \Delta\mathbf{q} \otimes \mathbf{q}. \tag{8.80}$$

Given two quaternions, $\Delta\mathbf{q} = [\Delta q_1 \ \Delta q_2 \ \Delta q_3 \ \Delta q_4]^T$, $\mathbf{q} = [\varepsilon_1 \ \varepsilon_2 \ \varepsilon_3 \ \eta]^T$, the quaternion product $\mathbf{q}_d = \Delta\mathbf{q} \otimes \mathbf{q}$ is defined as,

$$\mathbf{q}_d = \begin{bmatrix} q_{1d} \\ q_{2d} \\ q_{3d} \\ q_{4d} \end{bmatrix} = \begin{bmatrix} \varepsilon_1\Delta q_4 - \varepsilon_2\Delta q_3 + \varepsilon_3\Delta q_2 + \eta\Delta q_1 \\ \varepsilon_1\Delta q_3 + \varepsilon_2\Delta q_4 - \varepsilon_3\Delta q_1 + \eta\Delta q_2 \\ -\varepsilon_1\Delta q_2 + \varepsilon_2\Delta q_1 + \varepsilon_3\Delta q_4 + \eta\Delta q_3 \\ -\varepsilon_1\Delta q_1 - \varepsilon_2\Delta q_2 - \varepsilon_3\Delta q_3 + \eta\Delta q_4 \end{bmatrix} = \begin{bmatrix} \eta & \varepsilon_3 & -\varepsilon_2 & \varepsilon_1 \\ -\varepsilon_3 & \eta & \varepsilon_1 & \varepsilon_2 \\ \varepsilon_2 & -\varepsilon_1 & \eta & \varepsilon_3 \\ -\varepsilon_1 & -\varepsilon_2 & -\varepsilon_3 & \eta \end{bmatrix}\begin{bmatrix} \Delta q_1 \\ \Delta q_2 \\ \Delta q_3 \\ \Delta q_4 \end{bmatrix}. \tag{8.81}$$

Hence, it is expressed in matrix form as,

$$\mathbf{q}_d = \mathbf{C}_0\Delta\mathbf{q}, \ \mathbf{C}_0 = \begin{bmatrix} \eta & \varepsilon_3 & -\varepsilon_2 & \varepsilon_1 \\ -\varepsilon_3 & \eta & \varepsilon_1 & \varepsilon_2 \\ \varepsilon_2 & -\varepsilon_1 & \eta & \varepsilon_3 \\ -\varepsilon_1 & -\varepsilon_2 & -\varepsilon_3 & \eta \end{bmatrix}. \tag{8.82}$$

8.5.2 Relative Attitude Dynamics

Given the chaser spacecraft's current angular velocity vector $\boldsymbol{\omega}$, and that $\Delta\boldsymbol{\omega} = \boldsymbol{\omega}_d - \boldsymbol{\omega}$, the perturbed nonlinear Euler equations of the chaser spacecraft are,

$$\mathbf{I}\dot{\boldsymbol{\omega}} + \mathbf{I}\Delta\dot{\boldsymbol{\omega}} = \mathbf{M} + \mathbf{M}_{gg} + (\mathbf{M}_{ggd} - \mathbf{M}_{gg}) + \Delta\mathbf{M}_c + \mathbf{M}_d + \Delta\mathbf{M}_d - (\boldsymbol{\Omega} + \Delta\boldsymbol{\Omega})\mathbf{I}(\boldsymbol{\omega} + \Delta\boldsymbol{\omega}), \tag{8.83}$$

8.5 Nonlinear State Estimation for Spacecraft Rotation Rate Synchronization

where $\boldsymbol{\omega}$ satisfies Equation (8.68). The angular velocity perturbation is additive.

$$\boldsymbol{\omega} + \Delta\boldsymbol{\omega} = \begin{bmatrix} \omega_1 \\ \omega_2 \\ \omega_3 \end{bmatrix} + \begin{bmatrix} \Delta\omega_1 \\ \Delta\omega_2 \\ \Delta\omega_3 \end{bmatrix} = \begin{bmatrix} \omega_1 + \Delta\omega_1 \\ \omega_2 + \Delta\omega_2 \\ \omega_3 + \Delta\omega_3 \end{bmatrix}, \tag{8.84}$$

$$\boldsymbol{\Omega} + \Delta\boldsymbol{\Omega} = \begin{bmatrix} 0 & -\omega_3 - \Delta\omega_3 & \omega_2 + \Delta\omega_2 \\ \omega_3 + \Delta\omega_3 & 0 & -\omega_1 - \Delta\omega_1 \\ -\omega_2 - \Delta\omega_2 & \omega_1 + \Delta\omega_1 & 0 \end{bmatrix}. \tag{8.85}$$

If we let,

$$\Delta\mathbf{M}_{gg} = \mathbf{M}_{ggd} - \mathbf{M}_{gg}. \tag{8.86}$$

the chaser spacecraft's attitude dynamics relative to the target spacecraft may be linearized. Equation (8.83) may be expressed as,

$$\mathbf{I}\Delta\dot{\boldsymbol{\omega}} = \Delta\mathbf{M}_{gg} + \Delta\mathbf{M}_c - \{\boldsymbol{\Omega}\mathbf{I} - \mathbf{S}(\mathbf{h})\}\Delta\boldsymbol{\omega} + \Delta\mathbf{M}_d, \tag{8.87}$$

with,

$$\mathbf{h} = \begin{bmatrix} \sum_j I_{1j}\omega_j \\ \sum_j I_{2j}\omega_j \\ \sum_j I_{3j}\omega_j \end{bmatrix} = \begin{bmatrix} I_{11} & I_{12} & I_{13} \\ I_{12} & I_{22} & I_{23} \\ I_{13} & I_{23} & I_{33} \end{bmatrix} \begin{bmatrix} \omega_1 \\ \omega_2 \\ \omega_3 \end{bmatrix}, \quad \mathbf{S}(\mathbf{h}) = \begin{bmatrix} 0 & -I_{3j}\omega_j & I_{2j}\omega_j \\ I_{3j}\omega_j & 0 & -I_{1j}\omega_j \\ -I_{2j}\omega_j & I_{1j}\omega_j & 0 \end{bmatrix},$$

$$\tag{8.88}$$

and $\Delta\mathbf{M}_c$ is the vector of control torques. The equations are of the form,

$$\mathbf{I}\Delta\dot{\boldsymbol{\omega}} = \Delta\mathbf{M}_{gg} + \Delta\mathbf{M}_c - \mathbf{W}\Delta\boldsymbol{\omega} + \Delta\mathbf{M}_d, \tag{8.89}$$

where

$$\mathbf{W}\Delta\boldsymbol{\omega} \approx (\boldsymbol{\Omega} + \Delta\boldsymbol{\Omega})\mathbf{I}(\boldsymbol{\omega} + \Delta\boldsymbol{\omega}) - \boldsymbol{\Omega}\mathbf{I}\boldsymbol{\omega}, \quad \mathbf{W} = \begin{bmatrix} \mathbf{J}_1\boldsymbol{\omega} & \mathbf{J}_2\boldsymbol{\omega} & \mathbf{J}_3\boldsymbol{\omega} \end{bmatrix}. \tag{8.90}$$

The matrices $\mathbf{J}_1, \mathbf{J}_2, \mathbf{J}_3$ are defined as,

$$\mathbf{J}_1 = \begin{bmatrix} 0 & I_{13} & -I_{12} \\ -2I_{13} & -I_{32} & (I_{11} - I_{33}) \\ 2I_{12} & (I_{22} - I_{11}) & I_{23} \end{bmatrix},$$

$$\mathbf{J}_2 = \begin{bmatrix} I_{31} & 2I_{23} & (I_{33} - I_{22}) \\ -I_{23} & 0 & I_{12} \\ (I_{22} - I_{11}) & -2I_{12} & -I_{13} \end{bmatrix}, \tag{8.91}$$

$$\mathbf{J}_3 = \begin{bmatrix} -I_{21} & (I_{33} - I_{22}) & -2I_{23} \\ (I_{11} - I_{33}) & I_{12} & 2I_{13} \\ I_{23} & -I_{13} & 0 \end{bmatrix}.$$

The perturbation gravity-gradient moments $\Delta \mathbf{M}_{gg}$ may be expressed in terms of $\mathbf{c}(\mathbf{q}_d)$ and the corresponding cross-product operator $\mathbf{c}_\times(\mathbf{q}_d)$ and linearized as,

$$\Delta \mathbf{M}_{gg} \equiv 3n^2 \mathbf{c}_\times(\mathbf{q}_d)\mathbf{Ic}(\mathbf{q}_d) - 3n^2 \mathbf{c}_\times(\mathbf{q})\mathbf{Ic}(\mathbf{q}) \approx 3n^2 \times \mathbf{G}\frac{d\mathbf{c}(\mathbf{q}_d)}{d\Delta \mathbf{q}}, \quad (8.92)$$

with,

$$\mathbf{G} = [\mathbf{J}_1 \mathbf{c} \quad \mathbf{J}_2 \mathbf{c} \quad \mathbf{J}_3 \mathbf{c}] \quad \text{and} \quad \mathbf{c} = \mathbf{c}(\mathbf{q}). \quad (8.93)$$

Given the relative angular velocity vector of the chaser spacecraft $\Delta \boldsymbol{\omega}$, Equation (8.89) is reduced to,

$$\mathbf{I}\frac{d\Delta \boldsymbol{\omega}}{dt} = -\mathbf{W}\Delta \boldsymbol{\omega} + 3n^2 \times \mathbf{G} \times \mathbf{C}\Delta \mathbf{q} + \Delta \mathbf{M}_c + \Delta \mathbf{M}_d. \quad (8.94)$$

To obtain an expression for $\mathbf{C}$, in the above relation, $\mathbf{c}(\mathbf{q}_d)$ is first expressed in terms of relative attitude quaternion of the target. If we let the target's attitude quaternion relative to the chaser in the chaser's body frame be $\Delta \mathbf{q}$,

$$\Delta \mathbf{q} = [\Delta \varepsilon_1 \quad \Delta \varepsilon_2 \quad \Delta \varepsilon_3 \quad \Delta \eta]^T. \quad (8.95)$$

With no relative rotation, $\Delta \mathbf{q}$ is $\Delta \mathbf{q}_0$, which is, $\Delta \mathbf{q}_0 = [0 \quad 0 \quad 0 \quad 1]^T$. Then the matrix $\mathbf{C}$ is given by,

$$\mathbf{C} \equiv \frac{d\mathbf{c}(\mathbf{q}_d)}{d\Delta \mathbf{q}} = \frac{d\mathbf{c}(\mathbf{q}_d)}{d\mathbf{q}_d}\frac{d\mathbf{q}_d}{d\Delta \mathbf{q}} = 2\begin{bmatrix} \varepsilon_3 & -\varepsilon_2 & \varepsilon_1 & -\eta \\ \eta & \varepsilon_3 & \varepsilon_2 & \varepsilon_1 \\ -\varepsilon_1 & -\varepsilon_2 & \varepsilon_3 & \eta \end{bmatrix}\mathbf{C}_0. \quad (8.96)$$

The relative attitude quaternion of the target in the chaser's body frame satisfies a kinematic equation. The relative quaternion vector update equation relates the relative attitude quaternion $\Delta \mathbf{q}$ to the relative quaternion rate and to the angular velocity vector $\boldsymbol{\omega}$ of the chaser spacecraft. Thus, the relative quaternion vector update equation is,

$$\frac{d\Delta \mathbf{q}}{dt} = \frac{d}{dt}\begin{bmatrix} \Delta \boldsymbol{\varepsilon} \\ \Delta \eta \end{bmatrix} = \frac{1}{2}\mathbf{A}_\omega(\boldsymbol{\omega})\Delta \mathbf{q} + \frac{1}{2}\begin{bmatrix} \Delta \eta \mathbf{I}_{3\times 3} + \mathbf{S}(\Delta \boldsymbol{\varepsilon}) \\ -\Delta \boldsymbol{\varepsilon}^T \end{bmatrix}\Delta \boldsymbol{\omega}, \quad (8.97)$$

which may be written as,

$$\frac{d\Delta \mathbf{q}}{dt} = \frac{d}{dt}\begin{bmatrix} \Delta \boldsymbol{\varepsilon} \\ \Delta \eta \end{bmatrix} = \frac{1}{2}\mathbf{A}_\omega(\boldsymbol{\omega})\Delta \mathbf{q} + \frac{1}{2}\boldsymbol{\Gamma}(\Delta \mathbf{q})\Delta \boldsymbol{\omega}, \quad \boldsymbol{\Gamma}(\Delta \mathbf{q}) \equiv \begin{bmatrix} \Delta \eta \mathbf{I}_{3\times 3} + \mathbf{S}(\Delta \boldsymbol{\varepsilon}) \\ -\Delta \boldsymbol{\varepsilon}^T \end{bmatrix}. \quad (8.98)$$

The relative attitude quaternion also satisfies the well-known quaternion normalization relation.

8.5.3 Nonlinear State Estimation

When the angular velocity vector of the chaser spacecraft is measured by rate gyros and corrupted by biases and noise, the kinematics may be expressed in terms of the relative attitude quaternion $\Delta \mathbf{q}$ as,

8.5 Nonlinear State Estimation for Spacecraft Rotation Rate Synchronization

$$\dot{\mathbf{x}} = \begin{bmatrix} \Delta\dot{\mathbf{q}} \\ \dot{\mathbf{b}} \end{bmatrix} = \frac{1}{2}\begin{bmatrix} \mathbf{A}_\omega(\boldsymbol{\omega}) & \boldsymbol{\Gamma}(\Delta\mathbf{q}) \\ \mathbf{0} & \mathbf{0} \end{bmatrix}\begin{bmatrix} \Delta\mathbf{q} \\ \mathbf{b} \end{bmatrix} + \frac{1}{2}\begin{bmatrix} \boldsymbol{\Gamma}(\Delta\mathbf{q}) \\ \mathbf{0} \end{bmatrix}\Delta\boldsymbol{\omega} + \frac{1}{2}\begin{bmatrix} \boldsymbol{\Gamma}(\Delta\mathbf{q}) & \mathbf{0} \\ \mathbf{0} & 2\mathbf{I}_{3\times 3} \end{bmatrix}\begin{bmatrix} \boldsymbol{\eta}_1 \\ \boldsymbol{\eta}_2 \end{bmatrix}, \tag{8.99}$$

which is expressed in a state space matrix form as,

$$\dot{\mathbf{x}} = \tilde{\mathbf{A}}(\Delta\mathbf{q})\mathbf{x} + \tilde{\mathbf{B}}(\Delta\mathbf{q})\Delta\boldsymbol{\omega} + \tilde{\mathbf{B}}_d(\Delta\mathbf{q})\boldsymbol{\eta}, \tag{8.100}$$

where,

$$\tilde{\mathbf{A}} = \frac{1}{2}\begin{bmatrix} \mathbf{A}_\omega(\boldsymbol{\omega}) & \boldsymbol{\Gamma}(\Delta\mathbf{q}) \\ \mathbf{0} & \mathbf{0} \end{bmatrix}, \quad \tilde{\mathbf{B}} = \frac{1}{2}\begin{bmatrix} \boldsymbol{\Gamma}(\Delta\mathbf{q}) \\ \mathbf{0} \end{bmatrix}, \quad \tilde{\mathbf{B}}_d = \frac{1}{2}\begin{bmatrix} \boldsymbol{\Gamma}(\Delta\mathbf{q}) & \mathbf{0} \\ \mathbf{0} & 2\mathbf{I}_{3\times 3} \end{bmatrix}, \quad \boldsymbol{\eta} = \begin{bmatrix} \boldsymbol{\eta}_1 \\ \boldsymbol{\eta}_2 \end{bmatrix},$$

$$\mathbf{A}_\omega(\boldsymbol{\omega}) = \begin{bmatrix} 0 & \omega_3 & -\omega_2 & \omega_1 \\ -\omega_3 & 0 & \omega_1 & \omega_2 \\ \omega_2 & -\omega_1 & 0 & \omega_3 \\ -\omega_1 & -\omega_2 & -\omega_3 & 0 \end{bmatrix}, \quad \boldsymbol{\omega} = \begin{bmatrix} \omega_1 \\ \omega_2 \\ \omega_3 \end{bmatrix}, \quad \boldsymbol{\Gamma}(\mathbf{q}) = \begin{bmatrix} q_4 & -q_3 & q_2 \\ q_3 & q_4 & -q_1 \\ -q_2 & q_1 & q_4 \\ -q_1 & -q_2 & -q_3 \end{bmatrix}. \tag{8.101}$$

The quaternion normalization relation as applied to the relative attitude quaternion $\Delta\mathbf{q}$ is,

$$\Delta\mathbf{q}^T \cdot \Delta\mathbf{q} = 1. \tag{8.102}$$

The vector $\mathbf{b}$ is a rate-gyro drift rate bias, $\boldsymbol{\eta}_1$ is a Gaussian white noise process corrupting the angular velocity measurements, while the bias rate is assumed to be driven by another independent Gaussian white noise process, $\boldsymbol{\eta}_2$. We also assume that the quaternion rate satisfies the constraint,

$$\Delta\mathbf{q}^T \cdot \Delta\dot{\mathbf{q}} = 0. \tag{8.103}$$

To satisfy the constraints given by Equations (8.102) and (8.103), we assume a time step equal to Δt and discretize Equation (8.100) as,

$$\mathbf{x}(k+1) \approx \mathbf{F}\mathbf{x}(k) + \mathbf{G}\Delta\boldsymbol{\omega}(k) + \mathbf{G}_d\boldsymbol{\eta}(k), \quad \mathbf{F} = \exp\left(\tilde{\mathbf{A}}\Delta t\right), \quad \mathbf{G} = \tilde{\mathbf{B}}\Delta t, \quad \mathbf{G}_d = \tilde{\mathbf{B}}_d\Delta t. \tag{8.104}$$

As in the case of Equation (8.52), the evaluation of $\mathbf{F}$ is done in accordance with the approach recommended by Markley [30], which is equivalent to Equation (8.104) and given by the equation,

$$\Delta\mathbf{q}(k+1) = \mathbf{F}_1(\boldsymbol{\omega}_k + \Delta\boldsymbol{\omega}, \mathbf{b}_k + \boldsymbol{\eta}(k))\Delta\mathbf{q}(k), \tag{8.105}$$

which is expressed by the pair of equations as,

$$\Delta\mathbf{q}(k+1) \approx \mathbf{F}_1(\boldsymbol{\omega}(k), \mathbf{b}(k))\Delta\mathbf{q}(k) + \mathbf{G}_1(\boldsymbol{\omega}_d - \boldsymbol{\omega}(k)) + \mathbf{G}_{d1}\boldsymbol{\eta}_1(k), \tag{8.106}$$

$$\mathbf{b}(k+1) = \mathbf{b}(k) + \mathbf{G}_{d2}\boldsymbol{\eta}_2(k), \tag{8.107}$$

where,

$$\mathbf{F}_1(\boldsymbol{\omega}, \mathbf{b}) = \begin{bmatrix} \alpha_k \mathbf{I} & \beta_k(\boldsymbol{\omega} + \mathbf{b}) \\ -\beta_k(\boldsymbol{\omega} + \mathbf{b})^T & \alpha_k \end{bmatrix}, \quad \mathbf{G}_1(\Delta\mathbf{q}) = \mathbf{G}_{d1}(\Delta\mathbf{q}) = \frac{\Delta t}{2}\boldsymbol{\Gamma}(\Delta\mathbf{q}),$$

$$\mathbf{G}_{d2} = \Delta t \mathbf{I}, \quad \alpha_k = \cos\left(\frac{\Delta t}{2}\|\boldsymbol{\omega} + \mathbf{b}\|\right), \quad \beta_k = \sin\left(\frac{\Delta t}{2}\|\boldsymbol{\omega} + \mathbf{b}\|\right)/\|\boldsymbol{\omega} + \mathbf{b}\|. \tag{8.108}$$

Further, we assume that the relative attitude quaternion can be measured independently, and when this is done the measurement includes a quaternion rotation bias. Thus $\Delta \mathbf{q}$ is related to the measurement by,

$$\Delta \mathbf{q}_m = \Delta \mathbf{q}_\vartheta \otimes (\Delta \mathbf{q}_b \otimes \Delta \mathbf{q}), \qquad (8.109)$$

where $\otimes$ denotes quaternion multiplication and $\Delta \mathbf{q}_\vartheta$ is a random measurement rotation quaternion. The quantity $\Delta \mathbf{q}_b$ is a bias quaternion that satisfies,

$$\Delta \mathbf{q}_b(k+1) = \mathbf{v} \otimes \Delta \mathbf{q}_b(k), \qquad (8.110)$$

where $\mathbf{v}$ is another random rotation quaternion that drives the bias update.

The complete nonlinear equations for estimating the chaser spacecraft's angular velocity vector and attitude quaternion and the target spacecraft's attitude quaternion are obtained from the discrete equations for the chaser spacecraft's dynamics and are given by,

$$\mathbf{q}: \quad \mathbf{q}(k+1) \approx \mathbf{F}_1(\boldsymbol{\omega}(k), 0)\mathbf{q}(k), \qquad (8.111)$$

$$\boldsymbol{\omega}: \quad \mathbf{I}\boldsymbol{\omega}(k+1) - \mathbf{I}\boldsymbol{\omega}(k) = (\boldsymbol{\Omega}_d \mathbf{I} \boldsymbol{\omega}_d - \boldsymbol{\Omega} \mathbf{I} \boldsymbol{\omega})\Delta t - \Delta \mathbf{M}_{gg}(k)\Delta t - \Delta \mathbf{M}_c(k)\Delta t + \mathbf{I}\boldsymbol{\omega}_{noise}, \qquad (8.112)$$

$$\Delta \mathbf{q}: \quad \Delta \mathbf{q}(k+1) \approx \mathbf{F}_1(\boldsymbol{\omega}(k), \mathbf{b}(k))\Delta \mathbf{q}(k) + \mathbf{G}_1(\boldsymbol{\omega}_d - \boldsymbol{\omega}(k)) + \mathbf{G}_{d1}\boldsymbol{\eta}_1(k), \quad (8.113)$$

$$\mathbf{b}: \quad \mathbf{b}(k+1) = \mathbf{b}(k) + \mathbf{G}_{2d}\boldsymbol{\eta}_2(k), \qquad (8.114)$$

$$\Delta \mathbf{q}_b: \quad \Delta \mathbf{q}_b(k+1) = \mathbf{v} \otimes \Delta \mathbf{q}_b(k), \qquad (8.115)$$

where $\boldsymbol{\omega}_{noise}$ is a Gaussian white process noise vector. The measurement equations are,

$$\Delta \boldsymbol{\omega}: \quad \boldsymbol{\omega}_d = \boldsymbol{\omega}(k) + \Delta \boldsymbol{\omega}_d(k), \qquad (8.116)$$

where $\Delta \boldsymbol{\omega}_d(k)$ is measured to define $\boldsymbol{\omega}_d$,

$$\mathbf{q}: \quad \mathbf{q}_m = \mathbf{q}_\vartheta \otimes \mathbf{q}, \qquad (8.117)$$

$$\boldsymbol{\omega}: \quad \boldsymbol{\omega}_m(k) = \boldsymbol{\omega}(k) + \boldsymbol{\omega}_\vartheta, \qquad (8.118)$$

$$\Delta \mathbf{q}: \quad \Delta \mathbf{q}_m = \Delta \mathbf{q}_\vartheta \otimes (\Delta \mathbf{q}_b \otimes \Delta \mathbf{q}). \qquad (8.119)$$

The linear version of the first equation and the second are used for synthesizing the control law.

8.5.4 The Measurements

(a) **Measuring the Relative Attitude of a Target Body**

Since the attitude determination problem has been extensively investigated in numerous papers, we will assume that the attitude has been measured directly. There are three classes of methods (Yun, Bachmann, and McGhee [61], Shuster and Oh [7]) for estimating the attitude quaternion: (i) The TRIAD algorithm and its derivatives, (ii) Davenport's q method and its derivatives, and (iii) The QUEST algorithm and its derivatives. There are also methods proposed for estimating the attitude quaternion from distance-only measurements (see, for example, Trawny, Zhou, Zhou, and Roumeliotis [62]). Estimates from any of these algorithms, particularly the QUEST algorithm, can be applied to the case of relative attitude estimation, and these estimates can be used as pseudo measurements.

(b) Measuring the Relative Angular Velocity of a Target Body

As a rigid body moves, point A on it traces a circular path of radius $\vec{r}_{A/B} = |\vec{r}_{A/B}|$ relative to a point B on it, keeping the distance between the two unchanged. The angular velocity of this motion is simply the angular velocity ω of the rigid body. Then, for the time derivatives of rotating vectors we have,

$$\vec{v}_{A/B} = \frac{d}{dt}\vec{r}_{A/B} = \omega \times \vec{r}_{A/B}, \tag{8.120}$$

or,

$$\vec{v}_A = \vec{v}_B + \omega \times \vec{r}_{A/B}. \tag{8.121}$$

Observe that the expression reflects the decomposition of rigid body motion referred to previously. With B chosen as reference, the velocity of A is the vector sum of a translational portion $\vec{v}_B$ and a rotational portion $\omega \times \vec{r}_{A/B}$. The velocity of a point A of a body I with respect to the center of mass of the body is,

$$\vec{v}_{A,I} = \vec{v}_{CM,I} + \omega_I \times \vec{r}_{A,I/CM,I}. \tag{8.122}$$

Thus, the velocity of a point B of a body II with respect to the center of mass of the body is,

$$\vec{v}_{B,II} = \vec{v}_{CM,II} + \omega_{II} \times \vec{r}_{B,II/CM,II}. \tag{8.123}$$

In both cases, the frame of reference is assumed to be fixed in body I. The velocity of a point B of a body II with respect to a point A of a body I is,

$$\vec{v}_{B,II/A,I} = \vec{v}_{CM,II} - \vec{v}_{CM,I} + \omega_{II} \times \vec{r}_{B,II/CM,II} - \omega_I \times \vec{r}_{A,I/CM,I}. \tag{8.124}$$

Hence,

$$\vec{v}_{B,II/A,I} + \omega_I \times \vec{r}_{A,I/CM,I} = \vec{v}_{CM,II} - \vec{v}_{CM,I} + \omega_{II} \times \vec{r}_{B,II/CM,II}. \tag{8.125}$$

Considering another point D on body I,

$$\vec{v}_{D,I} = \vec{v}_{CM,I} + \omega_I \times \vec{r}_{D,I/CM,I}, \tag{8.126}$$

and another point E on body II,

$$\vec{v}_{E,II/D,I} + \omega_I \times \vec{r}_{D,I/CM,I} = \vec{v}_{CM,II} - \vec{v}_{CM,I} + \omega_{II} \times \vec{r}_{E,II/CM,II}. \tag{8.127}$$

Hence,

$$\vec{v}_{B,II/A,I} - \vec{v}_{E,II/D,I} + \omega_I \times \left(\vec{r}_{A,I/CM,I} - \vec{r}_{D,I/CM,I}\right) = \\ \omega_{II} \times \left(\vec{r}_{B,II/CM,II} - \vec{r}_{E,II/CM,II}\right). \tag{8.128}$$

or,

$$\vec{v}_{B,II/A,I} - \vec{v}_{E,II/D,I} + \boldsymbol{\omega}_I \times \left(\vec{r}_{A,I/CM,I} - \vec{r}_{D,I/CM,I} \right) =$$
$$\left(\vec{r}_{E,II/CM,II} - \vec{r}_{B,II/CM,II} \right) \times \boldsymbol{\omega}_{II}. \tag{8.129}$$

Introducing the relative angular velocity vector, $\Delta\boldsymbol{\omega} = \boldsymbol{\omega}_{II} - \boldsymbol{\omega}_I$, denoting the target body by the subscript "d" and the chaser by the subscript "c,"

$$\left(\vec{r}_{E,d/CM,d} - \vec{r}_{B,d/CM,d} \right) \times \Delta\boldsymbol{\omega}$$
$$= \vec{v}_{B,d/A,c} - \vec{v}_{E,d/D,c} + \boldsymbol{\omega}_c \times \left(\vec{r}_{A,c/CM,c} - \vec{r}_{D,c/CM,c} \right) - \left(\vec{r}_{E,d/CM,d} - \vec{r}_{B,d/CM,d} \right). \tag{8.130}$$

We have, from this relation,

$$\vec{r}_{E,d/B,d} \times \Delta\boldsymbol{\omega} = \vec{v}_{B,d/A,c} - \vec{v}_{E,d/D,c} + \boldsymbol{\omega}_c \times \left(\vec{r}_{A,c/D,c} - \vec{r}_{E,d/B,d} \right). \tag{8.131}$$

We assume the CM of the chaser spacecraft is precisely known. Thus, we measure the position vectors of two points on the target spacecraft, $\vec{r}_{B,d/CM,c}$, $\vec{r}_{E,d/CM,c}$. It follows that,

$$\vec{r}_{E,d/B,d} = \vec{r}_{E,d/CM,c} - \vec{r}_{B,d/CM,c}. \tag{8.132}$$

We also measure, $\vec{v}_{B,d/A,c}$, $\vec{v}_{E,d/D,c}$, and $\boldsymbol{\omega}_c$. Consequently, we can estimate $\Delta\boldsymbol{\omega}$ relative to the target body.

(c) **Other Measurements**

In addition, we assume that fiber optic gyro measurements of the chaser spacecraft's angular velocity vector and the chaser spacecraft's attitude quaternion are also available. The attitude quaternion can be accurately measured using state-of-the-art star trackers as discussed by Sun et al. [63].

8.5.5 The Controller Synthesis

The controller is synthesized in two phases. In the first phase, the attitude error between the actual attitude of the chaser spacecraft and its desired attitude is reduced to within acceptable limits. Once this objective is achieved, the controller is switched so the attitude of the chaser spacecraft is synchronized with that of the target spacecraft. The transition from the first control law to the second is achieved smoothly. We consider both these phases in the subsequent paragraphs.

(a) **Control for Rotation Rate Error Reduction and Near Synchronization**

In the absence of biases and noise, the relative attitude quaternion $\Delta\mathbf{q}$ update equations are,

$$\Delta\mathbf{q}(k+1) \approx \mathbf{F}_1(\boldsymbol{\omega}_k, 0)\Delta\mathbf{q}(k) + \mathbf{G}_1(\Delta\mathbf{q}(k))\Delta\boldsymbol{\omega}(k). \tag{8.133}$$

In the absence of disturbances, from Equation (8.111), the relative angular velocity update equations are,

$$\Delta\boldsymbol{\omega}(k+1) = \left(\mathbf{I}_{3\times 3} - \mathbf{I}^{-1}\mathbf{W}\Delta t\right)\Delta\boldsymbol{\omega}(k) + 3n^2\mathbf{I}^{-1}\mathbf{G}\mathbf{C}\Delta t\Delta\mathbf{q}(k) + \mathbf{I}^{-1}\Delta t\Delta\mathbf{M}_c, \tag{8.134}$$

where $\mathbf{I}_{3\times 3}$ is a 3×3 unit matrix. These equations may be used to construct the discrete time control law for attitude synchronization. The discrete-time linear quadratic regulator (LQR) with a finite final time and with magnitude constraints imposed on the state and control variables is first designed. A finite final time quadratic performance index could be used, so the $\mathbf{Q}$ and $\mathbf{R}$ weighting matrices will have to be chosen and a Riccati differential equation must be solved. With the presence of terminal constraints, the performance index is of the form,

$$J = \frac{1}{2}\sum_{k=0}^{N-1}\left(\mathbf{x}_k^T \mathbf{Q}\mathbf{x}_k + \mathbf{u}_k^T \mathbf{R}\mathbf{u}_k\right) + \mathbf{x}_N^T \mathbf{Q}_N \mathbf{x}_N. \qquad (8.135)$$

The solution for the discrete time optimal control law is obtained by solving the Riccati differential equation. However, with reference to the attitude controller design, while the methodology guarantees closed loop stability with full state feedback when the states are noise free, closed loop stability is not always possible when the states are noisy and the feedback is synthesized by an estimator. This is because the attitude dynamics are patently nonlinear, which under certain dynamic equilibrium conditions can result in uncontrollable linear models. For this reason, we only apply the discrete-time LQR controller over a finite time horizon.

(b) **Rotation Rate Synchronization Control and the UKF**
These topics were discussed in Chapter 4 and in Section 7.6.1, respectively, and will not be repeated here. An application example was also presented in Chapter 4.

8.6 Sensors for Localization

Apart from the classical sensors for navigation like gyroscopes, accelerometers, magnetic compasses, inclinometers that measure the direction of acceleration due to gravity, magnetometers that measure the Earth's magnetic field components, which are functions of the orbital position and the attitude of an Earth-orbiting satellite, and GNSS-based codeless satellite navigation attitude sensors, there are a number of ranging sensors that are used as sensors for localization. Classical navigation sensors such as accelerometers and rate-integrating gyroscopes are used to measure the acceleration or the attitude either in a navigation frame or in a body-fixed frame, and the corresponding equations of motion are integrated to obtain the position and orientation of a spacecraft. This is the principle behind the operation of inertial navigation systems discussed in detail in section 9.4 of Vepa [64]. Range and bearing sensors provide an alternative approach to navigation, which is particularly suitable for space and planetary navigation other than on the Earth.

Range sensing is generally based on measuring the round-trip travel time of an ultrasonic or narrow band electromagnetic wave to a target reflector and back. Ultrasonic range finders operate with very low power consumption and are particularly suitable for detection and ranging of landmarks from mobile vehicles. However, both lateral and depth resolution are limited by their wavelengths, which are generally of the order of millimeters. Short wavelength ultrasonic waves experience exponentially increasing attenuation in air. For this reason, ultrasonic range finders are not popular for high-precision three-dimensional ranging applications. RADAR and Light Detection and Ranging (LIDAR) use electromagnetic waves in radio and optical frequency bands, respectively, for ranging applications. Light waves have much shorter wavelengths than radio frequency waves; consequently, infrared (IR) and LIDAR-based rangers have a greater lateral resolution and depth precision than RADAR-based rangers, making them more suitable ranging applications. The most common architecture of a LIDAR-based ranging sensor measures the round-trip delay of a light pulse to and from a reflective target. Optical ranging using low-coherence interferometry has been in existence for more than two decades, as discussed by Danielson and Boisrobert [65]. Direct-detection LADAR systems based on LASERs have also been extensively developed, and have been reviewed by Richmond and Cain [66] and by Aboites and Wilson [67]. Infrared ranging and communication systems based on spread spectrum techniques have been developed for a variety of applications, as mentioned by Ueda and Mizui [68] and Kwak and Lee [69]. Charge-coupled imaging devices have also been developed for a variety of ranging applications that have been reviewed by Miyagawa and Kanade [70].

A LASER-based ranging sensor consists of a pulsed LASER or a pulsed light source, a detector, which is usually an avalanche-type semiconductor photo diode, and circuitry for comparing the time delay between the transmitted and received pulses. Considering the free-space speed of light is equal to 3×10^8 m/s, to achieve a precision range measurement below 10 μm, the electronic circuits must be able to measure the round-trip delay of the light pulse with a precision of approximately 70 femto(10^{-15})-secs(fs), which is well-nigh impossible in practice. Thus, one uses a frequency-modulated continuous wave or a frequency-modulated coded pulse, so the measurement of the delay reduces to the measurement of phase delay just as in the case of GPS code-less navigation sensors.

When the LIDAR cross-section of the target is narrow, one can relate the detected power P_D to the emitted power P_E and the background power P_B, in terms of opening angle γ_E, the atmospheric attenuation factor η_{atm}, the scatter at the target defined by the target cross-section σ (which represents an area across the beam direction), the target distance R, and the aperture of the receiver D_r and is given by,

$$P_D = P_E \frac{4}{\pi(\gamma_E R)^2} \frac{\sigma}{4\pi R^2} \frac{\pi(D_r)^2}{4} \eta_{atm} + P_B. \tag{8.136}$$

When the cross-section is not narrow, the returns are summed using the convolution integral, and ignoring the background power, the detected power from the ith target is expressed as,

$$P_{D_i} = \frac{(D_r)^2}{(\gamma_E R)^2} \frac{\eta_{atm}}{4\pi R^2} \int_{R_i-\delta}^{R_i+\delta} P_E(t - 2R/c)\sigma(R)dR, \tag{8.137}$$

where the differential cross-section of the target is 2δ. Thus, the detected power is inversely proportional to R^4. Pulse durations of the order of 2 to 10 *ns* are most commonly employed, and sampling times are usually of the order of 0.2 to 1 *ns*.

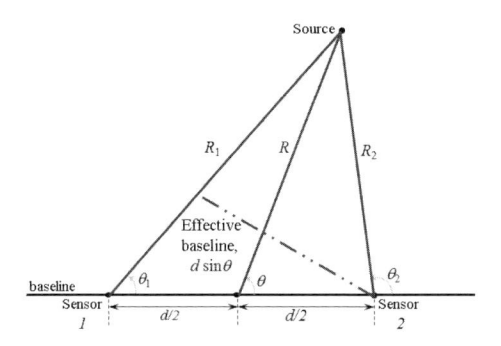

Figure 8.8 Typical geometry for ranging [71].

A variety of range scanners are currently available from various manufacturers. A scanning type of LIDAR ranger uses rotating mirrors to create a two-dimensional scanning platform able to cover a wider angle or a field of view. Laser scanners offer higher precision and directionality, and for these reasons they are used for map-building and localization applications.

Triangulation enables the determination of the range of the target using two ranging or imaging sensors located along a known baseline [71]. A typical geometry for ranging is illustrated in Figure 8.8. If we know the baseline d and angles of arrival (θ_1, θ_2) with sufficient accuracy, one can obtain the range R with simple geometric relationships,

$$\left(\frac{R}{d}\right)^2 = \frac{1}{2}\left\{\frac{\sin^2\theta_1 + \sin^2\theta_1}{\sin^2(\theta_2 - \theta_1)} - \frac{1}{2}\right\}. \tag{8.138}$$

The variance of the range error $\sigma_{\Delta R}$ is directly related to the variance $\sigma_{\Delta \theta}$ of the errors in the measurements of the angles θ_1 and θ_2, by the relation,

$$\sigma_{\Delta R} = \sqrt{2} \times \sigma_{\Delta \theta}\left(\frac{R}{d|\sin\theta|}\right). \tag{8.139}$$

The concept of photogrammetry for determining the range of features in visual imagery is also extensively used for purposes of ranging landmarks. Photogrammetry relies on the relative locations of common objects in each of two images obtained from two separated imaging sensors or cameras. The relative positions are used to solve for the desired unknowns related by a set of equations to the sensor positions, pointing angles, and the relative locations of the common objects in the image. Figure 8.9 illustrates a typical range computation geometry and the equation based on images captured on a camera's image plane.

8.7 Sensors for Navigation

In addition to the standard complement of sensors used for navigation discussed earlier, imaging cameras are also used as navigation sensors. The modeling of an imaging sensor as a typical pinhole camera is briefly discussed in the following section. Optical sensors for relative navigation are discussed by Fehse [72], in section 7.4.

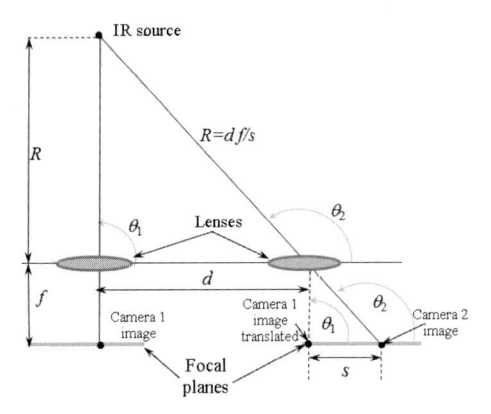

Figure 8.9 Typical range computation geometry and the range equation [71].

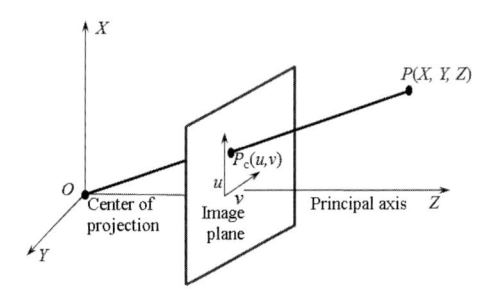

Figure 8.10 A pinhole camera projecting a point P onto an image screen.

8.7.1 Imaging Sensors and Cameras

Basic imaging sensors and cameras are often modeled as pinhole cameras. Modeling a pinhole camera involves setting up the relation that maps a three-dimensional point in space **P** onto a point P_c on a two-dimensional plane representing the imaging surface in the camera. When the origin of the two-dimensional image coordinates system does not coincide with the point where the Z axis intersects the image plane I_m, the point $\mathbf{P}_c$ must be translated to the two-dimensional frame in the imaging plane with the desired origin. For a more general case, the rectangle pixels on the image plane are assumed to represent a resolution p_u and p_v pixels/unit distance in u and v direction, respectively.

Figure 8.10 shows a model of a pinhole camera, where a point P is projected onto an imaging screen with the center of projection (COP) O and the principal camera axis parallel to the Z axis. The camera's imaging plane is assumed to be in front of the COP, and the distance of the plane from O is equal to the focal length f and in a direction away from O. The Z axis is perpendicular to the two-dimensional imaging plane $I_m \in \mathcal{R}^2$ and lies along the camera's optical axis, while the mutually perpendicular axes Y and X are defined toward the right and down, respectively, so the X, Y, and Z axes form a right-handed reference frame. The camera frame f_c attached to the camera has its origin at the COP of the camera. A three-dimensional point **P,** whose coordinates in the camera frame f_c are $\mathbf{P} = (X, Y, Z)$, is projected on the camera's image plane I_m with coordinates $\mathbf{P}_c = (u, \ v)$ in the image plane. This point P is imaged onto the camera's image plane I_m at distance $f = 1$ from the COP and at a point defined in homogeneous coordinates as,

$$\mathbf{p} = \frac{1}{Z}\mathbf{P}, \tag{8.140}$$

where

$$\mathbf{P} = [X \ Y \ Z]^{\mathrm{T}}. \tag{8.141}$$

To measure the position of the image point in the image plane $\mathbf{P}_c$ in pixels, its two-dimensional u and v coordinates must be multiplied by the scale factors p_u and p_v in pixels per unit distance, respectively. The complete transformation of point $\mathbf{P}$ onto the image plane may be expressed as,

$$\begin{bmatrix} u \\ v \\ 1 \end{bmatrix} = \begin{bmatrix} \alpha_x & s & u_0 \\ 0 & \alpha_y & v_0 \\ 0 & 0 & 1 \end{bmatrix} \mathbf{p} = \mathbf{KP}. \tag{8.142}$$

The transformation $\mathbf{K}$ is a 3×3 matrix usually called the intrinsic parameter matrix for the camera, as it only depends on the intrinsic parameters of the camera such as its focal length f and the distance of the center point of the image plane I_m in the camera frame f_c. The parameter s is the skew factor, which is zero for most of the cameras and characterizes the pixel distortion. The coordinates (u_o, v_o) define the principal point corresponding to the intersection between the image plane and the optical axis. In a strict sense, the matrix $\mathbf{K} = \mathbf{K}(\mathbf{P})$ does not represent a linear transformation, although t may be treated as one.

If the camera does not have its COP at (0, 0, 0), a translation $\mathbf{T}$ and a rotation $\mathbf{R}$ are needed to make the camera coordinate system coincide with the configuration in the figure. Let $\mathbf{T} = \begin{bmatrix} t_x & t_y & t_z \end{bmatrix}^T$ be the camera translation vector and let the rotation applied to co-locate the principal axis with the Z axis be given by the 3×3 rotation matrix $\mathbf{R}$. Then the matrix formed by first translating the object and then rotating to the camera frame is given by the 3×4 matrix,

$$\mathbf{E} = [\mathbf{R} \ \ \mathbf{RT}], \tag{8.143}$$

which is known as the extrinsic parameter matrix. The complete camera transformation can be expressed as,

$$\mathbf{KE} = \mathbf{K}[\mathbf{R} \ \ \mathbf{RT}] = [\mathbf{KR} \ \ \mathbf{KRT}] = \mathbf{KR}[\mathbf{I} \ \ \mathbf{T}]. \tag{8.144}$$

The vector $\mathbf{P}_c$ the projection image of $\mathbf{P}$, can be given by,

$$\mathbf{P}_c = \mathbf{KR}[\mathbf{I} \ \ \mathbf{T}]\mathbf{P} = \mathbf{CP}, \tag{8.145}$$

where $\mathbf{C}$ is a 3×4 matrix called the complete camera calibration matrix, which is formed by the three-dimensional homogeneous coordinates of $\mathbf{P}$, while $\mathbf{P}_c$, which is obtained from $\mathbf{CP}$, will be in two-dimensional homogeneous coordinates.

References

[1] Vepa, R. (2010) Spacecraft large attitude estimation using a navigation sensor. *Journal of Navigation*, 63(1): 89–104.

[2] Hatch, R. R. and Sharpe, T. (2001) A computationally efficient ambiguity resolution technique, in *Proceeding of ION GPS 2001*, Salt Lake City, UT, 11–14 Sept. 2001.

[3] Pervan, B., Cohen, C., and Parkinson, B. (1994) Integrity monitoring for precision approach using kinematic GPS and a ground-based pseudolite. *Navigation*, 41(2): 159–174.

[4] Black, H. D. (1964) A passive system for determining the attitude of a satellite. *AIAA Journal*, 2(7): 1350–1351.

[5] Davenport, P. R. (1965) Attitude Determination and Sensor Alignment via Weighted Least Squares Affine Transformations, NASA X-514–71-312.

[6] Davenport, P. R. (1968) A Vector Approach to the Algebra of Rotations with Applications, NASA TN-D-4696.

[7] Shuster, M. D. and Oh, S. D. (1981) Three-axis attitude determination from vector observations. *Journal of Guidance and Control*, 4(1): 70–77.

[8] Wahba, G. (1965) A least squares estimate of satellite attitude. *SIAM Review*, 7(3): 409–426.

[9] Farrell, J. L. (1970) Attitude determination by Kalman filtering. *Automatica*, 6: 419–430.

[10] Potter, J. E. and Vander Velde, W. E. (1968) Optimum mixing of gyroscope and star tracker data. *Journal of Spacecraft and Rockets*, 5(5): 536–540.

[11] Fujikawa, S. J. and Zimbelman, D. F. (1995) Spacecraft attitude determination by Kalman filtering of global positioning system signals. *Journal of Guidance, Control, and Dynamics*, 18(6): 1365–1371.

[12] Huang, G.-S. and Juang, J.-C. (1997) Application of nonlinear Kalman filter approach in dynamic GPS-based attitude determination. *Proceedings of the 40th Midwest Symposium on Circuits and Systems, 1997*, 2: 1440–1444.

[13] Lefferts, E. J., Markley, F. L., and Schuster, M. D. (1982) Kalman filtering for spacecraft attitude estimation. *Journal of Guidance, Control, and Dynamics*, 5(5): 417–429.

[14] Lerner, G. M. (1978) Three-axis attitude determination. In *Spacecraft Attitude Determination and Control*, J. R. Wertz and D. Reidel, eds., Dordrecht: D. Reidel Publishing Co., 420–428.

[15] Marins, J. L., Yun, X., Bachmann, E. R., McGhee, R. B., and Zyda, M. J. (2001) An extended Kalman filter for quaternion-based orientation estimation using MARG sensors, in *Proceedings of the 2001 IEEE/RSJ, International Conference on Intelligent Robots and Systems*, Maui, Hawaii, Oct. 29–Nov. 3, 2001.

[16] de Ruiter, A. H. J. and Damaren, C. J. (2002) Extended Kalman filtering and nonlinear predictive filtering for spacecraft attitude determination. *Canadian Aeronautics and Space Journal*, 48(1): 13–23.

[17] Vathsal, S. (1987) Spacecraft attitude determination using a second-order nonlinear filter. *Journal of Guidance, Control, and Dynamics*, 10(6): 559–566.

[18] Sage, A. P. and Melsa, J. L. (1971) *Estimation Theory with Applications to Communications and Control*, New York: McGraw-Hill, 106–156.

[19] Jazwinski, A. H. (1970) *Stochastic Processes and Filtering Theory*, London: Academic Press.

[20] Crassidis, J. L. and Markley, F. L. (1997) Predictive filtering for attitude estimation without rate sensors, *Journal of Guidance, Control, and Dynamics*, 20(3): 522–527.

[21] Crassidis, J. L. and Markley, F. L. (1997) Predictive filtering for nonlinear systems. *Journal of Guidance, Control, and Dynamics*, 20(4): 566–572.

[22] Julier, S. J. and Uhlmann, J. K. (2000) Unscented filtering and nonlinear estimation. *Proceedings of the IEEE*, 92: 401–422.

[23] Van Dyke, M. C., Schwartz, J. L., and Hall, C. D. (2004) Unscented Kalman Filtering For Spacecraft Attitude State And Parameter Estimation, *The American Astronautical Society, Paper* AAS-04–115.

[24] Kingston, D. B. and Beard, R. W. (2004) Real-Time Attitude and Position Estimation for Small UAVs Using Low-Cost Sensors, *AIAA 3rd Unmanned Unlimited Systems Conference and Workshop*, September, 2004, Paper no. AIAA-2004–6488.

[25] Creamer, G. (1996) Spacecraft attitude determination using gyros and quaternion measurements. *The Journal of the Astronautical Sciences*, 44(3): 357–371.

[26] Crassidis, J. L. and Markley, F. L. (2003) Unscented filtering for spacecraft attitude estimation. *Journal of Guidance, Control, and Dynamics*, 26(4): 536–542.

[27] Kraft, E. (2003) A quaternion-based unscented Kalman filter for orientation tracking. *Proceedings of the Sixth International Conference of Information Fusion*, 2003(1): 47–54.

[28] Cheon, Y.-J. and Kim, J.-H. (2007) Unscented filtering in a unit quaternion space for spacecraft attitude estimation, in *Proceedings of the IEEE International Symposium on Industrial Electronics, 2007,* ISIE 2007, 66–71.

[29] Markley, F. L. and Crassidis, J. L. (2014) *Fundamentals of Spacecraft Attitude Determination and Control*, New York: Springer-Verlag.

[30] Markley, F. L. (1978) Matrix and vector algebra. In *Spacecraft Attitude Determination and Control*, J. R. Wertz and D. Reidel, eds., Dordrecht: D. Reidel Publishing Co., 754–755.

[31] Hoots, F. R., Schumacher Jr., P. W., and Glover, R. A. (2004) History of analytical orbit modeling in the U.S. space surveillance system. *Journal of Guidance, Control, and Dynamics*, 27(2): 174–185.

[32] Julier, S. J., Uhlmann, J., and Durrant-Whyte, H. F. (2000) A New method for the nonlinear transformation of means and covariances in filters and estimators. *IEEE Transactions on Automatic Control*, 45(3): 477–482.

[33] Julier, S. J. (2002) The scaled unscented transformation. *Proceedings of the American Control Conference*, 6: 4555–4559.

[34] Farrenkopf, R. L. (1978) Analytic steady-state accuracy solutions for two common spacecraft attitude estimators. *Journal of Guidance, Control and Dynamics*, 1(4): 282–284.

[35] Wen, T.-Y. J. and Kreutz-Delgado, K. (1991) The attitude control problem. *IEEE Transactions on Automatic Control*, 36(10): 1148–1162.

[36] Bhat, S. P. and Bernstein, D. S. (2000) Topological obstruction to continuous global stabilization of rotational motion and the unwinding phenomenon. *Systems and Control Letters*, 39(1): 63–70.

[37] Fjellstad, O.-E. and Fossen, T. I. (1994) Position and attitude tracking of AUV's: A quaternion feedback approach. *IEEE Journal of Oceanic Engineering*, 19(4): 512–518.

[38] Fragopoulos, D. and Innocenti, M. (2004) Stability considerations in quaternion attitude control using discontinuous Lyapunov functions. *IEE Proceedings on Control Theory and Applications*, 151(3): 253–258.

[39] Dimarogonas, D. V., Tsiotras, P., and Kyriakopoulos, K. J. (2009) Leader-follower cooperative attitude control of multiple rigid bodies. *Systems and Control Letters*, 58(6): 429–435.

[40] Kang, W. and Yeh, H. H. (2002) Coordinated attitude control of multi-satellite systems. *International Journal of Robust Nonlinear Control*, 12(2–3): 185–205.

[41] Wang, P. K. C., Hadaegh, F. Y., and Lau, K. (1999) Synchronized formation rotation and attitude control of multiple free-flying spacecraft. *Journal of Guidance, Control, and Dynamics*, 22(1): 28–35.

[42] Pan, H. and Kapila, V. (2001) Adaptive nonlinear control for spacecraft formation flying with coupled translational and attitude dynamics, in *Proceedings of the 40th IEEE Conference on Decision and Control*, Orlando, FL.

[43] Nijmeijer, H. and Rodriguez-Angeles, A. (2003) *Synchronization of Mechanical Systems*, Singapore: World Scientific.

[44] Lawton, J. and Beard, R. W. (2002) Synchronized multiple spacecraft rotations. *Automatica*, 38(8): 1359–1364.

[45] Ren, W. and Beard, R. W. (2004) Decentralized scheme for spacecraft formation flying via the virtual structure approach. *Journal of Guidance, Control, and Dynamics*, 27(1): 73–82.

[46] Egeland, O. and Godhaven, J.-M. (1994) Passivity-based adaptive attitude control of a rigid spacecraft. *IEEE Transactions of Automatic Control*, 39(4): 842–846.

[47] Bai, H., Arcak, M., and Wen, J. T. (2008) Rigid body attitude coordination without inertial frame information. *Automatica*, 44(12): 3107–3175.

[48] Terui, F. (1998) Position and attitude control of a spacecraft by sliding mode control, in *Proceedings of the American Control Conference*, 217–221.

[49] Stansbery, D. T. and Cloutier, J. R. (2000) Position and attitude control of a spacecraft using the state-dependent Riccati equation technique, in *Proceedings of the American Control Conference*, 1867–1871.

[50] Singla, P., Subbarao, K., and Junkins, J. L. (2006) Adaptive output feedback control for spacecraft rendezvous and docking under measurement uncertainty. *Journal of Guidance, Control, and Dynamics*, 29: 892–902.

[51] Subbarao, K. and Welsh, S. (2008) Nonlinear control of motion synchronization for satellite proximity operations. *Journal of Guidance, Control, and Dynamics*, 31: 1284–1294.

[52] Xin, M. and Pan, H. (2010) Integrated nonlinear optimal control of spacecraft in proximity operations. *International Journal of Control*, 83: 347–363.

[53] Lizarralde, F. and Wen, J. (1996) Attitude control without angular velocity measurements: A passivity approach. *IEEE Transactions on Automatic Control*, 41(3): 468–472.

[54] Costic, B., Dawson, D., De Queiroz, M., and Kapila, V. (2000). A quaternion-based adaptive attitude tracking controller without velocity measurements, in *Proceedings of the 39th IEEE Conference on Decision and Control*, 3, Sydney, Australia, 2424–2429.

[55] Akella, M. R. (2001) Rigid body attitude tracking without angular velocity feedback. *Systems and Control Letters*, 42(4): 321–326.

[56] Sun, L. and Huo, W. (2015) Robust adaptive relative position tracking and attitude synchronization for spacecraft rendezvous. *Aerospace Science and Technology*, 41: 28–35.

[57] Lu, W., Geng, Y., Chen, X., and Zhang, F. (2011) Relative position and attitude coupled control for autonomous docking with a tumbling target. *International Journal of Control and Automation*, 4(4): 1–22.

[58] Brown, M. D. J. (2001). Continuous and Smooth Sliding Mode Control, chapter 8, *Ph.D. dissertation*, University of Alabama in Huntsville, Dept. of Electrical and Computer Engineering, 117–120.

[59] Wie, B. and Lu, J. (1995) Feedback control logic for spacecraft eigenaxis rotations under slew rate and control constraints. *Journal of Guidance, Control, and Dynamics*, 18: 1372–1379.

[60] Liu, H., Shi, X., Bi, X., and Zhang, J. (2016) Backstepping based terminal sliding mode control for Rendezvous and docking with a tumbling spacecraft. *International Journal of Innovative Computing, Innovation and Control*, 12(3): 929–940.

[61] Yun, X., Bachmann, E. R., and McGhee, R. B. (2008) A simplified quaternion-based algorithm for orientation estimation from Earth gravity and magnetic field measurements. *IEEE Transactions on Instrumentation and Measurement*, 57(3): 638–650.

[62] Trawny, N., Zhou, X. S., Zhou, K. X., and Roumeliotis, S. I. (2007) 3D relative pose estimation from distance-only measurements, in *Proceedings of the 2007 IEEE/RSJ International Conference on Intelligent Robots and Systems*, San Diego, CA, Oct. 29–Nov. 2, 1071–1078.

[63] Sun, T., Xing, F., Wang, X., You Z., and Chu, D. (2016) An accuracy measurement method for star trackers based on direct astronomic observation. *Scientific Reports*, 6: 22593. doi: 10.1038/srep22593.

[64] Vepa, R. (2009) *Biomimetic Robotics: Mechanisms and Control*, New York: Cambridge University Press.

[65] Danielson, B. L. and Boisrobert, C. Y. (1991) Absolute optical ranging using low coherence interferometry. *Applied Optics*, 30(21): 2975–2979.

[66] Richmond, R. D. and Cain, S. C. (2010) Direct-Detection LADAR Systems, SPIE Press, tutorial text, Bellingham.

[67] Aboites, V. and Wilson, M. (2018) Lasers. Chapter 16 in *Advanced Optical Instruments and Techniques*, Vol. 2, D. Malacara, D. M. Hernández, and B. J. Thompson, eds., Boca Raton, FL: CRC Press.

[68] Ueda, A. and Mizui, K. (2002) Vehicle-to-vehicle communication and ranging system using code-hopping spread spectrum technique with code collision avoidance algorithm, in *Proceedings of the Canadian Conference on Electrical and Computer Engineering*, 2002. IEEE CCECE 2002, No. 3, Winnipeg, Manitoba, Canada, May 12–15, 2002, 1250–1254.

[69] Kwak, J. S. and Lee, J. H. (2004) Infrared transmission for inter-vehicle ranging and vehicle-to-roadside communication systems using spread-spectrum technique. *IEEE Transactions on Intelligent Transport Systems*, 5(1): 12–19.

[70] Miyagawa, R. and Kanade, T. (1997) CCD-based range-finding sensor. *IEEE Transactions on Electron Devices*, 44(10): 1648–1652.

[71] Kelly, J. P., Klein, T., and Ilves, H. (1999) Design and demonstration of an infrared passive ranger. *Johns Hopkins APL Technical Digest*, 20(2): 220–235.

[72] Fehse, W. (2003) *Automated Rendezvous and Docking of Spacecraft*, Cambridge Aerospace Series, New York: Cambridge University Press.

Index

accelerometers 308, 339
adjoint 98
aerodynamic drag 123
aerodynamic forces 125
apoapse 34, 49
apoapsis 39, 47, 93
apogee 34–35, 40, 48, 50–51, 100, 110
Apogee Kick Motor 179
argument of perigee 34
artificial intelligence 26
artificial potential 181
ascending node 34, 43, 53, 55, 68, 71, 79, 86, 88, 99–100
attitude control 120, 129, 133–134, 136, 147, 149, 152, 158, 162, 164
attitude determination 312–313, 315, 319–320, 328, 336
attitude dynamics 26, 118, 123, 157
attitude estimation 314, 319, 324, 337
attitude quaternion 196
automatic control systems 24
autonomous 1, 4, 23–24, 26
autonomous control systems 24
autonomous robots 5
AUV 16
autonomy 25

Bayes rule 275–276
Bayesian estimation 277, 292
Bayesian inference 275
BLF 198–199, 202
body frame 119–120, 125, 129, 133–134, 156, 160, 164
body-fixed frame 119
bogie 242
burnout 109

camera 313, 341–343
Cartesian coordinates 183–184, 207
Cartesian space 181
center of mass 119, 123, 159–160
CP 123

centre of gravity 136, 145, 174–175, 188
characteristic velocity 110
CM 123, 183–184, 186, 188
CMG 149, 158–159, 161–163
conditional probability 269–270, 275, 278–279, 295
conditions for stability 131–132, 148
conservation of momentum 30
constraints 206–209, 221, 223–224
control laws 206
control moment gyros 134, 135, 137, 166

DH 183–184
D* algorithm 232, 254, 256
D'Alembert forces 214
D'Alembert's principle 178
DeBra-Delp diagram 193
DeBra-Delp 132, 138, 143, 148, 191–194
decoupled control 222
degrees of freedom 179, 183, 192–193
Delaunay elements 86
delta-vee 110
DH decomposition 167
de-orbit 14
descending node 34, 100
dilution of precision 262
disturbing potential 62, 65
docking 18
drag 35, 59, 62, 68, 102, 104–105, 108–109, 123, 125–126, 164
drawbar pull 243, 247

Earth's atmosphere 13
Earth-centered, Earth-fixed 105
eccentric anomaly 41–42, 46, 66
eccentricity 28, 31, 35, 38, 51, 54–55, 57, 59, 62, 70–71, 86–87, 92–93, 98–99
ECEF 105
ECI 43, 76, 104–105
electric propulsion 112–113, 115, 117
electromagnetic radiation 126
electrostatic ion thruster 10

349

ellipse 29, 31–32, 34–35, 38–39, 41–42, 49–50, 56–58, 62, 88, 93
elliptic orbit 33, 35, 37, 41, 49, 71, 77, 80
Enhanced Disturbance Map 180
environmental 123, 125, 129, 134
Epoch 34
equatorial plane 34, 65, 99, 103
equinoctial elements 87, 91, 93
Euler angle sequence 44, 79, 89, 91, 120, 129, 140, 238, 240
Euler angles 88, 120, 139, 152–153, 155
Euler-Lagrange equations 208, 214, 221
Euler rotation sequence 120
Euler-Lagrange 171, 178, 185, 209, 211
Euler-Newton formulation 178
European Space Agency 6, 8, 10, 13, 15, 21
extended Kalman filter 275, 288–289, 292, 298, 300, 304–305, 319–320, 324

feedback control 94, 97–98, 135, 151
foci 34, 57
free molecular flow 125
free-floating 180–181
free-flying 167

Gauss planetary equations 65, 99
Gauss-Newton 296–297, 303
geostationary 7, 13, 23
geostationary Earth orbit 47
geosychronous Earth orbit 47
geosynchronous 7
GTO 47
GPS 6
graph SLAM 300
gravitational forces 28, 59, 62
gravitational potential 64–65, 69, 103
gravity gradient 123–124, 131–132, 135, 137, 139, 143–144, 157–158, 163, 187–188, 191–192, 194
gyroscopes 308, 310, 313, 319, 339

Hamilton 86
Hill-Clohessy-Wiltshire 71, 73–75, 85
Hohmann transfer 49, 93–94, 115
homogeneous transformation 167
horizon sensors 312
Hough transform 313
hyperbolic orbit 31, 33

inclination 34, 43, 47–48, 53, 55, 62, 67–71, 79, 85–87, 92–93, 99–101, 115
induction-coil 311
inertial frame 118–119, 128, 146, 160
innovation 273, 295
interception 100–101
International Space Station 4–5, 17–18

inverse kinematics 180
Ion electric propulsion 11

Jacobian 168–170, 181, 204, 212, 289, 295, 297
JAXA 6, 11
joint angles 180
joint-space 180

Kalman filter 152, 251, 272–273, 275–276, 287, 290, 292, 295, 304, 319, 322–324, 339
Kane's method 178
Kepler's equation 42, 53, 66
Kepler's First Law 31
Kepler's Second Law 31–32
Kepler's Third Law 32
Keplerian 29, 34–35, 53, 62, 86, 88, 94
kinetic energy 171–172, 184, 190, 236–237, 240

LADAR 308, 340, 347
Lagrange brackets 63
Lagrange interpolation 247
Lagrange multiplier 208–209
Lagrange planetary equations 62
Lagrange points 60
Lagrange variables 98
Lagrange's formulation 179
Lagrangian 132, 170–172, 178, 183, 185, 236–237, 242–243
Lagrangian approach 170, 185
Lambert's problem 56, 101
Lambert's theorem 42–43, 51, 53, 58
LASER 287, 308, 340
law of gravitation 29, 35, 77, 124
Legendre 64, 102–103, 127
LEO 182
LERM 73
Levenberg-Marquardt 297, 303, 316
librations 131
LIDAR 308, 340–341
line of apsides 50, 86
line of nodes 34, 43, 53, 55, 88, 93
linear quadratic regulator 151, 339
localization 258–259, 263, 271, 275, 282–284, 289, 292, 296, 298–300, 306
Low Earth Orbit 125, 182
low thrust 94, 115
LVLH 71, 73, 77–79, 105
Lyapunov approach 206, 227
Lyapunov function 196, 198, 221
Lyapunov's method 96

magnetic dipole 123, 127–128, 164
magnetic field 126, 128–129, 135, 164–165, 311, 339
magnetic torquers 135

Index

Magnetic Torques 123, 126
manipulability 211–212, 222
manipulator 167–170, 172, 174, 178–181, 183, 185, 188–192, 194, 203–204, 206, 208–212, 215, 221–224, 226–228
Map 258
Markov process 269
Mars Exploration Rovers 9, 229, 255
Mars Express Orbiter 8
Mars Reconnaissance Orbiter 8–9
Mars Science Laboratory 9
mass expulsion 123
maximum likelihood 295
mean anomaly 35
mean motion 35
mission planning 230–231
mission profile 2
mobile platform 206, 208–209, 211–212, 215, 222, 224
Mohr-Coulomb 244
moment of momentum 121, 146–147, 149–150, 159, 161, 168–169, 185
moments of inertia 174, 191, 194
momentum wheel 134, 136–137, 143
Monte Carlo 193
Monte Carlo integration 277
Monte Carlo localization 285

NASA 2–3, 6, 8–10, 12, 15–16, 20, 23, 27
navigation 229–230, 233, 257–259, 261, 271, 286, 304, 306–307
NERM 73–74, 76, 82
nodes 34
non-holonomic 206, 208, 210, 212, 225
non-holonomic constraints 206, 208, 210, 212, 225, 233
nonlinear estimation 288, 292, 294–295
nutation 131, 135, 137, 139, 148, 164

oblateness 65, 69, 99, 103
observers 152
occupancy map 285, 287
optimal control 96, 98–100, 151, 164
optimal path 232, 254
optimization 247, 255
Orbital Inclination 34
Orbital mechanics 26
orbiting frame 118–119, 152, 157
osculating orbit 63

particle filter 277–278, 281–284, 286, 304–305
path following 231–233, 248, 251, 256
path planning 180, 231–233
performance 98, 115
periapse 34, 40, 42, 93
peri-focal 44–46

perigee 34
piezoelectric 309
planetary rovers 229–233, 243, 247, 255
Poisson matrix 63
posigrade 87
potential energy 171–172, 175, 179, 236–237, 241
probability density 264–265, 269–270, 273, 275–279, 281–282, 285–286, 295
probability distribution 269, 276, 302, 304
prograde 34, 47
pseudolite 314

q-method 316–318
quaternion 88–90, 92, 121, 152–156, 162–163, 197, 202, 313, 317–325, 328–331, 334, 336, 338, 345–346
QUEST algorithm 316, 318, 336

RAAN 88, 99–100
RADAR 308, 340
radiation pressure 126
reaction wheel 135, 137, 180–181
reference frame 118–119, 121, 146, 159
Remotely Operated Vehicles 16
rendezvous 100–101
retrograde 34, 87
Riccati 151
right ascension 34, 71, 79, 86, 88, 99
Robonauts 17
robot manipulator 167, 170, 194, 206
robotic arm 5, 12, 18, 21, 229
robotic manipulators 4, 26
rocker 235, 242
rocker-bogie suspension 242
ROSCOSMOS 6
route planning 257
rover 229–231, 233, 235, 237–238, 242, 253

screw motion 169
screw vector 167
sectoral 102
selenographic 70–71
semi-major axis 28, 31, 35, 38, 42, 53–56, 68, 86–88, 93, 98–101
serial manipulators 183, 209
SI 36
SLAM 258, 298–303, 305–307
sling shot maneuver 3
slip 231–233, 243, 246–248, 251, 253, 255
Smelt parameters 131, 193–194
solar radiation 123, 126
solar radiation pressure 102, 106
solar wind 126
SONAR 308
space debris 13–14, 27
space manipulators 181, 204

space robots 17
specific impulse 111, 115
spherical harmonic 127
stability 131, 206, 212, 215, 221, 223, 227–228
stability conditions 132, 191
stabilization 130–131, 133–135, 137, 139, 143–145, 147, 206, 227
staging 112
star sensors 313
station keeping 47
strapped-down navigation 308
Sun sensors 312
sun synchronous orbit 48

Terrain Adaptive Navigation 233, 255
tesseral 65, 69, 103
third body perturbation 106
three body problem 59
thrusters 126, 129, 134–137, 148–149, 180–181
traversability 232–233, 257–258
traversable 229, 231
TRIAD algorithm 316, 336

triangulation 263
trilateration 259, 295
Trojan points 60
true anomaly 31, 35, 38, 41–44, 46, 64–66, 71, 81, 87–88, 98
Tschauner-Hempel 81, 84, 116
two-link manipulator 236

unmanned systems 25
Unscented Kalman Filter 201, 275, 288, 290, 292–293, 298–300, 304–305, 320, 322–325, 330

vacant focus 33, 56
vectric 159
Virtual Manipulator 181, 204
visual odometry 231, 251, 255

Wheeled mobile manipulator 206, 209, 215
workspace 180, 204, 212, 222–224

zonal 65, 69, 99–100, 102–103